SOLUTE MODELLING IN CATCHMENT SYSTEMS

SOLUTE MODELLING IN CATCHMENT SYSTEMS

Edited by

STEPHEN T. TRUDGILL
University of Sheffield, UK

JOHN WILEY & SONS
Chichester · New York · Brisbane · Toronto · Singapore

Other Wiley Editorial Offices

John Wiley & Sons, Inc., 605 Third Avenue,
New York, NY 10158-0012, USA

Jacaranda Wiley Ltd, 33 Part Road, Milton,
Queensland 4064, Australia

John Wiley & Sons (Canada) Ltd, 22 Worcester Road,
Rexdale, Ontario M9W 1L1, Canada

John Wiley & Sons (SEA) Pte Ltd, 37 Jalan Pemimpin #05-04,
Block B, Union Industrial Building, Singapore 2057

Library of Congress Cataloging-in-Publication Data

Solute modelling in catchment systems / edited by Stephen T. Trudgill.
    p.   cm.
  Includes bibliographical references and index.
  ISBN 0-471-95717-8
  1. Landforms. 2. Solution (Chemistry) 3. Water chemistry.
I. Trudgill, Stephen T. (Stephen Thomas), 1947–
GB406.S578 1995
551.4′1—dc20   94-49356
                                                CIP

British Library Cataloguing in Publication Data

A catalogue record for this book is available from the British Library

ISBN 0-471-95717-8

Typeset in 10/12pt Times by MHL Typesetting Ltd, Coventry
Printed and bound in Great Britain by Bookcraft (Bath) Ltd.
This book is printed on acid-free paper responsibly manufactured from sustainable forestation,
for which at least two trees are planted for each one used for paper production.

Contents

List of Contributors

Tom Addiscott IACR-Rothamsted, Harpenden, Hertfordshire AL5 2JQ, UK

Mattias Alveteg Department of Chemical Engineering, Lund University, Box 124, Chemical Center, S-22100, Lund, Sweden

Adrian Armstrong ADAS Land Research Centre, ADAS Gleadthorpe, Meden Vale, Mansfield, Nottinghamshire NG20 9PF, UK

Jason Ball Department of Geography, University of Sheffield, Sheffield S10 2TN, UK

M.B. Beck Department of Civil Engineering, Imperial College of Science, Technology and Medicine, South Kensington, London, SW7 2BU, UK

K.J. Beven Centre for Research on Environmental Systems and Statistics, Institute of Environmental and Biological Sciences, University of Lancaster, Lancaster, LA1 4YQ, UK

Alex E. Blum US Department of the Interior, Geological Survey, MS 420, 345 Middlefield Road, Menlo Park, CA 94025, USA

Thomas D. Bullen US Department of the Interior, Geological Survey, Water Resources, MS 434, 345 Middlefield Road, Menlo Park, CA 94025, USA

T.P. Burt School of Geography, University of Oxford, Oxford, OX1 3TB, UK

Nils Christophersen Department of Informatics, Faculty of Mathematics and Natural Sciences, Informatics Building, University of Oslo, Gaustadalleen 23, PO Box 1080 Blindern, N-0316, Oslo, Norway

W. de Vries DLO Winand Staring Centre for Integrated Land, Soil and Water Research (SC-DLO), PO Box 125, 6700 AC Wageningen, The Netherlands

John Ewen Water Resource Systems Research Unit, Department of Civil Engineering, University of Newcastle upon Tyne, Newcastle upon Tyne, NE1 7TU, UK

J.E. Groenenberg DLO Winand Staring Centre for Integrated Land, Soil and Water Research (SC-DLO), PO Box 125, 6700 AC Wageningen, The Netherlands

M. Hornung Institute of Terrestrial Ecology, Merlewood Research Station, Grange-over-Sands, Cumbria, LA1 6JU, UK

Carol Kendall US Geological Survey, Water resources, MS 434, 345 Middlefield Road, Menlo Park, CA 94025, USA

M.J. Kirkby Department of Geography, University of Leeds, Leeds, LS2 9JT, UK

J. Kros DLO Winand Staring Centre for Integrated Land, Soil and Water Research (SC-DLO), PO Box 125, 6700 AC Wageningen, The Netherlands

Simon Langan Macaulay Land Use Research Institute, Craigiebuckler, Institute for Soil Research, Aberdeen, Scotland AB9 2QJ, UK

Peter Leeds-Harrison School of Agriculture, Food and Environment, Silsoe College, Cranfield University, Silsoe, Bedford MK45 4DT, UK

J.J. McDonnell College of Environmental Science and Forestry, State University of New York, Syracuse, NY 13210, USA

Jan Mulder Department of Soil Science and Geology, Agricultural University, PO Box 37, 6700 AA Wageningen, The Netherlands

C. Neal Institute of Hydrology, Wallingford, Oxfordshire OX10 8BB, UK

Malcolm Newson Department of Geography, University of Newcastle upon Tyne, Newcastle upon Tyne, NE1 7RU, UK

Tomas Pacés Czech Geological Survey, Malostranske Namesti 19, 110 00 Praha, Czech Republic

C.J. Reinds DLO Winand Staring Centre for Integrated Land, Soil and Water Research (SC-DLO), PO Box 125, 6700 AC Wageningen, The Netherlands

B. Reynolds Institute of Terrestrial Ecology, Bangor Research Unit, University of Wales Bangor, Deiniol Road, Bangor, Gwynedd, LL57 2UP, Wales, UK

W.H. Robertson Department of Statistics and Modelling Science, University of Strathclyde, Glasgow, G1 1XQ, Scotland, UK

A.J. Robson Institute of Hydrology, Wallingford, Oxfordshire, OX10 8BB, UK

Hans H. Seip Department of Chemistry, University of Oslo, PO Box 1033 Blindern, 0315 Oslo, Norway

Michael G. Sklash The Dragun Corporation, Farmington Hills, Michigan, USA

P.A. Stevens Institute of Terrestrial Ecology, Bangor Research Unit, University of Wales Bangor, Deiniol Road, Bangor, Gwynedd, LL57 2UP, Wales, UK

M.K. Stewart Institute of Geological and Nuclear Sciences Ltd, Lower Hutt, New Zealand

Harald Sverdrup Department of Chemical Engineering, Lund University, Box 124, Chemical Center, S-22100, Lund, Sweden

Geir Taugbøl Department of Chemistry, University of Oslo, PO Box 1033 Blindern, 0315 Oslo, Norway

Steve Trudgill Department of Geography, University of Sheffield, Sheffield, S10 2TN, UK

P.V. Unnikrishna Department of Civil and Environmental Engineering, Utah State University, Longan, UT 84322-5215, USA

C. van der Salm DLO Winand Staring Centre for Integrated Land, Soil and Water Research (SC-DLO), PO Box 125, 6700 AC Wageningen, The Netherlands

Michael Anthony Velbel Department of Geological Sciences, 206, Natural Science Building, Michigan State University, Each Lansing, MI 48824-1115, USA

Per Warfvinge Department of Chemical Engineering, Lund Institute of Technology, Box 124, Chemical Center, S-22100, Lund, Sweden

H.S. Wheater Department of Civil Engineering, Imperial College of Science, Technology and Medicine, South Kensington, London, SW7 2BU, UK

M.J. Whelan Department of Biological Sciences, University of Exeter, Exeter, EX4 4PS, UK

Art F. White US Geological Survey, MS 420, 345 Middlefield Road, Menlo Park, CA 94025, USA

P.G. Whitehead Department of Geography, University of Reading, Whiteknights, PO Box 227, Reading, Berkshire, RG6 2AB, UK

Preface

It is now nearly nine years since John Wiley published my edited *Solute Processes* (1986) book and much has happened in the field since then. However, the developments have largely been in the field of solute modelling and so I did not think that it would be appropriate merely to have an updated text with the same conceptual framework. I felt that this new book should now include particular advances in solute modelling and especially at the catchment scale. The present volume therefore stresses both the modelling and the catchment aspects with particular emphasis on the influences of weathering (Section II), ecosystem processes (Section III) and hydrological processes (Section IV), as well as the application of models (Section V), while Chapter 17 discusses management implications. Chapter 1 reviews selected solute models and Chapter 2 discusses modelling principles.

This book is an internationally authored research-level text, aimed at providing a synopsis of developments in solute modelling in catchment ecosystems in the last ten or so years — and also at indicating future needs and possibilities. It cannot pretend to be totally comprehensive, though Chapter 1 does attempt a review where many models are covered, albeit in abbreviated form. There are omissions of models in the book, in that not every existing model is covered by a chapter, and such omissions are due in no small part to the difficulties that many over-committed authors expressed in keeping to deadlines. The intended audience is research scientists in cognate fields and postgraduate research workers. It is especially aimed at those interested in catchment systems and the intention is that it will also enable specialists to grasp the essentials of topics in which they are not specialists. Thus, reviews and case studies are provided not only for specialists who wish for an overview of their own topic but also by specialists in related fields who wish to understand separate but cognate fields (e.g. hydrology for biologists, ecology for geochemists, geochemistry for hydrologists and so on). I feel that this is essential if we are to make real progress in providing integrated models.

While essentially a research-level text, the intention is that the volume will also be of use in a management context because successful water quality management actions are often dependent on the predictive success of models (Chapter 17). There will thus also be a context of how the models can be used to assess the sensitivity of solute production to environmental change and human influence such as land-use change (including manipulation of semi-natural ecosystems and changes in both land-use type and practices) and variation in atmospheric inputs. The intention is to focus on surface catchments, rather than to include much on the topics of groundwater and soil solutes, which are too large to also include in the volume. An emphasis is on describing models, their calibration, validation and application, with reported field results included in the contexts of model use. From the term "solutes", the volume focuses on the major divalent and monovalent cations and anions (calcium, magnesium, sodium, potassium, nitrate, phosphate, chloride, sulphate, etc.), together with pH and acidity.

I would like to thank a number of anonymous referees who responded to my requests to review chapters.

Stephen Trudgill   *November 1994*

Note: While the SI unit is M (mol l^{-1}) for solute concentrations, many other expressions are also commonly used. Conversion factors are given in the Appendix (p. 461).

SECTION I

OVERVIEWS

1

Overview of Solute Modelling

JASON BALL and STEVE TRUDGILL

Department of Geography, University of Sheffield, UK

1.1 EARLY MODELS

Modelling involves not just elements of observation, description and explanation but also of prediction — from the envisaged operation of a process we should be able to predict future patterns. In 1882 Charles Darwin in his book on vegetable mould and earthworms quoted (Darwin, 1882, p. 244) a Dr Gilbert as saying that ". . . several square yards on his lawn were swept clean, and after two or three weeks all the worm castings on the space were collected and dried. These were found to contain 0.35 (*sic*) of nitrogen". There then follows the key statement: "Supposing a quantity of (earthworm) castings equal to 10 tons in the dry state were annually deposited on an acre, this would represent a manuring of 78 lbs of nitrogen per annum. . . ." So is this one of the first models to quantitatively predict a solute input? Did this set the precedent that supposing that certain conditions and relationships operate, then solute production will equal . . .?

One assumes that the Dr Gilbert quoted by Darwin is he who worked with Lawes (Lawes *et al.*, 1882) on the composition of rain and drainage water at Rothamsted. In this paper were some of the first observations on solute production from soil and, while their work would hardly have been called a model, it is clear that they were able to generalise that solutes could be derived from soil material.

An early development in solute modelling was the observation of concentration–discharge relationships reported by Hem (1948), but by the mid 1960s one could assume from a review on hydrological models (More, 1967) that, apart from a reference to salinity, solute modelling did not exist and it is probably true to say that the early hydrological models were focused more upon the physical aspects of water resources including flooding. This much is clear from, for example, the establishment of the Coweeta watershed hydrological laboratory in 1934, when it is evident that all the early papers by Hursh in the 1940s were preoccupied with streamflow — the water quality, rather than quantity, programme did not really begin until 1970 (Gaskin *et al.*, 1983). However, there were clear signposts on the road to solute modelling, notably Johnson *et al.*, (1969) who, rather in the style of Lawes and Gilbert of observations followed by inferences, had made observations on the stream chemistry of the Hubbard Brook watershed established in the early 1960s and produced what they called a "working model" to explain the

Solute Modelling in Catchment Systems. Edited by Stephen T. Trudgill
© 1995 John Wiley & Sons Ltd

variation in stream water chemistry. This period also saw other papers, including Hall's important (1970, 1971) papers on dissolved solid-discharge relationships.

It is clear that by 1970 there was a good understanding of the geochemistry of surface waters at that time (Hem, 1970) and a quite well developed field of enquiry concerning the relationships between solute levels in runoff and bedrock geochemistry. For example, Clarke (1924) had made one such study in Bohemia and Conway (1942) had contrasted the (greater) amounts of solutes from basins on sedimentary rocks with those on igneous and metamorphic rocks. There was also a good realisation of the importance of atmospheric inputs (Gorham, 1955), and other essentially spatially distributed analyses had been undertaken dating back over a hundred years (Livingstone, 1963, and as reviewed by Walling and Webb, 1986, p. 251).

Hydrochemical processes were less well understood but many papers on solutes in streams had begun to focus on changes in stream water solute levels with discharge. Indeed, it soon seemed that every relevant research institute had a "V" notch weir and was measuring variations in solute concentration with discharge, but any "modelling" was often only really a descriptive and quantitative treatment of the data. This period is already well summarised by Gregory and Walling (1973, pp. 219−225) and by Walling and Webb (1986).

By 1980 there was a considerable wealth of relevant literature (see Gower, 1990, and Trudgill, 1986) but still little in the way of catchment modelling apart from Johnson *et al.*'s (1969) working model for Hubbard Brook. The relevant literature included the following:

1. Geochemical, aqueous chemistry and soil solute works (e.g. Garrels and Christ, 1965; Stumm and Morgan, 1970; Nye and Tinker, 1977).
2. Early (and continuing) engineering literature on flow and solutes (e.g. Danckwerts, 1953).
3. Many developments in groundwater hydrology (e.g. Freeze and Cherry, 1979).
4. A proliferation of data on river solutes (reviewed by Walling and Webb, 1986).
5. The development of two-component mixing models (e.g. Hall, 1970; Pilgrim *et al.*, 1979; Sklash *et al.*, 1976)
6. Other model developments in an agricultural context (though not at the catchment scale), especially CREAMS (Chemical, Runoff and Erosion from Agricultural Management Systems), (Knisel, 1980).
7. Many soil profile leaching models (e.g. Rose, 1973; de Wit and van Keulen, 1972; Frissel and Reininger, 1974), especially for nitrate (Burns, 1974; Addiscott, 1977; and the review of Wild and Cameron, 1980) together with many leaching studies (e.g. Cooke and Williams, 1970; Kolenbrander, 1975; Wild and Babiker, 1976).

There were also parallel developments in hydrological modelling during this period, especially in physically based variable contributing area models (TOPMODEL; see Beven and Kirkby, 1979, and the review of Beven, 1989). These, while not initially directed at solute modelling, were later to be developed to include solutes.

Despite the academic potential for synthesising this knowledge into solute models, solute model development only became widely facilitated by the recognition of practical problems. Some of the work on solute movement in porous media was, for example spurred on by the problems of disposal of waste, especially radioactive waste at Oak Ridge in Tennessee, USA (see, for example, Watson and Luxmoore, 1986). More significantly in our context, catchment-scale solute modelling only really began to "take off" in the mid 1980s under the spur of what is often called the "acid rain problem'. Here, it came to be realised that not only was it necessary to try to relate atmospheric inputs to soil and bedrock geochemical processes in order to understand stream runoff water quality, but also, and more importantly perhaps, that there might also be research money available. There was also a parallel development of soil leaching models, developing from the initial study of nutrient movement in the rooting zone in a plant growth context to the movement of nutrients beyond the soils and into streams, rivers and lakes in pollution and water quality contexts. Thus from the beginning of the 1980s to date there has been a great increase in the interest in the development of and the application of catchment-scale solute models to perceived environmental issues (Moldan and Černý, 1993).

1.2 SOLUTE MODELS IN THE 1980s and 1990s

The late 1970s and the early 1980s saw an increase in the application of predictive models. These models attempted to simulate solute behaviour from the relationships between input and output, given certain assumptions about the environmental characteristics. The models were essentially concerned with components and the flows between them. An important consequence of their development was that the models could be used to assess the sensitivity of solute behaviour in output to changes in atmospheric inputs and other catchment characteristics. This was in contrast to the earlier modelling attempts which sought to describe existing hydrochemical data and produce a "model" which was not much more than a set of retrospective generalisations (albeit quantitative) about hydrochemical behaviour.

There now exists a large array of solute models; Section 1.4 is a reference section which aims to describe some of the models in more detail. The text and accompanying tables attempt, where possible, to describe the hydrochemical processes employed, their data requirements, their temporal and spatial resolution and the degree to which spatial processes are lumped together. The aim of Section 1.4 is not to be an all-encompassing guide to solute modelling, but to provide a starting point for those wishing to pursue the subject more thoroughly. There are two tables to accompany the text: Table 1.1 is a guide to the input requirement, time step and output of the models and Table 1.2 attempts to show the internal representations of the hydrological and biogeochemical processes included. The tables and text provide necessarily abbreviated summaries of the models which may inevitably not do full justice to all aspects of the model or the modeller's work — this can only be achieved in detail by reference to the papers cited in the text.

The models often overlap in terms of their approaches, but they are grouped in Section 1.4 according to the context in which they were developed. Clearly, acidification and pollution have provided substantive contexts for model development and these are the two main headings under which the models may be grouped.

The acidification context includes short- and long-term models of chemical reactions, geochemical soil profile models and catchment-scale models. Where other models have been developed independently of this context, based on considerations of hydrological processes involving mixing (Sklash, 1990) and the role of topography (Beven and Kirkby, 1979), their actual application as solute models has often been in an acidification context. Thus, although they have other wider contexts and applications they are also included under the acidification heading.

The pollution context includes models of the transport of nutrients, pesticides and organic material through the soil profile. Particular developments include nutrient leaching models, especially of nitrate, at the soil profile, hillslope and catchment scale as well as the modelling of downstream changes in pollutant loadings (see James, 1993).

TABLE 1.1 The major inputs, processes and outputs for a selection of solute catchment models

Model	Input and parameters	Time step	Output
Birkenes	Rainfall, mean daily air temperature, wet and dry deposition of sulphate and chloride Hydrological routing and threshold parameters Gibbsite equilibrium constant, Gapon equation constants, sulphate adsorption/ desorption rate, carbonic equilibrium constant, organic acidity constant, partial pressure carbon dioxide in soil (fixed at ×10 atm) and weathering rate for base cations Measured dissolved organic carbon data also required for organic acid concentrations	Day or less	Discharge Mg^{2+} and Ca^{2+} (lumped as M^{2+} in earlier versions), H^+, Al^{3+}, Na^+, Cl^- and SO_4^{2-}, HCO_3^- and organic anion concentration (Rustad et al., 1986)
PULSE	Rainfall, mean air temperature, monthly standard potential evapotranspiration Recession coefficient for each compartment Alkalinity submodel with fixed minimum alkalinity value: version 1: empirical coefficient describing seasonal CO_2 variation and maximum alkalinity (at ∞); version 2: level of pulse in saturated zone, maximum alkalinity at bottom of aquifer in summer and winter (two parameters) Nitrogen leaching submodel: atmospheric nitrogen input rate, plant uptake rate, two soil moisture threshold parameters and seven empirical coefficients found by calibration	Day	Discharge Alkalinity and pH (Bergström et al., 1985) and NO_3–N (Bergström et al., 1987)
Allt a'Mharcaidh	Spatially distributed rainfall Evaporation in hourly time step Eleven parameters defining hydrological thresholds and recession characteristics of catchment	Hour	Alkalinity, discharge
EMMA	Soil water, groundwater and rainfall chemistry used to define end members, together with stream water chemistry data for any conservative solute	Event-based	Fraction of end member in total flow
TOPMODEL	Inputs include: spatially distributed rainfall; digital terrain map to define topographic index, channel reach lengths and catchment areas Parameters include: maximum interception storage, maximum litter layer storage, direct throughfall fraction, overland flow velocity, lateral transmissivity (at surface when saturated) and vertical hydraulic conductivity (at surface), 'field capacity', exponential decay for lateral transmissivity and vertical conductivity and two channel velocity constants	Hour or day	Discharge, soil moisture deficit

TABLE 1.1 *Continued*

Model	Input and parameters	Time step	Output
Ion balance model	Data on pH precipitation, non-marine fraction of calcium and magnesium (alkalinity from weathering) and non-marine sulphate (acidity input)	Long-term (year)	Alkalinity
ILWAS	Daily precipitation, dry and wet monthly deposition for Mg^{2+}, Ca^{2+}, K^+, Na^+, NH_4^+, H^+, SO_4^{2-}, NO_3^-, Cl^-, and alkalinity (monthly), daily maximum and minimum temperatures, chemical composition of leaves Parameters include: catchment parameters (area, lake area, aspect, slope); deposition parameters (collection efficiency, deposition velocity, gas adsorption velocity); canopy factors (maximum interception storage, leaf area index and density, foliar exudation rate); soil parameters (layer depths, hydraulic conductivity, cation exchange capacity, nitrification rate, field capacity and saturation water contents) channel and overland flow parameters (Manning's coefficient) Evaporation constants, nutrient uptake rate, litterfall rate, mineralisation and immobilisation rate constants, gibbsite rate constant, carbonic acid and aluminium complexation equilibrium constants, Gapon exchange, weathering rate and adsorption isotherm constants	Day or less	Mg^{2+}, Ca^{2+}, K^+, Na^+, NH_4^+, H^+, $Al(OH)$ species, SO_4^{2-}, NO_3^-, total P, Cl^-, organic ligands, total inorganic carbon and alkalinity Discharge, moisture contents and lake stage
Trickle Down and ETD	Rainfall, pan evaporation and air temperature Alkalinity of wet and dry deposition Sulphate and chloride wet and dry deposition in ETD Parameters include: catchment details (area, lake area, shorline length); snowmelt (melt rate factor and degree-day coefficient); soil and unsaturated zones (depth, lateral flow recession constant, vertical and horizontal (ETD only), hydraulic conductivity, porosity); lake permeability; ground water (hydraulic conductivity, porosity); evaporation pan coefficient; alkalinity parameters (base weathering rates, sum of bases, bulk density, sediment base transport coefficient), pH (partial pressure of CO_2, Henry's law constant, carbonic acid dissociation constants) ETD: sulphate sorption partition and reduction rate coefficients	Day	Alkalinity, pH, sulphate and chloride (ETD) Lake stage and discharge

TABLE 1.1 *Continued*

Model	Input and parameters	Time step	Output
MAGIC	Annual rainfall and runoff, atmospheric deposition of Cl^-, F^-, NO_3^-, SO_4^{2-}, NH_4^+, Ca^{2+}, Na^+, K^+ and Mg^{2+} Parameters include: cation exchange capacity, sulphate maximum adsorption capacity and half saturation constant, thermodynamic constants for Al^{3+} complexation reactions, bulk density, depth, deposition factors, base cation weathering rates, soil exchange selectivity coefficients for base cations For water and soil: Al^{3+} solubility coefficient, CO_2 partial pressure, organic matter content and pK constant for soil and water, enhancement factors for increased evapotranspiration, nutrient uptake and dry deposition by growing forests, nutrient uptake rates (calcium, magnesium, potassium and sulphate), (Cosby *et al.*, 1990)	Year	Soil and stream water chemistry for Mg^{2+}, Ca^{2+}, K^+, Na^+, Al^{3+}, SO_4^{2-}, Cl^-, F^-, NO_3^-, alkalinity and pH Organic anions
PROFILE and SAFE	Precipitation, runoff, soil temperature, deposition of cations (NH_4^+, Mg^{2+}, Ca^{2+}, Na^+, K^+ and Al^{3+}) and anions (Cl^-, SO_4^{2-}, NO_3^-) Base cation (Mg^{2+}, Ca^{2+} and K^+) and nitrogen uptake rates Parameters include: weathering rate coefficients, soil mineralogy, soil depth, moisture content, bulk density, specific surface area, soil CO_2 partial pressure, cation exchange capacity (for SAFE only), dissolved organic carbon and equilibrium constants for aluminium, organic and carbonate systems and nitrification rate factor	Year	Mg^{2+}, Ca^{2+}, K^+, Na^+, H^+, NH_4^+, Al^{3+} species, SO_4^{2-}, NO_3^-, Cl^-, dissolved organic carbon and alkalinity of soil and stream
SMART	Precipitation surplus Atmospheric input rates of sulphate, nitrate, ammonium and base cations, nitrogen uptake and nitrogen immobilisation rates, base cation weathering and uptake rates Parameters: soil layer depth, volumetric water content, bulk density, CO_2 partial pressure, cation exchange capacity, Henry's law constant, aluminium hydroxide and carbonate dissolution constants, cation exchange selectivity coefficients for H^+ and Al^{3+}, nitrification and weathering rate factors	Year	BC^{2+}, pH, Al^{3+}, SO_4^{2-}, NH_4^+, NO_3^- and HCO_3^- Base saturation and molar Al/BC ratio

TABLE 1.1 *Continued*

Model	Input and parameters	Time step	Output
RESAM	Annual precipitation, evapotranspiration, temperature Sulphur, nitrogen, base cation, ammonia and chloride dry and wet deposition rates Elemental content of stems, leaves and roots and initial concentrations of cations and anions in soil solution Parameters: soil layer depth, bulk density, volumetric water content, carbon dioxide partial pressure, Henry's constant, cation exchange capacity, sulphate adsorption rate and sulphate sorption capacity; aluminium hydroxide equilibrium and Elovich constants; carbonate dissolution rate and equilibrium constant; cation exchange equilibrium constants; weathering rate constants for base cations; protonation, nitrification, denitrification, root decay, litterfall, net growth and foliar exudation rate constants; root distribution data	Year	Major cations, anions, organic anion, pH and alkalinity of the soil solution
MIDAS	Precipitation, evaporation, deposition of base cations (Ca^{2+}, Mg^{2+}) and hydrogen Parameters: soil depth, soil water content, cation exchange capacity, aluminium hydroxide solubility coefficient, net weathering rates for base cations and hydrogen consumption Calibrated parameters: Gaines—Thomas selectivity coefficient, two cation exchange rate coefficients, net nutrient uptake rates	Year	Mg^{2+}, Ca^{2+}, H^+ and Al^{3+} of soil solution
LEACHN	Input: rainfall and rainfall nitrate and amonium concentrations, evapotranspiration, temperature Initial estimates of organic and soil mineral nitrogen pools and crop uptake rates Parameters: layer depth, soil retentivity and conductivity estimated from Campbell's (1974) empirical equations, dispersion coefficient for the CDE, partition coefficients for adsorbing chemicals, mineralisation, nitrification, denitrification and volatilisation rate constants with water and temperature correction factors, half-saturation constant (denitrification), maximum nitrate/ammonium ratio (nitrification), efficiency and humification factors and C/N ratios for organic pools	Day	Drainage and concentration of nitrate, ammonium and urea with depth

TABLE 1.1 *Continued*

Model	Input and parameters	Time step	Output
MACRO	Rainfall, evaporation, maximum moisture content in macropores, residual water content, saturated hydraulic conductivity of macro- and micropores, pore size distribution index, tortuosity factor for macro- and micropores, dispersivity in micropores, diffusion coefficient, initial solute concentration, mass-transfer coefficients for water and solute	Day, hour	Drainage, water table depth, concentration of solute or pesticide
Burns model	Rainfall, evaporation, water content at field capacity, initial solute concentration	Day, hour	Drainage and solute concentration at profile base
TETrans	Rainfall, irrigation, evapotranspiration, concentration of irrigation or rainfall, initial concentration and water content of soil For each layer: thickness, soil bulk density, mobility coefficient, water content at wilting point and field capacity Planting and harvesting date, days of maturity, plant root distribution	Day	Drainage, water content, solute concentration within the soil profile
Addiscott and Whitmore (1987)	Rainfall, open-water evaporation, soil temperature, mineral nitrogen in rainfall and fertiliser applications Initial mineral nitrogen concentration within the profile Parameters include: soil moisture characteristics, moisture content, layer depth, evaporative limit, maximum rainfall above which bypassing flow occurs, nitrification and mineralisation rate constants, Arrhenius coefficients, days between sowing and field capacity, sowing date, fraction of ammonium accessible to crop, rate constants for crop growth and nitrogen uptake, maximum crop growth and crop nitrogen uptake	Day	Drainage, mineral nitrogen concentration within the soil profile and leached at the profile base
TFM (White, 1987)	Rainfall Fractional transport volume derived from probability density functions of solute travel times from experimental data	Hour	Nitrate concentrations at base of profile

TABLE 1.1 *Continued*

Model	Input and parameters	Time step	Output
GLEAMS	Hydrology model requires: daily rainfall, mean monthly maximum and minimum temperatures, mean monthly solar radiation; monthly leaf area index, slope, drainage area, effective rooting depth, soil phosphorus sorption coefficient, soil temperature, SCS curve number, evaporation limit constant For each soil horizon: depth, porosity, bulk density, saturated hydraulic conductivity, field capacity, wilting point, organic matter content, soil pH, clay content, base saturation and calcium carbonate content Erosion model requires: rainfall erosivity, peak runoff rate, fractions of clay, silt and sand. Specific surface areas of clay, silt, sand and organic matter, overland profile characteristics USLE parameters: erodibility, crop and practice factors, channel dimensions, non-erodible layer depth, Manning's coefficient, enrichment ratio Pesticide model requires: pesticide half-life in soil and foliage, water solubility, organic carbon partition coefficient, initial foliar and soil concentration, coefficients for transformation and plant uptake, application rates Nutrient model requires: nitrogen concentration in rainfall, nitrate and labile phosphorus in irrigation, crop residue Initial contents of phosphorus and nitrogen pools including: crop residue (fresh organic nitrogen and phosphorus), organic humus phosphorus, organic nitrogen and phosphorus in animal waste, labile phosphorus and potential mineralisable nitrogen, active and stable mineral phosphorus Mineral and organic fractions of nitrogen and phosphorus in inorganic and organic fertilisers Application dates; tillage number, dates and type; crop planting and harvest date, C/N ratio, N/P ratio, dry matter ratio and crop yield for each crop	Day	Includes: annual, monthly and daily for surface runoff, percolation, evapotranspiration, sediment yield, pesticide loss in sediment, runoff and percolation Leaching loss of nitrate, ammonium, mineral phosphorus Sediment loss of organic nitrogen and phosphorus, ammonium and mineral phosphorus Runoff loss of nitrate, ammonium and mineral phosphorus Soil nitrogen and phosphorus content by layer

TABLE 1.1 *Continued*

Model	Input and parameters	Time step	Output
QUASAR	Inputs include: rainfall, water quality variable concentrations, chlorophyll-a concentrations, stream water temperature, number of sunlight hours, mean monthly air temperature, maximum air temperature for period, decay parameters, water quality and flow levels of inputs and abstractions along river reach, number of reaches and for each reach, channel cross-sectional area, length and slope	Hour or day	Nitrate, chloride, BOD, ammonium, ammonia, pH, DO, temperature and *Escherichia coli* Any conservative pollutant and inert material Downstream discharge
ADZ	Lumped parameters of the transfer function form of the ordinary differential equation determined using stochastic, recursive parameter estimation techniques Concentration data at upstream and downstream sites required for model calibration Concentration data at upstream site (e.g. pulse input) require for input	Hour or day	Concentration of conservative or non-conservative solute at downstream site
Export Coefficient Model	Lake and watershed areas Export coefficient for each land use is calculated from inorganic and organic application rates, farming practices, livestock types and numbers Other nutrient inputs: sewage treatment works (human population data)	Year	Nutrient loss from catchment area
SHETRAN	Inputs include: spatially distributed rainfall, meteorological data, vegetation type and cover, man-controlled inputs and abstractions, overland and channel dimensions, initial sediment and contaminant concentrations and inputs of sediment and contaminants from rainfall, bank erosion, etc. Parameters: evaporation rate constant, land-cover indices, canopy storage capacity, root distribution, aerodynamic and canopy resistance, overland and channel roughness coefficients, soil moisture characteristic and unsaturated hydraulic conductivity – soil moisture relationship for each soil horizon, porosity and saturated hydraulic conductivity for saturated zone, snowmelt degree-day factor Sediment size distribution, depth of loose soil and sediment, soil porosity, bed porosity, sediment longitudinal dispersion coefficient Contaminant dispersion coefficient for porous media and river channels, radioactive decay constant, fraction of adsorption sites and mobile water in porous media, partition coefficient for sediment and porous media, mass-transfer coefficient in mobile and immobile water, mass-transfer coefficients (2) between river, bed surface and deep bed layer and plant uptake efficiency factor	Varying time step	Discharge from all sections, soil moisture status; sediment and contaminant concentration (in adsorbed and dissolved phases)

TABLE 1.2 The major processes included in a selection of catchment models

Model	Hydrology	Chemistry
Birkenes	Two compartment hydrology Upper compartment representing organic soil horizons producing storm flow Lower compartment producing baseflow and representing mineral soil horizons Snowmelt processes by degree-day method can be included Evapotranspiration calculated from air temperature	Charge balance approach assuming mobile anion concept Cation exchange between H^+, and calcium and magnesium lumped as M^{2+} and sodium (Rustad et al., 1986) — Gapon equation, gibbsite equilibrium, sulphate adsorption/desorption described by linear isotherm, sulphate mineralisation by empirical equation, base cation weathering (Ca^{2+}, Mg^{2+}, Na^+) by first-order reaction, HCO_3^- and organic acid equilibrium (Rustad et al., 1986)
PULSE	(Semi)-spatially distributed model with different catchment areas represented by different storage compartments Each storage compartment consists of an unsaturated and a saturated zone which has it own recession characteristics through which a pulse of water is tracked	Cation exchange, weathering and the seasonal variation of CO_2 on weathering are considered by semi-empirical equations Seasonal variation of CO_2 governed by evapotranspiration data pH is determined from a relationship with alkalinity
Allt a'Mharcaidh	Spatially distributed five store model, each describing different soil types in the catchment	Stream alkalinity concentrations are a result of conservative mixing of three alkalinity sources or end members within the catchment No chemical mechanism is employed
EMMA	No hydrological model EMMA can be used to infer information on hydrological flowpaths	No chemical process is modelled All chemical reactions assumed fast enough for end members to be considered constant in time and space Mixing of end members is conservative either on the way to or in the stream
TOPMODEL	Semi-distributed model with topographic index used to calculate effects of local topography on catchment hydrological response, i.e. runoff generation from saturated contributing areas Evaporation, interception and throughfall zones, quick flow (overland flow, return flow and macropore flow), vertical percolation through a root zone to the water table and subsurface saturated flow are included	TOPMODEL has been used to predict stream ANC variations from mixing two end members assigned to different subsurface flow components identified by their depth of origin (Robson et al., 1991, 1992)

TABLE 1.2 *Continued*

Model	Hydrology	Chemistry
Ion balance model	No hydrological processes are described	Empirical relationship based on data from large number of lakes (Henriksen, 1980) No chemical process is described explicitly
ILWAS	Multi-compartment model dividing catchment into subcatchments and stream and lake segments Each subcatchment has a snow pack, canopy and soil layer module Processes simulated are throughfall, interception, evapotranspiration, soil freezing and thawing, snow accumulation and melting, overland, shallow subsurface and groundwater flow Lake is divided into 12 layers and lake stratification and ice formation are described Flow through the soil layer can be vertical percolation and/or lateral interflow	Charge balance approach Processes include organic matter decomposition, carbonic and organic acid and aluminium complexation equilibria, base cation (silicate) and gibbsite weathering, anion retention of sulphate, organic acid anions and H_2PO_4 (linear isotherm), cation exchange of Mg^{2+}, Ca^{2+}, K^+, Na^+, NH_4^+ and H^+ (Gapon equation), nitrification (Michaelis–Menten equation) in soil, nutrient uptake in lake and soil, mineralisation and immobilisation (of NH_4, SO_4^{2-}, Ca^{2+}, Mg^{2+}, K^+, Na^+), CO_2 production and exchange, foliar uptake/exudation and litterfall
Trickle Down and ETD	Multi-compartment model describing snow pack, soil, unsaturated zone, groundwater and lake processes In ETD greater description of overland flow, frozen ground processes and discharge from the lake are included Flow is routed through the stores assuming mass balance Flow can be both lateral and vertical Evapotranspiration is included	Cation exchange and chemical weathering described by first-order kinetic reactions pH is determined by a relationship relating the CO_2 system to alkalinity Sulphate sorption in terrestrial modules described by linear isotherm Sulphate reduction in lake module described by first-order reaction
MAGIC	One compartment representing vertical soil layer in which vertical flow only is considered Effects of evapotranspiration can be incorporated (Cosby *et al.*, 1990) Jenkins and Cosby (1989) have applied one-layer, two-layer and two-layer (with flow routing) versions	Charge balance approach based on mobile anion concept Processes include: cation exchange for Al^{3+}, Ca^{2+}, Mg^{2+}, K^+, Na^+ (Gaines–Thomas approach) anion retention of sulphate – Langmuir isotherm, base cation weathering, aluminium hydroxide and carbonic acid equilibria and Al^{3+} complexation with OH^-, SO_4^{2-} and F^- Net nutrient uptake for SO_4^{2-}, Ca^{2+}, Mg^{2+}, Na^+ and K^+ Organic acid speciation and nitrogen (NH_4^+ and NO_3^-) uptake by first-order rate reaction (Cosby *et al.*, 1990)

TABLE 1.2 *Continued*

Model	Hydrology	Chemistry
PROFILE and SAFE	Multi-layer compartment representing soil horizons with vertical and lateral water movement	Mass balance approach. Processes include: mineral weathering — kinetic reactions from soil geophysical data; cation exchange — Gapon mass transfer; nitrification (first order); base cation uptake and nitrogen uptake; soil solution submodel: aluminium, organic acid and carbonate equilibrium reactions and autoprotolysis of water
SMART	One lumped compartment representing vertical flow in soil layer. Can be subdivided into homogeneous soil horizons (De Vries *et al.*, 1989)	Charge balance approach based on mobile anion concept. Processes include: cation exchange — Gaines–Thomas equation (for BC^{2+}, H^+ and Al^{3+}); base cation weathering of silicate (model input); carbonate, aluminium hydroxide and carbonic acid equilibria. Also biogeochemical processes included as model inputs: net nitrogen immobilisation, nitrification (proportional to deposition) and net uptake of base cations and nitrogen
RESAM	Multi-layer soil compartment with vertical flow. Root water uptake and evapotranspiration effects are included	Processes include: foliar uptake (NH_3) and exudation (Ca^{2+}, Mg^{2+}, K^+); net growth; litterfall, root decay, mineralisation and root uptake (all for N, S, Ca^{2+}, Mg^{2+}, K^+); nitrification and denitrification (NO_3^- and NH_4^+), protonation of organic anions, carbonate (as $CaCO_3$) dissolution/precipitation, weathering of primary minerals containing aluminium and base cations (Mg^{2+}, Ca^{2+}, K^+, Na^+), aluminium hydroxide dissolution/precipitation, cation exchange — Gaines–Thomas equilibrium (H^+, Al^{3+}, Ca^{2+}, Mg^{2+}, Na^+, K^+, NH_4^+), sulphate adsorption/desorption — Langmuir isotherm, inorganic carbon dissolution equilibrium equation

TABLE 1.2 *Continued*

Model	Hydrology	Chemistry
MIDAS	One lumped compartment representing soil horizons Vertical water percolation only Evaporation effects are considered	Kinetic equations coupling processes of cation exchange (modified Gaines–Thomas equation, H^+ and Al^{3+} lumped as acid trivalent cations and Ca^{2+} and Mg^{2+} lumped as divalent base cations); weathering of base cations (input); aluminium hydroxide dissolution (equilibrium equation); net nutrient uptake combining nutrient uptake by plants and organic decomposition (input)
LEACHN	Water transport through soil matrix by the Richards equation Evapotranspiration and plant water uptake	Solute transport by convection–dispersion equation with additional terms to account for chemical adsorption to the soil, nitrogen uptake by plants and the nitrogen transformation processes: mineralisation, nitrification, denitrification and ammonium volatilisation Plant growth effects also included
MACRO	Soil profile divided into macropore (Darcy's law) and micropore (Richard's equation) phases with convective water transfer between the two Evaporation, irrigation, plant water uptake and surface runoff are also considered	Solute transport described by CDE in micropore region Solute transport in macropore region by convective mass flow Solute convection and dispersion between the two phases
Burns model	Infiltration, drainage to field capacity and evaporation represented by mass balance equations for each soil layer	Solute leaching from surface-applied (Burns, 1974, 1975) and soil-derived solute (Burns, 1976)
TETrans	Infiltration and drainage to field capacity, bypassing flow, plant water uptake and evapotranspiration represented by mass balance calculations over conceptual layered profile	Solute leaching between layers by mass balance equations with complete, instantaneous chemical equilibration within a layer after infiltration and again after removal of water by plants and evaporation
Addiscott and Whitmore (1987)	Infiltration in the mobile phase only and diffusion between mobile and immobile phases, evaporation and plant water uptake Infiltration between layers is modelled by mass balance equations	Nitrogen leaching, mineralisation, nitrification, crop growth, crop nitrogen uptake Incomplete solute equilibration between immobile and mobile phases

TABLE 1.2 *Continued*

Model	Hydrology	Chemistry
TFM — White (1987)	Drainage from soil profile modelled as a stochastic function	Nitrate leaching in profile is modelled as a stochastic function
GLEAMS	Multi-layer compartment model which allows for infiltration from rainfall or irrigation, storage, leaching losses to groundwater and evapotranspiration from each layer Overland and channel flow from surface layer, impoundment of surface runoff, crop growth and water uptake are considered Evapotranspiration is considered as the sum of soil evaporation and plant transpiration	Pesticide transformation and degradation to metabolites, pesticide partitioning between soil and solution phase in soil layers and extraction to runoff in surface layer, pesticide transport with sediment Nitrogen and phosphorus cycling, including nitrogen fixation by legumes, crop nutrient uptake, two-stage mineralisation (first order) of ammonification followed by nitrification, ammonium volatilisation, inorganic and organic fertiliser application Effects of crop type, planting, harvesting, tillage operations on nitrogen and phosphorus dynamics also considered Sediment yield from sediment detachment and transport in overland flow and channel flow Sediment deposition in overland, channel and impoundment elements Sediment concentration of each particle type
QUASAR	River flow variation in a reach is represented by a transfer function model of an ordinary, lumped parameter, differential equation of mass conservation Each reach can have possible inputs (tributaries, sewage outfall) or abstractions (industry or irrigation) River flow model coupled to stochastic rainfall – runoff model to account for soil moisture and evapotranspiration effects between upstream and downstream sites	Chemical advection and dispersion downstream is modelled using a transfer function Chemical decay and biological processes such as reaeration, effects on BOD of mass algae death and nitrate decay downstream are modelled as empirical relationships
ADZ	River flow variation in the river reach is represented by a transfer function model of an ordinary, lumped parameter, differential equation of mass conservation	Chemical advection, dispersion and decay (if relevant) within the river reach is represented by a transfer function model

TABLE 1.2 *Continued*

Model	Hydrology	Chemistry
Export Coefficient Model	Catchment hydrology ignored Considered only as a nutrient input to the catchment as rainfall and a nutrient loss as discharge	Chemical processes ignored Nutrient loss rates are calculated from all possible source input rates
SHETRAN	Fully distributed, physically based model Spatial distribution is represented by horizontal grid surface Each grid square is divided vertically to represent surface and subsurface processes Processes modelled include: snowmelt (energy budget method), canopy interception (Rutter accounting procedure; Rutter *et al.*, 1971), evapotranspiration (Penman–Monteith equation), overland and channel (St Venant's equation) flow, unsaturated (one-dimensional Richard's equation) and saturated (two-dimensional Boussinesq's equation) subsurface flow and stream/aquifer interaction	Sediment module: sediment transport and dispersion in overland and channel flow by convection–dispersion equation, soil erosion by raindrop impact, leaf drip impact and overland flow, sediment deposition on river bed and ground surface, overbank transport, bank and river bed erosion and infiltration of finer sediments into river bed Contaminant module: transport in dissolved and adsorbed forms by convection–dispersion equation in channel and overland flow. Also, adsorption on to sediment and soil, absorption into immobile phase, radioactive decay, atmospheric deposition, plant uptake and recycling, exchange between river and river bed, deposition of contaminated sediments and erosion of contaminated soils

The models that we have selected may be grouped under two headings as shown below (with the location in the reference section 1.4 indicated in brackets).

1.2.1 Acidification models

1. *Episodic event modelling*
   (a) Chemical reaction/storage compartment based:
   Birkenes: simulates episodic acidification from considering chemical reactions and hydrological processes in soil stores (1.4.1)
   PULSE: simulates alkalinity/pH in runoff during short-term "pulse" events from semi-empirical equations (1.4.2)
   (b) Mixing modelling based:
   SWAP/Allt a'Mharcaidh: uses mixing models; high-alkalinity groundwater mixes with low-alkalinity peat runoff to explain short-term variations in acidity (1.4.3)
   EMMA: end member mixing analysis — explains temporal stream water variations in acid neutralising capacity (ANC) by mixing, in varying proportions, contrasting chemical sources (1.4.4)
   TOPMODEL: a hydrological model incorporating a topographic index which links subsurface flow components to chemical end members derived from mixing models (1.4.5)

2. *Predicting long-term acidification effects*
   Ion balance: simulates changes in lake alkalinity from acid inputs/ion balances (1.4.6)
   ILWAS: Integrated lake water acidification study — semi-distributed process model, including biogeochemical (canopy and root zone) processes to explain lake acidification from acid deposition (1.4.7)
   Trickle Down (TD): simulates episodic events using a daily time step at a catchment scale to predict lake acidification (1.4.8)
   MAGIC: Model of Acidification of Groundwater in Catchments — uses soil chemical reactions, flow in soil layers and acid deposition to explain stream acidification (1.4.9)
   PROFILE/SAFE: soil acidification in forest ecosystems — uses soil mineralogy, texture and geochemical properties in the soil profile layers to simulate weathering and soil acidification (1.4.10)
   SMART: Simulation Model for Acidification of Regional Trends — uses atmospheric deposition and soil chemistry in a soil compartment to simulate soil acidification (1.4.11)
   RESAM: regional soil acidification model — similar to SMART but includes biological processes in multi-layer compartment modelling (1.4.12)
   MIDAS: model of ion dynamics and acidification of soil — simulates soil acidification using soil chemical reactions (1.4.13)

1.2.2 Models of the transport of nutrients and pollutants

1. *Soil profile*
   LEACHM/N: leaching of nitrate from soil nitrogen pools to groundwater (1.4.14)
   MACRO: transport of nitrate in structured soils (1.4.15)
   BURNS: leaching of nitrate in non-structured soils (1.4.16)
   TETrans: trace element transport in structured soils (1.4.17)

Addiscott and Whitmore: model for nitrate leaching and cycling in structured soils
(1.4.18)

TFM: stochastic transfer function model for nitrate leaching in soil profile (1.4.19)

2. *Field scale*
CREAMS: Chemical, Runoff and Erosion in Agricultural Management Systems —
simulates nutrient and pesticide leaching and nutrient, sediment and pesticide
concentration in surface runoff (1.4.20)

GLEAMS: Groundwater Loading Effects of Agricultural Management Systems — as for
CREAMS but includes percolation to groundwater (1.4.20)

3. *River water quality*
QUASAR: Quality Simulation Along Rivers (1.4.21)

ADZ: Aggregated Dead Zone model for downstream transport of pollutants (1.4.22)

4. *Catchment scale*
Export Coefficient: catchment scale export of nutrients in relation to land use in the context
of eutrophication of rivers and especially lakes (1.4.23)

SHETRAN: movement of contaminants in catchments and groundwater (1.4.24)

1.3 PERSPECTIVES ON SOLUTE MODELS

1.3.1 The practical issue

Research into solute processes and the subsequent development of models for use in prediction
is thus seen as primarily stimulated by the practical issues facing the environment. The "acid
rain" issue prompted funding for developments such as the surface waters acidification
programme (SWAP; Mason, 1990) and also the development of models such as MAGIC,
SMART, Birkenes and ILWAS. In an agricultural context, the economic and environmental
concern for the loss of pesticides and fertilisers into groundwater and streams encouraged the
development of models such as LEACHN, GLEAMS and Burns. More detailed reviews
concerning soil solute leaching models can be found in Addiscott and Wagenet (1985), Nielsen
et al. (1986), van Genuchten and Jury (1987), Feddes *et al.* (1988) and Bache (1990).
Movement of radioactive waste in the environment has also caused concern and models such as
SHETRAN have been developed with this problem in mind.

1.3.2 Solute model structure

The actual nature and structure of the model can be constrained by its objectives (see Section
2.1). In addition, the model can be entirely mechanistic, accounting for the internal processes
within the system by implementing the fundamental laws of physics, or it can be entirely data
based, using statistical techniques such as regression or time-series analysis to define a
relationship between two or more variables without considering the internal mechanisms at all.
In reality, many models fall between these two extremes, incorporating empirical and physics-
based formulations, possibly with simplifications. Also, in any given model the level of
complexity in the hydrology and chemical components may vary. For example, in the long-
term acidification models, hydrological processes are often minimised relative to
considerations of the soil chemical aspects.

Fully physically based models incorporate the greatest degree of realism (according to the current level of knowledge). They are usually relatively complex involving a high degree of parameterisation, but in theory require no calibration. They can be applied in any situation where input data exists to parameterise the model. Models of this type include SHETRAN and PROFILE.

More often, models incorporate simplifications into their structure. For example, many soil profile models simplify water and solute transport by using a series of mass balance equations to route water through a number of vertical layers, in what are usually called box models (e.g. the Burns, Addiscott and Whitmore, GLEAMS and many of the long-term acidification models). These models are less data intensive and easier to implement, which makes them ideal for management use. The Addiscott and Whitmore (1987) model is a good example of a management-oriented model, and likewise SMART and MAGIC, the former having been used to assess the impacts of different deposition scenarios in the RAINS decision support system (Alcamo *et al.*, 1990). Many catchment models (Birkenes, PULSE and ILWAS) also divide the catchment into a number of hydrological stores. Like the soil profile models, each store or layer is assumed to be chemically and hydrologically homogeneous.

Model simplifications may also occur due to lack of understanding and uncertainty. This is highlighted by the increased use of chemical end members derived from catchment chemistry data and simple mixing relationships and incorporated into hydrology codes (TOPMODEL and Allt a'Mharcaidh). Omitting a process because little is known about the subject is also common; e.g. in the long-term acidification models internal cycling of nutrients in the upper soil layers and the organic complexation reactions with aluminium (except ILWAS) are not included (Warfvinge and Sverdrup, 1992; De Vries *et al.*, 1989).

At the other extreme are the empirical and simple deductive models. These describe a relationship, which is usually statistical, between two or more variables. Examples of these are the concentration – discharge relationships derived from experimental data (e.g. Johnson *et al.*, 1969; Hall, 1970), mixing model relationships (e.g. Eshleman, 1988), the multiple regression models of episodic acidification (e.g. Lynch *et al.*, 1986; Gerritsen *et al.*, 1990) and the empirical charge balance lake acidification model of Henriksen (1980). A review of episodic acidification models can be found in Eshleman *et al.* (1992). Another technique known as time-series analysis can be used to describe input – output behaviour and is used to model rainfall runoff in the QUASAR package (Whitehead *et al.*, 1981). The relationship derived from the data in one situation are not applicable to other catchments or rivers because they do not take account of the internal mechanisms of the system — they are known as black box models.

Chemical processes

The complexity of the representation of chemical processes in solute models varies considerably from model to model. In some cases, only chemical transport is considered and chemical reactions are ignored. Leaching models such as Burns' model consider solute movement only through the soil profile as a series of mass balance equations. More physically based approaches are employed by LEACHN, MACRO and SHETRAN where the CDE (convection – dispersion equation) (see Nye and Tinker, 1977; Sposito *et al.*, 1986) is used to model solute movement. The inclusion of processes that affect solute transport, such as soil adsorption and desorption and plant uptake, can be included in the CDE by adding extra terms in the equation. SHETRAN, MACRO and LEACHN all allow for these extra processes.

Many of the long- and short-term acidification models (Birkenes, ILWAS and MAGIC) assume the charge balance principle, which incorporates the concept of the mobile anion (Seip, 1980; Reuss *et al.*, 1986). This assumes that the deposition and transport of strong acid anions through the catchment is generally unaffected by chemical reactions (although this is not always the case due to anion retention and biological influence). The negative charge associated with the anions is balanced by an equivalent positive charge from the base cations or, when they are depleted, by acid protons and aluminium.

The range of first-generation acidification models varies in the representation of internal soil geochemical processes and in how they are implemented. Soil chemical processes in these models include cation exchange, weathering, sulphate adsorption and desorption, carbon dioxide and gibbsite dissolution. These geochemical processes have little importance in the transport of nutrients and pollutants and are not included in the solute leaching models.

Cation exchange reactions can be either described by the Gaines—Thomas equations (Gaines and Thomas, 1953; as in MIDAS, RESAM, SMART and MAGIC) or by the Gapon equations (see Bohn *et al.*, 1979, p. 154; Greenland and Hayes, 1981, pp. 140 – 1; as in ILWAS, SAFE and Birkenes). Simplifications and assumptions about the relative importance of specific ions involved in cation exchange allow for further differentiation between models. ILWAS, MAGIC, RESAM and SAFE consider all the base cations in the exchange, whereas Birkenes, SMART and MIDAS consider base cations as a lumped divalent species representing calcium and magnesium. Conversely, for acid cations, Birkenes and ILWAS consider only hydrogen, whereas MAGIC considers aluminium; SMART and RESAM include both acid cations and MIDAS considers the acid cations as a lumped trivalent species assuming that cation exchange in forest soil is always influenced by aluminium. In contrast to the above, the TD/ETD models describe cation exchange using a first-order kinetic expression and the PULSE model uses a semi-empirical relationship to describe alkalinity production from cation exchange and weathering.

Methods for including mineral weathering also vary. It can be a model input (zero order; as in SMART and RESAM) of base cations, determined from input—output budget data or from the chemical analyses of soil samples such as strontium isotope techniques (Jacks *et al.*, 1989). As a model input it may be used as an adjustable parameter in calibrating the model against observed data (e.g. MAGIC). On the other hand, the ILWAS model describes weathering by a first-order, pH-dependent equation controlled by a rate constant. The PROFILE and SAFE models are unique in that they consider weathering in entirely mechanistic terms (employing transition state theory), using soil mineralogical data and other soil physical data as input. In contrast, the representation of aluminium hydroxide or gibbsite dissolution is similar in all acidification models. It is described by a lumped equilibrium equation which accounts for all reactions between hydrogen and aluminium. An exception to this is ILWAS where the reaction is assumed to be rate limited.

Sulphate adsorption and desorption is normally represented by the Langmuir isotherm (MAGIC and RESAM) or a linear adsorption isotherm (as Birkenes, ILWAS, TD and ETD), but is not included in all acidification models. In the application of SMART by De Vries *et al.* (1989), sulphate adsorption is assumed to be negligible. It is also assumed to be entirely mobile in the SAFE model.

In the first-generation acidification models (as Birkenes and ILWAS) soil chemical processes are represented using established theories of ion exchange and soil—soil solution equilibria. These are lumped into homogeneous compartments thought to represent chemically and hydrologically different soil layers. However, with the increasing availability of catchment chemistry and isotope data, uncertainty in model assumptions has emerged (Christophersen *et al.*, 1993).

Firstly, many of the chemical formulations used to represent reactions such as cation exchange did not match data collected in the field (Mulder *et al.*, 1989; Neal, 1992). Secondly, large variations in the soil chemistry data were observed within the soil horizons which in the models were assumed to be homogeneous (Campbell *et al.*, 1989). Finally and from a hydrological standpoint, data from conservative tracers such as oxygen-18 and chlorine led to the discovery that hydrological structures were overparameterised; i.e. a number of parameter sets gave equally good model fits and a unique set of parameter values could not be identified (Christophersen *et al.*, 1985; Lindström and Rodhe, 1986). Also, hydrograph separation using isotope data also conflicted with ideas on storm runoff generation, suggesting that the storm event was not always dominated by rainfall (Hooper and Shoemaker, 1986; Sklash *et al.*, 1986). A review of the state of science in hydrochemical modelling can be found in Christophersen *et al.* (1993).

This uncertainty has been reflected by simplifications in the chemical structure of hydrochemical models. In some cases, stream water chemistry could be described as a mixture of soil waters, or end members, and in a way that does not imply, or need description of, the chemical processes. The end members are assumed to be constant at the temporal scale of catchment response and are assumed to mix conservatively (Christophersen *et al.*, 1990; Hooper *et al.*, 1990). They have been used in conjunction with hydrology codes (Robson *et al.*, 1992) to enable inferences to be made about hydrological processes and therefore go some way towards defining a more identifiable hydrological structure. They do not, however, account for any chemical processes themselves; they are merely an end point or "quasi" end point in the chemical evolution of water travelling through the catchment. At the scale in question, little is known about how an end member acquires its chemistry or how it is transformed from one end member to another, if indeed it is. If these transformations occur they may be a function of the chemical reactions occurring in the substrate through which the end member flows or of the initial chemistry of the end member or of the residence time of the water in the system. Attempts to resolve this problem are addressed by Ferguson *et al.* (1994) who propose a combined mixing and uptake model to account for variations in stream chemistry.

Hydrological processes

The long-term soil acidification models (SMART, RESAM, MAGIC, MIDAS and SAFE) lump the hydrology into a one- or multi-layer compartment where vertical flow only is considered. Annual discharge, precipitation and evaporation data are used in water balance calculations.

There are also examples of soil profile and field-scale models where the soil horizon is represented by conceptual layers or compartments (e.g. Burns, TETrans, MACRO, LEACHN, Addiscott and Whitmore (1987) and GLEAMS). The conceptual layers may or may not represent actual soil layers and again only vertical movement through the soil profile is considered. Although LEACHN and MACRO consider the soil as layers, downward movement of water is modelled using the Richards equation (see Hillel, 1971, p. 109).

The distribution of water in the soil layers is often represented in two phases. This division is an attempt to account for the structured nature of soils. A mobile phase accounts for the water in larger pores where most flow occurs and the other immobile phase represents water held in the smaller pores of the soil matrix. Models that incorporate this include TETrans, Addiscott and Whitmore (1987) and MACRO. The exact definition of the phases varies from model to model. In the Addiscott and Whitmore model the division is fixed, being estimated from soil survey data. In TETrans the division is dependent on the current residual water content of a layer, where MACRO defines a micro- and macropore phase with flow being modelled by physical equations in both phases. All three models account for water movement between the phases.

Agricultural field-scale models are represented here by GLEAMS (Leonard *et al.*, 1987), but other models which are similar in structure to GLEAMS do exist, e.g. the pesticide root zone model (PRZM) developed by Carsel *et al.* (1985) and the agricultural non-point source (AGNPS) model (Young *et al.*, 1989). There are a number of similarities between these models, for instance, all three employ the soil conservation service (SCS) curve number method of estimating surface runoff and all three use a modified form of the Universal Soil Loss Equation (USLE) to estimate soil erosion yield. The AGNPS model and PRZM are less data intensive that the GLEAMS model. The PRZM is primarily concerned with estimating the fate of pesticides in the soil and a major difference between this and GLEAMS is in the characterisation of the surface layer of soil. In PRZM it is assumed to be much thicker, which makes the soil less responsive to rainfall in producing surface runoff and soil erosion. The inclusion of surface runoff is important in determining the sediment and the concomitant chemical load removed from the field.

Evapotranspiration processes are included in all the soil profile layer models, usually as a series of model inputs. GLEAMS uses meteorological data to calculate evaporation and transpiration separately.

Many of the catchment acidification models represent the hydrology with a number of stores or compartments which can be likened to different parts of the catchment or different hydrological processes such as interception, unsaturated soil water and groundwater movement. For example, Birkenes uses two compartments. The upper soil layer represents quick flow runoff from the upper soil layers and the lower compartment describes base flow from deeper soil layers. The Allt a'Mharcaidh model uses five conceptual stores but these represent different soil types in different parts of the catchment. The ILWAS model has a great number of stores for canopy, snowmelt, soil, unsaturated zone and groundwater processes. In both soil layer models and the catchment model using stores, the routing of water is controlled by mass balance equations with thresholds that control the maximum storage in a compartment and recession coefficients to determine the rate at which water is released from the catchment store. The incorporation of lateral flow processes into lumped compartment models in addition to vertical percolation increases the number of outputs from a store and therefore the number of parameters. An increase in the number of model parameters may result in problems in identifying a unique parameter set for the model (see Wheater and Beck, Chapter 11 of this volume).

The incorporation of channel routing into hydrological representations is important to account for changes in the magnitude, the speed of arrival and the shape of the flood wave at the basin outlet. A number of methods, both empirical and physically based, are available to account for channel routing effects, a detailed discussion of which can be found in Fread (1985). Of the models mentioned in this review, SHETRAN explicitly includes the effects of channel and overland flow routing by the numerical solution of the one-dimensional St Venant equations. More empirical methods are employed by TOPMODEL. Overland flow is routed using a time-delay histogram related to the current saturated contributing area and channel routing velocity is estimated by a non-linear algorithm derived from site discharge—velocity measurements.

Snowmelt is recognised as having an important role in acidification, resulting in the production of acidic pulses with the associated meltwater. This process is recognised as being important and is included in the Birkenes, ILWAS, TD and ETD models. The amount of meltwater produced is related to mean air temperature using the degree day method (Chen *et al.*, 1982; Schnoor *et al.*, 1984; Rustad *et al.*, 1986). In ILWAS the calculation is more complex as it takes into account the aspect of the subcatchment.

Biological processes

A further challenge also lies in integrating the biological components of the system with the hydrochemical components, as Swank (1986) has already done for streams. The Hubbard Brook ecosystem study (Likens *et al.*, 1977; Swank, 1986; and Velbel, 1992) suggests that there is a great deal of biotic regulation of solute losses from catchments (and see Section III, Chapters 6, 7 and 8 of this volume).

Some acidification models attempt to incorporate biological factors. For example, the ILWAS and RESAM models consider the effects of foliar exudation and uptake, litterfall, root decay and plant growth on solute dynamics in the catchment. They are, however, unique in acidification models which tend either to ignore the biological influence or reduce it to a net model input either to allow for model simplification or because not enough is known about the processes. The models SMART, MAGIC, SAFE and MIDAS all define net nutrient uptake as model input. Data for model input are estimated from input—output budget studies (SAFE) or the input is calibrated against observed soil and stream concentrations (MIDAS and MAGIC).

Nitrogen cycling is included explicitly in nutrient leaching models such as Addiscott and Whitmore, LEACHN, AGNPS and GLEAMS and also, with less complexity in soil acidification models such as MAGIC, ILWAS, RESAM and SMART. This primarily reflects the importance of the nitrogen cycle to the objectives of the models. In the nutrient leaching models it is important to account for all the transformations affecting the size of mineral nitrogen pools in the soil, whereas, in the soil acidification models, nitrogen mineralisation processes are just one source of acidity. Processes affecting nitrogen cycling include mineralisation, nitrification, denitrification, ammonia volatilisation and nitrogen uptake by plants. Not all these processes are included in these models. LEACHN and GLEAMS consider all the processes, whereas Addiscott and Whitmore's model considers all but the process of denitrification which is considered to be negligible in the cool winter periods for when it was designed to be used. Nitrogen uptake is generally incorporated into acidification models as an input, except in MAGIC where it is described by a first-order reaction. However, the rate constant for this reaction is determined by calibration against observed data. Nitrification is assumed to be a first-order reaction in SAFE and RESAM, but as a model input in SMART. RESAM and ILWAS have the most comprehensive descriptions of nitrogen cycling out of the acidification models including the effects of mineralisation and denitrification.

1.3.3 Incorporating spatial information

One of the problems facing solute modelling is to what degree do spatially distributed processes, both hydrological and chemical, have to be incorporated into model structures. This is partly defined by the model's objective. For example, if a model is to map a spatial process such as the movement of a contaminant plume in an aquifer, then some form of spatial representation is required. The potential of GIS (geographic information systems) techniques have yet to be fully realised in this context, but this is an actively developing field. However, if the model is only required to predict the output at a single point in space such as at the basin outlet then the question arises as to whether and, if so, at what level spatial information needs to be included to accurately model the system, or, to put it another way, will model predictions from a spatially lumped representation be significantly different from a more distributed structure?

Although models can be termed either spatially lumped or distributed, the reality is a continuum and even the most spatially distributed model is lumped at some scale. SHETRAN is probably one of the most distributed catchment models to be developed. SHETRAN imposes a horizontal grid over the catchment. Each quadrangle defined by the grid is divided into vertical layers to represent canopy and subsurface processes. Transport of solute and water within the catchment is tracked using physical equations and parameter values need to be estimated for these formulations for each grid node in space. However, a number of problems exist relating to the spatial heterogeneity of hydrological and chemical processes at the large scale. These are illustrated below with reference to SHETRAN and other models. These problems are applicable to many other lumped or disaggregated models and are discussed in more detail in a hydrological context by Beven (1989) and in a hydrochemical context by Christophersen *et al.* (1993).

Firstly, SHETRAN employs numerical solutions to physical equations. These equations were derived for small-scale homogeneous environments. Applying these equations to a highly heterogeneous large-scale system does not follow any theoretical framework, but they are applied without correction and with the same parameter values. This aggregation problem is also apparent for the leaching models such as MACRO and LEACHN. Although the scale here is more representative of that in which the equations were derived, the spatial variability observed in measured field parameters, often over two orders of magnitude, means that deriving representative parameter values is difficult (Nielsen *et al.*, 1973).

A second problem lies in relating measured parameters to the larger grid scale. Parameters such as infiltration capacity are measured on the relatively homogeneous decimetre scale, but they are assumed to be identical when lumped to the large, heterogeneous scale of the model. Again there is no framework for doing this and alternatively there are also no field techniques to measure these large-scale effective parameters. Estimating parameter values in lumped compartment models such as recession coefficients are prone to similar problems. The parameters do have some physical meaning, but they are not measurable using field techniques. Synthetic data analyses, such as that of Binley and Beven (1989), are addressing this problem. It is hoped these techniques will be able to determine the type and density of measurements required to uniquely identify and calibrate a model structure.

The long-term acidification models (SMART, RESAM, MAGIC and MIDAS) all have similar representations of spatially distributed processes. The hydrology is lumped into vertical flow in a one- or two-layer soil compartment operating on an annual time step. Geochemical and biochemical reactions operate within this compartment which represents soil types and different deposition input scenarios over a national or regional scale. Whether this level of lumping affects the accuracy of the results or whether it is justified to simplify the processes to this degree can in some way be answered by the following examples. Jenkins and Cosby (1990) tested three different structures of the MAGIC model; these were one-layer, two-layer and two-layer with flow routing versions. The results differed significantly from one structure to another, again highlighting the problem of relating measured parameter values to lumped and distributed model structures. However, these differences were small when compared to measurement error and for this particular application a one-layer model was assumed appropriate. The degree to which processes may be lumped may be site specific. Another application of MAGIC (Hooper and Christophersen, 1992) at Panola Mountain (Georgia, USA) suggests a two-layer model is more appropriate to predict the response to acid deposition. The "average" one-layer model was unable to account for the dominant influence of a thin organic horizon.

Spatial variability due to differing soil types and deposition inputs can be accounted for in long-term acidification models by including a stochastic element. This allows for variation in model parameter and input data. Such a technique is known as the Monte Carlo analysis. This requires information on the distribution of parameter and input data for measurements in the field. Parameter and input data are then randomly selected to produce an output value. This process is repeated for a large number of simulations until a distribution of output values can be obtained for the corresponding range in input data. For examples, see Kros *et al.* (1993) (RESAM) and Whitehead *et al.* (1981) (QUASAR).

So it seems that when incorporating spatial information a number of problems exist. Firstly, there is the problem of lumping physical equations derived for small-scale homogeneous environments to the large heterogeneous scale. Also, there is the problem of relating measured field data to the scale of the model. This is further confounded by the spatial heterogeneity observed in the field, although this can be accounted for to some degree by incorporating stochastic techniques. At least for the case of the MAGIC application (Jenkins and Cosby, 1990), it still seems that the lumping of spatial processes is valid. Even though varying the structure of the model to incorporate more spatial information did significantly affect the output, the differences were smaller than the measurement errors in the data. However, as the MAGIC application at Panola (Hooper and Christophersen, 1992) suggests, attention must be paid to the characteristics of the application site before specifying and applying a particular model structure.

The lumping procedure gains further support from synthetic data studies by Neal (1992) and Neal and Robson (1993) who have been concerned with applying non-linear chemical equilibria derived for homogeneous environments as lumped averages in heterogeneous environments. Their results suggest that average soil properties can be related to average soil chemistry using the same chemical equilibrium theory derived for homogeneous environments, but a quantitative assessment is still required (Christophersen *et al.*, 1993).

Problems in incorporating spatial heterogeneities, such as macropores in soil structure, could have important consequences for the transport of solutes. Indeed, the importance of flowpaths in determining stream chemistry has been noted (Neal *et al.*, 1988; Hooper *et al.*, 1990; Robson *et al.*, 1992), as have the importance of macropores in rapidly transferring water to the stream during an event (Buttle, 1994). Therefore it would seem necessary to include them in any modelling effort. However, their spatial characteristics make them difficult to represent in a model, i.e. they only occupy small areas of catchment, but can have a disproportionately large effect on the transport of water and solutes. Time-varying spatial heterogeneity could exacerbate the problem further. For example. a clay soil that dries in the summer may crack, creating macropores. These macropores will disappear once the clay begins to wet up and swells again. Will a lumped average parameter for an effectively "homogeneous" compartment accurately estimate any contribution from this spatially discrete source or is a more spatial representation required? If this is the case, can the occurrence and relative importance of these spatial heterogeneities be predicted without the need for intensive field investigations? To answer these questions more studies are needed which integrate field work and modelling efforts. Especially, there is a need to focus on how present methods of measuring parameters can be related to lumped model structures (Beven, 1989; Christophersen *et al.*, 1993) or whether new methods at a more appropriate scale must be developed.

Finally, an alternative approach is suggested by Kirchner *et al.* (1993) who state that "there is a need for techniques that predict or infer catchment geochemical behaviour, without requiring detailed data concerning catchment flowpaths". They go on to propose a multiple regression technique to predict the effects of acid anions on water quality by isolating geochemical and hydrological influences rather than integrating them, which so many of the hydrogeochemical models tend to do.

1.3.4 Model utility

It is worth reflecting on our view that development of solute models has been application driven: Bishop (1990) was able to conclude that governments would wait in vain if they wanted to use the range of SWAP acidification models as a basis for decision making. It is certain that scientific research often inevitably makes things more complicated (Trudgill and Richards, 1994), but it is also certain that while many models produced have been site specific, many have been able to give reasonable answers or predictions. The fundamental question here is "what is the simplest way a system may be modelled and still give reasonable answers that are useful to management" rather than saying how complicated any system is (Trudgill, 1988; Trudgill, 1994). There is perhaps a divergence between scientific effort in attempting to be more realistic or integrated and management which appears to want a simple answer. If models are to be judged in terms of their reliability and utility, it does not seem that substantial progress has been made but that there is some way to go. The achievement of the past two decades has perhaps been that of increasing awareness of what might be involved in solute modelling rather than of a consolidated consensus. Once the models have been described below in Section 1.4 and selected aspects discussed in subsequent chapters, the use of solute models in environmental management is considered in the final chapter of this book.

1.4 SUMMARIES OF SELECTED MODELS

Acidification models
1. Episodic event modelling
  (a) Chemical reaction/storage compartment based

1.4.1 Birkenes

The Birkenes catchment in southern Norway is in one of the areas most seriously affected by acidification. This prompted a study, commencing in 1971−72, which involved the collection of data on precipitation, runoff volume and chemistry. From these data a hydrological model was developed by Lundquist (1976, 1977). On to this a sulphate submodel was attached

(Christophersen and Wright 1981a, 1981b) and a cation submodel was added later (Christophersen *et al.*, 1982). The model is conceptual in nature with the hydrology being described by a two-store model. The upper store produces storm flow and can be likened to the organic soil horizons while the lower store produces base flow effectively from the mineral horizons. Evapotranspiration is calculated from temperature. A third store was added by Rustad *et al.* (1986) to describe snowmelt processes.

Initially the model did not simulate chloride and sodium, because their concentrations in rainfall and discharge were generally in balance and sodium is more mobile than other cations at Birkenes. Failing to simulate chloride delayed the discovery of an important model flaw. This was discovered when an attempt to model the conservative behaviour of the oxygen isotope (^{18}O) tracer by Christophersen *et al.* (1985) failed to reproduce the damped tracer response observed in the stream water. This deficiency was verified by Stone and Seip (1989) when they attempted to model chemically inert chloride. The hydrological submodel was subsequently modified to include piston flow (Seip *et al.*, 1985), which involved enhancing the mixing within the model by transferring water from the upper store to the lower store whilst removing an equivalent amount of water from the lower store.

The chemistry is based on the mobile anion concept, with sulphate being considered to be the dominant anion. Chemical processes include: cation exchange (Gapon approach as M^{2+}; see Bohn *et al.*, 1979, p. 154; Greenland and Hayes, 1981, p. 140), gibbsite weathering (equilibrium) in the upper compartment, gibbsite and base cation (first-order reaction) weathering in the lower compartment, sulphate adsorption/desorption (linear isotherm) and mineralisation. Sulphate mineralisation is modelled by an empirical equation and is dependent on water saturation and temperature. Rustad *et al.* (1986) included an equilibrium reaction for carbonic acid chemistry and an equilibrium reaction for organic acids ($RCOO^-$). The organic acids were calculated using a formula based on research by Oliver *et al.* (1983) from measured dissolved organic carbon concentrations and pH. Working on a daily time step or less, the model is designed for simulation of short-term events. Input data includes precipitation volume, sulphate and chloride wet and dry deposition and mean daily air temperature. Output includes discharge, sulphate, calcium, magnesium, pH, aluminium, bicarbonate, sodium, chloride and organic acids. A more detailed review and critical update of this pioneering model is given in Chapter 15.

1.4.2 PULSE

The PULSE model was primarily developed for the simulation of short-term variations ("pulses"), in runoff, alkalinity and pH on a daily time step in flowing waters. It is a conceptual model, semi-empirical in nature, operating on the catchment scale (Bergström *et al.* 1985; Lindström and Rodhe, 1986). The model considers water movement through the catchment, as a non-mixing pulse of known volume and hydrochemistry — hence the model name. A recent application of PULSE can be found in Sanner *et al.* (1993).

Hydrologically, the model has evolved from the HBV model (Bergström and Forsman, 1973; Bergström, 1975). Spatial variability in the hydrology is accounted for by having different storage compartments for different parts of the catchment. Snow and rainfall entering a compartment passes through an interception zone, a snowmelt zone, an unsaturated zone and on to a saturated zone. Within the unsaturated zone, there are routines to account for soil moisture and evaporation. Once in the saturated zone each pulse of rain or snowmelt is treated individually and drains by a recession coefficient relative to the level in that zone. This enables the age of a water pulse to increase with depth and the pulse volume to decrease exponentially with time. Each storage compartment has different characteristics, i.e. different recession coefficients, and is linked via routing procedures to produce a hydrograph at the basin output. Lakes can also be considered. Lindström and Rodhe (1986) modified the model by adding additional storage to it. This enabled the simulation of the dampened response of the isotope tracer, ^{18}O. Model inputs are daily precipitation, daily mean air temperature and monthly standard values of potential evapotranspiration.

Alkalinity is modelled with both physical and empirical relationships (Bergstrom *et al.*, 1985). Cation exchange, weathering processes and seasonal variation in carbon dioxide levels in the soil are represented empirically. Water entering the soil is assigned an initial fixed alkalinity value, assuming that any variation in rainfall alkalinity will level out in the organic soil horizon. Alkalinity increases due to cation exchange and weathering processes can be modelled by two methods: either by taking into account the residence time of the pulse, increasing from a minimum to a maximum value at time infinity, or relating alkalinity to the level of the pulse in the storage compartment, i.e. the deeper the pulse the higher the alkalinity. Both relationships are "semi-empirical" in nature. pH is determined from the alkalinity−pH relationship. Atmospheric chemical input is not considered, except for the pH of rainfall on to lake reservoirs.

Another version of PULSE incorporating inorganic nitrogen leaching also exists (Bergström *et al.*, 1987). Here, the authors have considered fertilisation, mineralisation, atmospheric fallout, plant uptake, leaching and denitrification to be the main processes involved. The model routines describing these processes are empirical in nature and have been influenced by agricultural studies and the chemical formulations in the following models: CREAMS (Knisel, 1980), the SOIL-N model (Johnsson *et al.*, 1987) and the NITCROS model (Hansen and Aslyng, 1984).

(b) Mixing modelling based

1.4.3 The surface waters acidification programme (SWAP) and the Allt a'Mharcaidh model

The surface water acidification programme (SWAP; Mason, 1990) was a five-year programme to study the physical, chemical and biological processes controlling the acidification of streams and lakes and to determine their effects on fisheries. A comprehensive field programme was initiated covering several different disciplines, including studies on land use, hydrology, soil, soil water and surface chemistry, acid deposition, palaeolimnology and effects on aquatic biota. A number of catchments at different stages of acidification were monitored, including the "intermediate" site at Allt a'Mharcaidh, the highly acidified Birkenes catchment and the pristine Høylandet site in northern Norway. Many studies were concentrated at the well-monitored Allt a'Mharcaidh site including the application of the acidification models MAGIC and Birkenes.

The Allt a'Mharcaidh catchment has a rapid hydrological response to rainfall events. Typically, the hydrograph can rise and fall within a few hours. Therefore in order to accurately simulate the transient pulses of acidity and aluminium associated with a storm event, a new hydrological model was needed which operated on a time step of one hour (Wheater *et al.*,

1990a). Detailed plot studies and catchment observations (Wheater *et al.*, 1990b, p. 121) led to the formulation of a conceptual model consisting of five stores corresponding to different catchment soil areas. There is an alpine store, two peaty podzol stores and two peat stores. Input is by spatially distributed (over the alpine and peat areas) rainfall and evapotranspiration.

To show the interaction of hydrological and chemical processes, a two-component mixing model was applied to stream data for alkalinity, total organic carbon (TOC), calcium, magnesium, sodium, chloride and sulphate (Kleissen *et al.*, 1990; Wheater *et al.*, 1990a). Results of this analysis suggest that event response is mainly due to low alkalinity, high TOC water with base flow originating from a high alkalinity source. In some cases, a short, high alkalinity response was observed during a storm, superimposed on the base flow component, indicating a third alkalinity source. Using the alkalinity sources identified by the mixing model and assuming they are conservative in nature, a simple hydrochemical model for alkalinity was developed. Essentially, a high alkalinity was assigned to base flow, a low alkalinity was assigned to quick flow from the peaty podzol soils and a further more ambiguous intermediate alkalinity from downslope drainage areas was also assigned. The main features of the alkalinity response was predicted (Wheater *et al.*, 1990a).

1.4.4 EMMA

EMMA (end member mixing analysis) emerged from the increasing availability of soil water chemistry data from catchment studies. EMMA was proposed by Neal and Christophersen (1989) and was developed further and applied by Christophersen *et al.*(1990) in Birkenes in southern Norway and Plynlimon in mid-Wales, at the Panola Mountain Research Watershed in the Piedmont Province, Georgia, USA, by Hooper *et al.* (1990) and at Panola in combination with the MAGIC model (Hooper and Christophersen, 1992).

EMMA is based on the concept that stream water is a mixture of different soil water classes or end members. An end member may be identified from any area within the catchment that has been identified as contributing towards stream chemistry such as groundwater, mineral horizons or hillslope areas. Any number of end members may be used to identify stream concentrations and any number of chemical species may be considered, provided that the soil end members are significantly different from one another. Solutes which are highly correlated with flow are usually the same ones that exhibit marked differences in concentration across the soil horizons and hence are appropriate for EMMA (Christophersen *et al.*, 1990). Thus, water originating from the upper organic horizons may be distinctly high in hydrogen and aluminium but low in base cations and can be remarkably different from water from mineral or groundwater horizons which are usually low in acid cations and high in base cations.

EMMA is based on two assumptions: firstly, that the chemical species involved are governed by fast reactions with the soil and therefore represent a constant end member for the horizon and, secondly, that the end members behave conservatively, i.e. there will be no further chemical reactions taking place on the way to the stream or in the stream. In other words, the end member is constant through space and time relative to stream water concentrations.

Provided that a set of end members can be identified which explain all the stream water variations, then the contribution of each end member to the stream is determined by least-squares methods (Hooper *et al.*, 1990). The end member contribution to total flow can provide information on hydrological pathways. EMMA has been applied to Birkenes (see Chapter 15) and used in conjunction with TOPMODEL (see Chapter 13).

1.4.5 TOPMODEL

TOPMODEL is a quasi-physical, semi-distributed hydrological model (Beven and Kirkby, 1979; Beven and Wood, 1983; Beven, 1987; Quinn *et al.*,1991). The model uses a topographic index to describe spatial variation in catchment topography and relate it to hydrological behaviour. The index is used in calculating the saturated contributing area, which is dynamic in nature, contracting and expanding according to soil moisture status and rainfall input. High indexes saturate first and indicate wet areas of the catchment such as convergent hollows, valley bottoms or flat areas which are unable to drain. Low index values indicate steep slopes. Grid squares with the same index value are assumed to be hydrologically similar and hence the topography can be summarised by a distribution of index values. Digital terrain maps are used to calculate the index (Quinn *et al.*,1991).

Input to TOPMODEL is by hourly, spatially distributed rainfall which is routed initially to an interception store and on to a litter layer store. A proportion of the rainfall bypasses the interception store and enters the litter layer store as direct throughfall (Hornberger *et al.*, 1985). Evapotranspiration can be removed from these and the root zone stores. Water leaving the interception store is routed vertically to the unsaturated zone towards the saturated zone. Vertical percolation in the unsaturated zone is determined by the hydraulic gradient to the local water table depth which, in turn, is linked to the topographic index. The unsaturated store is subdivided into an active and inactive store. This is effectively a root zone and gravity drainage zone respectively. Rain enters the root zone store and, when it fills, it spills over to the gravity drainage store and on to the saturated zone. The threshold between the two zones is effectively a "field capacity" value of the soil. As the water table rises and the zone becomes saturated or near saturated, quick flow is generated. Quick flow may encompass overland flow from saturated areas, return flow and macropore flow and together with subsurface saturated flow comprise stream flow. Subsurface saturated flow is described assuming Darcy's law and is proportional to the local slope. It decreases exponentially with depth and is controlled by a lateral transmissivity value (a value "determined" when the soil is saturated to the soil surface).

A recent study by Robson *et al.* (1991, 1992) shows that subsurface flow contributions identified by TOPMODEL and chemical hydrograph separation compare well. The subsurface flow components identified by TOPMODEL were separated by their depths of origin, whereas flow components identified by chemical hydrograph separation used a deep, well-buffered groundwater and an average soil water concentration for ANC (acid neutralisation capacity). This led to the combination of the two approaches with two ANC end members being assigned to different soil depths in TOPMODEL in order to predict short-term variations in stream ANC. A soil end member of low ANC was assigned to soil below 1 metre and another end member of high ANC to soil above 1 metre. Results indicated that flow from saturated contributing areas makes a significant contribution to total flow. A detailed description of this work is given in Chapter 13.

2. Predicting long-term acidification effects

1.4.6 Empirical ion balance models

One of the simplest models to assess the impacts of acid rain on surface water quality is the empirical, ion balance model of the kind used by Henriksen (1980). This is a steady-state, long-term model which predicts the alkalinity of a lake from the input of atmospheric sulphur and the alkalinity derived from chemical weathering (assumed to be independent of acid input and proportional to the sum of calcium and magnesium). It treats the process rather like a titration between an acid and alkali. The pH is predicted from the titration curve of the bicarbonate ion with a strong acid. The effect of different acid inputs on the acidity of the lake can be quantified by moving up and down the titration curve. The model can be used to assess the critical load of acid deposition on an ecosystem and is useful in the long-term predictions where the effects of ion exchange and other transient processes can be ignored.

1.4.7 ILWAS

The aim of the ILWAS (Integrated Lake Water Acidification Study) was to establish links between acid deposition and lake acidification (Goldstein *et al.*,1980). The project comprised field work and laboratory studies, culminating in a mathematical model (Chen *et al.*, 1982, 1983). ILWAS is a process-oriented, semi-distributed model comprising hydrology, canopy chemistry, snowmelt chemistry, soil chemistry, stream and lake water quality modules. It is an event-based model running on daily time-series data or shorter, but can be used to simulate long-term changes in lake acidification over a number of years. The application of the model is generally limited to research catchments (Chen *et al.*, 1985, 1988; Davis and Goldstein, 1988) since there are large data requirements (see Table 1.1).

The model conceptually divides the drainage basin into a number of subcatchments and stream and lake segments. Each subcatchment is composed of several vertical layers for canopy vegetation, snow pack and soil horizons. Hydrological processes modelled by ILWAS include: canopy interception, throughfall, evapotranspiration, freezing and thawing of soil, snow accumulation and melting. Flow is routed both vertically and laterally through the canopy and layered soil horizons to the streams, via surface, shallow subsurface and deep groundwater flow routes. Ultimately flow enters the lake compartment, which is divided into 12 layers. Subcatchment areas require geometrical information on area, slope and aspect. Land-use and vegetation effects are considered through the use of input data on the percentage land cover of deciduous, coniferous and open areas and monthly leaf indices are used to describe the seasonal variation in canopy cover.

Chemically, ILWAS employs the charge balance principle with the mobile anion concept. Biogeochemical processes include: organic matter decomposition (equilibrium reaction) nutrient uptake (model input), foliar uptake/exudation which is taken as being proportional to dry deposition and the elemental composition of the leaves, litterfall (model input), nitrification (Michaelis–Menten kinetic equation; see Nye and Tinker, 1977, p.105; Wild, 1988, p.161), mineralisation and immobilisation of ammonium, sulphate, calcium, magnesium, sodium and potassium (first-order reaction) and carbon dioxide production and exchange. Soil geochemical processes include: mineral weathering of silicate (first-order pH-dependent reaction), cation exchange (equilibrium reaction) and aqueous chemical equilibria. The aqueous chemical equilibria account for carbonic acid and aluminium complexation processes (with OH^-, F^- and SO_4^{2-}) and the effect of organic acid ligands on acid buffering. Aluminium hydroxide weathering is described by a rate-limited reaction dependent on pH. Anion retention rates for $RCOO^-$, sulphate and H_2PO_4 are described by linear adsorption isotherms.

Most major cations and anions are simulated together with total phosphorus, organic ligands, total inorganic carbon and alkalinity. Precipitation quantity and quality, dry deposition rates and daily maximum and minimum temperatures are model inputs. Model outputs also include lake stage and discharge and soil moisture contents for each compartment.

1.4.8 The Trickle Down and Enhanced Trickle Down models

An earlier version of the Trickle Down (TD) model was used to simulate episodic events and long-term annual averages by Schnoor *et al.* (1984). The hydrological submodel was added by Banwart (1983) and a snowmelt module by Carleton (1986). Primarily, the model is concerned with predicting the effects of acidification on seepage lakes, i.e. lakes that have no input or output from tributaries. Conceptual in nature, it is composed of three submodels: hydrology, alkalinity and aluminium speciation (Lin and Schnoor, 1986). It operates on a daily time step at the catchment scale.

Water flow is described by mass balance equations for the modules of snowmelt, soil and unsaturated soil zones, lake and groundwater. In the soil and unsaturated modules lateral flow and vertical percolation are included. Seepage between the lake and groundwater modules is described by Darcy's law and is assumed to follow an exponential distribution to account for greater discharge nearer the shoreline than towards the centre of the lake. Darcy's law also describes flow between the groundwater and soil and unsaturated compartments. Evapotranspiration effects are included using input data on pan evaporation. The hydrology module is calibrated by varying the lateral flow recession constants and vertical hydraulic conductivities against lake stage and groundwater levels.

Alkalinity is determined through a set of mass charge balance equations for the terrestrial and lake modules. This is calculated using kinetic reactions of cation exchange and chemical weathering rates as opposed to ion exchange equilibria (see Birkenes, MAGIC and ILWAS). Cation exchange rates are proportional to the remaining base concentration and weathering is dependent on soil surface area, flow rate and hydrogen ion concentration (Nikolaidis *et al.*, 1988). pH is determined from a relationship between alkalinity, total inorganic carbon content, carbonic acid dissociation constants, Henry's law and carbon dioxide partial pressure. Inputs required to run the model include precipitation, pan evaporation, temperature, dry and wet alkalinity loadings in rainfall (Lin and Schnoor, 1986). Outputs include discharge and lake stage, concentrations of pH and alkalinity. Simulated alkalinity is calibrated by varying the weathering rate constants against measured alkalinity values.

To adapt the model to drainage basins, an Enhanced Trickle Down (ETD) model was proposed by incorporating a number of new hydrological processes including more detailed representation of overland flow, frozen ground processes and surface water discharge from the lake. A chloride submodel was introduced in order to test the validity of the hydrological structure and a sulphate submodel was introduced to simulate sulphate adsorption in the terrestrial environment and sulphate reduction in the lake sediments. Sulphate sorption is modelled by a linear isotherm equation controlled by a partition coefficient and sulphate reduction is a first-order process subject to a rate constant that is dependent on temperature. Hence, for ETD, sulphate and chloride loading as dry and wet deposition are extra inputs. Applications of ETD include Georgakakos *et al.* (1989).

1.4.9 MAGIC

The model of acidification of groundwater in catchments (MAGIC; Cosby *et al.*, 1985a, 1985b) is designed to predict the long-term effects of acid deposition on soil and surface water quality. The model is a dynamic, process-oriented model which can operate on the catchment to regional scale on an annual time step. MAGIC has been frequently applied to many sites over Europe and North America, including the White Oak Run catchment in the Shenandoah National Park, Virginia, and the Panola Mountain Research Watershed in the Piedmont Province, Georgia (Cosby *et al.*, 1985a, 1985b; Hooper and Christophersen, 1992), Loch Chon and Kelty in central Scotland (Cosby *et al.*, 1990), a regional application to Wales (Whitehead *et al.*, 1990), the Dargall Lane catchment of the Loch Dee system in south-west Scotland (Whitehead, 1992) and several sites in Scandinavia (Warfvinge *et al.*, 1992). MAGIC is covered in greater detail in Chapter 14.

MAGIC lumps catchment soil and soil chemistry into one compartment. Within this compartment, hydrological description is limited to vertical percolation through the layer. The model requires annual precipitation and runoff volume as input. Evapotranspiration can also be considered (Cosby *et al.*, 1990). Other model structures have been applied by Jenkins and Cosby (1989). They compared the lumped one-layer version with a less lumped two-layer structure and a two-layer structure with flow routing. Results highlighted the problem in estimating parameter values in aggregated or distributed structures, because, although the three structures were capable of reproducing surface stream chemistry, there were significant differences in soil and soil water variables. However, these differences were small when compared to measurement errors.

The chemistry of stream and soil water is controlled via a soil—soil solution chemical equilibria in which the concentrations of the major ions are assumed to be governed by simultaneous reactions involving anion retention (of sulphate—Langmuir isotherm; see Greenland and Hayes, 1981, p. 139), cation exchange (of aluminium, calcium, magnesium, sodium and potassium using the Gaines—Thomas expression; see Gaines and Thomas, 1953), aluminium hydroxide equilibria, carbonic acid equilibria and aluminium complexation reactions (with hydroxide, fluoride and sulphate anions). Aluminium dissolution and precipitation is assumed to be controlled by an equilibrium with a solid phase of gibbsite. Input and output of the major ions, to and from the soil, by atmospheric deposition, mineral weathering, net uptake in biomass and loss to runoff is via a set of mass balance equations (Cosby *et al.*, 1985a, 1985b, 1990).

Information on the wet and dry deposition levels over time is needed for the major cations and anions. Base cation weathering input rates (assumed to be constant) and cation exchange selectivity coefficients are derived by calibration against observed data. Physical soil parameters (such as soil and water and carbon dioxide partial pressures, bulk density, depth and cation exchange capacity) are measured from different soil types within the catchment. These values can be "averaged" to produce a vertically aggregated value for each soil type. The vertically aggregated value is then weighted by percentage area of each soil type within the catchment to give weighted mean values for model input. Parameter values obtained from the field are subject to measurement error; hence, in MAGIC ranges of parameter values are chosen to account for this error and the spatial heterogeneity within the catchment (Cosby *et al.*, 1990). Output includes soil and stream concentrations for pH, alkalinity and all major ions, together with organic anions (Cosby *et al.*, 1990).

Cosby *et al.* (1990) included the process of organic acid speciation, treating them as diprotic acids and also refined nutrient uptake in MAGIC by including nitrogen uptake (nitrate and ammonium) as a first-order reaction with a rate constant determined by calibration, and net nutrient uptake rates were considered separately for base cations and sulphate. Rates were scaled and estimated from biomass concentrations and biomass accumulation rates (Miller and Miller, 1976). The effects of forest growth on atmospheric deposition, nutrient uptake and evapotranspiration ware included by using enhancement factors to modify deposition inputs.

1.4.10   PROFILE and SAFE

PROFILE is a steady-state, mechanistic model for the determination of long-term acidification of stream and soil, on a national to regional scale, from independent information on soil mineralogy, texture and geochemical properties (Sverdrup *et al.*, 1987; Sverdrup and Warfvinge, 1993; see Chapter 3 or this volume). It has been used to calculate critical loads of acid deposition in Sweden (Warfvinge and Sverdrup, 1992). SAFE (Soil Acidification in Forested Ecosystems; Warfvinge and Sverdrup, 1990; Warfvinge *et al.*, 1993) is the dynamic counterpart of PROFILE and can simulate changes in soil and stream chemistry over time, requiring time-series data as input. For an example of applications see Warfvinge *et al.* (1992).

PROFILE is a multi-layer model which represents the horizons of a soil profile. Water can flow through the soil layers to recharge groundwater, following vertical and horizontal flowpaths. Within each layer, the model combines a set of mass balance equations with equations to model uptake and release processes. These processes are cation and nitrogen uptake, mineral weathering, nitrification, aluminium, carbonate and organic acid (Oliver equation; see Oliver *et al.*, 1983) equilibria, autoprotolysis of water and cation exchange. All reactions take place via a soil solution submodel. Cation exchange is based on the Gapon mass-transfer kinetic equations and considers base cations as a lumped variable comprising calcium, magnesium and potassium. Mineral weathering is calculated from independent mineralogical data and transition state theory applied to the 42 most common primary and secondary minerals. Sulphate adsorption is ignored. Nitrification is governed by a kinetic reaction and is lumped together with immobilisation. Nitrogen uptake by biomass is by model input and base cation uptake is constrained to be less than the sum of base cations produced by weathering, input from the atmosphere or the overlying soil layer. Seasonal variations in soil moisture content, partial pressure of carbon dioxide and nitrification are ignored and cation exchange capacity (SAFE only) is assumed to be constant over long-term simulation periods. Temperature dependence is considered for all processes involved.

Model inputs include precipitation, runoff, mean annual soil temperature, anion (sulphate, chloride and nitrate) and cation (ammonium, aluminium, calcium, magnesium, potassium and sodium) deposition, nutrient uptake of nitrogen and base cations, together with data on soil texture and mineralogy. Output variables include alkalinity, pH, dissolved organic carbon and most major cations and anions.

1.4.11   SMART

The Simulation Model for Acidification's Regional Trends (SMART) is a forest soil acidity model designed to simulate the effects of soil acidification resulting from the dissociation of carbon dioxide and the deposition of nitrous oxides, sulphate and ammonia. As a submodel of the RAINS (Regional Acidification INformation and Simulation) project (Alcamo *et al.*, 1990) its primary objective is to account for the temporal and geographical extent of soil acidification on a European scale for alternative emission scenarios (De Vries *et al.*, 1989). The model is a process-oriented, dynamic model functioning on a time step of a year. The spatial distribution of soil type is incorporated into SMART by using transfer functions to relate soil survey data to model parameters (De Vries, 1989). An application of SMART can be found in Warfvinge *et al.* (1992).

Model inputs include atmospheric deposition of nitrous oxides, base cations, ammonia and sulphate. Soil chemistry incorporates the charge balance principle and is based on the mobile anion concept (cf. Birkenes, MAGIC and ILWAS). This governs the following chemical processes: base cation and aluminium production from silicate weathering (model input), carbonate weathering (equilibrium reaction), aluminium hydroxide and carbonic acid equilibria, cation exchange (between the lumped base cation (BC^{2+}), hydrogen and aluminium). The base cation weathering from silicate minerals is a model input and the aluminium weathering is proportional to this value. The inclusion of carbonate weathering and aluminium hydroxide dissolution allows SMART to simulate soil behaviour in all buffer ranges. Sulphate retention is ignored with input from deposition assumed to be equal to output (De Vries *et al.*, 1989), but it is included in later versions of the model (see Chapter 12), together with organic acid protonation and nitrogen dynamics.

Some biogeochemical processes are incorporated, but only as zero-order reactions (model input). Nutrient uptake of base cations and nitrogen are modelled, together with net nitrogen immobilisation (nitrification plus immobilisation). Nitrification is assumed to be proportional to deposition. Biological fixation of nitrogen and denitrification are assumed to be negligible in a forested catchment (De Vries *et al.*,1989). Base cations, pH, aluminium, inorganic nitrogen (nitrate and ammonium), sulphate and bicarbonate ions are simulated. Output also includes Al/BC ratios which provide a good indication of forest damage.

The hydrology is represented simply by a lumped, vertical flow, one-compartment model describing the soil layer. The soil can be further subdivided into homogeneous soil horizons, if required (De Vries *et al.*, 1989). Precipitation input is in the form of an annual precipitation surplus value, which is annual rainfall minus interception and evapotranspiration.

1.4.12 RESAM

The regional soil acidification model (RESAM; De Vries and Kros, 1989; Kros *et al.*, 1993) was developed to analyse the long-term response of forest soil to acid deposition and has been linked as a submodel in the overall framework of the integrated Dutch Acidification Simulation (DAS model; Olsthoorn *et al.*, 1990). In this context, RESAM has been used for predicting the environmental impacts of sulphur and nitrogen emissions to evaluate the effectiveness of abatement strategies. The model operates on a regional to national scale with a time base of a year. RESAM is a process-oriented, dynamic model and a more detailed account can be found in Kros *et al.* (1993) and Chapter 12, together with an account of a similar model, NUCSAM, which runs on a daily time step.

RESAM includes a range of biogeochemical processes occurring in the forest canopy, litter layer and root zone. These processes are foliar uptake and exudation, litterfall and root decay, mineralisation, root uptake, nitrification and denitrification and the protonation of organic anions, carbonate dissolution and precipitation, weathering of primary minerals containing aluminium and base cations (sodium, magnesium, potassium and calcium), aluminium hydroxide dissolution and precipitation, cation exchange of hydrogen, aluminium, base cations and ammonium, sulphate adsorption and desorption equilibrium, dissolution and speciation of inorganic carbon.

Foliar exudation, litterfall, root decay, net growth, nitrification and denitrification, organic anion protonation and base cation weathering are all determined by first-order reactions controlled by rate constants. Mineralisation is taken to be equal to the sum of litterfall and root decay. Foliar uptake is assumed to be a fraction of dry deposition and root uptake is calculated from the sum of litterfall, foliar exudation and root decay minus foliar uptake and net growth. Data on the elemental contents of the leaves, stems and roots of tree species are required for the biochemical calculations. Dissolution of calcium and aluminium from carbonates and hydroxides respectively are rate controlled when undersaturated, but are set to equilibrium reactions when supersaturation occurs. Cation exchange is controlled using the Gaines—Thomas expression (Gaines and Thomas, 1953) and sulphate adsorption/desorption is controlled by the Langmuir isotherm.

The soil layer is represented by a multi-layer compartment in which water movement is considered only as vertical flow and lateral flow is ignored. Information on soil layer thickness, bulk density, volumetric moisture content and cation exchange capacity are required. Water uptake by roots per soil layer is assumed to be proportional to transpiration for the soil layer.

Input is by atmospheric deposition, both wet and dry, of sulphate, ammonium and nitrate, base cations, ammonia and chloride. Dry deposition of base cations and sea salt sulphate are described by dry deposition factors. Annual precipitation and temperature data are also required. The output variables include major cations, anions, pH, alkalinity and organic anion concentrations of soil water.

1.4.13 MIDAS

The Model of Ion Dynamics and Acidification of Soil (MIDAS; Holmberg *et al.*, 1989) is a long-term, mechanistic, dynamic, catchment soil acidification model developed to simulate changes in forest soil chemistry due to the deposition of hydrogen ions and base cations, nutrient cycling in the soil and biomass, weathering of minerals and the dissolution of aluminium. Nutrient cycling incorporates nutrient uptake by vegetation and release due to litter decomposition. The model is based on a dynamic model of Oksanen *et al.* (1984). A modified version of MIDAS which accounts for the continuous vertical spatial distribution of soil moisture and ionic concentrations also exists (Holmberg *et al.*, 1985a, 1985b).

Soil chemical processes included in the model include cation exchange which is described by the Gaines—Thomas expression. The equation is slightly modified in that equivalent concentrations are used instead of activities and the cation exchange selectivity coefficients can be split into rate coefficients (Holmberg *et al.*, 1989). The acid cations, hydrogen and aluminium are lumped as a trivalent species and calcium and magnesium are lumped as divalent base cations. Bulking the acid cations as a trivalent species is acceptable only when aluminium is the main exchangeable acid cation and hence applicability of this assumption depends on the prevalent processes at the site in question. Aluminium dissolution is described by an equilibrium equation. Weathering rates of base cations are by model input. Net uptake rates of hydrogen and base cations, a sum of the processes of organic matter decomposition and plant nutrient uptake, are also model inputs. Kinetic equations relate the processes of transport, cation exchange, weathering and nutrient cycling together. The net nutrient uptake rates, the Gaines—Thomas selectivity coefficients and cation exchange rate coefficients are determined by calibration against collected data. Wet and dry deposition of hydrogen, calcium and magnesium are chemical inputs. The dynamics of base cations, hydrogen and aluminium in the soil solutions are predicted.

Soil horizons are lumped into one layer of a known depth. Hydrological input into the system is by annual rainfall, from which evaporation is subtracted. Water flow is by vertical percolation only.

Models of the transport of nutrients and pollutants
1. Soil profile

1.4.14 LEACHM and LEACHN

The LEACHM (Leaching Estimation and Chemical Model) series of models (Wagenet and Hutson, 1989) are deterministic, profile-scale models designed to simulate the movement of agricultural chemicals such as pesticides and nitrate through the soil into groundwater. The pesticide leaching version is described by Wagenet and Hutson (1989) and the version simulating nitrogen dynamics in the soil, known as LEACHN (Hutson and Wagenet, 1991), is described in more detail here.

Water and solute transport in the unsaturated soil is modelled by the physics-based Richards equation (Hillel, 1971, p. 109) and CDE (Sposito *et al.*, 1986) respectively. Included in these equations are terms for nitrogen and water uptake by plants (loss by evapotranspiration) and terms to account for loss and gain by chemical transformation. Adsorption of ammonium, urea and nitrate to the soil surface is modelled and is controlled by a linear adsorption isotherm. The amount adsorbed for a given chemical is regulated by a partition coefficient. The soil can be subdivided into layers to account for depth-varying soil characteristics such as bulk density and hydraulic conductivity. Although macropore flow is not explicitly included in the model it has been suggested by Jabro *et al.* (1994) that specifying a site measured value of infiltration rate as input (which is some measure of macropore flow) can adequately account for preferential flow.

Nitrogen cycling in LEACHN is represented by equations for the transformations between the three organic pools (likened to the litter, faeces and humus fractions), urea and mineral nitrate and ammonium pools. These transformations include the processes of mineralisation, nitrification, denitrification and volatilisation.

Mineralisation of the organic pools and the urea fraction are described by a first-order equation controlled by the amount of nitrogen in the humus and biomass pools. Mineralisation of the litter and faeces organic fractions can be via the humus fraction, as well as directly to the mineral state.

Ammonium volatilisation in the surface layer of the soil to produce ammonia is described by a first-order reaction. Nitrification of urea and ammonium is controlled by the first-order reaction controlled by a rate constant. This constant decreases as a user-defined maximum nitrate/ammonium ratio is approached. Denitrification is described by the Michaelis–Menten kinetic equation (see Nye and Tinker, 1977, p. 105; Wild, 1988, p. 161).

The organic pools are controlled by efficiency and humification factors and C/N ratios. If insufficient nitrogen is produced in the humus fraction by litter and faeces mineralisation then mineral nitrogen is immobilised. This brings about a net decrease in the mineralisation rates. Denitrification rates can be reduced if there is a lack of carbon substrate. Nitrification, mineralisation and denitrification are all dependent on water content and temperature with optimum rates being defined for all three rate constants. A detailed account of the nitrogen reactions used in LEACHN can be found in the description of the SOIL-N model (Johnsson *et al.*, 1987). The effects of plant growth are also included.

The model can run on a daily or hourly time step, with input on rainfall and evaporation. Initial input data on soil moisture retentivity and conductivity are required and can be obtained from the empirical equations of Campbell (1974). The amount of urea and mineral nitrogen leached from the soil profile are model outputs.

1.4.15 MACRO

The MACRO model (Jarvis, 1991) can currently be used to predict the short-term transport of non-reactive solutes and reactive pesticides in macroporous soils. Conceptually, MACRO divides the soil profile into a number of layers and each layer into two phases: the macro- and the micropore phase. Solute transport in the micropore phase is described by the physics-based CDE and Richards equations. Solute transport in macropore flow is by convection only and is related to Darcy's law (Jarvis, 1991) assuming a unit hydraulic gradient. A value of hydraulic conductivity is required by Darcy's law and this is estimated via a power law from the saturated hydraulic conductivity (Beven and Germann, 1981), which, in turn, is dependent on the swelling and shrinkage of the macropores and different management practices. If saturation occurs, excess water is redistributed upwards to the overlying layers; if the surface layer is reached then runoff occurs. Solute transfer between the two phases is based on an empirical approach (van Genuchten and Wierenga, 1976) describing convection and diffusion across a hydraulic and concentration gradient respectively.

The model runs on an hourly or daily time step and the model is driven by a rainfall input of known concentration, as well as evaporation. The effects of irrigation can be modelled provided input on irrigation dates, amounts and concentrations is available. Water uptake by plants (from the macropore phase) is described by an empirical relationship (Jarvis, 1989). Output includes water table depth, drainage and solute and pesticide concentration. The MACRO model is discussed in greater detail in Chapter 5.

1.4.16 Burns model

The solute leaching model of Burns (1974, 1975, 1976) can be used to describe the vertical movement of a surface-applied solute in the soil profile. Leaching of non-reactive species such as chloride can be simulated and the model has been applied to crop systems to simulate nitrate leaching (Burns, 1980; Burns and Greenwood, 1982). The model has been adapted (Burns, 1976) to describe the leaching of a solute initially spread uniformly throughout the soil.

Functional in nature, the model considers the soil profile as a number of layers of known depth. Water and solute entering the top layer is assumed to completely mix with water and solute already present in that layer. The infiltrating water causes a temporary increase in water content. If the water content is in surplus of field capacity then the excess water is displaced downwards. This process is repeated for each layer by mass balance calculations until either the bottom layer is reached or the infiltrating water does not cause the field capacity of a layer to be exceeded. Solute is considered only to be transported by displacement and all rainfall is assumed to infiltrate. Evaporation effects (Burns, 1974) are incorporated in the model by defining an evaporative limit. Water lost from the top layer reduces the water content until it reaches the evaporative limit. Below this limit water and solutes are drawn upwards by capillary rise from the underlying layers.

Recent work by Scotter *et al.* (1993) using a transfer function form of the Burns equation has suggested that the equation is able to describe solute transport with some preferential flow. This is consistent to the soil being likened to a set of independent flow tubes in which the soil solution travels at varying speeds and not as Burns suggested that soil solution and drainage water are completely mixed.

The Burns model requires the minimum of information: rainfall, evaporation, water content at field capacity and initial solute concentration.

1.4.17 TETrans

Several functional models have attempted to describe preferential flow in structured soils (Addiscott, 1977; van Ommen, 1985a, 1985b; and Corwin *et al.*, 1991). One such model developed by Corwin *et al.* (1991) and known as TETrans (Trace Element Transport) simulates the vertical movement with trace elements and non-volatile organic chemicals in the unsaturated zone. TETrans subdivides the soil into a number of layers. Each layer considers water in two phases: an immobile and a mobile phase. The water content of the immobile phase is not fixed, but is defined as a fraction of the current residual water content before input. This allows for the displacement of all solutes above the minimum water content which may be useful in long, low-intensity events. The model uses a mass balance approach to account for water and solute movement down the soil layers. If infiltrating water causes the field capacity of a layer to be exceeded then the excess is displaced or bypassed to the underlying layer. It is assumed that a fraction of all infiltrating water to a layer is subject to bypassing flow.

Bypassing flow is controlled by a mobility coefficient which is defined as the fraction of soil liquid phase subject to displacement. It is both spatially and temporally variable and is estimated from the deviation of measured solute concentrations (such as chloride) from concentrations assuming complete mixing and piston-type displacement. A depth-averaged and/or a time-averaged coefficient may be used to account for spatial variability in the prevalence of macropores.

Plant water uptake is included in TETrans and requires information on evapotranspiration and root distribution. Relative water uptake for a given layer is determined by the root distribution in the profile, which can be either linear or exponential. Root growth is taken to increase linearly from planting time to harvesting time. Water uptake and evaporation do not bring about the upward movement of solutes but instead cause concentration in a given layer. Water uptake is restricted by defining an evaporative limit (the water content at wilting point) below which no water can be removed. Instantaneous chemical equilibration between the mobile and immobile phases is also assumed: once after solute and water have infiltrated and again after plant water uptake. Solute uptake by plants is not considered in the model.

Input requirements are small using only rainfall/irrigation and evapotranspiration on a daily time step, plus layer depths, initial solute concentrations and water contents, physical soil properties of the soil layer and crop date such as the dates of harvest and planting, days of maturity and plant root distribution for each crop. Measurements of chloride or another conservative solute in the soil solution are needed to calibrate the mobility coefficient. Output includes water content, drainage and solute concentration over the soil depth.

1.4.18 Addiscott and Whitmore

The management-oriented, conceptual model developed by Addiscott and Whitmore (1987) is designed to simulate changes in soil mineral nitrogen and crop uptake of nitrogen in the soil profile. It can provide information to farmers on nitrogen leaching losses so that fertiliser applications may be adjusted accordingly. Primarily, it is designed for use in the autumn, winter and spring seasons when winter wheat is sown and harvested, although it can be extended to full season simulation periods. The model considers the processes in the following order: leaching, mineralization, nitrification and crop growth, crop uptake of nitrogen and evaporation. Denitrification is not included in this model version as the temperatures associated with the cooler seasons do not encourage it (Addiscott and Whitmore, 1987). Both ammonium and nitrate are treated individually. It operates on a daily time step, although other time steps are possible.

The soil profile is subdivided into conceptual layers, usually considered to be 5 cm thick, and each layer is further divided into two phases: a mobile and an immobile phase separated at a soil moisture potential of 2 bars. The amount of water in each phase is calculated from the soil moisture characteristic, which can be obtained from soil survey data. Drainage of water to field capacity takes place through the mobile phase only, displacing the resident water and solute to the next layer. Solute equilibration is allowed to occur between the two phases. However, the mixing process is incomplete because of retardation due to intra-aggregate diffusion. To allow for this, a portion of the mobile water is moved before equilibration takes place and the remaining portion after equilibration. Input of rainfall over a specified critical rate is assumed to bypass both soil phases to the base of the profile without displacing any solute.

The model considers three nitrogen pools: organic, ammonium and nitrate. Transformations between these pools is by mineralisation and nitrification respectively. Net mineralisation and nitrification are modelled by "pseudo-zero-order kinetics" (Addiscott and Whitmore, 1987) empirical equations and are dependent on a rate constant which is related to the moisture content and soil temperature through the Arrhenius equation. The mineralisation rate is assumed to decline with water content only when the soil moisture tension is below 0.33 bars. The mineralisation base rate (at $20°C$) is assumed to be constant within the plough layer, which is taken to be $20-25$ cm in depth, but below this the mineralisation base rate is assumed to decrease exponentially with depth. Crop growth is driven by degree-days of soil temperature and is limited only by water and nitrogen availability. Crop uptake of nitrogen is described by a simple equation relating to degree-days of soil temperature, sowing date and the date the soil returns to field capacity. The size of the crop determines how much water is transpired by plants and how much is evaporated from the soil. Evaporation from the soil surface brings about the subsequent upward movement of solute.

Data required for model input include daily rainfall, evaporation, soil temperature and mineral nitrogen of the rain if appropriate. Fertiliser application input can be included. Initial concentrations of mineral nitrogen within the soil layers also needs to be specified and can be input as concentrations for individual layers or as a uniformly distributed value for the whole profile. Output includes estimation of mineral nitrogen concentrations within the soil through time and the quantity of mineral nitrogen leached below the soil profile. See Chapter 5 for a discussion of a similar model, SLIM (Addiscott and Whitmore, 1991).

1.4.19 Stochastic transfer function models

An example of a stochastic, soil profile leaching model is that of the transfer function model (TFM) which has been used by Jury *et al.* (1986) and White (1987) to describe the effects of preferential flowpaths on solute transport. The TFM does not take into account any physical process, but represents solute transport within a soil profile of known depth as a stochastic function of the cumulative amount of water applied at the surface. It must be mentioned that any deterministic, linear model can be formulated as a transfer function. Examples of these are found in the Burns model (Scotter *et al.*, 1993) and the CDE (Sposito *et al.*, 1986; Jury and Roth, 1990).

White (1987) used a TFM to describe nitrate leaching in a mole and tile drainage system. To operate the model requires the calculation of an operationally defined transport volume for each rainfall event, which is obtained from the log normal probability density function of the travel times of a non-reactive solute through the system (nitrate and chloride; White, 1987). A large number of solute nitrate concentrations and drainflow records are required to parameterise the probability density functions of the solute travel times (White, 1987). The transport volume defines the volume of soil water that is effective in nitrate transport and also the volume in which any physical, chemical or biological reaction takes place and is much smaller value than field capacity (White *et al.*, 1986). Using the transport volume derived from the TFM, together with initial nitrate concentration and cumulative drainage during an event, an estimate of nitrate leaching can be obtained for the base of the profile.

2. Field scale

1.4.20 CREAMS and GLEAMS

GLEAMS (Groundwater Loading Effects of Agricultural Management Systems) (Leonard *et al.*, 1987) is a complex, physically-based model operating at the field scale on a daily time step. It is primarily concerned with simulating the effects of agricultural management practices on the movement of nutrients and pesticides in the plant root zone and their subsequent leaching losses to groundwater. GLEAMS is an extension and modification of the CREAMS (Chemicals, Runoff and Erosion from Agricultural Management Systems) model (Knisel, 1980).

GLEAMS consists of three main submodels: the hydrology, erosion and chemical components. The plant root zone is divided into seven conceptual layers and the land area can be divided into subareas to reflect the spatial distribution of crop types. Input by daily rainfall into the surface layer drives the hydrological component. The surface layer is 1 cm in depth and is assumed to be well mixed. If the surface layer becomes saturated then surface runoff is initiated. Surface runoff is estimated using the SCS curve number method. The SCS curve number for the soil moisture condition is a function of land use, management and hydrological factors. The production of runoff from snowmelt is also included in GLEAMS. Infiltration is by a storage routing equation and water percolation to the underlying layer occurs when storage is greater than field capacity. Storage is related to transit time which, in turn, is related to the field capacity and the saturated hydraulic conductivity of the layer. Water loss from soil evaporation and plant transpiration are considered separately and they are calculated from information on potential evaporation and the leaf area index for the crop. Potential evaporation is calculated from temperature, albedo and solar radiation data. Both soil evaporation and plant transpiration are reduced when soil moisture is limited. The removal of evapotranspiration (the sum of plant and soil evaporation) is distributed through the soil layers using information on rooting depth and root growth. It also initiates the upward movement of solute into overlying layers. Any water leaving the deepest layer is assumed to contribute to the vadose or saturated zones.

The erosion and sediment yield component in GLEAMS differs only slightly from CREAMS. As well as estimating sediment loss from a field, it is important in determining the loss of nutrient and pesticide adsorbed on to sediment and dissolved in runoff water. To achieve this, the field area is subdivided into a user-defined combination of overland, channel and impoundment elements. Sediment loss from each element is calculated, in turn, from the uppermost element to the downslope elements. The sediment yield is a function of sediment available for detachment and the transport-carrying capacity of the runoff. If the potential load is greater than the transport-carrying capacity then deposition occurs. If, however, the sediment load is less than the transport-carrying capacity then detachment occurs. This is affected by topography, soil characteristics, plant cover and rainfall/runoff rates and volumes.

Sediment detachment in rill and inter-rill areas of overland flow elements are predicted by a modified version of the USLE (Foster *et al.*, 1977). The sediment transport capacity of each particle type in overland and channel flow is described by the Yalin equation (Yalin, 1963) which is a function of slope, particle diameter and shear velocity. This allows plant type cover (or the lack of) to have an effect on transport capacity.

Channel elements consider spatially varied flow and flow concentration from overland areas. Detachment in concentrated flow is considered differently to overland elements, with consideration given to erosion of the erodible and also the non-erodible layers below the tillage layer. Impoundment elements collect flow, allowing it to drain between storms and also allowing the sediment to settle out with the concomitant nutrient and pesticide load. Slope topography of the field can be either linear, concave or convex or a combination of these three. For accuracy in estimating sediment yields, the slope is divided into a number of uniform segments.

Particle size distribution, important in determining the sediment type deposited and detached, is described by five sediment classes: primary clay, silt and sand, plus small and large aggregates. An enrichment ratio is calculated from the specific surface areas of organic matter, clay, silt and sand and is the ratio of the specific surface area of the sediment leaving the field to that of the parent soil. It is used in determining the amount of adsorbed chemical in the sediment yield.

The chemistry component includes a nutrient component and a pesticide component. The pesticide component in GLEAMS models the movement of pesticides in the surface and subsurface layers of the soil by considering the following processes: pesticide degradation and transformation into different metabolites (Leonard *et al.*, 1990), pesticide extraction to runoff, vertical flux through the profile, pesticide transport with sediment, upward movement of pesticide due to evaporation and plant uptake of systemic pesticides. GLEAMS can account for up to 10 pesticides and their metabolites. Degradation on plant foliage and the soil surface follows an exponential relationship controlled by half-life parameters. The rate of formation of metabolites is controlled by kinetic rate equations. The corresponding rate constants are determined from field and laboratory experiments (Leonard *et al.*, 1990). Pesticide extraction into runoff is from the 1 cm surface layer, which is assumed to be completely mixed, and takes place when the surface layer is saturated. The amount extracted to runoff is dependent on the concentration gradient (between the runoff and the soil phase) and the partition coefficient. The partition coefficient determines the pesticide distribution between the soil and solution phases and varies from each pesticide and is dependent on water solubility and organic carbon content. GLEAMS models the vertical flux of pesticide movement down the soil profile with water movement as a series of mass balance equations. Partitioning of the pesticide between soil and solution phases occurs before the pesticide and water are moved to the next layer. The pesticide mass transported with sediment is related to the sediment mass, the enrichment ratio and the pesticide concentration in the sediment phase. The enrichment ratio is a measure of the enrichment of clay and organic matter fractions onto which the pesticide preferentially adsorbs.

The plant nutrient component in GLEAMS differs largely from that used in CREAMS and is similar to that used in the EPIC model (Sharpley and Williams, 1990). Nitrogen and phosphorus dynamics are described by the processes: nitrification, ammonification, volatilisation of surface-applied animal waste and nitrogen fixation by legumes. These processes are controlled by rate equations and are affected by soil temperature and moisture. The model allows for applications of ammonium, nitrate and mineral phosphorus fertilisers and organic fertilisers such as animal wastes. Nitrogen mineralisation is a two-stage, first-order process of ammonification followed by nitrification (which is, in fact of zero order) and occurs in the organic animal waste, the crop residue, the ammonia and the active soil mineralisable pools. Mineral ammonium, nitrate and phosphorus are formed directly from the crop residue and animal waste pools, as well as via the active soil mineralisable pool. Tillage operations are important in mixing surface residues and the active soil mineralisable nitrogen and phosphorus. Twenty-two tillage operations are allowed for in GLEAMS. On the harvesting of a crop, nutrients and crop residues are apportioned to either crop yield or mixed into the surface and root zone layers. GLEAMS can manage 78 crop types via the internal database. Percolation of mineral nitrogen and phosphorus is subject to the water storage routing procedure and adsorption on to the soil surface. Crop uptake of nitrogen is described and leaching losses of nitrogen and phosphorus in surface runoff are simulated. Soluble forms are extracted via partition coefficients into runoff and insoluble forms (phosphorus, ammonia) are adsorbed to sediment particles and lost by sediment transport.

Estimation of the many parameters required is either in the form of site-specific measured data such as soil samples or, if not all the required data can be measured, from general estimations by internal algorithms. The entry for GLEAMS in Tables 1.1 and 1.2 will serve only to give a general idea of model requirements.

3. River water quality

1.4.21 QUASAR

Quality simulation along rivers, or QUASAR, has been developed at the UK Institute of Hydrology (Whitehead and Hornberger, 1979; Whitehead *et al.*, 1981; Whitehead, 1992), the precursor to this being the classical dissolved oxygen sag model of Streeter and Phelps (see James, 1993, p. 141). Its primary objective is to simulate the dynamic behaviour of flow and water quality along the river system. It is a stochastic, dynamic model combining rainfall−runoff, river flow and water quality submodels (Whitehead and Hornberger, 1979; Whitehead *et al.*,1981). The model can operate on an hourly or daily time step.

Flow routing along the river network in QUASAR employs a multi-lag approach. The river is composed of a number of reaches and each reach contains a number of compartments or lags. The number of lags or compartments in a reach is a measure of the advection and dispersion in the reach and is determined from tracer studies. Stream flow variation employs the laws of mass conservation in the form of an ordinary, lumped parameter, differential equation represented by a transfer function. Mass balance calculations are performed sequentially down the river reach to determine stream flow output. Empirical equations to account for velocity − flow relationships are included.

The river reach model can be coupled to a stochastic, rainfall − runoff model which accounts for flow variations between upstream and downstream gauging sites. The transfer function model incorporates the effects of evapotranspiration (long-term) and soil moisture (short-term) by "filtering" the rainfall input. Parameters are estimated by recursive techniques (Young, 1974). Inputs to the rainfall − runoff model include time-series data on rainfall, mean monthly air temperature and maximum temperature.

Eight water quality variables are simulated: nitrate, biochemical oxygen demand (BOD), ammonium, ammonia, pH, dissolved oxygen (DO), temperature and *Escherichia coli* levels, together with any other conservative pollutant or inert material in solution. Source inputs can include tributaries, groundwater, direct runoff, effluents or storm water. Abstractions from industry and irrigation are also accounted for. As with the stream flow component, water quality variations are described by mass balance relationships for each reach. Additional terms, such as decay parameters to account for the chemical delay and biological behaviour of the variable are incorporated into the model. These are in the form of empirical equations. For example, calculation of dissolved oxygen levels takes into account the oxygen saturation level (which in turn is dependent on stream water temperature), chlorophyll-a concentration and sunlight hours. Additional data requirements include a catchment structure incorporating a river map and reach parameters consisting of data specific to that reach, such as channel cross-sectional area.

There are two modes of operation: stochastic and deterministic. The stochastic mode is used in planning the level of abstractions and effluent consent conditions for a given water quality objective within the river. This mode employs a technique known as Monte Carlo simulation which calculates a distribution histogram for each water quality variable at a point downstream by repeatedly running the model using different inputs defined by the mean, standard deviation and probability distribution type of the input variables. This requires statistical data on water quality and flow for the top reach and any inputs or abstractions downstream.

The deterministic of dynamic mode enables real-time forecasting of quality levels from information on inputs and abstractions in the upper river reaches. This requires time-series water quality and flow data as input. The water quality output variables are calculated, sequentially, for each reach along the river. Applications include the simulation of heavy metal transport in the River Tawe in south Wales and nitrate and algae in the Thames (Whitehead and Williams, 1982; Whitehead and Hornberger, 1984).

1.4.22 ADZ model

The Aggregated Dead Zone (ADZ) model (Young and Wallis, 1986; Wallis *et al.*, 1989) is a model to describe solute dispersion and transportation in a river reach. It employs the concept of dead or storage zones (Valentine and Wood, 1977). Dead zones are areas of low-velocity water flow caused by obstructions in the river. They can be found in pool-riffle sequences, hollows in the river bank and bed or may be due to the generation of turbulent eddies. These dead zones cause dispersion of a pollutant or solute and any number may be found within the reach. The ADZ model lumps these dead zones together and treats them as a single aggregated dead zone.

The ADZ model is essentially a transfer function form of a lumped parameter, ordinary differential equation of mass conservation. This representation has greater numerical stability, can predict observed dispersion patterns better and is easier to formulate than the distributed form of the traditional advection−dispersion equation. The effects of a nonconservative solute can be incorporated by employing a decay term. Time-series analysis, notably recursive instrumental variable techniques (Young, 1985), are used to identify model structure (order) and to estimate parameters. Concentration time-series data from tracer studies at upstream and downstream sites are required to calibrate the model parameters.

4. Catchment scale

1.4.23 Export coefficient models

The export coefficient model developed by Vollenweider (1968, 1975, 1986) and Jørgenson (1980) and applied by Johnes and O'Sullivan (1989) is designed to determine nutrient losses from catchments. The model is entirely empirical in nature, operating on an annual time step. To determine nutrient loss from an area, information on land-use area has to be collected. Export coefficients for each land use are applied to give an areally weighted measure of catchment nutrient loss in kg ha^{-1} a^{-1}. No hydrological or chemical processes are included in the approach; the model output is simply the sum of the relative contributions from different non-point sources. Point nutrient sources and inputs from precipitation can also be added.

Johnes and O'Sullivan (1989) used the technique to calculate nitrogen and phosphorus loadings to a small lake from the surrounding catchments. In their study, lake nutrient loadings were determined from precipitation input into the lake, organic and inorganic nutrient losses from different land types and inputs from other sources such as sewage treatment works.

Information on land use is required for model input. Such data include a proportion of temporary and permanent grassland, cereal and vegetable cultivation, rough pasture, market gardening, urban areas and woodland. Details of the human population and livestock numbers are also required. To determine nutrient losses from the various land uses, data are collected on inorganic and organic fertiliser application rates, as well as information on farming practices, e.g. the proportion of the year that cattle are grazing outside or are kept in holdings. This enables the determination of an export coefficient for each land-use type for each nutrient. In reality, coefficients may be determined partly from collected data and partly from literature.

1.4.24 SHETRAN

SHETRAN (Bathurst and Purnama, 1991) is a physically-based, spatially distributed model which is designed to simulate water, sediment and contaminant transport at the river basin scale. SHETRAN is an extension of the hydrological model, SHE (Système Hydrologique Européen) (Abbot *et al.*, 1986a, 1986b), and the sediment transport model, SHESED (Wicks,

1988). It runs on a time step which varies from two hours in base flow periods to 15 minutes during the storm period (see Ewen, Chapter 16 of this volume) and incorporates the processes of snow pack accumulation and snowmelt, canopy interaction, evapotranspiration, overland, overbank and channel flow, unsaturated and saturated subsurface flow and stream/aquifer interaction. Conceptually, the river basin is seen as a horizontal grid on to which the spatial distribution of basin characteristics, hydrological response and precipitation input are superimposed. Each grid square is divided vertically into horizontal layers and represents one-dimensional vertical flow in the unsaturated zone and root zone, one-dimensional channel flow and two-dimensional overland flow. Saturated, horizontal groundwater flow is represented in two dimensions. The grid size can vary and is usually smaller around river channels to account for complex flowpaths. The majority of flow processes are represented by finite difference representations of physics equations. Processes modelled in this way include one-dimensional subsurface flow (Richards equation). Other processes such as snowmelt are represented by empirical equations.

The sediment transport component in SHETRAN accounts for erosion by raindrop impact, leaf drip impact and overland flow, sediment transport by overland and channel flow, sediment deposition on the river bed and loose sediment deposition and storage on the ground surface, overbank transport, bank and river bed erosion and infiltration of finer sediments into river bed. Sediment discharge is a function of the flow capacity and the incoming available sediment load and is calculated for each sediment size fraction with the cohesive fraction being considered as a single fraction. Consideration of a cohesive fraction allows for the preferential adsorption of contaminants on to finer sediment sizes (Bathurst and Purnama, 1991).

The SHETRAN contaminant component concentrates on the transport of radionuclide contaminants in both dissolved and particulate form. Transport can be as flow in either overland, channel, unsaturated or saturated zones. Transport of the contaminant is represented by a form of the convection−dispersion equation (CDE) with additional terms, depending on which flow process is being described, to account for adsorption on to soil and sediment, absorption into the immobile dead-space in rock and soil, radioactive decay, atmospheric deposition, erosion of contaminated sediment, sediment deposition, plant uptake, lateral convection between columns and exchanges between the river and river bed. The contaminant load is partitioned into adsorbed (on to sediment) and dissolved forms by the Freundlich adsorption isotherm (see Greenland and Hayes, 1981, p. 139). Absorption of contaminant into the dead space or the immobile phase is allowed and a plant model is included to account for uptake, storage and recycling for different plant types. Transport in channel flow is divided into three zones: the flow zone which carries dissolved and particulate contaminants, the bed surface layer where rapid bed/flow interactions occur and the deep bed layer which accounts for longer-term subsurface storage of contaminants. Mass-transfer coefficients allow for the exchange of contaminants between flow and bed sediment. Loss and gain of contaminant in the channel can occur due to bank erosion, radioactive decay and generation of daughter products, input from tributaries, plant uptake, overland flow, atmospheric wet and dry deposition, and exchanges between the flow zone, bed-surface layer and the deep bed layer.

REFERENCES

Abbott, M. B., Bathurst, J. C., Cunge, J. A., O'Connell, P. E. and Rasmussen, J. (1986a). "An introduction to the European Hydrological System — Système Hydrologique Européen, 'SHE', 1: history and philosophy of a physically based distributed modelling system", *J. Hydrol.*, **87**, 45−59.

Abbott, M.B., Bathurst, J. C., Cunge, J. A., O'Connell, P. E. and Rasmussen, J. (1986b). "An introduction to the European Hydrological System — Système Hydrologique Européen, 'SHE', 2: structure of a physically-based distributed modelling system", *J. Hydrol.*, **87**, 61−77.

Addiscott, T. M. (1977), "A simple computer model for leaching in structured soils", *J. Soil Sci.*, **28**, 554−563.

Addiscott, T. M. and Wagenet, R. J. (1985). "Concepts of solute leaching in soils: a review of modelling approaches", *J. Soil Sci.*, **36**, 411−424.

Addiscott. T. M. and Whitmore, A. P. (1987). "Computer simulation of changes in soil mineral nitrogen and crop nitrogen during autumn, winter and spring", *J. Agric. Sci.*, **109**(4), 149−188.

Addiscott, T. M. and Whitmore, A. P. (1991). "Simulation of nitrogen in soil and winter wheat crops, a management model that makes the best use of limited information," *Fertilizer Res.*, **27**, 305−312.

Alcamo, J., Shaw, R. W. and Hordijk, L. (1990). *The RAINS Model of Acidification: Science and Strategies in Europe*, Kluwer Academic Publishers, Dordrecht, The Netherlands.

Bache, B. W. (1990). "Solute transport in soils", in *Process studies in Hillslope Hydrology* (Eds. M. G. Anderson and T. P. Burt), Ch. 13, pp. 437−462, John Wiley, Chichester.

Bathurst, J. C. and Purnama, A. (1991). "Design and application of a sediment and contaminant transport modelling system" in *Sediment and Stream Water Quality in a Changing Environment: Trends and Explanation.* (Eds. N. E. Peters and D. E. Walling), IAHS Publication No. 203, pp. 305−313.

Banwart, S. A. (1983). "Development of a time-variable hydrological submodel for acid precipitation simulation", Thesis presented to the University of Iowa, in partial fulfilment of the requirements for the degree of Master of Science.

Bergström, S. (1975). "The development of a snow routine for the HBV-2 model", *Nord. Hydrol.*, **6**, 73−92.

Bergström, S. and Forsman, A. (1973). "Development of a conceptual deterministic rainfall-runoff model", *Nord. Hydrol.*, **4**, 147−170.

Bergström, S., Carlsson, B., Sandberg, G. and Maxe, L. (1985). "Integrated modelling of runoff, alkalinity and pH, on a daily basis", *Nord. Hydrol.*, **16**, 89−104.

Bergström, S., Brandt, M. and Gustafson, A. (1987). "Simulation of runoff and nitrogen leaching from two fields in southern Sweden", *Hydrol. Sci. J.*, **32**, 191−205.

Beven, K. (1987). "Towards the use of catchment geomorphology in flood frequency predictions", *Earth Surf. Proc. Landf.*, **12**, 69−82.

Beven, K. (1989). "Changing ideas in hydrology — the case of physically-based models", *J. Hydrol*, **105**, 157−172.

Beven, K. and Germann, P. F. (1981). "Water flow in soil macropores. II. A combined flow model", *J. Soil Sci.*, **32**, 15−29.

Beven, K. and Kirkby, M. J. (1979). "A physically-based variable contributing area model of basin hydrology", *Hydrol. Sci. Bull.*, **24**, 43−69.

Beven, K. and Wood, E. F. (1983). "Catchment geomorphology and the dynamics of runoff contributing areas", *J. Hydrol.*, **65**, 139−158.

Binley, A. and Beven, K. (1989). "Modelling heterogeneous Darcian headwaters", in *British Hydrological Society, 2nd National Symposium*, University of Sheffield, 4−6 September, Institute of Hydrology, Wallingford, Oxon, UK, pp. 1.17 −1.22.

Bishop, K. H. (1990). "General discussion", in *The Surface Waters Acidification Programme* (Ed. B. J. Mason), p. 507, Cambridge University Press.

Bohn, H., McNeal, B. and O'Connor, G. (1979). *Soil Chemistry*, John Wiley, New York.

Burns. I. G. (1974). "A model for predicting the redistribution of salts applied to fallow soils after excess rainfall or evaporation", *J. Soil Sci.*, **25**, 165−178.

Burns, I. G. (1975). "An equation to predict the leaching of surface applied nitrate", *J. Agric. Sci.*, **85**, 443−454.

Burns, I. G. (1976). "Equations to predict the leaching of nitrate uniformly incorporated to a known depth or uniformly distributed throughout a soil profile", *J. Agric. Sci.*, **86**, 305−313.

Burns, I. G. (1980). "A simpler model for predicting the effects of leaching of fertilizer nitrate during the growing season on the nitrogen fertilizer needs of crops.", *J. Soil Sci.*, **31**, 175−202.

Burns I. G. and Greenwood, D. J. (1982). "Estimation of the year variations in nitrate leaching in different soils and regions of England and Wales", *Agriculture and Environment*, **7**, 35−45.

Buttle, J. M. (1994). "Isotope hydrograph separations and rapid delivery of pre-event water from drainage basins", *Progress in Physical Geography*, **1**, 16–41.

Campbell, G. S. (1974). "A simple method for determining unsaturated hydraulic conductivity from moisture retention data", *Soil Sci.*, **117**, 311–314.

Campbell, D. J., Kinniburgh, D. G. and Beckett, P. H. T. (1989). " The soil solution chemistry of some Oxfordshire soil temporal and spatial variability", *J. Soil Sci.*, **40**, 321–339.

Carleton, J. E. (1986). "An acid retention simulation model with emphasis on snowmelt", Thesis presented to the University of Iowa, in partial fulfilment of the requirements for the degree of Master of Science.

Carsel, R. F., Mulkey, L. A., Lorber, M. N. and Baskin, L. B. (1985). "The pesticide root zone model (PRZM): a procedure for evaluating pesticide leaching threats to groundwater", *Ecological Modelling*, **30**, 49–69.

Chen, C. W., Dean, J. D., Gherini, S. A. and Goldstein, R. A. (1982). "Acid rain model: hydrologic model", *J. Env. Eng. Div. ASCE*, **108**, 455–472.

Chen, C. W., Hudson, R. J. M., Gherini, S. A., Dean, J. D. and Goldstein, R. A. (1983). "Acid rain model: canopy module", *J. Env. Eng. Div. ASCE*, **109**, 585–603.

Chen, C. W., Gherini, S. A., Goldstein, R. A. and Clesceri, N. L. (1985). "Effect of ambient air-quality on throughfall acidity", *J. Env. Eng. Div. ASCE*, **111**, 364–372.

Chen, C. W., Gherini, S. A., Munson, R., Gomez, L. and Donkers, C. (1988). "Sensitivity of Meander Lake to acid deposition", *J. Env. Eng. Div. ASCE*, **114**, 1200–1216.

Christophersen, N. and Wright, R. F. (1981a). "Sulphate budget and a model for sulphate concentrations in stream waters at Birkenes, a small forested catchment in southernmost Norway", *Water Resour. Res.*, **17**, 377–389.

Christophersen, N. and Wright, R. F. (1981b). *Sulphate budget and a model for sulphate concentrations in stream waters at Birkenes, a small forested catchment in southernmost Norway*, Report IR 70/80, SNSF Project, Norwegian Institute for Water Research, Oslo.

Christophersen, N., Seip, H. M. and Wright, R. F. (1982). "A model for stream water chemistry at Birkenes, Norway", *Water Resour. Res.*, **18**, 977–996.

Christophersen, N., Kjaernsrød, S. and Rodhe, A. (1985). "Preliminary evaluation of flow patterns in the Birkenes catchment using natural ^{18}O as a tracer", in *Hydrological and hydrogeochemical mechanisms and model approaches to the acidification of ecological systems* (Ed I. Johansen), Nordic IHP-Report No. 10, Oslo, pp. 29–40.

Christophersen, N., Neal, C., Hooper, R. P., Vogt, R. D. and Andersen, S. (1990). "Modelling stream water chemistry as a mixture of soil water end-members — a step towards second generation acidification models", *J. Hydrol.*, **116**, 307–320.

Christophersen, N., Neal, C. and Hooper, R. P. (1993). "Modelling the hydrochemistry of catchments: a challenge for the scientific method", *J. Hydrol.*, **152**, 1–12.

Clarke, F. W. (1924). "The data of geochemistry", *US Geol. Survey Bull.*, **770**, 841 pp.

Conway, E. J. (1942). "Mean geochemical data in relation to oceanic evolution", *Royal Irish Acad. Proc.*, **B48**, 119–159.

Cooke, G. W. and Williams, R. J. B. (1970). "Losses of nitrogen and phosphorous from agricultural land", *Water Treatment and Examination*, **19**, 253–276.

Corwin, D. L., Waggoner, B. L. and Rhoades, J. D. (1991). "A functional model of solute transport that accounts for bypass", *J. Environ. Qual.*, **20**, 647–658.

Cosby, B. J., Hornberger, G. M., Galloway, J. N. and Wright, R. F. (1985a). "Modelling the effects of acid deposition: assessment of a lumped parameter model of soil water and stream water chemistry", *Water Resour. Res.*, **21**, 51–63.

Cosby, B. J., Hornberger, G. M. and Galloway, J. N. (1985b). "Modelling the effects of acid deposition: estimation of long term water quality response in a small forested catchment", *Water Resour. Res.*, **21**, 1591–1601.

Cosby, B. J., Jenkins, A., Ferrier, R. C., Miller, J. D. and Walker, T. A. B. (1990). "Modelling stream acidification in afforested catchments: long-term reconstructions at two sites in central Scotland", *J. Hydrol.*, **120**, 143–162.

Danckwerts, P. V. (1953). "Continuous flow systems (distribution of residence times)", *Chemical Engineering Science*, **2**, 1–13.

Darwin, C. (1882). *The Formation of Vegetable Mould through the Action of Worms with Observations on Their Habits*, John Murray, London.

Davis, J. E. and Goldstein, R. A. (1988). "Simulated response of an acidic Adirondack Lake watershed to various liming mitigation strategies", *Water Resour. Res.*, **24**, 525–532.

De Vries, W. (1989). "Philosophy, structure and application methodology of a soil acidification model for the Netherlands", in *Impact Models to Assess Regional Acidification* (Ed. J. Kämäri), Kluwer Academic Publishers, Dordrecht, The Netherlands.

De Vries, W. and Kros, J. (1989). "The long-term impact of acid deposition on the aluminium chemistry of an acid forest soil", in *Regional Acidification Models* (Eds. J. Kämäri, D. F. Brakke, A. Jenkins, S. A. Norton and R. F. Wright) pp. 113–128. Springer-Verlag, Berlin, Heidelberg.

De Vries, W., Posch, M. and Kämäri, J. (1989). "Simulation of the long term soil response to acid deposition in various buffer ranges", *Water, Air and Soil Pollut.*, **48**, 349–390.

de Wit, C.T. and van Keulen, H. (1972). *Simulation of transport processes in soils*, Centre for Agricultural Publishing and Documentation, Wageningen, The Netherlands.

Eshleman, K. N. (1988). "Predicting regional episodic acidification of surface water using empirical models", *Water Resour. Res.*, **24**, 1118–1126.

Eshleman, K. N., Wigington, P. J., Davies, T. D. and Tranter, M. (1992). "Modelling episodic acidification of surface waters: the state of science", *Environmental Pollution*, **77**, 287–295.

Feddes, R. A., Kabat, P., van Bakel, P. J. T., Bronswijk, J. J. B. and Halbertsma, J. (1988). "Modelling soil water dynamics in the unsaturated zone — state of the art", *J. Hydrol.*, **100**, 69–111.

Ferguson, R. I., Trudgill, S. T. and Ball, J. (1994). "Mixing and uptake of solutes in catchments: model development", *J. Hydrol.*, **159**, 223–233.

Foster, G. R., Meyer, L. D. and Onstad, C. A. (1977). "A runoff erosivity factor and variable slope length exponent for soil loss estimates", *Transactions of the American Society of Agricultural Engineers*, **20(4)**, 683–687.

Fread, D. L. (1985) "Channel routing", in *Hydrological Forecasting* (Eds M. G. Anderson and T. P. Burt), John Wiley, Chichester.

Freeze, R. A. and Cherry, J. A. (1979). *Groundwater*, Prentice Hall, New York.

Frissel, M. J. and Reininger, P. (1974). *Simulation of Accumulation and Leaching in Soils*, Centre for Agricultural Publishing and Documentation, Wageningen, The Netherlands.

Gaines, G. L. and Thomas, H. C. (1953). "Adsorption studies of clay minerals, II. A formulation of the thermodynamics of exchange adsorption", *J. Chem. Phys.*, **21**, 714–718.

Garrels, R. M. and Christ, C. L. (1965). *Solutions, Minerals and Equilibria*, Harper and Row, New York.

Gaskin, J. W., Douglas, J. E. and Swank, W. T. (1983). *Annotated bibliography of publications on watershed management and ecological studies at Coweeta Hydrologic Laboratory, 1934–1984*, USDA, Forest Service, Southeastern Forest Experiment Station, General Technical Report SE-30.

Georgakakos, K. P., Valle, G. M., Nikolaidis, V. S. and Olem, H. (1989). "Lake acidification studies — the role of input uncertainty in long-term predictions", *Water Resour. Res.*, **25**, 1511–1518.

Gerritsen, J., Dietz, J., Wilson, H. T. Jr and Janicki, A. J. (1990). *Prediction of episodic acidification in Maryland coastal stream plains*, Maryland Department of Natural Resources, Report AD-90-1, Annapolis.

Goldstein, R. A., Chen, C. W., Gherini, S. A. and Dean, J. D. (1980) "A framework for the integrated lake-watershed study", Proceedings of International Conference on *The Ecological Impact of Acid Precipitation*, Sandefjord, Norway, 11–14 March 1980, pp. 252–253.

Gorham, E. (1955). "On the acidity and salinity of rain", *Geochim. Cosmochim, Acta*, **7**, 231–239.

Gower, A. M. (Ed.) (1990) *Water Quality in Catchment Ecosystems*, John Wiley, Chichester.

Greenland, D. J. and Hayes, M. H. B. (1981). *The Chemistry of Soil Processes*, John Wily, Chichester.

Gregory, K. J. and Walling, D. E. (1973). *Drainage Basin Form and Process. A Geomorphological Approach*, Arnold, London.

Hall, F. R. (1970). "Dissolved solids–discharge relationships: 1. Mixing models", *Water Resour. Res.*, **6**, 845–850.

Hall, F. R. (1971). "Dissolved solids–discharge relationships: 2. Applications to field data", *Water Resour. Res.*, **7**, 591–601.

Hansen, S. and Aslyng, H. C. (1984). *Nitrogen balance in crop rotation simulation model NITCROS*,

Hydrotechnical Laboratory, The Royal Veterinary and Agricultural University, Copenhagen.

Hem, J. D. (1948). "Fluctuations in concentration of dissolved solids of some southwestern streams", *Trans. Am. Geophys. Union*, **29**, 80–83.

Hem, J. D. (1970). "Study and interpretation of the chemical characteristics of natural water", US Geological Survey Water-Supply Paper 1473.

Henriksen, A. (1980). "Acidification of fresh waters — a large scale titration", in *Ecological Impact of Acid Precipitation* (Eds. D. Drablos and A. Tollan, pp. 68–74, SNSF project, Oslo-Ås.

Hillel, D. (1971). *Soil and Water. Physical Principles and Processes*, Academic Press, New York.

Holmberg, M., Heikkilä, E., Hari, P. and Ilvesniemi, H. (1985a). "Modelling solute and water dynamics in forest soil", Paper presented at IUFRO Symposium on *Water and Nutrient Movement in Forest Soils*, Hampton Beach, New Hampshire.

Holmberg, M., Makela, A. and Hari, P. (1985b). "Simulation model of ion dynamics in forest soil under acidification", in *Air Pollution and Plants* (Ed. C. Troyanowsky), pp. 236–239, VCH Publishers, Weinheim, FRG.

Holmberg, M., Hari, P. and Nissinen, A. (1989). "Model of ion dynamics and acidification of soil: application to historical soil chemistry data from Sweden", in *Regional Acidification Models* (Eds. J. Kämäri, D. F. Brakke, A. Jenkins, S. A. Norton and R. F. Wright), pp. 229–240, Springer-Verlag, Berlin, Heidelberg.

Hornberger, G. M., Beven, K. J., Cosby, B. J. and Sappingten, D. E. (1985). "Shenandoah watershed study: calibration of a topography-based, variable contributing area hydrological model to a small forested catchment", Water Resour. Res., **21**, 1841–1850.

Hooper, R. P. and Christophersen, N. (1992). "Predicting episodic acidification in the South-eastern United States: combining a long-term acidification model and the end-member mixing concept", *Water Resour. Res.*, **28**, 1983–1990.

Hooper, R. P. and Shoemaker, C. A. (1986). "A comparison of chemical and isotopic hydrograph separation", *Water Resour. Res.*, **22**, 1444–1454.

Hooper, R. P., Christophersen, N. and Peters, N. E. (1990). "Modelling stream water chemistry as a mixture of soil water end-members — an application to the Panola Mountain catchment, Georgia, USA", J. Hydrol., **116**, 321–343.

Hutson, J. L. and Wagenet, R. J. (1991). "Simulating nitrogen dynamics in soils using a deterministic model", *Soil Use Management*, **7**, 74–78.

Jabro, J. D., Lotse, E. G., Fritton, D. D. and Baker, D. E. (1994). "Estimation of preferential movement of bromide tracer under field conditions", *J. Hydrol.*, **156**, 61–71.

Jacks, G., Berg, G. and Hamilton, P. J. (1989). "Calcium budgets for catchments as interpreted by strontium isotopes", *Nordic Hydrol.*, **20**, 85–96.

James, A. (Ed.) (1993). *An Introduction to Water Quality Modelling*, John Wiley, Chichester.

Jarvis, N. J. (1989). "A simple empirical model of root water uptake", *J. Hydrol.*, **107**, 57–72.

Jarvis, N. J. (1991). "MACRO — a model of water movement and solute transport in macroporous soils", Swedish University of Agricultural Sciences, Department of Soil Sciences, Uppsala, 38pp.

Jenkins, A. and Cosby, B. J. (1989). "Modelling surface water acidification using one and two soil layers and simple flow routing", in *Regional Acidification Models* (Eds. J. Kämäri, D. F. Brakke, A. Jenkins, S. A. Norton and R. F. Wright), pp. 253–266, Springer, New York.

Johnes, P. J. and O'Sullivan, P. E. (1989). "The natural history of Slapton Ley nature reserve XVIII. Nitrogen and phosphorous losses from the catchment — an export coefficient approach", *Field Studies*, **7**, 285–309.

Johnson, N. M., Likens, G. E., Bormann, F. H., Fisher, D. W. and Pierce, R. S. (1969). "A working model for the variation in stream water chemistry at the Hubbard Brook experimental forest in New Hampshire", *Water Resour. Res.*, **5**, 1353–1363.

Johnsson, H., Bergström, L., Jansson, P.-E. and Paustian, K. (1987). "Simulated nitrogen dynamics and losses in a layered agricultural soil", *Agriculture, Ecosystems and Environ.*, **18**, 333–356.

Jørgensen, S. E. (1980). *Lake management*, Pergamon.

Jury, W. A. and Roth, K. (1990). *Transfer Functions and Solute Movement through Soil: Theory and Applications*, Birkhauser Verlag, Basel, Switzerland.

Jury, W. A., Sposito, G. and White, R. E. (1986). "A transfer function model of solute movement through soil. I. Fundamental concepts", *Water Resour. Res.*, **22**, 243–247.

Kirchner, J. W., Dillon, P. J. and LaZerte, B. D. (1993). "Separating hydrological and geochemical influences on runoff acidification in spatially heterogeneous catchments", *Water Resour. Res.*, **29**, 3903−3916.

Kleissen, F. M., Wheater, H. S., Beck, M. B. and Harriman, R. (1990). "Conservative mixing of water sources: analysis of the behaviour of the Allt a'Mharcaidh catchment", *J. Hydrol.*, **116**, 365−374.

Knisel, Walter G. (Ed.) (1980). *CREAMS: a field-scale model for Chemicals, Runoff and Erosion from Agricultural Management Systems*, US Department of Agriculture, Conservation Research Report No. 26, 640 pp.

Kolenbrander, G. J. (1975). "Nitrogen in organic matter and fertilizer as a source of pollution", IAWPR Specialised Conference, Copenhagen.

Kros, J., De Vries, W., Janssen, P. H. M. and Bak, C. I. (1993). "The uncertainty in forecasting trends of forest soil acidification", *Water, Air and Soil Pollut.*, **66**, 29−58.

Lawes, J. B., Gilbert, J. H. and Warington, R. (1882). "On the amount and composition of rain and drainage waters collected at Rothamsted", William Clowes & Sons Ltd, London.

Leonard, R. A., Knisel, W. G. and Still, D. A. (1987). "GLEAMS: groundwater loading effects of agricultural management systems", *Trans. ASAE*, **30**, 1403−1418.

Leonard, R. A., Knisel, W. G., Davis, F. M. and Johnson, A. W. (1990). "Validating GLEAMS with field data for fenamiphos and its metabolites", *J. Irrigation and Drainage Engng*, **116**, 24−35.

Likens, G. E., Bormann, F. H., Pierce, R. S., Eaton, J. S. and Johnson, N. M. (1977). *Biogeochemistry of a Forest Ecosystem*, Springer−Verlag, New York.

Lin, J. C. and Schnoor, J. L. (1986). "Acid precipitation model for seepage lakes", *J. Env. Eng. Div. ASCE*, **112**, 677−694.

Lindström and Rodhe (1986). "Modelling water exchange and transit times in till basins using oxygen-18", *Nord. Hydrol.*, **17**, 325−334.

Livingstone, D. A. (1963). "Chemical composition of rivers and lakes", US Geological Survey, Professional Paper, 440G, 64 pp.

Lundquist, D. (1976). *Simulation of the hydrologic cycle*, Report IR 23/76 (in Norwegian), SNSF Project, Norwegian Institute for Water Research, Oslo, 28 pp.

Lundquist, D. (1977). *Hydrochemical modelling of drainage basin*, Report IR 31/77 (in Norwegian), SNSF Project, Norwegian Institute for Water Research, Oslo, 27 pp.

Lynch, J. A., Hanna, C. M. and Corbett, E. S. (1986). "Predicting pH, alkalinity and total acidity in stream water during episodic events", *Water Resour. Res.*, **22**, 905−912.

Mason, B. J. (Ed.) (1990). *The Surface Waters Acidification Programme*, Cambridge University Press.

Miller, H. G. and Miller, J. D. (1976). "Effects of nitrogen supply on net primary production in Corsican pine", *J. Appl. Ecol.*, **13**, 249−256.

Moldan, B. and Černý, J. (1993). *Chemistry of Small Catchments. A Tool for Environmental Research. SCOPE 51*, John Wiley, Chichester.

More, R. J. (1967). "Hydrological models and geography", in *Physical and Information Models in Geography*, (Eds. R. J. Chorley and P. Haggett), Ch. 5, Methuen.

Mulder, J., van Breemen, N. and Eijck, H. C. (1989). "Depletion of soil aluminium by acid deposition and implications for acid deposition", *Nature*, **337**, 247−249.

Neal, C. (1992). "Describing anthropogenic impacts on streamwater quality: the problem of integrating soil water chemistry variability", *Sci. Tot. Environ.*, **115**, 207−218.

Neal, C. and Christophersen, N. (1989). "Inorganic aluminium−hydrogen ion relationships for acidified streams: the role of water mixing processes", *Sci. Tot. Environ.*, **80**, 195−203.

Neal, C. and Robson, A. R. (1994). "Integrating soil water chemistry variation at the catchment level with a cation exchange model", *Sci. Tot. Environ.*, **144**, 93−102.

Neal, C., Christophersen, N., Neale, R., Smith, C. J., Whitehead, P. G. and Reynolds, B. (1988). "Chloride in precipitation and stream water for the upland catchment of River Severn, mid-Wales: some consequences for hydrochemical models", *Hydrol. Proc.*, **2**, 156−165.

Nielsen, D. R., Biggar, J. W. and Erh, K. T. (1973). "Spatial variability of field measured soil water properties", *Hilgardia*, **43**, 215−260.

Nielsen, D. R., van Genuchten, M. Th. and Biggar, J. W. (1986). "Water flow and solute transport processes in the unsaturated zone", *Water Resour. Res.*, **22**, 89S−108S.

Nikolaidis, N. P., Rajaram, H., Schnoor, J. L. and Georgakakos, K. L. (1988). "A generalised soft water acidification model", *Water Resour. Res.*, **24**, 1983−1996.

Nye, P. H. and Tinker, P. B. (1977). *Solute Movement in the Soil−Root System*, Studies in Ecology 4, Blackwell.

Oksanen, T., Heikkilä, E., Mäkelä, A. and Hari, P. (1984). "A model of ion exchange and water percolation in forest soils under acid precipitation", in *State and change of forest ecosystems − indicators in current research*, (Ed. G. I. Agren, Swedish University Agricultural Science Department Ecology and Environmental Research, Report 13, pp. 293−302.

Oliver, G. G., Thurman, E. M. and Malcom, R. L. (1983). "The contribution of humic substances to the acidity of coloured natural water", *Geochim. Cosmochim. Acta.*, **47**, 2031−2035.

Olsthoorne, T. N., Jaarsveld, J. A., Knopp, J. M., Van Egmond, N. D., Mülschegel, J. H. C. and Van Duijvenbooden, W. (1990), "Integrated modelling in The Netherlands", in *Environmental Models: Emissions and Consequences*, (Eds. J. Fenhann, H. Larsen, G. A. Mackenzie G. A. and B. Rusmussen), pp. 461−479, Elsevier Science Publishers b.v., Amsterdam.

Pilgrim, D. H., Huff, D. D. and Steele, T. D. (1979). "Use of specific conductance and contact time relations for separating flow components in storm runoff", *Water Resour. Res.*, **15**, 329−339.

Quinn, P., Beven, K., Chevallier, P and Planchon, O. (1991). "The prediction of hillslope flow paths for distributed hydrological models using digital terrain models", *Hydrol. Process*, **5**, 59−80.

Reuss, J. O., Christophersen, N. and Seip, H. M. (1986). "A critique of models for freshwater and soil acidification", *Water, Air and Soil Pollut.*, **30**, 909−931.

Robson, A. J., Jenkins, A. and Neal, C. (1991), "Towards predicting future episodic changes in stream chemistry", *J. Hydrol.*, **125**, 161−174.

Robson, A., Beven, K. and Neal, C. (1992). "Towards identifying sources of subsurface flow: a comparison of components identified by a physically-based runoff model and those determined by chemical mixing techniques", *Hydrol. Process*, **6**, 194−214.

Rose, D. A. (1973), "Some aspects of the hydrodynamic dispersion of solutes in porous materials", *J. Soil Sci.*, **24**, 284−293.

Rustad, S., Christophersen, N., Seip, H. M. and Dillon, P. J. (1986). *Can. J. Fish. Aquat. Sci.*, **43**, 625.

Rutter, A. J., Morton, A. J. and Robins, P. C. (1971). "A predictive model of rainfall interception in forests. I. Derivation of the model from observations on a plantation of Corsican pine", *Agric. Met.*, **9**, 367−384.

Sanner, H., Wittgren, H. B., Carlsson, B. and Holmstrom, I. (1993). *Vatten*, **49**, 17−23.

Schnoor, J. L., Palmer, W. D. Jr and Glass, G. E. (1984). "Modelling impacts of acid precipitation of north-eastern Minnesota", in *Modelling of Total Acid Precipitation Impacts*, (Ed. J. L. Schnoor), Butterworth, Stoneham, Mass.

Scotter, D. R., White, R. E. and Dyson, J. S. (1993). "The Burns leaching equation", *J. Soil Sci.*, **44**, 25−33.

Seip, H. M. (1980). "Acidification of freshwater − sources and mechanisms", in *Ecological Impacts of Acidic Deposition*, (Eds. D. Drabløs and A. Tollan, pp. 358−366, Norwegian Institute for Water Research, Oslo.

Seip, H. M., Seip, R., Dillon, P. J. and de Grosbois, E. (1985). "Model of sulphate concentration in a small stream in the Harp Lake catchment, Ontario", *Can. J. Fish. Aquat. Sci.*, **42**, 927−937.

Sharpley, A. N. and J. R. Williams (Eds.) (1990). *EPIC − Erosion-Productivity Impact Calculator: 1. Model Documentation*, US Department of Agriculture, Technical Bulletin No. 1768, 235 pp.

Sklash, M. G. (1990). "Environmental isotope studies of storm and snowmelt runoff generation", in *Process Studies in Hillslope Hydrology*, (Eds. M. G. Anderson and T. P. Burt). Ch. 12, pp. 401−435, John Wiley, Chichester.

Sklash, M. G., Farvolden, R. N. and Fritz, P. (1976). "A conceptual model of watershed response to rainfall, developed through the use of oxygen-18 as a natural tracer", *Can. J. Earth Sci.*, **13**, 271−283.

Sklash, M. G., Stewart, M. K. and Pearce, A. J. (1986). "Storm runoff generation in humid catchments − 1. A case study of hillslope and low order stream response", *Water Resour. Res.* **22**, 1273−1282.

Sposito, G., White, R. E., Durrah, D. D. and Baker, D. E. (1986). "A transfer function model of solute transport through soil. 3. The convection−dispersion equation". *Water Resour. Res.*, **22**, 255−262.

Stone, A. and Seip, H. M. (1989). "Mathematical models and their role in understanding water acidification: an evaluation using the Birkenes model as an example', *Ambio*, **18**, 192−199.

Stumm, W. and Morgan, J. J. (1970). *Aquatic chemistry. An Introduction Emphasising Chemical Equilibria in Natural Water*, John Wiley, New York.

Sverdrup, H. and Warfvinge, P. (1993). "Calculating field weathering rates using a mechanistic geochemical model — PROFILE", *Appl. Geochem.*, **8**, 273 – 283.

Sverdrup, H., Warfvinge, P. and von Bromssen, U. (1987). "A mathematical model for acidification and neutralisation of soil profiles exposed to acid deposition", in *Air Pollution and Ecosystems*, Reidel, Dordrecht.

Swank, W. T. (1986). "Biological control of solute losses from forest ecosystem', in *Solute Processes*, (Ed. S. T. Trudgill), Ch. 3, John Wiley, Chichester.

Trudgill, S. T. (Ed.) (1986). *Solute Processes*, John Wiley, Chichester.

Trudgill, S. T. (1988). "Hillslope solute modelling", *Modelling Geomorphological Systems*, (Ed. M. G. Anderson), Ch. 11, John Wiley, Chichester.

Trudgill, S. T. (1994). "Barriers to a better environment: an overview of water quality. Integrated measures to overcome barriers to minimizing harmful fluxes from land to water", *Proceedings of Stockholm Water Symposium*, Stockholm Vatten, pp. 31 – 41.

Trudgill, S. T. and Richards, K. R. (1994) "The global or the local environment: is environmental policy incompatible with environmental science?", in *Proceedings of BGRG/ERG Symposium, IBG*. (Eds. K. R. Richards, S. Owens and T. Spencer), John Wiley, Chichester (in press).

Valentine, E. M. and Wood J. R. (1977). "Longitudinal dispersion with dead zones", *J. Hydraul. Div. ASCE*, **103**, 975 – 980.

van Genuchten, M. Th. and Jury, W. A. (1987). "Progress in unsaturated flow and transport modelling", *Rev. Geophys.*, **25**, (2), 135 – 140.

van Genuchten, M. Th. and Wierenga, P. J. (1976). "Mass transfer studies in porous sorbing media. I. Analytical solutions", *Soil Sci. Soc. of Am. J.*, **40**, 473 – 480.

van Ommen, H. C. (1985a). "Systems approach to an unsaturated – saturated groundwater quality model, including adsorption, decomposition and bypass", *Agric. Water Management*, **10**, 193 – 203.

van Ommen, H. C. (1985b). "Calculating the quality of drainage water from non-homogeneous soil profiles with an extension to an unsaturated – saturated groundwater quality model including bypass flow", *Agric. Water Management*, **10**, 293 – 304.

Velbel, M. A. (1992). "Geochemical mass balances and weathering rates in forested watersheds of the southern Blue Ridge, III. Cation budgets in an amphibolite watershed and weathering rates of plagioclase and hornblende", *Am. J. of Sci.*, **292**, 58 – 78.

Vollenweider, R. A. (1968) *Scientific fundamentals of stream and lake eutrophication, with particular reference to nitrogen and phosphorus*, OECD Technical Report No. DAS/DST/88, 182 pp.

Vollenweider, R. A. (1975). "Input – output models with special reference to the phosphorus loading concept in limnology", *Schweizerisches Zeitschrift für Hydrologie*, **37**, 58 – 83.

Vollenweider, R. A. (1986) "Phosphorus: the key element in eutrophication control", in Proceedings of the International Conference on *Management Strategy for Phosphorus in the Environment*, (Eds. J. Lester and D. W. W. Kirk), pp. 1 – 10, Selper Ltd, London.

Wagenet, R. J. and Hutson, J. L. (1989) *LEACHM: Leaching Estimation and Chemistry Model — A Process Based Model of Water and Solute Movement, Transformation, Plant Uptake and Chemical Reactions in the Unsaturated Zone. Continuum*, Vol. 2, Water Resources Institute, Cornell University, Ithaca, New York.

Walling, D. E. and Webb, B. W. (1986). "Solutes in river systems", in *Solute Processes*, (Ed. S. T. Trudgill), Ch. 7, John Wiley, Chichester.

Wallis, S. G., Young, P. C. and Beven, K. J. (1989). "Experimental investigation of the aggregated dead zone model for longitudinal solute transport in stream channels", *Proc. Instn. Civ. Engrs, Part 2*, **87**, 1 – 22.

Warfvinge, P. and Sverdrup, H. (1990). "The SAFE (Soil Acidification of Forested Ecosystem) model", in *Soil Acidification Models Applied on Swedish Historic Soils Data*. (Ed. B. Andersen), Nordic Council of Ministers.

Warfvinge, P. and Sverdrup, H. (1992). "Calculating critical loads of acid deposition with PROFILE — a steady state soil chemistry model", *Water, Air and Soil Pollut.*, **63**, 119 – 143.

Warfvinge, P., Holmberg, M., Posch, M. and Wright R. F. (1992). "The use of dynamic models to set target loads". *Ambio*, **21**, 369 – 376.

Warfvinge, P., Falkengren-Grerup, U., Sverdrup, H. and Andersen, B. (1993). "Modelling long-term cation supply in acidified forest stands", *Environ. Pollut.*, **80**, 209–221.

Watson, K. W. and Luxmoore, R. J. (1986). "Estimating macroporosity in a forest watershed by use of a tension infiltrometer", *Soil Sci. Am. J.*, **50**(3), 578–582.

Wheater, H. S., Kleissen, F. M., Beck, M. B., Tuck, S., Jenkins, A. and Harriman, R. (1990a). "Modelling short-term flow and chemical response in the Allt a'Mharcaidh catchment", in *The Surface Waters Acidification Programme*. (Ed. Sir B. J. Mason), pp. 455–466, Cambridge University Press.

Wheater, H. S., Langan, S. J., Miller, J. D., Ferrier, R. C., Jenkins, A., Tuck, S. and Beck, M. B. (1990b). "Hydrological processes on the plot and hillslope scale", in *The Surface Waters Acidification Programme*. (Ed. Sir B. J. Mason), pp. 121–135, Cambridge University Press.

White, R. E. (1987). "A transfer function model for the prediction of nitrate leaching under field conditions", *J. Hydrol.*, **92**, 207–222.

White, R. E., Dyson, J. S., Haigh, R. A., Jury, W. A. and Sposito, G. (1986). "A transfer function model of solute transport through soil. 2. Illustrative application", *Water Resour. Res.*, **22**, 248–254.

Whitehead, P. G. (1992). "Examples of recent models in environmental impact assessment", *J. of Inst. Water Environ. Management*, **6**, 475–484.

Whitehead, P. G. and Hornberger, G. M. (1979). "A systems model of stream flow and water quality in the Bedford Ouse River — I. Stream flow modelling", *Water Res.*, **3**, 1155–1169.

Whitehead, P. G. and Hornberger, G. M. (1984). "Modelling algal behaviour in the River Thomas", *Water. Res.*, **18**, 945–953.

Whitehead, P. G. and Williams, R. J. (1982). "A dynamic nitrate balance model for river basin", in *Proceedings of IAHS Exeter Conference*, IAHS Publication No. 139.

Whitehead, P. G., Beck, M. B. and O'Connell, E. (1981). "A systems model of flow and water quality in the Bedford Ouse River system — II. Water quality modelling", *Water Res.*, **15**, 1157–1171.

Whitehead, P. G., Jenkins, A. and Cosby, B. J. (1990), "Long term trends in water acidification", in *The Surface Waters Acidification Programme* (Ed. Sir B. J. Mason), pp.431–443, Cambridge University Press.

Wicks, J. M. (1988). "Physically based mathematical modelling of catchment sediment yield", Unpublished PhD thesis, University of Newcastle upon Tyne, UK, 238 pp.

Wild, A. (Ed.) (1988). *Russell's Soil Conditions and Plant Growth*, Longman.

Wild, A. and Babiker, I. A. (1976). "The asymmetric leaching pattern of nitrate and chloride in a loamy sand under field conditions", *J. Soil Sci.*, **27**, 46066.

Wild, A. and Cameron, K. C. (1980). "Soil nitrogen and nitrate leaching", in *Soils and Agriculture*. (Ed. P. B. Tinker), Society of Chemical Industry, pp. 35–70, Blackwell.

Yalin, Y. S. (1963). "An expression for bed load transportation", *J. Hydraulics Div., Proc. ASCE*, **89**(HY3), 221–250.

Young, P. C. (1974). "Recursive approaches to time-series analysis", *Bull. IMA*, **10**, 209–224.

Young. P. (1985). "Recursive identification, estimation and control", in *Handbook of Statistics*, Vol. 5, *Time Series in the Time domain*, (Eds. E. J. Hannan, P. R. Krishnaiah and M. M. Rao), Ch. 8, Elsevier, Amsterdam.

Young, P. and Wallis, S. G. (1986). "The aggregated dead zone (ADZ) model for dispersion in rivers", paper presented at the BHRA International Conference on *Water Quality Modelling in the Inland Natural Environment*, Bournemouth, England, 10–13 June 1986.

Young, R. A., Onstad, C. A., Bosch, D. D. and Anderson, W. P. (1989). "AGNPS: a non-point source pollution model for evaluating agricultural watersheds", *J. Soil Water Conserv.*, **44**, 168–173.

2

Basic Principles of Frequently Used Models

Per Warfvinge

Department of Chemical Engineering, Lund University, Sweden

2.1 INTRODUCTION

The objective of this chapter is to give an overview of the challenges one faces in modelling the solute composition in catchments. It is written as a personal view of the topic rather than a state-of-the art review of current catchment-scale models. For an overview of current models, refer to Chapter 1 of this volume.

Catchment solute models are basically used for two purposes, to predict and to explain. With the first object, it is satisfactory to develop a tool that offers a relationship between a dependent variable and one or more independent variables. Examples are NO_3^--leaching as dependent on N loading (fertilization) and pH in stream water as a function of acid input. The structure of a purely predictive tool does not need to be constrained by limited knowledge about the causal relations in the catchment, but can be based on purely statistical relations. Such empirical models can be very useful, but may not answer questions involving *why* and *how*. Catchment models with a high degree of explanatory power are used to say something about the inner working of the system, in our case the catchment. They also tend to involve more elements of natural sciences such as physics, e.g. fluid mechanics, heat mechanics and chemistry, e.g. geochemistry and physical chemistry. Apparently, such models contain less empiricism than many tools, such as regression models, developed for prediction only. Of course, the explanatory power of solute models relies on the predictive capability of the model. Models with explanatory capacity can be used to test scientific hypotheses. Such an hypothesis could be, "Is the current set of equations representing a set of processes sufficient to reproduce the patterns and variability in the observations?"

While some models certainly have less explaining potential than others, we should remind ourselves about Neils Bohr's statement: "It is wrong to believe that the objective of physics is to find out what nature is. Physics deals with what we can *say* about nature." Consequently, the predictive power is always the main objective with all modelling work, but the attitude of natural science is that we can increase the predictive power if we base our models on tested physical and chemical principles.

The wide range of catchment solute models and the different ways of modelling different processes shows that there is a significant element of empiricism even in models that claim to be based on fundamental physical and chemical principles. This is true with respect to the

Solute Modelling in Catchment Systems. Edited by Stephen T. Trudgill

mathematical formulation of individual processes, but especially the selection of parameter values that tie the magnitude of one chemical component or chemical process to another.

The wide range of catchment solute models has emerged for several reasons. Important factors that have led to this variety are:

1. Management objective
2. Dependent and independent variables chosen
3. Availability of data allowing a model to be validated
4. Differences in catchment characteristics
5. Scientific background of the modeller

The first item is of the utmost importance. Virtually all models described in this book were developed with financial support from a body with a clearly defined management responsibility. For example, the models dealing with soil and water acidification were created with support from the environmental ministries to provide a scientific platform to assess the effects of long-range transported air pollutants on terrestrial and aquatic ecosystems. With this objective, the independent variables in such models always become the rate of deposition of acidifying substances, and the dependent variable would be the biologically most relevant quantity such as stream pH and/or Al concentration. Examples of such models are MAGIC (Chapter 14), SAFE (Chapter 3) and RESAM (Chapter 12). If the concern is eutrophication of freshwater, the independent variable will be factors affecting the nutrient input to the aquatic system, as influenced by land use. One such model is by Whelan *et al.* (Chapter 6).

To guarantee usefulness, catchment solute models have to be validated. This calls for a good data set with which the predictive capacity can be assessed. Indeed, many models were developed in close connection with a certain set of data. For example, the development of ILWAS (integrated lake water acidification study) model (Chen *et al.*, 1983) relied heavily on the data that was generated within the catchment studies at Woods Lake and Panther Lake in the Adirondacks. With the high level of ambition within the field project, it was feasible to design a model containing a large number of processes (and parameters) and still maintain the possibility to parameterize and calibrate the model. The MAGIC model was developed with access to the 1000 lakes survey in Norway and a survey of 4000 streams, and the whole structure of the model, use and calibration was designed to make maximum use of available data. The way MAGIC was built allows both modelling on a catchment scale, as ILWAS does, and on a regional scale with available data. Both modelling exercises thus share a common objective, to assess the impact of acid deposition on runoff chemistry, but the resulting models are different.

The catchment characteristics are, of course, also very important for how a model is made. Different processes may be of different relative importance in different geochemical regions. For example, in a catchment dominated by calcareous soils, Al chemistry would be relatively unimportant, while it would be crucial to calculate Al concentrations accurately in catchments in a final phase of acidification.

Well, what is a catchment? According to Webster's dictionary, a catchment is the area from which (point in) a river draws its water supply. Therefore, a water sample taken in running water integrates a vast number of chemical and physical features of the catchment. It is therefore natural that some modellers may focus on the flux of water and yet another on the catchment itself rather than the water that leaves it.

Therefore it is unavoidable that the background of the modeller has a great impact on the resulting model structure, even if the objective is clearly stated. Two catchment solute models,

PULSE and TOPMODEL, can serve as an example of this. Both were basically hydrological models, which were later equipped with hydrochemical routines. While the approach in the solute transport component of the models is relatively sophisticated, the chemical parts are merely phenomenological. Those models do therefore contain an explanatory possibility for hydrology, but are only predictive with regard to hydrochemistry. A counter-example is SAFE which in its basic form has certain detailed hydrochemical parts and does not include hydrological processes (Warfvinge *et al.*, 1993),but has to be linked with a detailed hydrological model if the influence of hydrological dynamics on solute concentrations should be addressed, as done in Warfvinge and Sverdrup (1992b).

Does the diversity of catchment models reflect a waste of effort? Do we need so many models all addressing the same issues? Maybe not, but it is important to recognise that in all multi-disciplinary fields of work, such as ecosystem research, different views will always be represented. Different individuals will, literally, direct their consciousness to different parts of the ecosystem and search for explanations to various phenomena within their own area of specialization. As long as individuals are different, so will their models. Therefore, I will give some thought to the question, "What is a model?"

2.2 WHAT IS A MODEL?

Certainly, people with different background, interests and points of reference would put different meanings into the word model. For example, an architect occupied with the design of a new building creates a cardboard model in a scale that is appropriate to visualize how the building will function aesthetically and fit into the landscape. This is an example of a physical model. The soil scientist involved in catchment research may use a soil column to perform soil reaction experiments, from which conclusions about the catchment's response to perturbations are drawn.

In both cases, however, we can apply the following general meaning to the concept "model":

A model is a system that reproduces important features of another system.

The basic idea behind the use of models is thus that the properties of the model system should be so similar to the real system that they can be used to draw conclusions about the real system, despite the fact that the model includes fewer elements and relationships than the real system. The model may be a physical model, a subsample of the real system or a mathematic model implemented as computer code. For example, an architect does not bother to show every nail, the insulation or the electrical installation because they are not "important features" at the stage when the design of the building is to be decided upon. The soil scientist may use the soil column as a model substitute because manipulations of an entire catchment may be too tedious to carry out.

Another definition of "model" is "a model is a source of data" (Cellier, 1991). This definition is quite interesting, since it automatically brings our measurement systems into what is defined as a model. For example, a lysimeter collecting water from a soil does, with this broad definition, qualify as a model. This indeed seems reasonable, not only because it is a source of data but also because the lysimeter is a system that represents the solute component of the soil system. The lysimeter is thus a model with both definitions. When we compare the data from a mathematical model with "real data" we do in reality therefore compare two models with each other!

We can also see that the concept "model" is indivisibly linked to the concept "system". One possible definition of this concept is:

A system is a number of components connected to a whole.

This definition brings our attention to the fact that there are often causal relationships between components of the system. For example, in this monograph, the system in focus is the catchment. The components in the catchment are always the soil with its solid, gaseous and liquid, and the vegetation, or at least the most important features of the biological components. With this process-oriented perspective, we arrive at one definition of "modelling":

Modelling means the process of organizing knowledge about a given system.

The art of modelling systems has its basis in a skill to determine which components are important features of the system and how they are interrelated. The difficulty in modelling is, however, certainly not to find enough processes/equations to include; the difficulty lies in the problem of careful selection of the proper processes/equations to meet the objectives of the modelling exercise. Of course, we can use statistical means to organize our knowledge by using the observations from the catchment, and this is also an example of organizing the knowledge, although it is not done from the view that a system consists of different components.

Regardless of the modelling approach, modellers use their models to make simulations. A simulation has been defined as:

A simulation is an experiment performed on a model.

In the case of our architect, simulations can be made by illuminating a cardboard model with a lamp in the angle of the setting sun, while the architect leans over the model to examine how the evening light will fall through the window. In a similar fashion, the catchment modeller will exert a mathematical model to different boundary conditions and compare the data that the model generates with the response by the real system to similar conditions. Through this analysis, the modeller will be able to extend knowledge about the system's behaviour and answer questions about the (real) system that otherwise could be impossible to obtain for practical, theoretical or economical reasons. We can therefore justly put modelling as performed in this volume on an equal footing with experimental mathematics.

It is obvious that modelling will always be controversial since every model by definition includes simplifications of the original system. Also, in the model applications presented in this monograph, the models will sometimes be brought to the limit of their validity. Outside this limit, experiments made with the model will distort our perception of the real system's behaviour, rather than providing new insights. Therefore, it is a challenge for the modellers to define the frame of "experimental" conditions within which the models are valid. A criterion for what constitutes a good model can thus be that it is valid for a wide range of experimental conditions, i.e. it has a wide *experimental frame*. In management applications this is vital, since models are used to assess future effects of different management practices of which we know very little. Assessing the rate of recovery of acidified streams when acid input to the catchments is reduced over a period of several decades is a good example of such an application.

To summarize, we can state that in the context of mathematical modelling a model consists of a set of equations, generally solved with a numerical technique by a computer. The model system has in most cases been designed to mirror a conceptual idea about the interactions between different components in the "real" system. Special attention is given to ensure that the model will respond to specific variations in boundary conditions, such as changes in the climatic and chemical environment.

2.3 WHAT IS MODEL COMPLEXITY?

One aspect of modelling that is frequently discussed in modelling papers is what level of complexity is required. The discussion is generally focused on *structural complexity*. A model has high structural complexity if the number of state variables and parameters are large. The other aspect of complexity is "behavioural complexity". A model with high behavioural complexity will give us many surprises and its output is difficult to foresee, and it will generate highly non-linear responses to linear perturbations. Chaotic models comprise a class of models with high behavioural complexity which can be obtained with a very low structural complexity.

Process-oriented models are often developed with a particular ambition to keep the level of complexity to a minimum while still matching a model's anticipated field of application. When the catchment acidification model MAGIC was developed, it was an outspoken objective to keep the number of parameters down. This was done by lumping the parameters and letting one single compartment represent the entire catchment (Cosby *et al.*, 1985). The alternative to a lumped structure would have been to use a smaller grid size, i.e. a multi-layer catchment model or some other multi-grid structure to increase the spatial resolution. Such a structure would better represent the physical and chemical heterogeneity of the catchment in more detail, but result in higher structural complexity.

However, is the number of state variables and parameters, as suggested by Van Oene and Ågren (1995), an unambiguous measure of structural complexity? As an example, we can consider how geochemical models are parameterized with respect to the pool of exchangeable cations. Changes in size and composition of this pool have a large impact on solute concentrations. For example, the pool of exchangeable Ca at the Hubbard Brook Experimental Forest, New Hampshire, amounts to *ca.* 700 mmol m^{-2}, while the flux of Ca from the catchment is *ca.* 3 mmol m^{-2} yr^{-1} (Kirchner, 1990). This corresponds to less than 0.5% of the exchangeable pool. In most catchment studies the size of the exchangeable pool is determined by analysing the soil horizon by horizon. With multi-compartment models with a spatial resolution corresponding to the resolution in the measurements, the parameterization is straightforward. However, when a single- compartment model is parameterized it is less obvious how the measured values, with their often high vertical variability, should be utilized for parameterization. If the model is used to simulate the response to a change in acid load, the effect will move slowly down through the soil, and it becomes important how the exchange capacity is distributed. If there is horizontal spatial variability, additional problems will arise for model parameterization, as discussed by Reuss (1990).

Consequently I would like to state the following:

> The parameterization of the mathematical model is an integrated part of the total model structure.

We should thus be aware that when we lump original data into fewer elements, we apply an unstated model that may have significant consequences for the results of simulations. When original data are lumped we reduce the information and add another element of complexity that does not show in the mathematical model, i.e. the set of equations that form the whole. We cannot therefore say *a priori* that a spatially and temporally more lumped model is less complex than a model with a finer resolution. On the contrary, one may argue that the fine-grid structure model system can be implemented with a less subjective treatment of data.

Rather, the problem with models that require extensive parameterization is that they easily become overparameterized, i.e. inclusion of parameters in a model whose values cannot be uniquely determined by the calibration data (Hooper *et al.*, 1988). This is true also for models with outspoken ambitions to be mechanistic, i.e. NUCSAM (Chapter 12) and SAFE. Despite that the model parameters are not lumped and with a unique physical/chemical interpretation it is always possible to find a vast number of sets of parameters within the range of uncertainty of these parameters that yield, in principle, identical results. In such cases, it is necessary to perform an uncertainty analysis that demonstrates the variability in model output that can result from variability in parameter values. Parameters that have little influence on model output will contribute less to overparameterization than those with a large influence.

Behaviour complexity is often caused by feedback mechanisms. One such feedback mechanism could be between the forest stand and its nutrient uptake and solute chemistry. Another could be between P concentrations in water and algal growth and sedimentation. Obviously, such feedback mechanisms where the change in a state variable depends on the state itself creates non-linear, and therefore non-trivial, behaviour. It is tempting to agree with Van Oene and Ågren (1995), who concluded that increased (structural) complexity is warranted only if it entails new feedback mechanisms.

2.4 ASSUMPTIONS AND PRINCIPLES OF CATCHMENT MODELS

The catchment models that are described in this volume have been developed with the common scope of predicting the chemical composition of the outflow of the catchment. There are two lines of development of catchment models. One line has resulted in a class of models generally referred to as process-oriented models. These represent the reductionistic school in science, since the behaviour of a catchment is viewed as the integrated response to subprocesses that are ruled by basic scientific, testable laws. The whole is the result of the details.

The models that do not fall into this class are much more diverse. One common denominator is, however, that they make use of the observation at the lower boundary to say something about the catchment, while the process-oriented models do the opposite. We therefore find more holistic approaches in this category, i.e. models where certain rules apply to the catchment as a whole.

In the following I will try to give some examples of the principles that are shared by catchment models, with emphasis on the process-oriented models.

2.4.1 Process-oriented models

The most dominating class of catchment solute model contains the process-oriented models. This class of model ties an input to the catchment, i.e. atmospheric input, fertilization, etc., to a set of output responses, i.e. fluxes of components in a stream leaving a catchment through chemical and physical processes. The process-oriented models thus take their starting point as the basic properties of the catchment and the laws of the sub- systems.

The basic assumption of all process-oriented catchment models is that the law of continuity is applicable on a catchment scale. In words, the law of continuity states that what is produced of element i within a control volume must either leave the volume or be retained within it, i.e. either diverge or accumulate. In mathematical terms, this is written as

$$\frac{\partial C_i}{\partial t} = r_i - \nabla \cdot N_i \tag{1}$$

where C_i denotes the total concentration of component i, r_i the rate of production of i per unit volume and time, while N_i is a molar flux per unit area of the control volume. This basic law is usually called mass balance, and models based on this law of mass conservation are called biogeochemical models (Jörgensen, 1992).

In principle, we can calculate the concentration and fluxes of any solute with this equation. In practice, however, it is impossible to model every chemical aspect of every possible solute component. It is therefore necessary to restrict the level of ambition to what is absolutely necessary to meet the overall objectives of the modelling work.

We can also see that the law of continuity is quite clear about which possibilities the modeller has access to for simplifications. These are:

1. Spatial resolution and temporal resolution
2. Components included
3. Chemical reactions included

Although the equation of continuity can be solved analytically for a number of special cases, even inclusion of very simple chemistry leads to considerable mathematical difficulties (Parlange *et al.*, 1982). Therefore, no "real" biogeochemical systems can be modelled at a catchment scale based on analytical solutions of the law of continuity. Inevitably, the law of continuity must be discretised, i.e. the catchment must be divided into one or more discrete compartments, and the time span of the simulation must be divided into an appropriate number of time steps. It is also necessary to make decisions on which components are necessary, considering interactions between components, and the rate of different reactions.

As a minimum, the number of equations needed in a process- oriented model equals the number of components, but this applies only to systems where no chemical reactions take place, i.e. a purely physical system. In such cases the minimum set of equations can be used, together with appropriate initial values and boundary conditions. Of course, no interesting catchment system consists only of inert tracers. In most catchment solute models, it is therefore necessary to include more equations to account for chemical reactions that transform one component to another, i.e. NH_4^+ to NO_3^-. Because of this, process- oriented catchment models consist of a set of differential and algebraic equations. The total number of equations will increase to equal the number of components plus the number of independent chemical reactions. On top of this, parameter values must be assigned.

Two types of equations that are very common in solute models are the charge balance and equilibrium equations. Both of these types of equation replace mass balances. The charge balance can be formed as the sum of all mass balances, and thus does not add anything to the law of continuity. The equilibrium equations reflect the fact that the concentration of one component is totally dependent on the concentration of another. This causes the mass balances to be coupled, and one must be taken away. The method of substituting n mass balances for one charge balance and $n - 1$ equilibrium equations is often referred to as the "mobile anion concept'.

Dynamic versus static models

One important divide in model structure goes between dynamic models and static models. While the dynamic model is designed to predict the change in state variable as a function of time, the static model calculates the steady-state conditions for a set of boundary conditions. The dynamic models thus include both sides of eqn (1), while static models set the left-hand side of eqn(1) equal to 0. One example of a process-oriented static catchment model is PROFILE (Warfvinge and Sverdrup, 1992a).

Within the range of hydrochemical acidification models represented in this volume, we can recognize models that combine static and dynamic elements. Several models neglect temporal variations on a time-scale of an order of magnitude less than the other important forcing functions in the catchment. For example, the seasonal variations in nutrient uptake and biomass degradation are neglected in most catchment models that use annual average values for atmospheric inputs of nutrient and eutrifying/acidifying substances. This is also common for hydrological variations in catchments.

Presently most attention is given to dynamic model approaches, but the catchment models developed by Kirchner (1990) and Hooper *et al.* (1988) represent an interesting line of development in static modelling. Also, it is noteworthy that the static model PROFILE has been more widely used for its final objective, environmental management within the context of critical loads of acid deposition, than any dynamic model so far. The Henriksen steady-state water chemistry model (Sverdrup *et al.*, 1990) has also been very successful in providing a tool for environmental policy decisions, by giving input to the new sulphur emissions reduction protocol signed in Oslo in June 1994 under the LRTAP convention.

Spatial and temporal resolution of dynamic models

One of the most striking differences between dynamic catchment solute models is the differences in temporal and spatial resolution. The models presented in this volume will show that some models calculate solute chemistry with a temporal resolution ranging from hours to years, while the spatial resolution varies between 50 by 50 m and several square kilometres.

When the mass balance (eqn (1)) is discretised, it follows that the modeller assumes that the state variables are constant within the interval in time and/or space considered. The modeller should therefore present fair support to show that this could be valid.

The key questions that determine the resolution of process-oriented catchment solute models are:

1. How does the catchment integrate the chemistry and physics of its subsystems?
2. How does the biological component affected by the discharge quality (such as fish in a stream) integrate solute chemistry?

Obviously, nature does integrate properties on different scales. One example is the formation of distinct soil horizons during podsolisation. The horizons are chemically (macroscopically) homogeneous and often form a very well defined boundary to the next layer. With respect to vertical discretisation, the soil horizon appears to be a good starting point. For simulation of short-term variations in solute chemistry, applications of end member mixing analysis (Hooper *et al.*, 1990) supports this view, both with respect to chemistry and to hydrology. For long-term

TABLE 2.1 Classification of temporal and spatial resolution of models described in this volume

Spatial resolution	Temporal resolution	
	Year	Day/hour
Catchment	MAGIC SMART	TOPMODEL
Two-dimensional	RESAM SAFE	MACRO NUCSAM
Three-dimensional		SHETRAN-UK

simulations of acidification of catchments with a lateral flow component, distributed models indeed yield different results from spatially lumped models (Wright *et al.*, 1991).

To illustrate the second question, let us refer to a comparative study where four catchment-scale models, MAGIC, MIDAS, SAFE and SMART, were used to calculate the stream water chemistry at three Scandinavian sites for different loadings of acidifying atmospheric input (Warfvinge *et al.*, 1992) (see Table 2.1). The overall objective was to determine the acid load that would not cause damage to fish within a certain time frame. All models worked with an annual time resolution in input (deposition, hydrology), so the models could only predict annual mean values of stream chemistry. Yet it is known that one important cause of damage to fish population is acid episodes during critical periods of the hatching. The data available to relate the stream chemistry to damage on fish populations also showed the risk of damage and extinction of fish populations as a function of annual mean water chemistry (Henriksen *et al.*, 1990), expressed as the acid neutralizing capacity (ANC). This biological data therefore integrated the temporal variability of the annual mean level of acidity on the biota. Therefore the catchment solute models matched the data available and were appropriate for predictive purposes.

In this context, it is appropriate to be reminded of the distinction between the *temporal resolution* and *numerical time-step*. While the former refers to the time interval between input/output data, the latter refers to the number of subdivisions in this time interval that is necessary to solve the underlying equations without errors. The time step is therefore always shorter than the time resolution.

Components and reactions

The law of continuity (eqn (1)) allows the modeller to involve any number of components (solutes) in a model, but the challenge is to find the minimum amount of components while still obtaining the link between system input and the desired output. As the number of components (state variables) increase, so does the number of mass balance equations, the number of parameters and the structural complexity of the model.

This section will give two examples of how some classes of biogeochemical reactions have been modelled in different process-oriented catchment solute models. The two processes, cation exchange and chemical weathering, are included in all catchment models addressing soil and water acidification, and the examples illustrate the range of ideas that have been applied in current models.

The process-oriented models take as their starting point the assumption that the science has sorted out how different reactions proceed with respect to reactants, rate control and stoichiometry. Based on these ideas, the modeller creates a mathematical model in a similar fashion as an engineer builds a car, assembling parts from different suppliers. The parts are generally chemical principles taken from applied chemistry, developed and tested on a laboratory scale rather than on a catchment scale (Hauhs, 1990). As mentioned above in connection with the issue about model parameterization, we have as yet no theory for how we shall deal with the scaling problem.

Cation exchange and chemical weathering have been very much in focus in acid rain research and the subject of considerable research efforts. We are, however, still very much in the dark regarding some chemical aspects related to these reactions (Neal *et al.*, 1990).

Cation exchange In many catchments, cation exchange reactions are very important, and in most catchment solute models that treat Ca, Mg, K and NH_4 cation exchange reactions are included. As the example from Hubbard Brook Experimental Forest shows, the pool of exchangeable cations — the cation exchange capacity (CEC) — is very large as compared to the annual cation fluxes from the catchment. A small change in pool size will therefore exert a large influence on solute concentration in discharge. Many dynamic, process-oriented catchment models rely on a correct assessment of element fluxes between the soil solution and the exchange matrix for its predictive capacity.

Cation exchange reactions are known to take place on clays and solid organic matter in the soil. The submodels for cation exchange that are used in current catchment-scale models are based on models deduced from theories of cation exchange on discrete sites on clays (cf. Harmsen, 1982). Examples of such models are the Gaines–Thomas and the Gapon exchange isotherms. The cation exchange models applied by catchment modellers are based on three critical assumptions:

1. The CEC is constant.
2. The parameter in the cation exchange model, the selectivity coefficient, is a true constant.
3. The exchange between the soil and the exchange matrix is so rapid that equilibrium rules.

The first assumption is very critical; if the CEC changes over a period of several decades, the fluxes of ions from the exchange may be much greater than if CEC is constant, and possibly in other proportions than predicted by the equilibrium equation. Here the data are ambiguous. Some recent work also shows that the CEC may increase due to acidification, because Al blocking the exchange sites may be dissolved (Wesselink and Mulder, 1995), while a decrease in the CEC has been observed elsewhere (Sverdrup *et al.*, 1995).

The second assumption that selectivity coefficients are constant has never been supported. If they are not constant, the predicted buffering rate will be affected, just as different exchange isotherms result in different responses to hydrochemical changes. Several researchers have shown that the selectivity coefficients are not constant over a wide range of soil solution composition. After a review of published selectivity coefficients, Bruggenwert and Kamphorst (1982) concluded that ". . . authors have, in some cases, been shown to be guided by wishful thinking with regard to the constancy of their preferred coefficient . . .".

In lumped models with a very low spatial resolution, with, for example, one compartment representing the entire catchment, such as MAGIC, the catchment model will tend to predict a smaller change in soil solution composition and exchanger composition over time than the more distributed model. The assumption in such models with respect to cation exchange should thus be more realistic because of the smoother response to changes, as compared to multi-layer models. Therefore, one may argue that the inaccuracy of classic cation exchange models causes least damage in non-distributed models.

The problems of characterizing cation exchange are shared with SO_4^{2-} adsorption in soils, both with respect to estimating the pool of sites available for absorption and the laws regulating solution – solid interactions.

For use in practice, these shortcomings of prevailing methods of dealing with cation exchange may not be so alarming after all. For short-term changes, the change in the CEC is probably negligible and the soil solution composition is stable enough to make exchange isotherms an acceptable simplification. For long-term simulations >50 years, it does not matter if we are a decade off from the "true" scenario in our predictions, as long as the general trends are sufficiently reliable to serve the purposes for policy decisions.

The assumption that cation exchange reactions are instantaneous and reversible, and thus possible to describe with equilibrium equations, can also be challenged. In batch experiments, Sposito (1989) showed situations in which they were controlled by mass transfer. This implies that the third assumption above is actually equal to assuming that the rate of transport of reactants from the solution to the exchanger is greater than the convective flux of solutes through the soil. In fact, the question concerning exchange equilibria has so far not been analysed critically.

One interesting approach to modelling cation exchange has been proposed by Tipping and Hurley (1988). In the model CHAOS, cation exchange is modelled on a very detailed level, considering near- surface processes on a molecular level. Although this model may be too structurally complex to include in current hydrochemical models, it may be possible to identify certain key elements of CHAOS that could be linked to catchment models.

Chemical weathering One key process in catchment is chemical weathering. While cation exchange is a process with a limited pool available, chemical weathering of primary minerals is an almost inexhaustible source of cations, such as Na, K, Mg and Ca, and a sink for acidity. In the long-term perspective, weathering is a key process in ecosystems since it provides mineral nutrients and acts as a buffering mechanism. All predictions regarding the effects of acidification on solute chemistry with process-oriented models therefore include weathering as an important process.

To illustrate the importance of reasonable estimates of weathering rates for overall model performance, let us consider a catchment where the weathering rate is 100 meq m^2 yr^{-1} and the runoff is 0.4 m yr^{-1}. Then differences in model predictions of 50% result in a difference of 125 meq m^{-3} in runoff alkalinity, a significant difference indeed.

The mechanism behind chemical weathering involves a dissolution reaction that takes place at the mineral – solute interface and the chemical composition at the interface that determines the weathering rate. The historic problem has been that rate laws linking solute concentration to

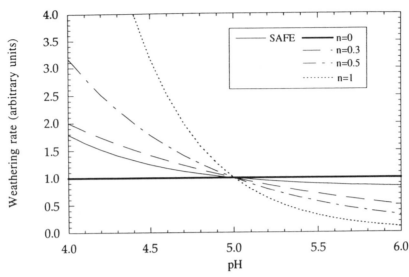

FIGURE 2.1   Comparison of pH dependence of chemical weathering with different model formulations

mineral dissolution have not been available and no scheme has existed showing how the mineral phase should be characterized. Issues related to the complex interactions between ecosystem processes and weathering are covered in Chapter 7 of this volume.

Despite the historic lack of scientific support, catchment modellers have simply been forced to assess weathering rates. Three main approaches have been taken:

1. Assignment of a constant weathering rate, determined either by calibration or by using an estimate for the actual catchment or some neighbouring site
2. Assignment of a standard rate, either arbitrary or determined as a function of soil properties, modified by the solution pH
3. Application of a geochemical weathering model

The first two approaches can be summarized by the following simple rate equation:

$$W = k \cdot [H^+]^n$$

where W is the weathering rate, k is a rate constant and n is the reaction order with respect to H^+. The constant weathering rate approach corresponds to $n = 0$, which is the submodel used in MAGIC. In SMART and RESAM, $n = 0.5$. With the geochemical approach, as implemented in SAFE (see Chapter 3), W is a function of solute composition (Al, base cations, DOC, DIC) and geochemical properties such as mineralogy and soil texture (Sverdrup and Warfvinge, 1992). A comparison of the different approaches is illustrated in Figure 2.1, where the models have been normalized at pH 5. We can see that, with $n = 0.5$, the weathering rate increases three-fold for a change in 1 unit of pH, a change that is in the range of actual decline in pH due to acidification over a period of 50 years (Falkengren-Grerup, 1987), and two-fold if $n = 0.3$. The geochemical model SAFE includes rate equations for the actual dissolution reaction with $n = 0.5$, but, due to the chemical interactions in the model, the geochemical model suggests even less change between pH 4.5 and 5.5 (40%) than an $n = 0.3$ model.

We can also see that in a vertically lumped model, an "average" H^+ concentration will be used to drive chemical weathering. Therefore, the higher the reaction order n and the greater the pH gradient through the soil, the more difficult is it to conceptualize the effect of spatial lumping on the calculated weathering rate. We can thus conclude that models with a high value of n will have larger behavioural complexity than models with lower reaction order for chemical weathering.

2.4.2 Other model types

Besides the process-oriented models, there is an important class of models where the output is not driven by explicit disturbances at the upper boundary. Many of these models are holistic, which means that the ecosystem works as a unit with properties of its own. An example of such models is AGNPS (agricultural non-point source), a model that is used to predict N leaching from farm land. It is an export coefficient model (see Chapter 1) and a specific N leaching is assigned to sub areas with a certain land use.

Another example of holistic modelling was introduced by Kirchner (1990). In its simplest form, the proposed model predicts variations in the concentration of one solute as a function of the concentration of another. The only assumption is that heterogeneous equilibrium rules in the catchment and that the parameters governing these equilibria are constant. One application has been to predict the concentrations of base cations in drainage water as a function of the acid anion concentrations, which, in a sense, reflects the conditions at the upstream boundary of the catchment (Kirchner *et al.*, 1992).

Another holistic model is the end member approach (EMMA) to predict stream-water variations in chemistry on a short-term (weekly) basis. EMMA is based on the hypothesis that stream water is a mixture of water from different distinct sources with a different, time-invariant composition. An underlying idea is that variations in hydrological flowpaths are responsible for these chemical variations. Indeed, examples show that the back-calculated hydrograph produced by EMMA may agree with the actual hydrograph (Hooper *et al.*, 1990). Thus, EMMA is dynamic with respect to hydrology but is static with respect to the internal catchment chemistry. The dynamics in stream water chemistry are entirely due to dynamics in the hydrological pathway.

The same basic idea was implemented by Bergström *et al.* (1985). There, the hydrochemical properties of runoff from different soil compartments were determined by a fitted, time-variant function, rather than experimentally confined end members. Since this category of model is based on a time-invariant end member composition, such models can only be applied to short-term variations, where the hydrological variations are greater than the changes in chemical properties of the catchment. They can therefore be applied not to make prognoses regarding long-term trends in catchment acidification but only to variations in stream water chemistry.

2.4.3 Model implementation

Why is it then such an effort to create a model? When broken down into "pure" equations, most models turn out to be very simple. They consist of a few differential equations, some equilibrium constraints and time series to feed the model, calculation loops — that is generally all. Despite this simplicity, the modeller may need thousands of lines of computer code to implement the model and solve the equations. There are several factors that make model implementation so difficult and time consuming. Firstly, the resulting systems of both differential equations and algebraic equations need to be solved in the correct order. Many times the model implementations are performed by the scientist who created the conceptual model as well as the mathematical model, and the code will mirror the causality of the processes, which may be very inefficient. Secondly, few models make use of high-level programming environments that allow the model to concentrate on the science rather than the implementation. Examples of powerful high-level simulation environments are STELLA, MatLab/SimuLink and DYMOLA. Finally, the modeller cannot see the contours of the "final" model when the work is started. Components may be added and deleted and time-invariant forcing functions (such as hydrology) may suddenly become time-variant. All of these changes call for modifications in the computer code. This can result in very patchy programming, far from object-oriented programming.

It is desirable that existing catchment solute models be distributed widely, preferably with commercial quality implementation and documentation. This would propagate the interest for use of models in ecosystem research and lead to better use of field data. The experience also shows that users who are not emotionally attached to a certain model are extremely valuable in the everlasting process of model development, verification and validation.

2.5 CONCLUDING REMARKS

This paper presents a number of concerns and obstacles associated with catchment solute modelling. This should not be interpreted as pessimism. Indeed, mathematical modelling of catchments is necessary in order to understand (explain) catchment behaviour and predict catchment responses to perturbations. Catchment modelling (organization of knowledge) must be performed in response to environmental pollution, and mathematics is the most powerful tool available to reduce the field data in a defined manner. There is thus no alternative to modelling to obtain guidance for policy decisions. The problem is that our understanding of catchment processes is not complete, and never will be. The reductionist modellers will therefore continue their search for a set of equations that can be validated by an increasing number of observations. The holistic modellers still need theories for the catchment scale, and will continue to risk being criticised by the reductionist for not including their "pet processes". The above discussion leads to some general conclusions regarding solute modelling in catchments:

1. A model can never be identical to the real system, and therefore models should not be criticised for being too simplistic, but modellers could be criticised for applying their model outside their experimental frame.
2. The usefulness of a model should be judged from its predictive capacity, not by its quantitative content of "laws" developed within other branches of science.
3. Processes that cannot be parameterised with confidence should either be kept out of the model, although this is the same as saying that the process is unimportant, or be subject to an uncertainty analysis with respect to its parameters.
4. Modellers should realise that high structural complexity will prevent others from taking an interest in the modelling work, simply because the complexity prevents them from gaining personal experience of the general behaviour of the model.

REFERENCES

Bergström, S., Carlsson, B., Sandberg, G. and Maxe, L. (1985). "Integrated modelling of runoff, alkalinity and pH on a daily basis", *Nord. Hydrol.*, **16**, 86−104.

Bruggenwert, M. G. M and Kamphorst, A. (1982). *Survey of Experimental Information on Cation Exchange in Soil Systems*, Elsevier Scientific Publishing Company, Amsterdam.

Cellier, F. E. (1991). *Continuous System Modelling*, Springer-Velag, New York.

Chen, C. J., Gherini, S., Hudson, R. M. and Dean, S. (1983). *The integrated lake−watershed acidification study*, Final Report EPRI EA-3221, Electrical Power Research Institute, Palo Alto, California.

Cosby, B. J., Wright, R. F., Hornberger, G. M. and Galloway, J. N. (1985). "Modeling the effects of acid deposition: assessment of a lumped parameter model for soil water and stream water chemistry", *Water Resour. Res.*, **21**, 51−63

Falkengren-Grerup, U. (1987). "Long-term changes in pH forest soils in southern Sweden", *Environmental Pollution*, **43**, 79−90.

Harmsen, K. (1982). "Theories of cation adsorption by soil constituents: discrete-site models", in *Soil Chemistry; B. Physio-Chemical Models* (Ed. G. H. Bolt), pp. 77−139, Elsevier Scientific Publishing company, Amsterdam.

Hauhs, M. (1990). "Ecosystem modelling: science or technology?", *J. Hydrol.*, **116**, 25−33.

Henriksen, A., Kämäri, J., Posch, M., Lövblad, G., Forsius, M. and Wilander, A. (1990). *Critical Loads to Surface Waters in Fennoscandia*, Nordic Council of Ministers, Copenhagen, Miljørapport 1990, p. 17.

Hooper, R. P., Stone, A., Christophersen, N., de Grosbois, E. and Seip, H. M. (1988). "Assessing the Birkenes model of stream acidification using multisignal calibration methodology", *Water Resour. Res.*, **24**, 1308−1316.

Hooper, R. P., Christophersen, N. and Peters, N. E. (1990). "Modelling streamwater as a mixture of soil water as a mixture of soil water end-members — an application to the Panola Mountain catchment, Georgia, USA", *J. Hydrol.*, **116**, 321−343.

Jørgensen, S. E. (1992). *Integration of Ecosystem Theories: A Pattern*, Kluwer Academic Publishers, Dordecht.

Kirchner, J. W. (1990). "Heterogeneous geochemistry of catchment acidification", *Geochim. Cosmochim. Acta*, **56**, 2311−2327.

Kirchner, J. W., Dillon, P. J. and LaZerte, B. D. (1992). 'Predicted response of stream chemistry to acid loading in Canadian catchments", *Nature*, **358**, 478−481.

Neal, C., Mulder, J., Christophersen, N., Neal, M., Waters, D., Ferrier, R. C., Harriman, R. and McMahon, R. (1990). "Limitations to the understanding of ion-exchange and solubility controls for acidic Welsh, Scottish and Norwegian sites", *J. Hydrol.*, **16**, 11−23.

Parlange, J.-E., Starr, J. L., Barry, D. A. and Braddock, R. D. (1982). "A theoretical study of the inclusion of boundary conditions and transport equations for zero-order kinetics", *J. Soil Sci. Soc. of Am.*, **46**, 701−704.

Reuss, J. O. (1990). *Analyses of soil data from eight Norwegian catchments*, NIVA Report 0-89153.

Sposito, G. (1989). *The Chemistry of Soils*, Oxford University Press, New York.

Sverdrup, H. and Warfvinge, P. (1992). "Calculating field weathering rates using a mechanistic geochemical model — PROFILE", *Applied Geochemistry*, **8**(3), 273−283.

Sverdrup, H. U., de Vries, W. and Henriksen, A. (1990). *Mapping Critical Loads*, Miljørapport 1990, p. 98, Nordic Council of Ministers, Copenhagen.

Sverdrup, H., Warfvinge, P., Goulding, K. and Blake, L. (1995). "Modeling recent and historical soil data from Rothamsted Experimental Station, England using SAFE", *Agriculture, Ecosystems and Environment*, **53**, 161−177.

Tipping, E. and Hurley, M. A. (1988). "A model of soil−solution interactions in organic soil based on the complexation properties of humic substances", *J. Soil Sci.*, **39**, 505−519.

Van Oene, H. and Ågren, G. I. (1995). "What do we need to know about nutrient cycling when modelling effects of acid deposition on forest growth? Complexity versus simplicity", *Ecological Bulletin*, **44**, (in press).

Warfvinge, P and Sverdrup, H. (1992a). "Calculating critical loads of acid deposition with PROFILE — a steady state soil chemistry model", *Water, Air and Soil Pollut.*, **63**, 119–143.

Warfvinge, P and Sverdrup, H. (1992b). "Hydrochemical modeling', in *Modeling Acidification of Groundwater*, pp. 79–114, (Eds. P. Sanden and P. Warfvinge), Swedish Meteorological and Hydrological Institute (SMHI), Norrköping.

Warfvinge, P., Falkengren-Grerup, U. and Sverdrup, H. (1993). "Modeling long-term base cation supply to acidified forest stands", *Environmental Pollut.*, **80**, 209–220.

Warfvinge, P., Holmberg, M., Posch, M and Wright, R. F. (1992). "The use of dynamic models to set target loads", *Ambio*, **5**, 369–376.

Wesselink, L. G. and Mulder, J. (1995). "Modeling aluminium solubility control in an acid forest soil, Solling, Germany", *Ecological Modelling* (in press).

Wright, R. F., Holmberg, M., Posch, M. and Warfvinge, P. (1991). *Dynamic models for predicting soil and water acidification: Application to three catchments in Fennoscandia*, NIVA, Blindern, Report 25/1991.

SECTION II

WEATHERING AND SOILS IN SOLUTE MODELLING

3

Biogeochemical Modelling of Small Catchments Using PROFILE and SAFE

HARALD SVERDRUP, MATTIAS ALVETEG

Department of Chemical Engineering, Lund University, Sweden

SIMON LANGAN

Macaulay Land Use Research Institute, Aberdeen, UK

and

TOMAS PAČÉS

Czech Geological Survey, Praha, Czech Republic

3.1 INTRODUCTION

3.1.1 Modelling Ecosystems

Ecological change can always be modelled as long as the kinetic rules and the boundary conditions can be defined and the properties of the ecosystem measured. Where these conditions can be met (and therefore the model is tightly constrained) it is possible to apply the model widely and run it over any time-scale. Models are especially important in research not because they produce results in their own right but because they allow complex and non-linear systems to be investigated and data from such systems to be interpreted. When forced to describe the subject of study in the form of equations and set values to coefficients, it is possible to formally test our understanding of the system. In a model formal understanding of the system is tested. There are no "maybe's" in modelling, as all parameters are assigned quantitative values according to unique and precise rules. Models can be seen as a systematic synthesis of established research knowledge, a synthesis often beyond what can be produced by empirical extrapolation used to produce two-dimensional linear regression plots. Modelling provides a powerful method for connecting and comparing experimental and theoretical results with the results of others. When processes on the catchment scale are investigated, then models become an almost inevitable tool for untangling all connections and feedbacks within the system, as well

---

Solute Modelling in Catchment Systems. Edited by Stephen T. Trudgill
© 1995 John Wiley & Sons Ltd

as a platform for testing whether the understanding of the system will stand up to comparison with observations.

There are a variety of geographical scales at which the modelling of ecosystems can be undertaken. The most appropriate scale for the model will depend on the object of the study and the data obtainable. If, for example, our objective is to predict the ecological response to changes in acid deposition the most appropriate scale of model will be that based on the function of soil processes in buffering acid inputs. However, in order to predict the responses at a different scale it is necessary to ensure that the model incorporates all of the relevant processes. For example, modelling at the catchment scale must utilise information on the various connections, interactions and feedbacks between atmospheric inputs, vegetation, soil and hydrology.

3.1.2 Objectives and scope

Before embarking on a modelling exercise it is vital that the objectives of the work are clearly defined. Typical modelling objectives are:(a) to make predictions about the future, (b) reconstructing past events and (c) for interpreting the processes controlling the changes in our observations.

The objective of this study is primarily to demonstrate the use and capabilities of the biogeochemical models PROFILE and SAFE. This is to be illustrated by application of PROFILE to sites in southwestern Sweden and Scotland. SAFE is applied to the extremely polluted Most catchment in the Czech Republic in order to illustrate how SAFE may be used to study the development over time, a problem not addressed by PROFILE.

3.1.3 Model types

We can distinguish between three basic types of models which fulfil different purposes in terms of prediction and catchment process representation:

1. Regression models
2. Process-oriented models
3. Mechanistic models

All of these models can be used for making different types of predictions, but only the last type can be used for interpreting processes on the catchment level. In soil modelling two varieties of the models listed above are used: steady-state models, where the path to a future change is bypassed and the final state calculated directly, and dynamic models where the whole path of change is described. In Figure 3.1 we have illustrated the basic difference between the two types of model, as compared to popular perception of models.

Experimental data are fundamental to the development and testing of models. Basically there are two different ways of utilizing data. The simplest approach is to use the data directly, and by trying to relate one type of observation directly to another, correlations are sought for. Such models are regression models.

Regression models utilise the information described by patterns in the data to construct correlations between the observed parameters. Such models require much data on which to

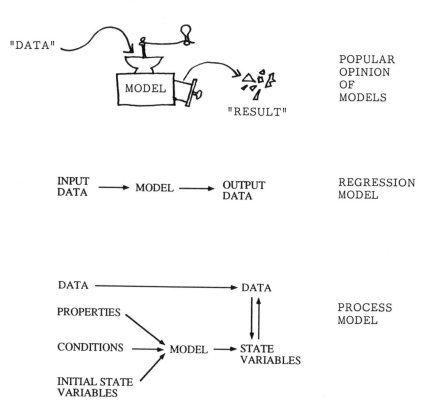

FIGURE 3.1 The figure illustrates how regression models explore relations between the observed parameters. Regression models require large amounts of data on which to calibrate, and have applicability only within the system for which they are calibrated. Process models operate on mathematical representation of fundamental principles and properties of the system. Data are divided into state variables and information used directly by the model such as boundary conditions, properties and initial states. The data state variables bypass the model and are only used to verify whether the model output is consistent with observations

calibrate, and they have applicability only within the system for which they are calibrated. Certain sets of equations may be contained in a regression model and by tuning key parameters to make the model output resemble the data, the model is calibrated. Extrapolation outside the range of the dataset will often be invalid.

Process models or mechanistic models operate on mathematical representation of fundamental principles and properties of the system. For such models "the data" is subdivided into different types of information: properties, boundary conditions and initial states are kept separate, as they will be important for determining the conditions from which the processes will be able to act. The input from a mechanistic model will consist of state variables which change with time (pH, concentrations etc.). With such a model it is possible to calculate synthetic state variables, "modelled data". Observations of state variables are used to compare with the calculated ones. If they agree, then it could be argued that the model contains the relevant processes, and our understanding of the system is correct. If however, there is a mismatch between modelled and observed data it is possible to conclude that the misfit is a quantification of the deficiency in our understanding of the system processes. Such quantification of how much we miss is very valuable when the model is being developed. It can be used to account for the relative importance of different processes. Once our understanding of the system as represented in the model formulation is acceptable, then it is possible to test the model. The ultimate test for the model and modeller is the ability to transfer a model to a similar system, without the need to recalibrate; only properties and boundary conditions are changed, as dictated by the new system.

An alternative approach is where the model is formulated from an *a priori* understanding of the system. This understanding is expressed in mathematical relations. Process models or mechanistic models operate on mathematical representation of fundamental principles and properties of the system. In this case data are used to verify whether the mathematical representation is basically correct or not. The basic properties of the system determine how the principles are acted out. For example, catchment output such as runoff chemistry can be calculated from our *a priori* knowledge of properties such as soil type, texture, mineralogy and conditions such as the amount of acid input and amount of rainfall. If our understanding of the catchment is basically correct and formulated in a similar manner within the model, then model output should be similar to the observed data. Any principles, processes or errors in our understanding of catchment processes will be highlighted as differences between our calculations and the observed data. Provided that all other processes are correctly represented, it is possible to calibrate one variable on one process whilst maintaining a high degree of freedom. This will allow the model to be used as a measurement instrument for that particular variable or process. Mechanistic models have good general applicability and good predictive capacity, where good quality input data are available.

Sometimes models are called "process oriented". This implies that they contain certain processes in mathematical representation but operate with regression polynomials for other processes. These models are hybrids and share both advantages and weaknesses of both types.

3.1.4 Calibrating acidification models

In modelling, the "cult of success" is very strong. The "cult of success" implies that the calculated output is expected to pass through all the observation points. If not, the user of the model might soon hear "The model is wrong . . ." or "It is obvious that the model does not work". However if it is obvious that the model was calibrated, critics will soon remark "You can probably make the thing fit anything . . .". In the following section we will discuss why calibration sometimes is necessary, but also why it must be used with great caution, in order not to prevent interpretations.

If a process model is calibrated on more than one or two processes, it becomes much more likely that errors in the model formulation can be effectively covered over. This may result in the calculations looking correct, but if it is not possible to distinguish between artefacts of the calibration and the effect of correct principles in the model, then the good fit will be of no help in understanding the system. Thus it is always relevant to reflect on what was the objective of the model application in the first place. It is often helpful to remember as a general rule that our understanding of system processes and functionality increases as a result of model misfit rather than when the model fit is good.

If one is certain that all the other processes except the one used for calibration are correct, then the model will serve as a measurement instrument for determining the coefficients of that uncertain process.

If more than three processes are calibrated simultaneously there will be several ways to model a dataset and fit the observations. Then any conclusion on catchment process will be meaningless. In this case the model is no longer a mechanistic model but has become a non-linear regression. Any conclusion based on such a model formulation will be subject to a high degree of uncertainty. Thus, before a model is calibrated, care should be taken to consider what kind of information may be lost in the calibration procedure. For the purpose of testing process formulations, generally calibration on more than one degree of freedom must be completely avoided. Calibrated runs ignoring important processes can be made to look like runs including full process representation.

Dynamic soil chemistry models are best calibrated on two parameters: the weathering rate and the initial base saturation. In all cases, with the exception of sites at Rothamsted (UK) (Sverdrup *et al.*, 1995) and a few Swedish sites (Falkengren and Eriksson, 1990; Warfvinge *et al.*, 1993), no historical values are available, and only one point in time (present) is available for testing the model. For short-term soil chemistry responses more data are often available.

Considering an acidification model, adjusting the weathering rate will change the slope of the line, the rate of change of acidification, as well as influence the final alkalinity level. The weathering rate has often been considered very difficult to estimate for field conditions; as a result several contemporaneous acidification models use it for calibration. The effect of using the weathering rate for model calibration is illustrated in Figure 3.2.

The figure shows that the initial base saturation determines the level from which the system starts. The higher the initial base saturation, the more must be leached, and this influences the time and rate of change required to reach the final state, i.e.the curve shifts up or down.

The gibbsite coefficient is also often used for calibration of acidification models. The constants estimated in laboratory experiments are useless and values like $pK_G = 6.5, 7.5, 8.5, 9.5$ are generally used for the O, E, B and C layers respectively. Gibbsite in European soil is extremely rare. If calibration of base saturation and the weathering rate is combined, it will always be possible to calibrate the model to go through the present time observation. This is illustrated in Figure 3.2. If further calibration is given on additional processes, like uptake, decomposition of organic matter or ion exchange, the model will be able to fit very diverse datasets.

3.1.5 The integration of biological response

The terrestrial flora and fauna of a catchment are (in most cases) very important for the cycling of nutrients, such as base cations, phosphorous and nitrogen. They are also of great importance for the balance of acidity in a catchment; acidity may be a strong driving force in many soil processes. Major biological processes in this context are the growth of plants, decomposition of organic matter, nitrification and denitrification. An example of how such a feedback loop can be coupled into a model is shown in Figure 3.3. Large changes in soil chemistry will imply that the rate of these processes will change, something to remember when the model does not include such feedbacks. For example, under severe acidification, plant growth will be inhibited by a low ratio of base cations necessary for growth and aluminium which can interfere with

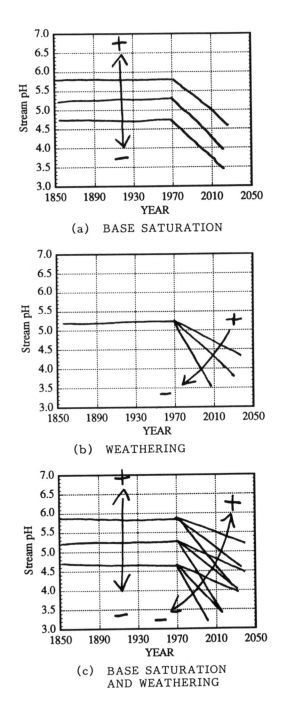

FIGURE 3.2 A dynamic model can be calibrated using (a) the initial base saturation, (b) the weathering rate or (c) by adjusting both the initial base saturation and the weathering rate

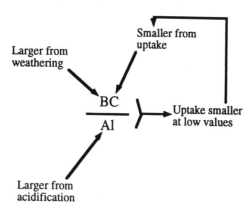

FIGURE 3.3   Example of a feedback loop between soil solution chemistry and tree growth in a catchment. The diagram illustrates how this has been implemented in the PROFILE model

uptake. This is sometimes expressed using the base cation to aluminium ratio ("BC/Al ratio"). A reduction in uptake of base cations will feed back on the ratio, causing the decrease in the ratio value to slow down. If uptake is less, then the base cations will stay in the soil solution, and the concentration will be changed by the diminished ability of the plants for growth. Decomposition or nitrification can be affected by chemistry and affect the chemistry in the same way.

In the present formulation, PROFILE has a limited amount of chemical feedback built in, although the effect of soil chemistry on the growth and nitrification is included. Similar feedback on denitrification, N immobilization and decomposition is currently under development.

3.2   MODEL DESCRIPTION

PROFILE and SAFE are the same model, but in different forms. PROFILE is a steady-state version, PROFILE bypasses the changes in soil state over time and calculates the final steady state directly, whereas SAFE calculates the change with time (Sverdrup and Warfvinge, 1991; Sverdrup *et al.*, 1992).

The model is based on a conceptual model of a forest soil; the soil may represent a profile or the whole catchment. They include the following chemical subsystems:

1. Deposition, leaching and accumulation of dissolved chemical components
2. Chemical weathering reactions between soil solution and minerals
3. Cation exchange reactions
4. Nitrification, immobilization and denitrification
5. Biological uptake of nutrient cations, with chemical feedback
6. Solution equilibrium reactions involving CO_2, Al and organic acids

In PROFILE the differentials for change in the soil solution and ion exchange reservoirs are set to zero, and the final base saturation is determined from equilibrium with the soil solution.

The PROFILE model is structured in different compartments in order to represent the natural vertical differences in soils (horizons), which have distinctly different chemical properties as a result of pedogenical processes. The internal coupling in each horizon is illustrated in Figure 3.4.

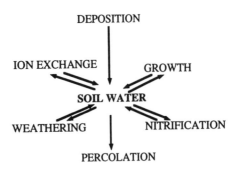

FIGURE 3.4   Coupling of the processes in the steady-state model PROFILE and the dynamic model SAFE

The change in soil chemistry and the subsequent change in the distribution of elements on the cation exchange matrix is calculated by means of mass balance equations. The cations Mg, Ca and K are lumped together into a divalent component, (BC). The hydrogen ion is treated as dependent on the variable acid neutralizing capacity (ANC). In the model, each soil horizon is assumed to be homogeneous.

In the model, the buffering in the liquid phase is controlled by the CO_2 – carbonate system, the acid – base reactions of an organic acid and an aluminium submodel. The latter is based on the assumption that the concentration of Al species is governed by the dissolution and precipitation of a solid gibbsite phase, $Al(OH)_3$. In the models dissolved organic carbon (DOC) is modelled as a monovalent organic acid. The dissociation of the acid functional groups of the DOC is quantified using a simple equilibrium between pH, total DOC and dissociated organic acid; this is called the Oliver equation.

The outstanding difference between PROFILE and other comparable soil chemistry models is that the weathering rate is calculated from independent geophysical properties of the soil system and the transition state theory applied to silicate minerals. This reduces the degrees of freedom in the model and consequently also the need for data available to calibrate on. Several chemical reactions between the mineral and constituents in the liquid solution contribute to the base cation release rate from the chemical weathering of silicate minerals, and the total will be the sum of the rates of the individual elementary reactions. The elementary reactions are (a) with H^+, Al and cations of the parent mineral, (b) with H_2O and Al, (c) with CO_2 and (d) with organic acids. The rate of dissolution is inhibited by the reaction products due to complexing on active dissolution sites (Sverdrup, 1990). The rate coefficients themselves are functions of the base cation and aluminium concentrations, as discussed in Sverdrup (1990). In soils the availability of moisture may be limited, causing the activity of the exposed mineral surface to be less than unity, with the degree of surface wetting depending on soil moisture saturation.

The dynamic version SAFE has been developed with the objective of studying the effects of acid deposition on soils and groundwater. It calculates the values of different chemical state variables as a function of time. It can therefore be used to study the process of acidification and recovery, as effected by deposition rates, soil parameters and hydrological variations.

PROFILE calculations require little computation time, whilst applying SAFE takes considerable computation time, typically a couple of hours for one run comprising four soil layers and a 200 year time period.

3.2.1 Initial conditions

Unique for SAFE is the need for temporal information. Whereas PROFILE ignores time, SAFE will require a starting point; with conditions at that starting point. As well as time series for some of the parameters, PROFILE will require a long-term average. To make sure that the simulations reflect changes due to variations in external and internal loads, rather than just instable initial conditions, SAFE is initiated with steady-state initial conditions calculated with PROFILE (Warfvinge and Sverdrup, 1992a, 1992b; Sverdrup and Warfvinge, 1993a, 1993b).

Although PROFILE calculates the exchangeable fractions on the cation exchange complex in the soil, the output steady-state base saturation is a function of the values chosen for the selectivity coefficients. The SAFE model is calibrated by adjusting the initial value of the base saturation of each soil horizon independently.

All dynamic models used for calculating changes in soil chemistry must be triggered from an equilibrium situation. They should also be expected to converge towards a steady-state solution if the acid deposition is brought to a constant value. The calculation with PROFILE cannot be calibrated. As can be seen from the input and output data displays in Figures 3.5 to 3.12, all inputs can be determined by measurement; therefore no parameter calibration is necessary.

3.3 APPLICATION OF THE PROFILE MODEL

In order to illustrate the utility and output from the model we have selected three catchment studies in which the model has been used to explore different aspects of acidification. In the first study the model has been used to calculate the supply of base cations from the chemical weathering of catchment soils. The determination of the release of base cations from mineral decomposition and dissolution is fundamental to assessing the sensitivity of soils and catchments to acidification.

Present assessments of the vulnerability of ecosystems to acidification are being based on the critical load hypothesis in which the critical load is equal the point at which the supply of base cations from chemical weathering plus those from atmospheric input are equal to the loss of base cations through biological uptake and soil water leaching at a biologically acceptable limit (Sverdrup and Warfvinge, 1993a, 1993b). The first case study at Lake Gårdsjön makes use of the PROFILE model. This study illustrates how the model output from the catchment soils can be used to predict surface water quality which in turn can be linked to the survival of fish within the catchments watercourses.

In the second example the PROFILE model is used to calculate weathering rates at Allt a'Mharcaidh in Scotland. Here the calculated weathering rate is compared with values from other estimates.

The third case study is used to examine how the rate of supply of base cations can be related to the decline in the productivity of forest ecosystems, using PROFILE at a highly impacted catchment in the Czech Republic.

3.4 THE GÅRDSJÖN CATCHMENT, SWEDEN

3.4.1 Catchment description

The data used for the demonstration is from the F1 research catchment at Lake Gårdsjön. The catchment is covered with a 60–100 year old Norway spruce stand; the trees are showing a significant degree of needle loss at present (1989–93, 38%). The lake was acidified during the 1960s and all fish were extinct by 1970 (Hultberg, 1985). The catchment is located close to the city of Göteborg in southern Sweden. The characteristics of the catchment are apparent from the PROFILE input data cards presented in Figures 3.5 to 3.8.

On the first input card (Figure 3.5) deposition of different elements, nitrification rate and climatic data should be given. There are several possible ways of defining deposition in the PROFILE model. What should be kept in mind is that sources included in the deposition figures must also be included in the uptake figures. For the case where all circulation is included, we define total input as the sum of wet deposition, dry deposition, canopy exchange, litterfall and net mineralization. Total uptake is defined as the sum of growth of stems, branches and coarse roots, litterfall and canopy exchange.

To determine the dry deposition of base cations from measurements of throughfall and wet precipitation we define a scaling factor:

$$F_{Na} = \frac{T_{Na} - D_{Wet,Na}}{D_{Wet,Na}} \tag{1}$$

where T_{Na} is throughfall and $D_{Wet,Na}$ is the wet deposition of sodium. The dry deposition of calcium is then given by

$$D_{Dry,Ca} = D_{Wet,Ca} \cdot F_{Na} \tag{2}$$

Sodium is used for scaling because it is conservative with respect to the vegetation.

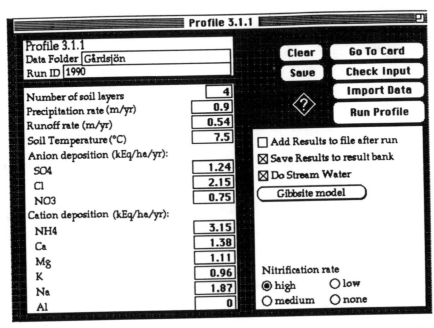

FIGURE 3.5   The run ID input card, the first card in the PROFILE stack. Data from the Swedish research catchment F1 at Lake Gårdsjön have been used

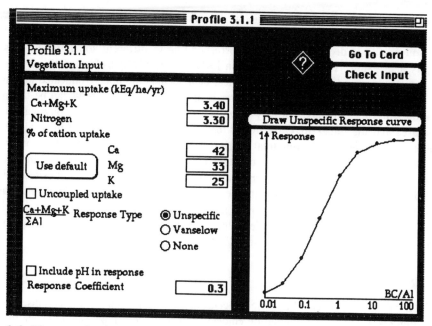

FIGURE 3.6   The vegetation input card. Data from the Swedish research catchment F1 at Lake Gårdsjön have been used

It is worth noticing that deposition is a function of canopy area and therefore also of tree growth. This feedback loop is not included in the PROFILE model. The specified deposition figures are used unchanged throughout the calculation. The card in Figure 3.5 shows the input values for 1990.

As can be seen in Figure 3.6, PROFILE needs information regarding the desired uptake of different nutrients. The tree growth submodel calculates uptake by:

$$\text{Uptake} = f_{Al} \cdot \min \text{ (input uptake, nutrients)} \qquad (3)$$

where f_{Al} is the uptake response factor:

$$f_{Al} = \frac{[BC]^n}{[BC]^n + K \cdot [Al]^m} \qquad (4)$$

where the Al variable is dependent on aluminium in different forms, but also other undesired ions adsorbing to the plant root surface:

$$[Al] = 3([Al^{3-}] + 2[Al(OH)^{2+}] + [Al(OH)_2{}^+])/3 + n \cdot [H^+] \qquad (5)$$

BC is the sum of Ca, Mg and K. Two different uptake response functions are included in the model: the unspecific response, where $n = m = 1$, and the Vanselow response, where $n = 3$ and $m = 2$. The inclusion of $n \cdot H^+$ in the response functions is optional. The use of f_{Al} as feedback is optional, and can be turned off.

Since all trees need more than one of the nutrients to survive, uptake is limited by the least available nutrient. If, for example, only 60% of the desired amount of calcium is available while there is an excess of all other nutrients, then uptake of each and all of the nutrients will be decreased to 60%. The coupling of nutrient uptake is optional, and can be turned off.

The soil profile is divided into layers corresponding to different soil horizons of similar chemistry and texture. For each soil horizon PROFILE needs hydrologic data, physical data such as mineralogy and density, and biological data such as the uptake allocated to that layer (Figure 3.7). The hydrologic data, soil water content and the water flow can be taken from simulation with other models, but they may also be estimated. Among the physical data, the surface area and the mineralogy are by far the most important factors. If mineralogy is not available it may be estimated from total analysis of the soil, following the method of Sverdrup et al. (1990). The gibbsite solubility constant (Figure 3.8) is included to obtain a relation between pH and aluminium. It is very unlikely that the solubility of gibbsite influences aluminium concentrations in soils, but it gives a reasonably good fit and a kinetic alternative is under development. The kinetic aluminium model at present is included as an option, but all sources of aluminium are not yet included. PROFILE therefore gives too low concentrations of aluminium when the kinetic model is selected.

3.4.2 Results

PROFILE calculates pH, ANC and the concentration of major anions and cations, as can be seen in Figures 3.9 and 3.10. Output on chloride and sulphate can be used to check the water balance, since they are treated as conservative by the model. Silica concentrations should not be taken too seriously, since there is no sink for silica when the gibbsite model is used.

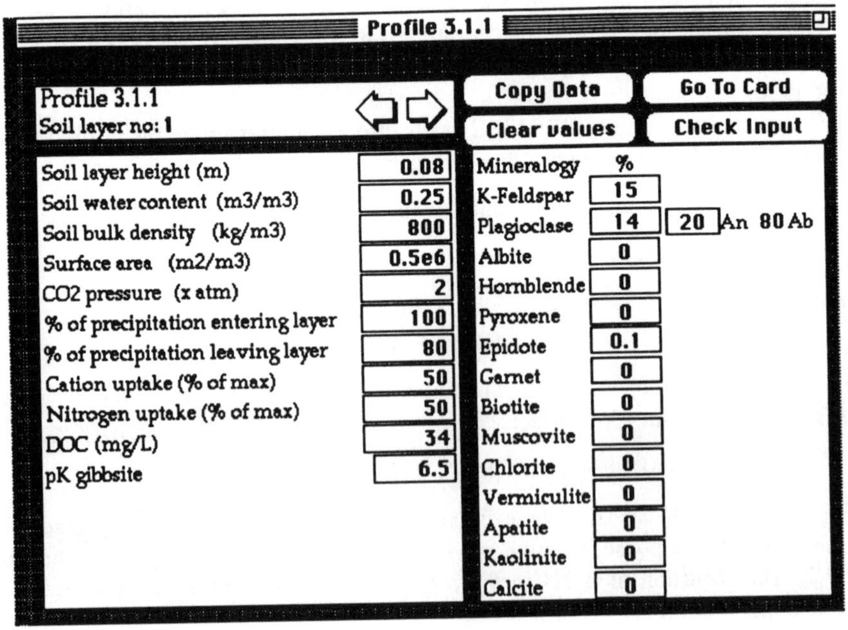

Profile 3.1.1				Copy Data	Go To Card
Soil layer no: 1				Clear values	Check Input

			Mineralogy	%	
Soil layer height (m)	0.08		K-Feldspar	15	
Soil water content (m3/m3)	0.25		Plagioclase	14	20 An 80 Ab
Soil bulk density (kg/m3)	800		Albite	0	
Surface area (m2/m3)	0.5e6		Hornblende	0	
CO2 pressure (x atm)	2		Pyroxene	0	
% of precipitation entering layer	100		Epidote	0.1	
% of precipitation leaving layer	80		Garnet	0	
Cation uptake (% of max)	50		Biotite	0	
Nitrogen uptake (% of max)	50		Muscovite	0	
DOC (mg/L)	34		Chlorite	0	
pK gibbsite	6.5		Vermiculite	0	
			Apatite	0	
			Kaolinite	0	
			Calcite	0	

FIGURE 3.7   A layer input card. There is one input card for each layer in PROFILE. Data from the Swedish research catchment F1 at Lake Gårdsjön has been used

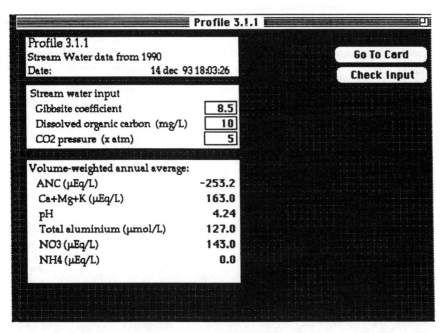

FIGURE 3.8 The stream water card. This card contains both input and output. Data from the Swedish research catchment F1 at Lake Gårdsjön have been used

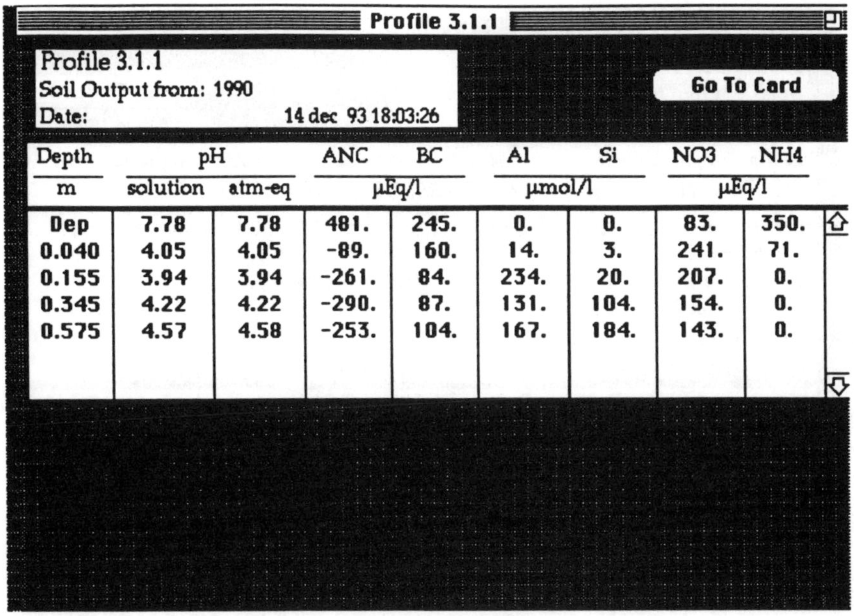

Depth	pH		ANC	BC	Al	Si	NO3	NH4
m	solution	atm-eq	µEq/l		µmol/l		µEq/l	
Dep	7.78	7.78	481.	245.	0.	0.	83.	350.
0.040	4.05	4.05	-89.	160.	14.	3.	241.	71.
0.155	3.94	3.94	-261.	84.	234.	20.	207.	0.
0.345	4.22	4.22	-290.	87.	131.	104.	154.	0.
0.575	4.57	4.58	-253.	104.	167.	184.	143.	0.

FIGURE 3.9 The first output card in the PROFILE model. This show the concentrations in the soil solution at steady state with the deposition in the input

Profile 3.1.1								
Profile 3.1.1								
Detailed Soil Output from:1990							**Go To Card**	
Date:			14 dec 93 18:03:26					
Depth	Ca	Mg	K	Na	CL	SO4	Al-org	Al-inorg
m				μEq/l			μmol/l	
0.040	93.	77.	75.	260.	299.	172.	12.	3.
0.155	39.	34.	48.	350.	398.	230.	180.	54.
0.345	50.	45.	39.	360.	398.	230.	48.	82.
0.575	67.	51.	45.	377.	398.	230.	76.	90.

FIGURE 3.10   The second output card in PROFILE. Here base cations are given as individual ions

Profile 3.1.1								
Profile 3.1.1								
Uptake Output from:1990							**Go To Card**	
Date:			14 dec 93 18:03:26					
Depth	Uptake		N Uptake response			BC/Al	Al precip.	Denitrif.
m	N	BC	Al	deficiency	total		kEq/ha,yr	
0.040	1.643	1.692	0.995 *	1.000 =	0.995	65.864	-	0.011
0.155	1.114	1.148	0.844 *	1.000 =	0.844	1.620	-	0.014
0.345	0.260	0.268	0.789 *	1.000 =	0.789	1.120	-	0.028
0.575	0.000	0.000	- *	- =	-	1.325	-	0.059
Totals:	3.017	3.108	0.914 *	1.000 =	0.914		-	0.111

Acidity produced by N-uptake and N-reactions   3.172   kEq/ha,yr
Atmospheric deposition of acidity   -4.330   kEq/ha,yr
Total potential acidity   1.970   kEq/ha,yr
Alkalinity leaching (acidity leaching if negative)   -1.367   kEq/ha,yr

FIGURE 3.11   On this card, results on tree growth are shown. The uptake response will only be valid for nitrogen if uptake is specified as uncoupled

The effect of chemical feedback on uptake and hence growth is shown on the vegetation response card, as shown in Figure 3.11. The card shows how this affects different horizons. The effect of Al is separated from the effect of nutrient limitations, and the total effect is calculated from those two elements.

In the weathering card, the total weathering rate is displayed, in addition to the weathering rate for individual base cations and different soil horizons. The weathering rate card is shown in Figure 3.12.

If the stream water chemistry is to be calculated, the key assumption is that the total thickness of the layers in PROFILE correspond to the average soil depth in the catchment. PROFILE needs the gibbsite solubility constant, the dissolved organic carbon and the carbon dioxide pressure for the stream water (Figure 3.8). PROFILE calculates ANC, pH and the concentration of nitrate, ammonium, aluminium and base cations for the stream water. When running PROFILE to find initial values for SAFE, it is not necessary to calculate stream water chemistry.

Fish survival was calculated using the Brown−Baker−Schofield model, parameterised from the experiments of Baker and Scofield (1982) and Brown (1982):

$$S = \frac{[BC^{2+}]}{[BC^{2+}] + k_{Al} \cdot [Al^{3+}] + k_{H} \cdot [H^{+}]} \tag{6}$$

where k_{Al} has the value 5 and k_{H} the value 15. Applying this model to the results on the output card yields a low survival index for Lake Gårdsjön. The fish were extinct by 1970, implying that survival of young fish was too low to keep up with natural mortality plus fishing. The same fish survival model was also applied to similar output from SAFE; results have been displayed in Figure 3.13.

Profile 3.1.1								
Profile 3.1.1 Weathering output from 1990 Date: 14 dec 93 18:03:26						Go To Card		
Depth	Weathering (kEq/ha/yr)							
m	BC+Na	Ca	Mg	K	Na	Al	Si	PO4
0.040	0.009	0.002	0.000	0.002	0.004	0.027	0.089	0.000
0.155	0.054	0.021	0.011	0.006	0.016	0.115	0.350	0.004
0.345	0.394	0.173	0.146	0.022	0.053	0.722	1.816	0.013
0.575	0.245	0.093	0.033	0.029	0.090	0.534	1.715	0.029
Totals:	0.702	0.289	0.190	0.059	0.164	1.398	3.970	0.047

FIGURE 3.12 The calculated weathering rates are available for individual base ions released

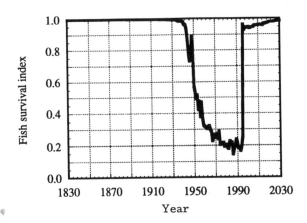

3.5   ALLT A'MHARCAIDH CATCHMENT, SCOTLAND

3.5.1   Catchment description

The Allt a'Mharcaidh catchment is situated in the Cairngorm mountains of north-east Scotland.
Detailed catchment soil and vegetation surveys and hydrochemical monitoring were undertaken
in the late 1980s as part of the Royal Society's Surface Water Acidification Project. A detailed
account of the work undertaken as part of this project and in particular on the Mharcaidh
catchment can be found in Mason (1990). Annual precipitation in the catchment is 925 mm,
approximately 20 mm of which falls as snow during the winter months (Ferrier and Harriman,
1990). Runoff is 860 mm, the average annual temperature 5.5°C. Excess sulphur inputs to the
catchment (1986−8) were 15 μeq l^{-1}, on the basis of which Ferrier and Harriman (1990) have
classified the catchment as transitional. The annual acid precipitation loading in 1988 was 0.25
kg H$^+$ ha^{-1} yr^{-1} (Patrick *et al.*, 1991).

   Total deposition was estimated for sulphur to be 0.7 keq ha^{-1} yr^{-1}, nitrogen 0.26
keq ha^{-1} yr^{-1} and base cations 0.55 keq ha^{-1} yr^{-1}. Base cation deposition is primarily from
maritime sources. Catchment altitude ranges from 330 metres at its outflow to 1111 m at its
highest point. The total catchment area is 10.14 km^2.

   Vegetation ranges from alpine azalea heath on the uppermost slopes through to heather moor
and blanket bog across much of the rest of the catchment. Some of the lower slopes also have a
broken cover of native Scots pine.

   The underlying geology is biotite granite, although the valley floor has a deep deposit of
glacial till. Soils of the upper area are dominated by freely drained alpine soils whilst the mid
and lower slopes are occupied by peaty and peaty gleyed podzols. On the mid slope shoulder
these mineral soils give way to an accumulation of organic blanket peat. Wheater *et al.* (1991)
have suggested that streamflow mirrors the twin flow dynamics generated by these three soil
components. For each of the soils identified from the soil survey soil pits were dug and each soil
profile sampled. Both physical and chemical analyses were carried out together with
mineralogy determinations for each sample. A typical soil profile representation and analysis of

the soils is given in Table 3.1, which was taken at Most, Czech Republic. Mineralogical analysis was undertaken using quantitative XRD methods as described by Bain *et al.* (1990a, 1990b). The mineralogy of the soil is dominated by quartz (30%) and feldspars (25% K-feldspar, 40% plagioclase), typical of many soils derived from granite. The texture is coarse.

3.5.2 Results

This information was used as an input to three PROFILE model runs, one for each soil component identified from the soil survey. The input data and output from these runs are shown in Figure 3.14. The weathering rates calculated fall within a narrow band with a mean for the three runs of 0.32 keq ha^{-1} yr^{-1}. These weathering rates calculated by PROFILE are in close agreement with those calculated on the same soils by Bain *et al.* (1993) (0.23 keq ha^{-1} yr^{-1}), using the more traditional method of stable isotopes. Similarly, the results are in agreement with those allocated to the soils by Langan and Wilson (1994) as part of the provisional British soils critical load map. At a catchment scale Whitehead *et al.* (1990) have suggested the weathering rate to be approximately twice this rate (0.62 keq ha^{-1} yr^{-1}), which indicates that there is some additional catchment source of base cations presently unidentified. The result of this will be to offer the stream water a greater buffer to acid inputs than that provided by the soils analysed here. Comparison of the supply of base cations from mineral weathering of the soil against the atmospheric acid loading suggests base cations are being lost at a rate which is in excess of their supply. Therefore if atmospherically deposited acidity is not reduced the soil exchange complex will become depleted with a consequent reduction in the base cation to aluminium ratio and subsequent loss of vegetation productivity. To examine this in more detail we will now utilise the data from the highly impacted catchment in the Czech Republic.

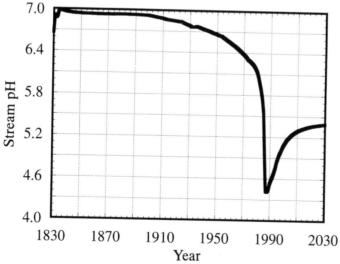

FIGURE 3.14   The calculations suggest that the transition from non-acidified to acidified will occur very rapidly at Allt a'Mharcaidh. Due to uncertainties in the input data such as catchment soil depth and the exact value of CEC, the timing of the change may be $+/-$ a decade

TABLE 3.1   Example of SAFE and PROFILE model input data for catchment X-14 at Most, Czech Republic

Parameter	Unit	Most, X-14			
		1	2	3	4
Morphology characterization		O	E	B	C
Soil layer thickness	m	0.05	0.15	0.6	0.6
Moisture content	$m^3\ m^{-3}$	0.2	0.2	0.2	0.25
Soil bulk density	$kg\ m^{-3}$	800	1200	1400	1600
Specific surface area	$10^{-6}\ m^2\ m^{-3}$	0.55	2	2	1.5
Cation exchange capacity	$keq\ kg^{-1}$	118	60	60	40
CO_2 pressure	times ambient	5	10	20	30
Dissolved organic carbon	$mg\ l^{-1}$	10	5	2	2
log Gibbsite equation constant	$kmol^2\ m^{-3}$	6.5	7.5	8.5	9.2
Inflow	% of precipitation	100	90	80	60
Percolation	% of precipitation	90	80	60	53
Mg+Ca+K uptake	% of total max	75	10	10	5
N uptake	% of total max	75	10	10	5
Mineral	% of total				
K-feldspar		8	9	9	9
Oligoclase		15	20	20	20
Albite		0	0	0	0
Hornblende		0	0	0	0
Pyroxene		0	0	0	0
Epidote		0	0	0	0
Garnet		0	0	0	0
Biotite		0	2	2.5	2.5
Muscovite		5	5	5	5
Chlorite		0	2	2.5	2.5
Vermiculite		5	5	7	7
Apatite		0	0.1	0.2	0.3
Kaolinite		0	0.1	0.2	0.3

3.6   THE MOST CATCHMENT, CZECH REPUBLIC

3.6.1   Catchment description

The catchment X-14 is situated near Vysoke Pec, in the Krusne Hory Mountains, near the city of Most in the northern part of the Czech Republic. Large coal-fired power plants and one of the largest open-pit mining operations for brown coal in the world are located at the foot of the hill with X-14 on top (Pačés, 1985).

The catchment used to be covered by Norway spruce and beech, but due to fumigation from the power plant and other industry in the region, nearly 70% of the spruce forest died in 1970–80, leaving grasslands. The size of the drainage area is 2.7 km², with the top of the catchment located 924 m above sea level. pH of the rain in the area was pH 4.2 in 1987; pH of the runoff was pH 4.9 in 1987. The catchment received 812 mm of rain; runoff was 430 mm. Input data to the model have been listed in Table 3.1.

X-14 belongs to a region which has received some of the highest sulphur depositions that have ever occurred in the world. A maximum was reached in 1979 when approximately 125 kg S^+ ha^{-1} yr^{-1} were deposited. At the same time atmospheric dust inputs generally base rich in nature were large; thus the depletion of the ion exchanger and the occurrence of low BC/ Al values were less drastic than what would have been the case if dust collectors had been installed at an early date. The deposition levels used in the model are shown in Figure 3.15. In the 1970s most of the trees in the catchment were killed by the onslaught of pollution. This is also reflected in the input data to the model, as seen in Figure 3.16.

3.6.2 Results

The models were used to calculate soil solution composition, base saturation and the weathering rate at Most. PROFILE was used to estimate the approximate initial conditions. By using present base saturation and soil solution concentrations, the selectivity coefficients could be calculated. These were assumed to be constant in time and used with the PROFILE model and the historic deposition to estimate a range for the historic base saturation.

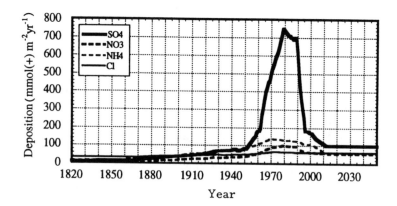

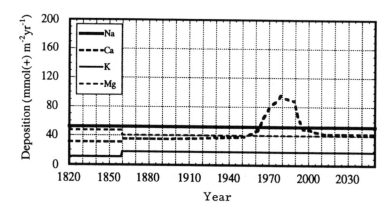

FIGURE 3.15 The deposition input data used in the calculations. The Northern part of the Czech Republic received extreme depositions of acidity during the time period 1970–90 due to use of large amounts of high-sulphur brown coal for power generation

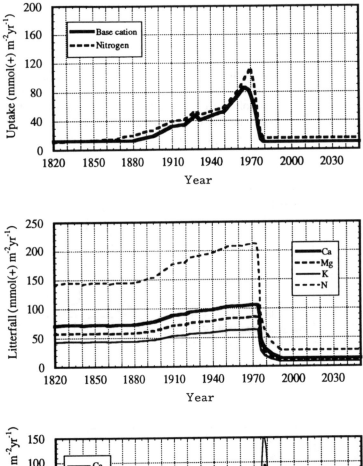

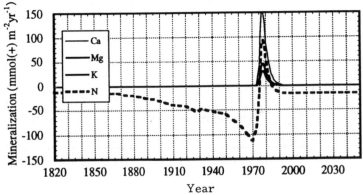

FIGURE 3.16   Uptake, litterfall, immobilization and decomposition input data used in the calculations.
During the 1970s, approximately 70% of the forest in the catchment was killed by pollution

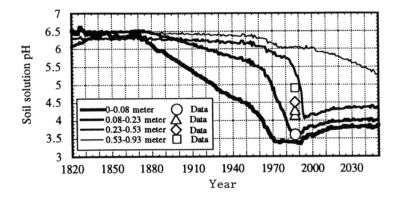

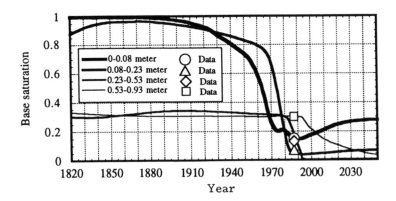

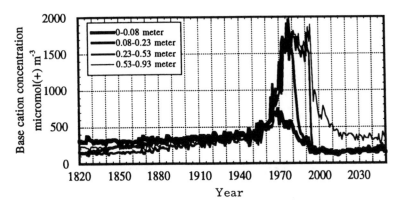

FIGURE 3.17   Calculated soil solution pH and base saturation at X-14 in Most, Czech Republic. The dots represents observed values in 1990

The calculated pH, base saturation and base cation concentration in the soil solution are shown in Figure 3.17. The dots represent observations. The severe acid deposition causing soil acidification in X-14 mobilises large amounts of base cations within the soil profile. This can be seen in the strong increase in soil solution base cation concentrations, starting in the 1960s.

This will persist until 2000 due to the large amount of acidity stored in the soil profile, which is moving downward.

The calculated soil solution BC/Al ratio at X-14 in Most, Czech Republic, is shown in Figure 3.18. The calculated soil pH and development of base saturation are shown in Figure 3.17. The dots represent the observations based on soil analysis.

The BC/Al ratio was used to calculate forest survival in the bottom diagram in Figure 3.18. It can be seen that the forest would have declined at Most even without direct fumigation effects, but soil acidification was insufficient to cause the amount of tree damage observed on such a large scale. The excess decline must accordingly be due to the combination with other types of pollution and stresses. The observed decline was calculated as the ratio between recorded tree

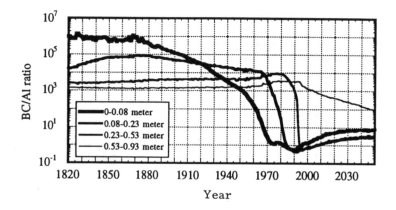

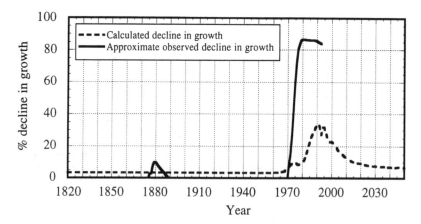

FIGURE 3.18   Calculated soil solution BC/Al ratio at X-14 in Most, Czech Republic, in the upper panel. The BC/Al ratio was used to calculate forest decline in the bottom diagram. It can be seen that the forest would have declined at Most even without direct fumigation effects, but soil acidification was not enough in 1975 to kill the forest on such a large scale as actually happened then. The excess decline must accordingly be due to the combination with other types of pollution and stresses. SO_2 concentrations in the air are known to have been extreme from 1960 to 1992

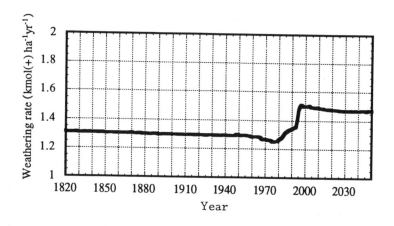

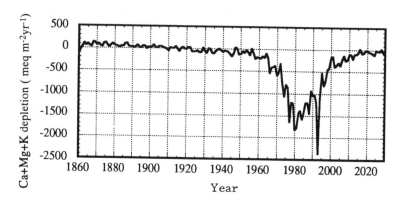

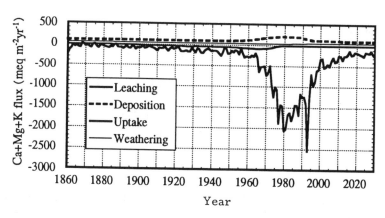

FIGURE 3.19 The model output was used to make mass balances for Ca+Mg+K for the soil at Most. The analysis shows that leaching due to acid deposition has been the most overwhelming cause of soil acidification

growth and the maximum possible set by the nitrogen load and the availability of base cations:

$$\text{Decline} = \frac{\text{actual growth}}{\text{possible growth}} \qquad (7)$$

Possible growth was estimated from what the maximum would be if nitrogen and base cations were the only limitations:

$$\text{Possible growth} = \min (N_{\text{flux}}, BC_{\text{flux}}) \qquad (8)$$

where N_{flux} and BC_{flux} are the maximum fluxes of nutrients that the plant can get from the system. The reduction in forest growth must thus not necessarily imply a net loss of growth. Growth will have increased due to better forestry practices and increased availability of N. Effects of soil acidification will turn this growth increase down to a smaller increase in growth if the acidification impact is modest and, stagnation or a net decrease in growth if the impact of acidification is very severe. As a result, we can have what appears to be a paradox: the forest grows better than ever before and acidification is at the same time significantly damaging growth. This occurs because without acidification, growth on fertile soils would have been even better.

Figure 3.19 shows the weathering rate in the catchment. Pačés (1985) estimated the weathering rate to 1.51 keq ha^{-1} yr^{-1} based on mass balance. Pačés (1985, 1986a, 1986b) maintained that soil acidification in the Czech Republic has been so severe that the weathering rate has significantly increased, which the present PROFILE calculations support.

The results were used to construct a mass balance for base cations. They were grouped as follows into sources of base cations:

$$\text{Sources} = \text{deposition} + \text{weathering} \qquad (9)$$

and sinks for base cations:

$$\text{Sinks} = \text{uptake} + \text{leaching} \qquad (10)$$

The depletion can be calculated as the difference:

$$\text{Depletion} = \text{sources} - \text{sinks} \qquad (11)$$

There is a net depletion when eqn (11) has a negative value. For a system in balance, the average value of depletion should be zero. It can be seen how the depletion term is the most important term for creating an unbalance. The mass balance is shown in Figure 3.19. The procedure would be similar for a mass balance on acidity.

3.7 FINAL REMARKS

The models demonstrated here are built in a compartmentalised way, and new modules for other processes may easily be added. At present, better descriptions of the nitrogen cycle and the interaction between growth, litterfall and decomposition is being developed. Including the release of heavy metals would be relatively simple. The models are available free of charge from the authors.

REFERENCES

Bain, D. C., Mellor, A. and Wilson, M. (1990a). "Nature and origin of an aluminous vermiculite weathering product in acid soils from upland catchments in Scotland", *Clay Minerals*, **25**, 467–475.

Bain, D. C., Mellor, A., Wilson, M. J. and Duthie, D. M. L. (1990b). "Weathering in Scottish and Norwegian Catchments", in *The Surface Waters Acidification Programme* (Ed. B. Mason), pp. 29–30, Cambridge University Press.

Bain, D. C., Mellor, A., Robertson-Rintoul, M. S. E. and Buckland, S. T. (1993). "Allt a'Mharcaidh, what a place!!", *Geoderma*, **57**, 275–293.

Baker, J. P. and Schofield, C. L. (1985). "Aluminium Toxicity to Fish in Acidic Waters", *Water, Air and Soil Pollution*, **18**, 289–309.

Brown, D. (1982). "The effect of pH and calcium on fish and fisheries", *Water Air and Soil Pollution*, **16**, 343–351.

Falkengren-Grerup, U. and Erikson, H. (1990). "Changes in soil, vegetation and forest yield between 1947 and 1988 in beech and oak sites of southern Sweden", *Forest Ecology and Management*, **38**, 37–53.

Ferrier, R. C. and Harriman, R. (1990). "Pristine, transitional and acidified catchment studies in Scotland", in *The Surface Waters Acidification Programme* (Ed. B. Mason), pp. 9–18, Cambridge University Press.

Langan, S. J. and Wilson, M. J. (1994). "Critical loads of acid deposition on Scottish soils", *Water, Air and Soil Pollution*, **75**, 177–191.

Mason, B.J. (Ed.) (1990). *The Surface Waters Acidification Programme*, Cambridge University Press.

Pačés, T. (1985). "Sources of acidification in central Europe estimated from elemental budgets in small basins", *Nature*, **315**, 31–36.

Pačés, T. (1986a). "Rates of weathering and erosion derived from mass balance in small drainage basins", in *Rates of Chemical Weathering of Rocks and Minerals*, (Eds. S. Coleman and D. Dethier), pp. 531–550.

Pačés, T. (1986b). "Weathering rates of gneiss and depletion of exchangeable cations in soils under environmental acidification", *J. Geological Soc.*, Lond., **143**, 673–677.

Patrick, S., Waters, D., Juggins, S. and Jenkins, A. (1991). "The United Kingdom acid waters monitoring network: site descriptions and methodology report, Published by ENSIS, 26 Bedford Way, London, WC1H 0AP.

Sverdrup, H. (1990). *The Kinetics of Chemical Weathering*, Lund University Press.

Sverdrup, H. and Warfvinge, P. (1991). "On the geochemistry of chemical weathering", in *Chemical Weathering under Field Conditions*, (Ed. K. Rosen), pp. 79–118, Department of Forest Soils.

Sverdrup, H. U., de Vries, W. and Henriksen, A. (1990). *Mapping Critical Loads*, Nordic Council of Ministers, Copenhagen.

Sverdrup, H. and Warfvinge, P. (1993a). "Calculating field weathering rates using a mechanistic geochemical model — PROFILE", *J. Appl. Geochem.*, **8**, 273–283.

Sverdrup, H. and Warfvinge, P. (1993b). *Soil acidification effect on growth of trees, grasses and herbs, expressed by the (Ca+Mg)/Al ratio*, Reports in Environmental Engineering and Ecology, **2**, 1–123, Institute of Technology, Lund University, Sweden.

Sverdrup, H. *et al.* (1992). "Mapping critical loads and steady state stream chemistry in the state of Maryland", *Environ. Pollut.*, **77**, 195–203.

Sverdrup, H., Warfvinge, P., Blake, L. and Goulding, K. (1995). "Modeling recent and historical soil data from Rothamsted Experimental Station, England using SAFE", *Agriculture, Ecosystems and Environment*, **53**, 161–177.

Warfvinge, P. D and Sverdrup, H. (1992a). "Calculating critical loads of acid deposition with PROFILE — a steady-state soil chemistry model", *Water, Air and Soil Pollut.*, **63**, 119–143.

Warfvinge, P. and Sverdrup, H. (1992b). "Hydrochemical modeling", in *Modeling Acidification of Groundwater*, (Eds. P. Warfvinge and P. Sandén), SMHI, Norrköping.

Warfvinge, P., Falkengren-Grerup, U. and Sverdrup, H. (1993). "Modeling long-term base supply to acidified forest stands", *Environ. Pollut.*, **80**, 209–221.

Wheater, H. S., Langan, S. J., Brown, A. and Beck, M. B. (1991). "Hydrological response of the Allt a'Mharcaidh catchment — inferences from experimental plots", *J. Hydrol.*, **123**, 163–199.

Whitehead, P. G., Jenkins, A. and Cosby, B. J. (1990). "Allt a'Mharcaidh, modelling, measuring and monitoring", in *The Surface Waters Acidification Programme*, (Ed. B. Mason), pp. 19–29, Cambridge University Press.

4

Climatic Effects on Chemical Weathering in Watersheds: Application of Mass Balance Approaches

ART F. WHITE and ALEX E. BLUM

United States Geological Survey, Department of the Interior, Menlo Park, California, USA

4.1 INTRODUCTION

Weathering rates of primary minerals in soils and regolith are of considerable interest in understanding the biogeochemistry of natural and perturbed watersheds. Besides the transport of some of the most abundant elements of the earth's surface, issues include the impacts of weathering on soil development, nutrient cycling and atmospheric acid deposition. Recently, interest has focused on the interrelationship between chemical weathering and global climate change (e.g. Volk, 1987; Berner, 1991, 1994; Brady, 1991). Silicate weathering is the most important long-term sink for protons in the lithosphere and can potentially buffer atmospheric concentrations of greenhouse gases such as CO_2. Inorganic nutrients, such as base cations and phosphorus, are released by weathering in soils and impact fertility, and therefore productivity and the global cycling of organic carbon. Finally, soil weathering is important in regulating hydrologic processes such as evapotranspiration and runoff which effect the exchange of moisture between the atmosphere and hydrosphere.

The origin of the term "weathering" implies that chemical weathering is strongly affected by climate. The climatic controls on chemical weathering are related principally to moisture and temperature (Ollier, 1984). Moisture is influenced by the total amount, intensity and seasonality of precipitation, humidity, evapotranspiration, runoff and infiltration. Thermal effects include average air temperatures, seasonal temperature variations and thermal gradients in soils. Changes in these climatic parameters are expected to directly impact chemical weathering.

Watersheds represent an important resource in discerning the interconnection between climate and chemical weathering. Watersheds are upstream inputs to larger continental and oceanic hydrochemical systems, but are less encumbered by complex and poorly constrained regional meteorologic, biological and geochemical processes. In climatic studies, watersheds can also be selected to minimize variables such as geology, geomorphology, agricultural impacts and regional atmospheric contamination, and to accentuate climatically controlled

Solute Modelling in Catchment Systems. Edited by Stephen T. Trudgill
© 1995 John Wiley & Sons Ltd

biogeochemical interactions. Watersheds can provide important information on chemical weathering under different climatic conditions existing today, as well as serving as a potential harbinger of climatic changes in the future.

The first part of this chapter will briefly review fundamental geochemical processes that both affect weathering rates and are potentially sensitive to climatic differences. Subsequent sections will review existing watershed weathering data in the context of the impacts of precipitation, evapotranspiration and temperature. These impacts can be investigated from three prospectives: (a) temporal chemical changes in a specific watershed related to historical climatic changes (b) comparison of soil weathering in different watersheds related to long-term climate differences and (c) comparison of solute fluxes from different watersheds related to present climatic differences. The historical perspective is the most direct approach. However, as in many types of climate studies, the historical record is too short, and the magnitudes of the chemical changes are too small to directly correlate with any obvious systematic change in the climatic record of a specific watershed (Driscoll *et al.*, 1989).

The present chapter will therefore focus on the last two approaches, i.e. correlating soil and solute chemistries in watersheds with different climates. Techniques used to make these comparisons involve chemical mass balances for soils and solutes. The chapter will utilize existing watershed data in this synthesis. Intercomparison of chemistry based on climatic differences is a relatively new approach in watershed studies and climate-specific data are often lacking. A benefit of the present review includes recognition of data needed to make more detailed climatic comparisons between watersheds.

4.2 FUNDAMENTAL CONTROLS ON CHEMICAL WEATHERING

Many climatic factors tend to vary sympathetically in watersheds, making the isolation of individual variables very difficult. For example, higher rainfall, greater primary plant productivity, increased soil P_{CO_2} and higher dissolved organic acids all tend to vary sympathetically in many watersheds. In this section, we will examine experimental and theoretical constraints on weathering rates of silicate minerals, with an emphasis on isolating these variables.

4.2.1 The effects of temperature

The most simplistic description of the effect of temperature on chemical reaction rates is the Arrhenius relationship. The rate of a reaction r (mol·cm^{-2}·s^{-1}) can be described by the equation:

$$r = A \, e^{-E_a/(RT)} \tag{1}$$

where E_a (kJ mol^{-1}) is the reaction activation energy, T is temperature (K), R is the gas constant and A is a pre-exponential factor. Experimentally measured activation energies for mineral dissolution are generally poorly constrained. Values vary from 35 kJ mol^{-1} for calcite to 85 kJ mol^{-1} for Ca and Mg silicates such as augite and forsterite (Brady, 1991). There have been several measurements of activation energy for feldspars, common primary minerals contained in watershed rocks and soils. Knauss and Wolery (1986) and Brantley *et al.* (1992) suggested values of 55 and 60 kJ mol^{-1} for albite respectively. Sverdrup (1990) reports activation energies of 64 and 80 kJ mol^{-1} for albite and oligioclase respectively. Lasaga (1984) suggested 60 kJ mol^{-1} as the most reasonable activation energy for silicates for which there are

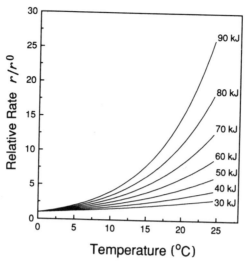

FIGURE 4.1 The effect of temperature on reaction rates as predicted by the Arrenhius relationship (eqn (2)). The reference temperature (T_0) is 0°C and the rate is expressed as the ratio to the rate at T_0 (r_0). The rate increases significantly over the range of activation energies (30–90 kJ) proposed for silicate dissolution

not experimental measurements, and this value has been adopted in several models for the feedback between consumption of atmospheric CO_2 by weathering and global temperature (Brady, 1991; Berner, 1994).

The Arrhenius relationship (eqn (1)) assumes that chemical reactions occur via a single reaction pathway, that there is no change in reaction mechanism and that nearly constant differences exist in the chemical potential between the reactants and products. Clearly, chemical weathering in a watershed does not represent such ideal conditions. Solution composition, temperature, soil hydrology and vegetation may all influence weathering reactions involving many different mineralogical phases. Therefore, a rigorous thermodynamic basis for the Arrhenius relationship describing the weathering process is not strictly valid. However, if the multitude of reactions involved in weathering are each individually controlled by an Arrhenius behaviour, and the activation energies of the contributing reactions do not vary greatly, the sum of the individual rates will increase exponentially with temperature. Equation (1) may then be used to approximate the temperature dependence of the weathering process.

The ratio of reaction rates r and r_0 at temperatures T and T_0 respectively can be predicted by the expression

$$\frac{r}{r_0} = \exp\left[\frac{E_a}{R}\left(\frac{1}{T_0} - \frac{1}{T}\right)\right] \qquad (2)$$

Calculating the ratio of the rates at two different temperatures does not require knowledge of either the absolute reaction rates or the pre-exponential factor (eqn (1)). Figure 4.1 shows the change in reaction rates as a function of temperature for activations energies between 30 and 90 kJ mol^{-1}. Clearly the ratio of the rates increases significantly between 0 and 25°C, which is the temperature range encountered in most watersheds. Velbel (1993) recently used the above approach to calculate an activation energy of 77 kJ mol^{-1} for weathering of a metamorphic terrain in North Carolina, USA, based on weathering fluxes in two adjacent watersheds with a temperature difference of 1°C.

4.2.2 The Effect of pH

Solution pH is also a significant factor in controlling experimental silicate dissolution rates. This is because both H$^+$ and OH^{-1} are important participants in dissolution mechanisms and fluctuate more widely in concentration than any other natural solute. Figure 4.2 shows measured dissolution rates for albite as a function of pH at temperatures near 25°C. As indicated, dissolution rates are at a minimum between pH 5 and 8, and increase in both the acidic and basic regions. This pattern is common to all feldspars and many other silicates (e.g. Helgeson *et al.*, 1984; Lasaga, 1984; Murphy and Helgeson, 1987).

The above relationship strongly suggests at least two different dissolution mechanisms dependent on pH: a proton-promoted mechanism in the acid region and a hydroxyl-promoted mechanism in the basic region (Brady and Walthers, 1989; Blum and Lasaga, 1991). Several workers (Chou and Wollast, 1985; Knauss and Wolery, 1986; Mast and Drever, 1987) have suggested a third dissolution mechanism for feldspars that dominates in the neutral pH region and is pH independent. The slope of the rate data at low pH is proportional to [H$^+$]$^{0.5}$. These data suggest that weathering rates would change by a factor of ~ 7 between pH 4 and 5, a range typical of acid soils. However, the majority of soil and surface waters are between pH 5 and 8, a range over which experimental dissolution rates are not very pH sensitive.

The partial pressure of CO$_2$ (P_{CO_2}) in soils is a factor of 5–100 times higher than atmospheric P_{CO_2}. Experimental dissolution of feldspars indicates that CO$_2$ does not directly influence the kinetics of weathering reactions (Busenburg and Clemency, 1976; Brady and Carroll, 1993, 1994). However, CO$_2$ will depress and buffer the pH of soil solutions, and thus increase silicate dissolution rates by increased H$^+$ activities (Figure 4.2). Climatic variables such as temperature and precipitation can affect biologic activity and CO$_2$ production in the soils. In contrast, increases in atmospheric CO$_2$ from anthropogenetic sources are not expected to directly impact weathering rates because soil CO$_2$ is already elevated far above atmospheric levels.

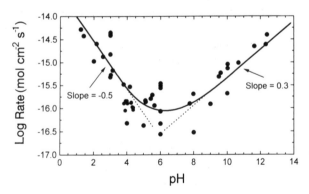

FIGURE 4.2   Compilation of experimental dissolution rates for albite as a function of pH at temperatures near 25°C (after Blum, 1994)

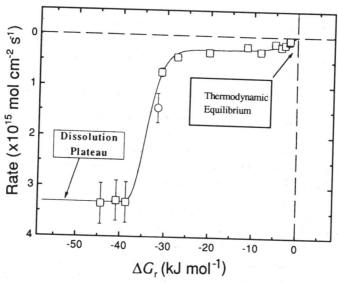

FIGURE 4.3   Albite dissolution rate versus ΔG_r of the reaction at pH 8.8 and 80°C (from Burch *et al.*, 1993). The dissolution plateau is the region far from equilibrium where the dissolution rate is independent of solution composition. At equilibrium, both ΔG_r and the dissolution rate are zero

Organic acids produced by biological activity can depress and buffer the soil solution pH, accelerating proton-induced weathering. Organics may also bind on the mineral surface where they may either catalyze or retard the dissolution reaction. Finally, organics in solution may also complex with dissolved species, most notably Al and Fe, decreasing the concentrations of the free ions and, thus, decreasing the saturation state. Dissolved organics are produced by biological activity and are therefore dependent on temperature and precipitation.

4.2.3   Effect of reaction affinities

The net dissolution rate observed in solution is the difference between the amount of dissolution and precipitation at the mineral surface. In infinitely dilute solutions, there is no precipitation, and the net dissolution rate will be at a maximum. At equilibrium, dissolution and precipitation rates at the surface are equal, and there is no net reaction. Between these two well-defined end points, there will be a progressive decrease in the net dissolution rate as equilibrium is approached from infinitely dilute solutions. A common assumption is that the net dissolution rate decreases proportionally to the decrease in the thermodynamic driving force (see Nagy *et al.*, 1991). This can be expressed in terms of the free energy of reaction (ΔG), the change in chemical potential between components in the solid and solution ($\Delta\mu$) or the solution saturation state (ion activity product/equilibrium constant), all of which are essentially equivalent treatments.

Burch *et al.* (1993) experimentally observed a strong dependence of albite dissolution on the free energy of reaction (Figure 4.3). The decrease in rate is much greater than that predicted by the thermodynamic driving force, and suggests that the influence of solution composition on the formation of etch pits at dislocations controls the net dissolution rate (see Lasaga and Blum, 1986; Blum *et al.*, 1990). Another possible explanation is that dissolution is inhibited by Al reabsorbed on to the mineral surface from solution. In any case, Figure 4.3 indicates that the accumulation of solutes in soil solutions may significantly retard the weathering reaction. Evapotranspiration (ET) may further concentrate solutes during wetting and drying cycles, enhancing the inhibitory effects of solute concentrations on mineral dissolution.

4.3 EFFECT OF CLIMATE ON SOIL WEATHERING

Soils represent the residual product of long-term chemical weathering within watersheds. Many workers (e.g. Peltier, 1950; Büdel, 1980; Ollier, 1984) have developed semi-quantitative relationships between the degree of weathering and temperature and precipitation. Mass balance calculations can be used to compare present and past weathering rates in soils (White *et al.*, 1992). Solid-state chemical and mineralogical changes within a soil reflect the integrated weathering rate over the entire period of soil development. This average weathering rate can be compared to present-day weathering rates calculated from solute fluxes, as discussed in following sections. The differences between short- and long-term weathering rates may reflect differences between the past and present climates.

Several workers have used this approach to investigate differences between present and past weathering rates in specific watersheds. Sverdrup (1990) compared past chemical weathering rates in 12 ka soils in the Gårdsjön watershed with present watershed fluxes and found comparable rates. White *et al.* (1992) also found comparable past and present weathering rates for 10 – 3000 ka soils in California, and concluded that past climates may have been similar to present conditions. April *et al.* (1986) estimated long-term weathering rates in the Adirondack Mountains, USA (Woods Lake and Panther Lakes; see Table 4.1) by quantifying mineral and base cation losses from 14 ka soils. These authors found that past weathering rates in the two watersheds were similar, but that present rates vary as a result of watershed acidification, and possibly changes in climate. Kirkwood (1989) compared cation losses in soils in the Plastic Lake watershed with solute fluxes. He concluded that present-day weathering rates are a factor of two faster than historical rates, and attributed the increases to acid deposition.

4.3.1 Mass balance approaches

Mass balances in soils can be used to quantitatively compare the long-term weathering rates within soils forming under different climatic conditions. Researchers have long applied mass balances in soils to estimate the extent of chemical weathering (Merrill, 1906; Barth, 1961; Brimhall and Dietrich, 1987). The basic approach involves comparing the ratio of weatherable (or mobile) and non-weatherable (or conservative) constituents currently present in a soil with the ratio of the same constituents in the initial parent rock or sedimentary deposit. The proportion of an element remaining in a soil after weathering (ΔM) can be defined as

$$\Delta M = \frac{M_w/M_p}{C_c/C_p} \tag{3}$$

where M is the wt % of the mobile soil component and C is the wt % of the conservative component. The subscript p refers to the initial or parent material and the subscript w refers to the component in the weathered soil. The apparent concentration of a soil component compared with its initial concentration is also dependent on volumetric changes that have occurred in the soil as a result of weathering. The volumetric change (ΔV) in the soil can be determined from

TABLE 4.1 Chemical, climatological and physical characteristics of watersheds on granitic-tupe rocks. Silica fluxes (mol ha⁻¹ yr⁻¹) are calculated for weathering based on eqns (5) and (6). Si concentrations (μM) and Si fluxes are discharge-weighted annual averages. Evapotranspiration (%) is calculated from the difference between precipitation and runoff. Temperature is annual mean air temperature. Elevations are averages of maximum and minimum reported temperatures

Watershed	Location	Si flux	Si concentration	pH	Precipitation (mm)	Runoff (mm)	ET (%)	Temperature (°C)	Area (ha)	Elevation (m)	Reference
Loch Vale	Colorado, USA	154	26	6.3	1104	504	45	0	860	3560	Mast et al. (1990)
Rusape	Zimbabwe	220	251		663	87	87	17.1	733	1550	Owens and Watson (1979)
Rawson Lake NW	Ontario, Canada	387	135	6.0	803	277	66	2.4	62	260	Schlinder et al. (1976)
Pond Branch	Maryland, USA	392	155		933	169	82	12.3	38	152	Cleaves et al. (1970)
Emerald Lake	California, USA	392	28		1830	1410	23	6.0	120	3108	Williams and Melack (1991)
Martinell	Colorado, USA	401	26	6.4	1295	1507	-10	0.0	8	3490	Reddy (1988)
Sage Hen	California, USA	404	283		621	141	47	1.5	274	3140	Marchard (1971)
Tharp's Creek	California, USA	425	227	6.8	887	187	79	7.2	14	2161	Williams et al. (1993)
Log Creek	California, USA	466	128	6.7	887	363	59	7.2	50	2232	Williams et al. (1993)
Bear Brook	Maine, USA	485	55	5.5	1400	881	37	5.0	10	370	Norton et al. (1994)
Hartviko	Czechoslovakia	507	108	6.8	781	467	40	6.0	98	734	Paces (1985)
Filson Creek	Minnesota, USA	517	177	6.2	680	270	60	3.5	2520	603	Siegel and Pfannkurch (1984)
Vocadio	Czechoslovakia	535	327	6.2	736	171	77	6.5	59	321	Paces (1986)
Salacova Lhota	Czechoslovakia	542	428	7.1	685	128	81	6.5	168	260	PACES (1986)
Rawson Lake E	Ontario Canada	544	196	6.0	803	277	66	2.4	170	260	Schlinder et al. (1976)
Rawson Lake NE	Ontario Canada	589	212	6.0	803	277	66	2.4	10	657	Schlinder et al. (1976)
Hubbard Brook	New Hampshire, USA	628	78	4.9	1300	800	38	5.0	3076	400	Likens et al. (1977)
Glendye	Scotland	641	93	6.5	789	817	-4		4125	2973	Creasey et al. (1986)
Rabbit Ears	Colorado, USA	659	104		1132	617	45	0.7	200	252	N.E. Peters (1994)[a]
Panola	Georgia, USA	676	204	5.0	1149	338	71	16.3	41	295	N.E. Peters (1994)[a]
Hanley 3	British Columbia, Canada	728	70	6.7	2040	1040	49	9.2	44	150	Feller and Kimmins (1984)
Mundberry Brook	Massachusetts, USA	820		6.7	1428			7.0			Yuretich et al (1993)
Fort River	Massachusetts, USA	837	156	6.1	1080	507	53	8.4	1053	213	Yuretich and Batchelder (1988)
Cadwell Creek	Massachusetts, USA	849		5.8	831	793	45	7.0	139	245	Yuretich et al. (1993)
Hanley 2	British Columbia, Canada	930	75	6.7	1990	1240	38	9.2	68	295	Feller and Kimmins (1984)
Juliasdale	Zimbabwe	980	245		705	400	43	13.8	91	1900	Owens and Watson (1979)
Hanley 1	British Columbia, Canada	989	98		1820	1010	45	9.2	23	295	Feller and Kimmins (1984)
Pont Donar	France	1062	316	6.7	976	379	61		24	25	Bouchard (1983)
Peatfold	Scotland	1096	188	6.6	1160	580	50		200	485	Creasey et al. (1986)
Bathalde	Germany	1193	86		2000	1395	30			2000	Stahr et al. (1980)
Silver Creek	Iadaho, USA	1270	371		915	342	65	4.2	109	1585	Clayton (1985)
Jamieson Creek	British Columbia, Canada	1521	49	6.4	4541	3668	19	3.4	299	792	Zeman (1978)
Indian River	Alaska, USA	1896	69		3300	2824	14	5.7	800	754	Stednick (1981)
Rio Icacos	Puerto Rico	8066	305		4300	3680	16	22.0	326	708	McDowell and Astbury (1994)

[a] Personal communication.

the densities (ρ) and concentrations of the conservative component (C) in the soil and parent material such that

$$\Delta V = \left(\frac{\rho_p}{\rho_w}\right)\left(\frac{C_p}{C_w}\right) - 1 \qquad\qquad (4)$$

The terms ΔM and ΔV are ratios, and are therefore independent of the units used. When $\Delta M = 1$ there is no loss or gain of a soil component, while a value of zero indicates a total loss. As discussed by White *et al.* (1995), the above equations have equally applicability to mineral and elemental ratios. Conservative components used in the calculation of volumetric changes are most often Zr, Ti, rare earth elements and quartz.

An application of the above mass balance approach to soils is illustrated for a deep sapolitic weathering profile in the Rio Icacos watershed in the Luquillo Maintains of Puerto Rico (Blum *et al.*, 1993). This watershed (Table 4.1) is situated in a humid tropical rain forest. Required data for the calculations included the chemical composition and density of the initial quartz diorite and the chemistry and bulk density within the 7 m profile of soils and saprolite.

The calculation of volumetric changes (ΔV) in the saprolite (eqn. (4)) was performed using several possible conservative elements including Zr, Ti and the rare earth elements Nb and Y. Although such elements are generally considered immobile with respect of weathering, their conservation under any specific geochemical condition is not guaranteed. Therefore, the best strategy is to consider a suite of such elements to establish the extent of volumetric changes undergone by a soil during weathering. As indicated for the Rio Icacos soil, Zr, Ti and Nb produce consistent estimates for volume changes (ΔV) which centre close to zero (Figure 4.4(a), vertical dashed line). This lack of a significant volume change during weathering is consistent with a soil porosity of nearly 50% and the preservation of primary igneous textures in the saprolite. Use of Y in eqn (4) produces calculated volume increases of up to a factor of three in the saprolite. Such unrealistic increases in volume reflect the loss of Y from the soil during weathering.

Fractional mass loss (ΔM, eqn (3)) of non-conservative major cations and silica in the saprolite relative to the initial quartz diorite are shown in Figure 4.4(b). The extent of weathering under these extreme climatic conditions is evidenced by nearly complete loss of Na and Ca from the soil. Approximately 80% of Mg and 50% of SiO_2 are also lost. These elemental distributions correspond to the observed mineralogy. Primary plagioclase, alkali feldspars and hornblende are totally depleted during weathering, which explains the losses of Na, Ca and SiO_2. The only remaining primary minerals are hydrobiotite, which accounts for the residual Mg, and quartz, which along with secondary kaolinite accounts for the remaining SiO_2. The extent of elemental and mineralogical losses from weathering in the upper 7 m of the soil profile are constant. Current weathering occurs within a narrow interface (< 10 cm) directly above the underlying quartz diorite (Figure 4.4).

4.3.2 Comparison of weathering rates in soils

The mass balance approach can be used to compare the extent of weathering of soils exposed to different climatic conditions. Unfortunately, data for mass losses in watershed soils are not generally available in the literature. However, the approach can be illustrated using three datasets, including the Rio Icacos soil and two chronosequence soils in California, USA. (A chronosequence is a group of soils that differ in age but have similar parent material; Jenny, 1941). The Merced chronosequence is comprised of soils developed on alluvial terraces formed from glacially derived granitic outwash from the Sierra Nevada, and deposited along the Merced River in central California (Pavich *et al.*, 1985; Harden, 1987). The present climate is of a Mediterranean type, with hot summers and wet cold winters. Estimated ages of these deposits (0.2−3000 ka) were obtained by ^{14}C, uranium trend analysis and K−Ar dating methods. Mass balance reconstructions, using eqns (3) and (4), were performed using quartz as the conservative component (White *et al.*, 1995). Mineralogical data show a progressive loss of plagioclase, K-feldspar and hornblende with time and an increase in secondary kaolinite.

The Mattole soil chronosequence is composed of marine terrace deposits and situated near the mouth of the Mattole River in northern California. Soil ages are based on radiocarbon and sea-level stands, and range from 3.6 to 240 ka (Chadwick *et al.*, 1990; Merritts *et al.*, 1992). The dominant parent sedimentary material was arkosic sandstone with some siltstone and shale from the Jurassic Francisian formation. This site has a seasonal Mediterranean-type climate characterized by mild wet winters and a prolonged cool but dry summer. In the mass calculations, Merritts *et al.*, (1992) assumed that Zr behaved as a conservative component. Results show that elemental losses of Na, K, Ca, Mg and SiO_2 increased progressively with time.

Residual SiO_2 ratios (M_w/M_p, eqn (3)) were chosen as indicators of the extent of weathering within the soils. These ratios for Merced and Mattole soils are approximated by exponential decay functions (dashed lines) when plotted against time (0−3000 ka) in Figure 4.5. The exponential decrease in weathering rates with time is attributed to chemical selectivity during

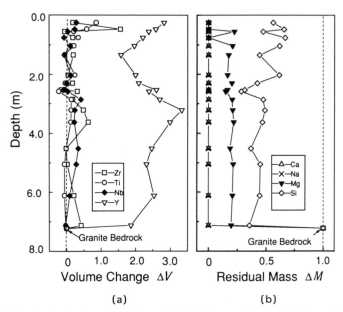

FIGURE 4.4   Calculated volume changes (ΔV, eqn (4)) and elemental losses (ΔM, eqn (3)) as functions of depth in the Rio Icacos soils of Puerto Rico. The elemental compositions are unity at the interface with the parent granite bedrock

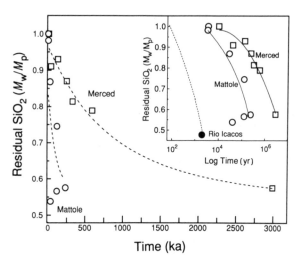

FIGURE 4.5 Residual SiO$_2$ concentrations (eqn (3)) versus age for the Merced and Mattole chronosequences. The dashed line is an exponential fit to data. The insert displays residual SiO$_2$ plotted against log time and incorporates the Rio Icacos data

weathering. The most reactive minerals, including plagioclase and mafic silicates such as hornblende, dominate the initial weathering processes and result in rapid SiO$_2$ loss in younger soils. In the older soils, only the more chemical-resistant minerals are present. Further decreases in SiO$_2$ are therefore much slower. Both the oldest Merced and Mattole soils asymptotically approach an SiO$_2$ content of between 50 and 60%, indicating that most of the residual SiO$_2$ is contained in quartz and kaolinite, two minerals highly resistant to additional chemical weathering.

The above weathering rates can also be compared to those for the Rio Icacos watershed. The denudation rates in the Rio Icacos watershed are estimated to be between 20 mm ka^{-1}, based on *in situ* [10]Be production (Brown *et al.*, 1993), and 75 mm ka^{-1} estimated from sediment transport rates (Larson and Sánchez, 1992). The soil chemical profiles (Figure 4.4) indicate that residual SiO$_2$ reaches a constant content of $\sim 50\%$ within 10 cm of the underlying diorite surface. Assuming an average steady-state weathering rate of 50 mm ka^{-1} leads to the conclusion that the bedrock is completely altered to saprolite in ~ 2 ka. Residual SiO$_2$ for the Rio Icacos soil is plotted along with the Merced and Mattole soils on a log−time scale in the Figure 4.5 insert. For a comparable age, the residual SiO$_2$ contents of soils strongly decrease in the order Merced > Mattole > Rio Icacos.

The differences in observed weathering rates for the soils correlate with differences in climate. The Merced site receives approximately 300 mm yr^{-1} of rain, the Mattole site receives 1100 mm yr^{-1} and the Rio Icacos rain forest receives 4300 mm yr^{-1}. Annual mean temperatures for the Merced and Mattole soils are 16 and 13°C respectively. The tropical Rio Icacos site is warmer at 22°C. The range in precipitation at the three sites exceeds an order of magnitude while the maximum temperature difference is only 9°C. The Arrhenius relationship, using an activation energy of 60 kJ mol^{-1}, indicates the maximum difference in reaction rates for the soils should be $\sim 215\%$ (eqn (2) and Figure 4.1), and the cooler Mattole site (13°C) has weathered considerably faster than the warmer Merced site (16°C). The effect of temperature, therefore, appears to be less significant than the effect of precipitation in producing the extreme differences observed in weathering rates of these soils. The preceding comparison assumes that present climatic conditions are indicative of past climates. The fact that a direct correlation exists between present precipitation and long-term weathering suggests that this assumption may be correct.

The preceding soil data demonstrates two important aspects related to chemical weathering and mass fluxes in watersheds. Weathering is clearly faster in watershed soils that are wetter and/or warmer, a correlation that has been previously documented in many soil studies. However, the mass balance approach provides a method to quantify soil weathering rates and solute fluxes in watersheds and to compare them in terms of climatic factors such as temperature and precipitation. The soil data also show that present chemical weathering rates are dependent on weathering history. For soils derived from similar parent material, rates of weathering decrease significantly with age. This is due to selective weathering of the most reactive mineral phases. Therefore, weathering rates in different watersheds will be expected to vary due to their geomorphic and climatic histories. Watersheds with very old stable soils should have lower solute fluxes than watersheds with younger soils. These latter watersheds generally have high relief and/or have been exposed to recent physical erosion such as glaciation. The above considerations are important in attempting to correlate present watershed chemical weathering rates with present climatic differences.

4.4 THE EFFECT OF CLIMATE ON SOLUTE CHEMISTRY

Solute mass balances in small watersheds provide an important opportunity to assess the impacts of climatic conditions, such as precipitation and temperature, on chemical weathering. Small watersheds on a scale < 5000 ha (50 km^2) have advantages over larger scale river and basin systems in that they often represent single climatic zones and may contain homogeneous rock and vegetation types. Watersheds are also usually chosen to be closed systems with minimum extraneous chemical inputs from regional groundwater sources. The disadvantages of watersheds are related to their limited geography that can be affected by site-specific processes such as atmospheric deposition, which perturb apparent weathering rates. Therefore, a statistically large watershed data set, in addition to detailed chemical and hydrologic information for each site, are required to make valid correlations between climate and chemical weathering.

4.4.1 Mass balance calculations

Quantitative reaction rates in watersheds are ideally defined based on individual mineral phases at known solid-state and solution compositions. Such natural rates, in turn, can be compared to laboratory-determined rates based on experimental and theoretical results, such as shown in Figures 4.2 and 4.3. However, as discussed by Schnoor (1990), White and Peterson (1990) and Sverdrup (1991), the number of watershed studies that have calculated quantitative reaction rates is very limited. Such rate determinations require detailed evaluations of mineralogy, grain size distributions, estimates of reactive surface areas and evaluation of hydrologic pathways, all in addition to elemental mass balances (Paces, 1983; Velbel, 1985; Swoboda-Colberg and Drever, 1993).

An alternative approach is to consider net solute fluxes from watersheds to be representative of chemical weathering rates. Although chemical fluxes do not incorporate detailed physical, chemical and hydrologic information, they represent a means of empirically comparing the weathering rates in watersheds with different climates. This approach is only valid if climatic effects are of paramount importance in weathering relative to other biogeochemical factors. Chemical fluxes are much more commonly reported in the literature and several workers have compared chemical fluxes to regional differences in climate. Dethier (1986) calculated SiO_2 and cation fluxes for river systems in the Pacific Northwest, USA. Suchet and Probst (1993) tabulated dissolved inorganic carbon fluxes from watersheds in France. Bruijnzeel (1990) tabulated nutrient fluxes including Ca, Mg, and K from tropical catchments. However, no comprehensive effort has been attempted to correlate existing world-wide watershed data with climatic parameters.

Operationally a solute flux Q for a chemical species i in a watershed is calculated as

$$Q_i = C_i \frac{V}{A \cdot t} \tag{5}$$

where C_i is the chemical concentration, V is the fluid mass, A is the aerial extent of the watershed and t is time. Q_i is reported in the literature in a wide array of units, but is standardized to mol ha^{-1} yr^{-1} in the present discussion. Although Q_i has the same dimensions as the kinetic rate (r in eqn (1)), Q_i represents a chemical flux and not a reaction rate.

The chemical mass balance used to define the extent of chemical weathering in a watershed can be determined operationally by the relationship

$$Q_w = Q_{dis} - Q_{precip} - Q_{dry} \pm Q_{bio} \pm Q_{exch} \tag{6}$$

where

Q_w	= net solute flux due to chemical weathering from soils and bedrock
Q_{dis}	= output flux based on chemical concentrations and stream discharge
Q_{precip}	= input flux based on wet precipitation
Q_{dry}	= input flux based on dry fall from aerosols and soluble particulates
Q_{bio}	= input or output flux based from biological activity
Q_{exch}	= input or output flux from ion exchange

The chemical weathering flux Q_w can be calculated for a watershed if the fluxes on the right-hand side of eqn (6) can be determined. Other potential fluxes, such as extraneous groundwater and anthropogenic inputs from agriculture, are assumed to be negligible.

The first two terms on the right-hand side of eqn (6), Q_{dis} and Q_{precip}, are commonly determined in watersheds based on measured chemical concentrations and volumes of discharge and precipitation (eqn (5)). Measurement of dry fall Q_{dry} is technically much more difficult, and is reported as a separate component in a more limited number of studies. Q_{bio} and Q_{exch}, the fluxes based on biological activity and ion exchange, can either represent input or output fluxes. A degrading ecosystem produces a positive biological flux because organic decomposition releases nutrients such as K, Ca and Mg to the abiotic environment. In an aggrading ecosystem, Q_{bio} is negative because increasing biomass sequesters such nutrients. This biological term is considered quantitatively in only a limited number of studies (see Velbel, Chapter 7 of this volume) and its significance is open to debate (Likens *et al.*, 1977; Stauffer and Wittchen,

1991). In most watersheds with a mature and/or steady-state ecosystem, the net biological flux is most often assumed to be zero.

The exchange flux Q_{exch} represents the sorption or desorption of cations or anions between soil waters and ion exchange sites, principally on clays and ferric oxyhydroxide minerals. In a geologic time frame, the net flux should be positive due to the progressive formation of these minerals and the increasing exchange capacity of the soil profile. However, in terms of annual chemical budgets in undistributed watersheds at steady state, the net flux due to ion exchange should approach zero. This is because ion exchange reactions are much more rapid than chemical weathering and, therefore, at equilibrium (Cresser and Edwards, 1987). However, for disturbed watersheds, such as those impacted by acid deposition, sorbed species represent very large chemical reservoirs, which, when released, can significantly affect chemical outputs from a watershed and therefore alter the apparent weathering rate (eqn (6)).

An extensive literature exists on acid deposition and its effect on the chemical budgets of watersheds. Leaching of base cations, such as Ca and Mg, from ion exchange sites by strong acid anions has been the focus of several watershed acidification models documenting the importance of such leaching on watershed output budgets (Cosby *et al.*, 1985); Christophersen and Neil, 1990). However, as noted by Johnson (1984) and April *et al.* (1986), there is little direct evidence that acidification significantly affects chemical weathering rates. The pH ranges of soil and surface waters found in most watersheds are greater and less variable than the pHs of incoming precipitation. As previously indicated from experimental studies, dissolution rates exhibit very little change as a function of pH in the near neutral range (Figure 4.2). Therefore, fluxes of species least impacted by exchange processes, notably SiO_2, should not be significantly affected by acidification. For example, steady-state SiO_2 concentrations have been shown to remain nearly constant in stream water over a 25 year period at Hubbard Brook. This occurred despite significant decreases in base cations and sulphate concentrations resulting from decreases in atmospheric deposition (Driscoll *et al.*, 1989).

4.4.2 Geologic effects

Differences in the soil and bedrock type must also be considered in any attempt to quantitatively compare weathering fluxes between watersheds. As discussed by Bricker and Rice (1989), differences in mineralogy are of paramount importance in controlling weathering rates. The relative reactivities of minerals decrease in the order

$$Carbonates > mafic silicates > feldspars > quartz$$

Less obvious in most watersheds is the correlation between the dominant rock mineralogy and the type and amount of minerals readily accessible to weathering. Much ongoing weathering does not involve bedrock but occurs in the soil zone. Soils tend to have significantly different mineralogy than bedrock due to long- term selective weathering. This effect was previously shown by the mass balance calculations that documented significant decreases in soil weathering rates with time (Figure 4.5).

Stallard and Edmond (1983) differentiated such mineral selectivity in terms of weathering-limited and transport-limited regimes. In a weathering-limited regime, mechanical erosion is faster than chemical weathering. Therefore, the most reactive phases will always be available for weathering. For example, calcite dissolution dominates Ca and alkalinity fluxes in the high-altitude Loch Valle watershed (Table 4.1), which has been impacted by recent glaciation (Mast *et al.*, 1990). This occurs even though calcite is present in only trace amounts in the igneous and high-grade metamorphic rocks. In a transport-limited regime, erosion is less intense and weathering is less selective. All weatherable minerals ultimately contribute to the solute load in proportion to their abundances in the bedrock. Under such a scenario, trace calcite would be rapidly lost and weathering would become dominated by dissolution of silicate phases such as feldspars and hornblende. This is the case for the complete weathering of primary aluminosilicates in the Rio Icacos soils (Figure 4.4). Weathering selectivity implies that current chemical fluxes are dependent not only on the bulk rock mineralogy but also on the past history of physical and chemical weathering in the watershed.

4.4.3 Watershed data

In an ongoing study of the relationship between watershed weathering and climate, a literature review was conducted of chemical mass balances for well-characterised watersheds (White and Blum, 1995). For the purposes of the present discussion, only silica fluxes will be considered. The data are tabulated in Table 4.1 for 34 watersheds in order of increasing SiO_2 weathering fluxes. Other information includes mean annual SiO_2 discharge concentrations and pH, annual precipitation and runoff, mean air temperatures and mean watershed elevations. Additional unpublished data were contributed by a number of authors, as indicated in the acknowledgements.

Table 4.1 includes only watersheds underlain by acidic crystalline rocks, consisting of either intrusive granitoid rocks or high-grade metamorphic gneiss. This classification was used to guarantee at least a minimum similarity in composition of the parent mineralogy. Silica may be the solute species most indicative of chemical weathering rates (eqn (6)). Silica is present in low concentrations in precipitation, is not considered a major nutrient and is not influenced significantly by exchange reactions. However, SiO_2 is not a conservative solute because it is incorporated into secondary mineral phases during weathering. As such, SiO_2 fluxes are dependent on the stoichiometry of the weathering reactions involving both primary and secondary phases.

There are several simplifying assumptions that can be made concerning the stoichiometry of weathering reactions of granitoid rocks. Soils developed on such rocks form predominately kaolinite, which has an Si/Al ratio of 1:1. The predominant granitoid minerals, albite, K-feldspar, hornblende and biotite have Si/Al ratios of 3:1. Consequently, weathering of these minerals of kaolinite will release ~2/3 of the total SiO_2 as a net flux out of the watershed. The anorthite component of plagioclase has an Si/Al ratio of 1:1, and weathering to kaolinite will not result in any net Si flux. The dissolution of kaolinite, with the resultant mobilization of Al or formation of gibbsite, could also increase the observed SiO_2 fluxes. However, all of these watersheds have significant topography, and most clays are probably physically eroded before a significant proportion of the clays are chemically dissolved.

The geographic distribution of the watersheds listed in Table 4.1 is limited, with all but three in northern Europe and North America (Figure 4.6). The watersheds generally encompass alpine to temperate climatic conditions. Weathering of granitoid rocks under truly tropical conditions occurs only in the Rio Icacos watershed in the El Verde rain forest in Puerto Rico (McDowell and Asbury, 1994). Clearly, in terms of an adequate representation of global climate, a more world-wide distribution of watershed data would be desirable. In spite of limited geographical representation, the tabulated watersheds do encompass a significant range in chemical variability, with SiO_2 fluxes ranging from 150 to 8000 mol ha^{-1} yr^{-1} (Table 4.1). In turn, annual precipitation ranges between 620 and 4500 mm yr^{-1}, runoff between 90 and 3700 mm yr^{-1} and mean annual temperatures between 0 and 22°C. The following discussion will attempt to correlate these chemical fluxes to the corresponding climatic characteristics.

4.4.4 Runoff versus precipitation

Chemical fluxes are dependent on the magnitude of runoff (eqn (5)). In a hydrologically closed watershed, runoff can be related to precipitation (*P*) and evapotranspiration (ET) by the relationship

$$R = P - \text{ET} \tag{7}$$

Precipitation is a significant variable that is extremely important in global climate models. ET encompasses evaporation from surface waters and shallow soils, in addition to respiration by plants. ET is therefore dependent on a number of climatic factors including temperature, humidity, wind velocity and extent and type of vegetative cover. The above hydrologic balance (eqn (7)) forms a basis for relating chemical fluxes, which are dependent on runoff, to the climatic variables precipitation and evapotranspiration.

The relationship between runoff and precipitation in watersheds is shown in Figure 4.7. Solid points represent watersheds with alpine/temperate climates tabulated by White and Blum (1995) for which data in Table 4.1 are inclusive. Additional precipitation and runoff data for tropical watersheds, tabulated by Bruijnzeel (1990), are plotted as the open symbols in Figure 4.7. The alpine/temperate and tropical data produce linear and generally mutually exclusive correlations between precipitation and runoff. As expected, the lower temperature, alpine/temperate watersheds plot at higher runoff to precipitation ratios than the warmer tropical watersheds. However, tropical watersheds with very high precipitation, such as Rio Icacos (Table 4.1), tend

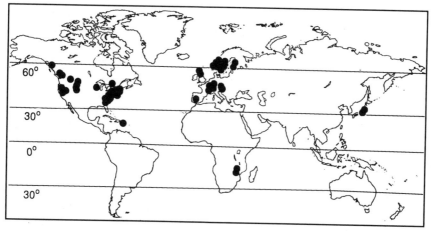

FIGURE 4.6 Geographic distributions of watersheds tabulated in Table 4.1

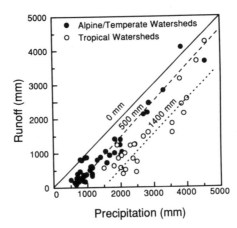

FIGURE 4.7 Distributions of runoff versus precipitation for alpine/temperature (•) and tropical (○) watersheds. The solid line corresponds to a one-to-one ratio, signifying no evapotranspiration (ET). The average ETs for the watersheds are equivalent to the intercepts of the linear regression fits (eqns (8) and (9)), 500 mm for the alpine/temperate data (dashed line) and 1400 mm for the tropical data (dotted line)

to more closely approximate the alpine/temperate ratios. The solid diagonal line (Figure 4.7) denotes a direct correlation between runoff and precipitation. Most runoff falls before this line, denoting the effects of ET (eqn (7)). However, runoff from several watersheds plot above this line (such as Glendye and Martinell in Table 4.1), implying extraneous recharge from groundwater or glacial ablation.

Linear correlations between runoff and precipitation can be approximated for the alpine/temperate (dotted line) and tropical (dashed line) data sets by the relationships

$$R = -535 + 1.04 \, P_{(temperate)} \; (r^2 = 0.90) \quad \text{and} \quad R = -1425 + 1.05 \, P_{(tropical)} \; (r^2 = 0.82) \quad (8)$$

These regression fits indicate that runoff is directly proportional to precipitation (i.e. the slopes of eqn (8) equal 1.04 and 1.05). The average ET is, therefore, generally independent of the annual precipitation and can be estimated by the respective intercepts of the alpine/temperate and tropical data trends (535 mm and 1425 mm respectively). This occurs because the annual precipitation in the tabulated watersheds generally exceeds the maximum potential annual ET. In such cases, ET is dependent on the temperature and humidity but not precipitation. The exception would be for watersheds in arid climates where precipitation is less than the potential ET. In such situations, the runoff would approach zero except during heavy precipitation events, and the actual ET would be dependent on the amount of precipitation.

Although the absolute water loss from ET in watersheds with similar climates is relatively constant, the proportions of runoff to ET are not. Thus ET loss will be much higher in watersheds with low precipitation than in watersheds with high precipitation. This is shown by the range in ET percentages tabulated in Table 4.1 (0−85%). Water loss by ET increases solute concentrations and decreases the thermodynamic affinities of the weathering reactions (Figure 4.3). High ET may also result in drying of the soil zone, enhancing the precipitation of secondary clays, decreasing permeability and thus decreasing chemical and hydrologic transport rates.

4.4.5 Silica concentrations as a function of ET

The solute flux is equal to the product of runoff and solute SiO_2 concentration (eqn (5)). Silica concentrations (Table 4.1) correspond to the mean annual concentrations weighted with respect to discharge. As indicated by Figure 4.8, these SiO_2 concentrations decrease asymptotically with increasing precipitation and runoff. The offset between precipitation and runoff (Figure 4.7) is attributed to water losses by ET which concentrate SiO_2. The relatively constant offset in the relationships of SiO_2 and precipitation and runoff is attributed to the effect of similar absolute water losses by ET (Figure 4.7).

In addition to weathering rates, the major control on SiO_2 concentrations in many watershed streams appears to be evapotranspiration. The increase in SiO_2 concentrations with the percentage of water loss by ET is clearly demonstrated in Figure 4.9. The trend in the SiO_2 concentration can be modelled on the simple relationship

$$C = C_0 \left(\frac{P}{P - \mathrm{ET}} \right) \tag{9}$$

where C is the measured concentration of SiO_2 and C_0 is the initial concentration produced by weathering. In watershed balances, ET is calculated from the difference between precipitation (P) and discharge (R) (eqn (8)). Therefore, the term in parentheses on the right-hand side of eqn (9) is equal to the ratio of P to R. Equation (9) defines concave lines that describe the evolution of SiO_2 concentrations produced by evaporation of initial solutions given by the left-hand intercept in Figure 4.9. As indicated, all measured SiO_2 concentrations for the watersheds (Table 4.1) fall within the envelope of curves assuming values of C_0 between 20 and 150 μM.

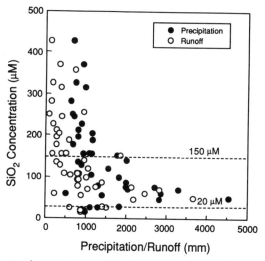

FIGURE 4.8 Silica concentrations versus precipitation and runoff for the watersheds shown in Table 4.1. The horizontal dashed lines correspond to SiO_2 concentrations of 20 and 150 M, the range of silica concentrations attributed to chemical weathering based on the fit of eqn (10), and which is shown in Figure 4.9. Silica concentrations above this range have been impacted significantly by evapotranspiration

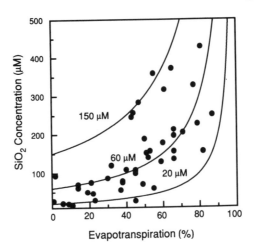

FIGURE 4.9 Watershed SiO$_2$ concentrations versus percentage of evapotranspiration. The lines correspond to the trajectories for the concentration of silica as the solution volume decreases as a result of evapotranspiration (eqn (10)). The labels indicate the initial solution composition at % ET = 0. The SiO$_2$ concentrations in the range of 20–150 μM at % ET = 0 are attributable to weathering

Silica concentrations have a negative correlation with precipitation and runoff (Figure 4.8) because of concentration by ET. Although the absolute amount of ET does not vary with either precipitation or runoff, the relative percentage of ET increases with decreases in both parameters. The initial SiO$_2$ concentration range shown in Figure 4.9 (20–150 μM) is represented by the parallel dashed lined in Figure 4.8. For watersheds with high annual precipitation ($P > 1500$ mm yr^{-1}, SiO$_2$ concentrations generally fall within this envelope, representing soil water compositions least impacted by ET. Most of the watersheds with an annual precipitation < 1500 mm yr^{-1} have SiO$_2$ concentrations which plot above this envelope, indicating significant concentration by ET. Due to the impact of variable ET on watershed chemistry, the use of SiO$_2$ concentrations as a means of comparing weathering rates is generally not a valid approach.

The extent of chemical concentration by ET also has important ramifications for the nature of the weathering processes in watersheds. Presumably, the initial chemical concentrations fitted by eqn (9) correlate with soil chemistries dominated by weathering. This chemistry is later subjected to the effects of ET. Although minor ET is associated with surface water flow during discharge, most ET within watersheds occurs either by evaporation or plant respiration in the shallow soil zone. To be subsequently affected by ET, the preponderance of chemical weathering must also occur in this environment. If weathering occurs at depth, for example in a groundwater system, the observed ET effects would not be generated.

4.4.6 Silica fluxes as a function of precipitation

SiO$_2$ fluxes for the tabulated watersheds do not exhibit a strong correlation with the percentage of ET. This occurs because the chemical flux is proportional to the product of runoff and chemical concentration. As runoff is decreased by ET, a proportional increase occurs in the chemical concentration (eqn (5)).

SiO_2 fluxes generally increase with runoff and precipitation (Figure 4.10). This is in contrast to SiO_2 concentrations which decrease with these parameters (Figure 4.8). Increasing SiO_2 fluxes implies that precipitation and runoff must increase weathering rates. The relationship between silica fluxes and the amount of precipitation and runoff (excluding the Rio Icacos data) can be approximated by the linear regressions

$$Q_{SiO_2} = 324 + 0.31\ P\ (r^2 = 0.45) \quad \text{and} \quad Q_{SiO_2} = 476 + 0.32\ R\ (r^2 = 0.42) \quad (10)$$

The slopes of the lines (Figure 4.10 insert) are nearly identical (0.31 and 0.32), as would be expected from the linear relationship between precipitation and runoff (Figure 4.7). The slope of the linear regression fit to the flux data with respect to runoff is equivalent to an initial SiO_2 concentration (i.e. before ET) of 32 mol l^{-1}. This SiO_2 concentration falls within the concentration envelope defined by the evaporation curves (Figure 4.8), confirming that most watershed SiO_2 concentrations are affected by ET. These watershed data can be compared to those of Dethier (1986), who found a slope of 0.85 ($r^2 = 0.30$) for SiO_2 fluxes in runoff from coastal rivers of the western USA. The significantly higher slope and poorer correlation of Dethier's data can be attributed to more rapid weathering of diverse rock types including a preponderance of young volcanic flows.

As shown in Figure 4.10, the silica fluxes for the Rio Icacos watershed, situated in a tropical rain forest, is a factor of four times greater than the fluxes obtained from temperate North American rain forests (such as Jamesion Creek and Indian River in Table 4.1). This occurs in spite of similar annual precipitation, dense vegetation (although very different species) and very steep topography. This difference indicates that under extreme climatic conditions, other factors, such as temperature, are important in controlling weathering rates. The variation in silica fluxes at low and moderate amounts of precipitation can also be related to other watersheds characteristics. For example, the Loch Vale, Emerald Lake and Martinell

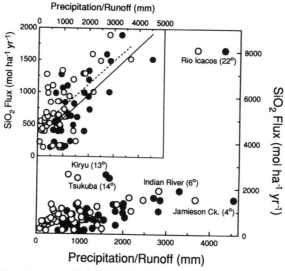

FIGURE 4.10 Silica fluxes versus precipitation (closed symbols) and runoff (open symbols). The outer figure shows data inclusive of the Rio Icacos and the insert shows an expanded flux scale with the linear regression fits to the alpine/temperate watershed data (eqns (12) and (13))

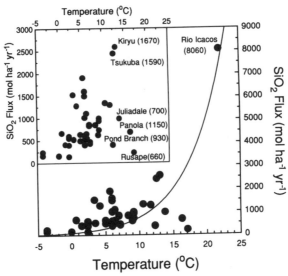

FIGURE 4.11   Silica fluxes versus temperature. The outer figure is inclusive of the Rio Icacos data. The dashed line shows the optimized fit to the Arrhenius expression (eqn (1)). The insert contains an expanded view of the alpine/temperate data, showing an apparent lack of correlation between SiO_2 fluxes and temperature

watersheds (Table 4.1) have low SiO_2 fluxes and are situated in high-elevation alpine basins with minimum soil development and vegetation. The relatively short hydrologic residence time, in additional to low temperatures, may contribute to low weathering rates in these watersheds.

4.4.7   Silica flux as a function of temperature

Silica fluxes are plotted as a function of mean annual air temperature in Figure 4.11. Diurnal and seasonal fluctuations are highly moderated in soils, and annual air temperature is a reasonable approximation of soil temperature (Velbel, 1993). Air temperatures are usually monitored at base elevations near stream outflow. With large elevation gradients, this temperature may vary by several degrees from the higher reaches of the watershed. In the few cases where temperatures were measured at elevations significantly different from base elevations, the temperature was adjusted using a thermal gradient of 0.0065 degree m^{-1} elevation. The Arrhenius equation (eqn (1) was fit to the data using an optimizing routine to minimize the sum of the residuals $\Sigma(Q-Q')^2$, where Q is the actual SiO_2 flux and Q' is the predicted value.

The resulting fit (dashed line, Figure 4.11) is dominated by the high-temperature Rio Icacos data and yields values of $A = 10^{27.4}$ and $E_a = 133$ kJ mol^{-1}. An activation energy of 133 kJ mol^{-1} is absurdly high, over twice as large as determined experimentally for silicate minerals, and much higher than suggested from global climate models and field studies. The Rio Icacos is the highest temperature site at 22°C, but also has very high precipitation (4300 mm yr^{-1}). Excluding the Rio Icacos (22°C), the alpine/temperature watersheds (0−16°C) show no clear trend between silica flux and temperature (insert, Figure 4.11), indicating that precipitation is a much stronger influence on weathering rates in these watersheds. To effectively evaluate the isolated effect of temperature and precipitation, it is necessary to consider the effect of these variables simultaneously.

4.4.8 Silica flux as a function of both precipitation and temperature

We can propose a function that will describe the simultaneous dependence of the weathering flux of SiO_2 in watersheds on both precipitation and temperature such that

$$Q = (A+BP) \exp\left[-\frac{E_a}{R}\left(\frac{1}{T} - \frac{1}{T_0}\right)\right] \tag{11}$$

The pre-exponential term on the right-hand side of eqn (11) assumes a linear correlation between precipitation and the SiO_2 flux where A is the intercept and B is the slope. The form of this term is based on the approximate linear relationship between fluxes and precipitation for the alpine/temperate watersheds (eqn (10). The exponential temperature term in eqn (11) is equivalent to eqn (2), which describes the variation in fluxes as a function of the difference between T and T_0. At the reference temperature (T_0), the SiO_2 flux equals the pre-exponential term ($A+BP$). As the temperature deviates from T_0, the rate will scale as a function of the activation energy (E_a) contained in the Arrhenius relationship.

Equation (11) can be solved numerically for the values of A, B and E_a which will minimize the residual between the predicted and measured values for the SiO_2 flux as functions of temperature (T) and precipitation (P). Equation (11) requires the selection of a reference temperature (T_0), but the solution for the activation energy (E_a) and the ratio of the intercept (A) to the slope (B) are insensitive to the choice of T_0. We choose a value of $T_0 = 5°C$ close to the mean of the temperature data in Table 4.1. In fitting eqn (11) to the watershed data, precipitation and temperature are treated as independent variables. In our dataset, the correlation between P and T has $r^2 = 0.10$, suggesting that this is a valid assumption.

The best fit of eqn (11) to the data yields values of $A = 35.0$, $B = 0.403$ and $E_a = 60.3$ kJ mol^{-1}. The three-dimensional surface describing the dependence of the SiO_2 flux on temperature and precipitation is shown in Figure 4.12. At any fixed temperature, the SiO_2 flux increases linearly with precipitation. However, the slope of the line describing the precipitation dependence increases with increasing temperature, varying from 0.25 at 0°C to 0.62 at 10°C, and 1.53 at 20°C. Therefore, we would expect a linear correlation between precipitation (or runoff) and weathering rates for watersheds with similar temperatures, such as the data of Meybeck (1986) and Dethier (1986). The slope of that correlation, however, would depend upon the mean temperature in the region. From eqn (11), it is clear that, at any fixed precipitation, the rate has a simple Arrhenius relationship, which is scaled linearly by the magnitude of the precipitation.

The standard deviation of the measured data from eqn (11) is 325 mol ha^{-1} yr^{-1}, or 34% of the mean SiO_2 flux. A useful way to visualize the goodness of fit is to "correct" the measured data to a standard temperature or precipitation. Figure 4.13 compares the uncorrected SiO_2 fluxes (open circles) with the same data corrected to the standard T_0 of 5°C (closed circles) using the best-fit activation energy of 60.3 kJ mol^{-1} (eqn (11)). The temperature correction pulls the high SiO_2 flux of the Rio Icacos watershed on to the general precipitation trend. To some extent, this reflects the dominant influence of the Rio Icacos watershed on the determination of the fitted activation energy. The temperature corrections to the other data points are all much smaller. Only in high-temperature environments does the Arrhenius equation predict that temperature will have a dramatic influence on the weathering rate, which will dominate over precipitation (Figure 4.1). After the temperature correction, the linear fit to the precipitation data increases to $r^2 = 0.60$, a considerably better fit than the uncorrected data, either with or without the Rio Icacos watershed.

The SiO_2 flux, normalized to a precipitation of 1000 mm yr^{-1}, is plotted versus temperature in Figure 4.14, with the dashed line showing the best fit from eqn (11). Figure 4.14 can be compared with Figure 4.11 showing the fit to the data uncorrected for precipitation. The data point at 22°C (Rio Icacos) still has a large influence on the activation energy, but a consistent trend in the entire fitted dataset is now apparent. The residual scatter (Figure 4.14) is still too large to differentiate whether an Arrhenius relationship is the most appropriate function. However, the best-fit activation energy of $E_a = 60$ kJ mol^{-1} is in the range of experimental results for silicate dissolution experiments, and is consistent with the activation energies, which yield reasonable predictions in global mass balance models. This supports the appropriateness of using the Arrhenius relationship for modelling the temperature dependence of the overall weathering process. Nevertheless, there is a distinct need for more data from warm climates to better define the temperature dependence.

4.4.9 Other factors controlling watershed weathering rates

The resulting correlation coefficients for the preceding model indicate that approximately 60% of the variability in SiO_2 fluxes for the tabulated watersheds can be explained by differences in annual precipitation and temperature. However, the residual variability is still considerable, and is dependent on several other factors that may or may not be related to climate.

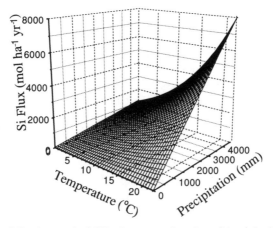

FIGURE 4.12 Optimized fit of watershed SiO_2 fluxes as a function of precipitation and temperature. At a constant temperature, the flux is a linear function of precipitation. For constant precipitation, the flux is an exponential function of temperature

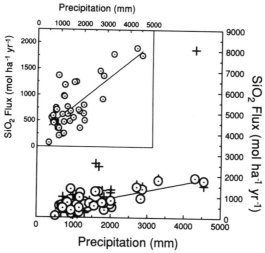

FIGURE 4.13   Comparison of uncorrected and temperature-corrected SiO$_2$ flux data normalized to 5°C. The line corresponds to a cross section through the best-fit surface shown in Figure 4.12

Bedrock lithology

The preceding comparison includes watersheds underlain by granitoid and gneissic bedrock, in which the primary minerals weathering are K-feldspars, sodic plagioclase (An$_{\approx 30}$) hornblende and micas. Even this constraint leaves room for significant variability in watersheds, including the proportions and compositions of these minerals, the extent of jointing, the degree of secondary hydrothermal alteration and the presence of other minor minerals and lithologies. These factors may have large influences on weathering rates and SiO$_2$ fluxes, and probably account for a significant proportion of non-climatic variability.

Age of soil cover

Chronosequence data showed a significant decrease in the chemical weathering rates of soils over time due to the preferential removal of reactive minerals. However, this process is not as obvious when comparing SiO$_2$ fluxes for different watersheds. Soil ages are difficult to determine and quantify. However, there is no clear differentiation between alpine watersheds that have experienced recent glaciation (such as Loch Vale and Emerald Lake), those that experienced Pleistocene glaciation (such as Hubbard Brook and Rawson Lake) and those that have not been glaciated in at least 250 ka (such as Panola and Rusape). It appears that the effects of soil age are secondary to precipitation and temperature in controlling weathering rates. This also implies that the Pleistocene continental glaciations did not have an overwhelming effect on chemical weathering rates.

Thickness of the soil cover

Chemical fluxes are only normalized with respect to watershed area and not to the soil volume or mineral surface areas. Data on average soil thicknesses are not available in most watershed

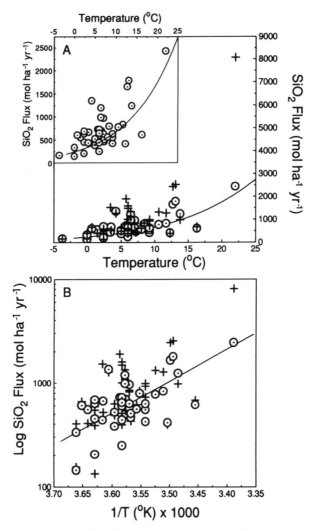

FIGURE 4.14 Silica fluxes as a function of temperature normalized to a constant precipitation (1000 mm). The line is a cross section through the best fit surface shown in Figure 4.13

studies. The significance of soil thickness in weathering is suggested, however, by low chemical fluxes in the high-elevation alpine basins with thin soils. Lower weathering rates may be attributed both to lower total reactive surface areas of minerals and to less moisture retention within the soil profile.

Topography

Topographic relief is required to hydrologically define a small closed watershed. Within the range of hilly to mountainous topography represented (20−2000 m change in elevation, with the majority in the 100−800 m range), there is no clear evidence that slope steepness affects chemical weathering rates. Obviously, flat basins are totally unrepresented. In these watersheds with at least minimal topographic relief, erosion of weathering products such as clays, oxides and residual quartz does not severely limit the extent of chemical weathering.

Vegetation

The tabulated data are heavily biased towards alpine and forested watersheds. Grasslands and other ecosystems are under-represented. Vegetation varies from alpine tundra and bare outcrops to coniferous forest, hardwood deciduous forest and one tropical forest site. There is no clear trend that would suggest that vegetation is a major factor in controlling net weathering rates. However, the type of vegetation tends to vary sympathetically with precipitation and temperature, and any effect of vegetation may be incorporated into these trends.

The preceding parameters impact the variations observed in the weathering rates of these watersheds to an extent comparable to the effects of precipitation and temperature. However, unlike precipitation and temperature, these parameters are much more difficult to quantify. Additional advances in climatic comparisons of watersheds await advances in such techniques.

4.4.10 Application to climate models

A number of climate models have considered the mechanism that stablizes global climate over geologic time due to the feedback between atmospheric CO_2 and silicate weathering (Walker *et al.*, 1981; Berner *et al.*, 1983; Volk, 1987; Marshall *et al.*, 1988; Brady, 1991; Berner, 1994). The major uncertainty in such models is the relative sensitivity of weathering rates on climatic variables, principally temperature, precipitation and runoff. Efforts to calibrate such models have focused both on experimental dissolution and natural weathering studies.

The effect of temperature on the ratios of weathering fluxes (Q_T/Q_0) predicted by eqn (11) for constant values of precipitation (750–4000 mm) are represented by the solid lines in Figure 4.15. The standard flux Q_0 is that defined for 5°C and global average precipitation (750 mm). These results, except at high precipitation and low temperature, are bracketed by Arrhenius relationships (Figure 4.15) describing temperature effects based on experimental activation energies proposed by Brady (1991) for Ca silicates (48 kJ mol^{-1}) and Mg silicates (115 kJ mol^{-1}). Global carbon models most often assume that the dominant reactions that consume CO_2 involve Ca and Mg silicates (Berner *et al.*, 1983; Volk, 1987). This assumption is based on the relatively large global export of Ca and Mg relative to Na and K, in addition to reaction stoichiometry which requires that alkali earth elements consume twice as many protons as alkali metals. The calculated activation energy based on watershed SiO_2 fluxes (59 kJ mol^{-1}) falls between these experimental values. Also included in Figure 4.15 is the predicted temperature dependence based on the activation energy of 77 kJ mol^{-1} obtained by Velbel (1993) from the Coweeta watersheds.

Recently Berner (1994) in the GEOCARB II model considered the coupled effects of temperature and runoff on apparent weathering fluxes. This model was calibrated based on the estimated activation energy of 62 kJ mol^{-1} (Brady, 1991) and river runoff/concentration data presented by Dunne (1978) and Peters (1984). Observed trends in solute concentrations were assumed to result from the combined effects of dilution and solute residence times in the

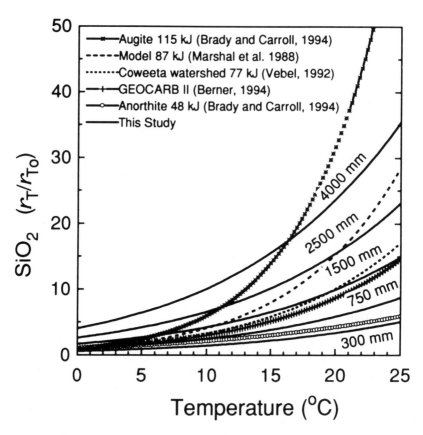

FIGURE 4.15 Comparison of silica flux ratios (solid lines) as functions of temperature at indicated constant precipitation. Q_0 is at 5°C and 750 mm of precipitation (global average). Dotted lines are temperature trends based on experimental activation energies for Mg and Ca silicates (Brady and Carroll, 1994) and solute fluxes at Coweeta (Velbel, 1993). The dashed line is trend predicted by the GEOCARB II model (Berner, 1994)

regiolith. As in previous models, the effects of ET on solute concentrations were not considered. This model also assumes that the reaction rates are not kinetically influenced by moisture content, as is suggested in the present study. Figure 4.15 indicates that weathering rates produced by GEOCARB II and the present coupled approach (eqn (13)) produce comparable rates for moderate precipitation environments (1000–2000 mm), but that GEOCARB II weathering would significantly underestimate watershed weathering rates in wet tropical environments.

4.5 CONCLUSIONS

A strong interconnection exists between chemical weathering and climate. Climate influences weathering principally through precipitation and temperature, and can potentially impact global climate via CO_2 consumption, soil fertility, carbon cycling and hydrological processes including evapotranspiration and runoff. This chapter demonstrates that the effects of precipitation and temperature can be investigated by the use of chemical mass balances which relate the chemistry of soils and streams to the rates of weathering in watersheds with different climates but similar lithologies.

The chemistry of soils of different ages may be used to quantify the amount of chemical weathering that has occurred over geologic time. By comparing the ratios of reactive and conservative elements of minerals in soils with the initial bedrock, the total loss of elements or primary minerals over the time span of soil development can be quantified. Results demonstrate the importance of long-term differences in precipitation and temperature on the rates of soil profile development. Mass balances in soil chronosequences indicate that rates of weathering decrease significantly with soil age as a result of the preferential loss of more reactive minerals during the early stages of weathering. This observation implies that present rates of weathering within watersheds should be dependent on the previous weathering history, including changes in climate during soil development, history of glaciation and tectonic regime.

The effect of climate on chemical weathering was also investigated by comparing solute weathering fluxes in watersheds with climatic parameters. This approach is detailed for SiO_2 fluxes in watersheds underlain by granitoid rocks. The results demonstrate the interrelationship between watershed hydrology and SiO_2 chemistry. In watersheds with similar temperatures (i.e. alpine/temperate or tropical), the *absolute* amount of ET is fairly constant and independent of both the amount of precipitation and runoff. However, the *proportion* of precipitation lost through ET increases significantly as precipitation decreases. This increases SiO_2 concentrations in watershed streams through evaporative concentration and implies that the range of SiO_2 concentrations in soil solutions involved in weathering is much lower and narrower than the SiO_2 concentrations of the stream waters. The use of solute chemical concentrations, therefore, is not a viable tool in making wide-scale climatic comparisons of watershed weathering rates, or for modelling the chemical controls on weather rates in soils.

In contrast, net watershed SiO_2 fluxes are not susceptible to ET effects and can be successfully used to quantify the effects of precipitation and temperature on chemical weathering rates. In watersheds with temperate climates, there is an apparent linear trend between SiO_2 fluxes and precipitation. In contrast, there is no obvious correlation between SiO_2 fluxes and temperature. This indicates that for alpine/temperate climates, precipitation is a much stronger control on chemical weathering rates than is temperature. However, in tropical climates, both precipitation and temperature strongly influence SiO_2 fluxes. Therefore, in investigating global climatic differences, the approach of independently considering the effects of temperature and precipitation is clearly inappropriate.

This resulted in the development of a model that describes weathering as the product of a linear dependence on precipitation and an exponential (Arrhenius) dependence on temperature. The calculated activation energy for this temperature dependence was 60 kJ mol^{-1}. This activation energy is highly dependent on a single tropical watershed, but is comparable to experimental silicate dissolution data and values used in current global climate change models. Statistical analysis indicates that approximately 60% of the observed variation in watershed SiO_2 fluxes is attributable to precipitation and temperature. The remaining variations can be attributed to other watershed characteristics, including bedrock lithology, age and thickness of soil cover, topography and vegetation. More detailed climatic analysis awaits a more diverse distribution of watersheds, as well as methods to quantify these other important processes relative to climatic effects.

ACKNOWLEDGEMENTS

The authors would like to thank the following individuals who contributed additional information to the watershed tabulation: Paul P. Campbell, Emery T. Cleaves, D. A. Everson, J. P. Kimmins, George H. Leavesley, Alisa M. Mast, Norman E. Peters, Michael M. Reddy and Richard F. Yuretich.

REFERENCES

April, R., Newton, R. and Coles, L. T. (1986). "Chemical weathering in two Adirondack watersheds: past and present rates", *Geol. Soc. Am. Bult.*, **97**, 1232–1238.

Barth, T. F. (1961). 'Abundance of the elements, aerial averages and geochemical cycles", *Geochim. Cosmochim. Acta*, **23**, 1–8.

Berner, R. A. (1991). "Model for atmospheric CO_2 over Phanerozoic time", *Am. J. Sci.*, **291**, 339–376.

Berner, R. A. (1994). "GEOCARB II: A revised model of atmospheric CO_2 over Phanerozoic time", *Amer. J. Sci.*, **294**, 56–91.

Berner, R. A., Lasaga, A. C. and Garrels, R. M. (1983). "The carbonate silica geochemical cycle and its effect on atmospheric carbon dioxide and climate", *Amer. J. Sci.*, **283**, 641–683.

Blum, A. E., (1994). "Feldspars in weathering", in *Feldspars and Their Reactions* (Ed. I. Parsons), NATO Advanced Study Workshop Series C, Kluwer, The Netherlands, 595–630.

Blum, A. E. and Lasaga, A. C. (1991). "The role of surface speciation in the dissolution of albite", *Geochim. Cosmochim. Acta*, **55**, 2193–2201.

Blum, A. E., Yund, R. A. and Lasaga, A. C. (1990). "The effect of dislocation density on the dissolution rate of quartz", *Geochim. Cosmochim. Acta*, **54**, 283–297.

Blum, A. E., White, A. F., Bullen, T., Schulz, M. S. and Larson, M. (1993). "Chemical weathering of silicate minerals in a mountainous tropical rain forest, Puerto Rico", *Geol. Soc. Am. Abst.*, **25**, A255.

Bouchard, M. (1983). "Bilan géochimique et origine d'élements dissous dans un bassin-versant granitique breton (France)", *Revue Géol. Dynam. Géograph. Phys.*, **24**, 363–379.

Brady, P. V. (1991). "The effect of silicate weathering on global temperature and atmospheric CO_2", *J. Geophys. Res.*, **96**, 18101–18106.

Brady, P. V. and Carroll, S. A. (1993). "Geochemical weathering and global climate control", *Trans. Am. Geophys. Union*, **74**, 267.

Brady, P. V. and Carroll, S. A. (1994). "Direct effects of CO_2 and T on silicate weathering: possible implications for climate control", *Geochim. Cosmochim. Acta*, **58**, 1853–1856.

Brady, P. V. and Walthers, J. V. (1989). "Controls on silicate dissolution in neutral and basic pH solutions at 25°C", *Geochim. Cosmochim. Acta*, **53**, 2823–2830.

Brantley, S. L., Stillings, L. L. and Voigt, D. E. (1992). "Effect of oxalic acid on albite dissolution", *Abstracts of Papers, 203th ACS National Meeting*, GEOC 4.

Bricker, O. P. and Rice, K. (1989). "Acidic deposition to streams", *Environ. Sci. Technol.*, **23**, 379–385.

Brimhall, G. H. and Dietrich, W. E. (1987). "Constitutive mass balance relations between chemical composition, volume, density, porosity, and strain in metasomatic hydrochemical systems: results on weathering and pedogenesis", *Geochim. Cosmochim. Acta*, **51**, 567–587.

Brown, E. T., Stallard, R. F., Larsen, M. C., Raisbeck, G. M. and Yiou, F. (1993). "Denudation rates based on accumulation of *in situ* produced [10]Be compared with watershed balance results in the Luquillo Experimental Forest, Puerto Rico", *Trans. Am. Geophys. Union*, **74**, 295.

Bruijnzeel, L. A. (1990). *Hydrology of Moist Tropical Forests and Effects of Conversion: A State of Knowledge Review*, UNESCO International Hydrological Programme, Free University, Amsterdam, 224 pp.

Büdel, J. (1980). "Climatic geomorphology", *Zeit. Geomorph. Suppl. Bult.*, **36**, 1–8.

Burch, T. E., Nagy, K. L. and Lasaga, A. C. (1993). "Free energy dependence of albite dissolution kinetics at 80°C and pH 8.8", *Chem. Geology*, **105**, 137–162.

Busenberg, E. and Clemency, C. V. (1976). "The dissolution kinetics of feldspars at 25°C and 1 atm CO_2

pressure", *Geochim. Cosmochim. Acta*, **40**, 41−50.

Chadwick, O. A., Brimhall, G. H. and Hendricks, D. M. (1990). "From a black to a grey box — a mass balance interpretation of pedogenesis", *Geomorphology*, **3**, 369−390.

Chou, L. and Wollast, R. (1985). "Steady state kinetics and dissolution mechanisms of albite", *Am. J. Sci.*, **285**, 963−993.

Christophersen, N. and Neal, C. (1990). "Linking hydrologic, geochemical, and soil chemical processes on a catchment scale: an interplay between modelling and field work", *Water Resour. Res.*, **26**, 3077−3086.

Clayton, J. L. (1985). "An estimate of plagioclase weathering in the Idaho Batholith based upon geochemical transport rates", in *Rates of Chemical Weathering* (Eds. S. M. Colemen and D. P. Dethier), pp. 453−466, Academic Press, Orlando, Florida.

Cleaves, A. T., Godfrey, A. E. and Bricker, O. P. (1970). "Geochemical balance in a small watershed", *Geol. Soc. Am. Bull.*, **81**, 3013−3032.

Cosby, B. J., Wright, R. F., Hornberger, G. M. and Galloway, J. N. (1985). "Modelling the effects of acid deposition: estimation of long term water quality responses in a small forested catchment", *Water Resource. Res.*, **21**, 1591−1601.

Creasey, J., Edwards, A. C., Reid, J. M., Macleod, D. A. and Cresser, M. S. (1986). "The use of catchment studies for assessing chemical weathering rates in two contrasting upland areas of Northeast Scotland", in *Rates of Chemical Weathering of Rocks and Minerals* (Eds. S. M. Colman and D. P. Dethier), pp. 467−501, Academic Press, Orlando, Florida.

Cresser, M. and Edwards, A. (1987). *Acidification of Freshwaters*, Cambridge University Press, London, 112 pp.

Dethier, D. P. (1986). "Weathering rates and chemical fluxes from catchments in the Pacific Northwest USA", in *Rates of Chemical Weathering of Rocks and Minerals* (Eds. S. M. Colman and D. P. Dethier), pp. 503−528, Academic Press, Orlando, Florida.

Driscoll, C. T., Likens, G. E., Hedlin, L. O., Eaton, J. S. and Borman, F. H. (1989). "Changes in the chemistry of surface waters", *Environ. Sci. Technol.*, **23**, 137−142.

Dunne, T. (1978). "Rates of chemical denudation of silicate rocks in tropical catchments", *Nature*, **274**, 244−246.

Feller, M. C. and Kimmins, J. P. (1984). 'Effects of clearcutting and slash burning on streamwater and watershed nutrient budgets in southwestern British Columbia", *Water Resour. Res.*, **45**, 1421−1437.

Harden, J. W. (1987). "Soils developed in granitic alluvium near Merced, California", *US Geol. Survey Bull.*, 1590-A, 135 pp.

Helgeson, J. C., Murphy, W. M. and Aagaard, P. (1984). "Thermodynamic and kinetic constraints of reaction rates among minerals and aqueous solution. II. Rate constants, effective surface area, and the hydrolysis of feldspar", *Geochim. Cosmochim. Acta*, **48**, 2405−2432.

Jenny, H. (1941). *Factors of Soil Formation*, McGraw-Hill, New York, 281 pp.

Johnson, N. M. (1984). "Acid rain neutralization by geologic materials", *Geologic Aspects of Acid Rain* (Ed. O. P. Bricker), pp. 37−53, Butterworth Publishing, Massachusetts.

Kirkwood, D. E. (1989). "Long term chemical and mineralogical weathering within soils of Plastic Lake, Ontario", *Ms. thesis*, Department of Geology, University of Western Ontario, 174 pp.

Knauss, K. G. and Wolery, T. J. (1986). "Dependence of albite dissolution kinetics on pH and time at 25°C", *Geochim. Cosmochim. Acta*, **50**, 2481−2497.

Larson, M. C. and Sánchez, A. J. T. (1992). "Landslides triggered by Hurricane Hugo in eastern Puerto Rico, September 1989", *Caribbean J. Sci*, **28**, 113−125.

Lasaga, A. C. (1984). "Chemical kinetics of water-rock interactions", *J. Geophys. Res.*, **89**, 4009−4025.

Lasaga, A. C. and Blum, A. E. (1986). "Surface chemistry, etch pits and mineral-water reactions", *Geochim. Cosmochim. Acta*, **50**, 2363−2379.

Likens, G. E., Bormann, F. H., Pierce, R. S., Eaton, J. S. and Johnson N. M. (1977). *Biogeochemistry of a Forested Ecosystem*, Springer Verlag, Berlin, 146 pp.

McDowell, W. H. and Asbury, C. E. (1994). "Export of carbon, nitrogen, and major ions from three tropical montane watersheds", *Limin. Oceanogr.*, **39**, 111−125.

Marchard, D. E. (1971). "Rates and modes of denudation, White Mountains, Eastern California", *Am. J. Sci.*, **270**, 109−135.

Marchard, D. E. (1971). "Rates and modes of denudation, White Mountains, Eastern California", *Am. J. Sci.*, **270**, 109−135.

Marshall, H. G., Walker, J. C. G. and Kuhn, W. R. (1988). "Long-term climate change and the geochemical cycle of carbon", *J. Geophys. Res.*, **93**, 791−801.

Mast, M. A. and Drever, J. L. (1987). "The effect of oxalate on the dissolution rates of oligoclase and tremolite", *Geochim. Cosmochim. Acta.*, **51**, 2559−2568.

Mast, M. A., Drever, J. I. and Baron, J. (1990). "Chemical weathering in Loch Vale Watershed, Rocky Mountain National Park, Colorado", *Water Resour. Res.*, **26**, 2971−2978.

Merrill, G. P. (1906). *A Treatise on Rocks, Rock Weathering, and Soils*, Macmillan, New York, 400 pp.

Merritts, D. J., Chadwick, O. A., Hendricks, D. M., Brimhall, G. H. and Lewis, C. J. (1992). "The mass balance of soil evolution on late Quaternary marine terraces, northern California", *Geol. Soc. Am. Bull.*, **104**, 1456−1470.

Meybeck, M. (1986). "Composition chimique des ruisseaux non pollués de France", *Sci. Geol. Bull.*, **39**, 3−77.

Meybeck, M. (1987). "Global chemical weathering of surfical rocks: effect of temperature", *Am. J. Sci*, **287**, 401−428.

Murphy, W. M. and Helgeson, H. C. (1987). "Thermodynamic and kinetic constraints on reaction rates among minerals and aqueous solution. III. Activated complexes and pH dependence on rates of feldspar, pyroxene, wollastonite and olivine hydrolysis", *Geochim. Cosmochim. Acta*, **51**, 3137−3153.

Nagy, K. L., Blum, A. E. and Lasaga, A. C. (1991). "Dissolution and precipitation kinetics of kaolinite at 80°C and pH 3: the dependence on solution saturation state", *Am. J. Sci*, **291**, 649−486.

Norton, S. A., Kahl, J. S., Fernandez, I. J., Rustad, L. E., Scofield, P. and Haines, T. A. (1994). "Response of the West Bear Brook Watershed, Maine to the addition of $(NH_4)_2SO_4$ − Three year results", *Forest Ecology and Management* (in press).

Ollier, C. (1984). *Weathering*, Longman, London, 270 pp.

Owens, L. B. and Watson, J. P. (1979). "Rates of weathering and soil formation on granite in Rhodesia", *Soil Sci. Soc. Am. J.*, **43**, 160−166.

Paces, T. (1983). "Rate constants of dissolution derived from the measurements of mass balance in hydrologic catchments", *Geochim. Cosmochim. Acta*, **47**, 1855−1863.

Paces, T. (1985). "Sources of acidification in Central Europe estimated from elemental budgets in small basins", *Nature*, **315**, 31−35.

Paces, T. (1986). "Weathering rates of gneiss and depletion of exchangeable cations in soils under environmental acidification", *J. Geol. Soc. London*, **143**, 673−677.

Pavich, M. J., Brown, L., Harden, J., Klein, J. and Middleton, R. (1986). "[10]Be distribution in soils from Merced River terraces, California", *Geochim. Cosmochim. Acta*, **50**, 1727−1735.

Peltier, L. (1950). "The geographic cycle in periglacial regions as it is related to climatic geomorphology", *Ann. Assoc. Am. Geol.*, **40**, 214−236.

Peters, N. (1984). "Evaluation of environment factors affecting yields of major dissolved ions of streams in the United States", *US Geol. Sur. Water Supply Paper*, **2228**, 39 pp.

Reddy, M. M. (1988). "A small alpine basin budget: Front Range, Colorado", in *International Mountain Watershed Symposium, Subalpine Processes and Water Quality* (Eds. I. G. Poppoff, C. R. Goldman, S. L. Loeb and L. B. Leopold, pp. 370−385. Tahoe Resource Conservation District, South Lake Tahoe, California.

Schlinder, D. W., Newbury, R. W., Beaty, K. G. and Cambell, P. (1976). "Natural water and chemical budgets for a small Precambrian Lake Basin in Central Canada", *J. Fish. Res. Board Canada*, **33**, 2526−2543.

Schnoor, J. L. (1990). "Kinetics of chemical weathering: a comparison of laboratory and field weathering rates", in *Aquatic Chemical Kinetics* (Ed. W. Stumm, pp. 475−533, Wiley-Interscience, New York.

Siegel, D. I. and Pfannkurch, H. O. (1984). "Silicate dissolution influence on Filson Creek chemistry, northeastern Minnesota", *Geol. Soc. Am. Bull*, **95**, 1446−1453.

Stahr, K., Zottl, H. W. and Hadrich, R. (1980). "Transport of trace elements in the ecosystems of the Barhalde watershed in the southern Black Forest", *Soil Sci.*, **130**, 217−224.

Stallard, R. F. and Edmonds, J. M. (1983). "Geochemistry of the Amazon: 2. The influence of the geology and weathering environment on the load", *J. Geophys. Res*, **88**, 9671−9688.

Stauffer, R. E. and Wittchen, B. D. (1991). "Effects of silicate weathering on chemistry in a forested, upland, felsic terrain of the USA", *Geochim. Conmochim. Acta*, **55**, 3253−3271.

Stednick, J. D. (1981). "Hydrochemical balance of an alpine watershed in southeast Alaska", *Arctic Alpine Res.*, **13**, 431−438.

Suchet, P. A. and Probst, J. L. (1993). "Modelling of atmospheric CO_2 consumption by chemical weathering of rocks. application to the Garonne, Congo, and Amazon basins", *Chem. Geol.*, **107**, 205−210.

Sverdrup, H. U. (1990). *The Kinetics of Base Cation Release Due to Chemical Weathering*, Lund University Press, Lund, Sweden, 245 pp.

Swoboba-Colberg, N. G. and Drever, J. I. (1993). "Mineral dissolution rates in plot-scale field and laboratory experiments", *Chem. Geol.*, **105**, 51−69.

Velbel, M. A. (1985). "Geochemical mass balances and weathering rates in forested watersheds of the southern Blue Ridge", *Am. J. Sci*, **285**, 904−930.

Velbel, M. A. (1993). "Temperature dependence of silicate weathering in nature: how strong of a negative feedback on long term accumulation of atmospheric CO_2 and global greenhouse warming?", *Geology*, **21**, 1059−1062.

Volk, T. (1987). "Feedbacks between weathering and atmospheric CO_2 over the last 100 million years", *Am. J. Sci.*, **287**, 763−779.

Walker, J. C. G., Hays, P. B. and Kashing, J. F. (1981). "A negative feedback mechanism for the long-term stabilization of the earth's temperature", *J. Geophys. Res.*, **86**, C 10, 9776−9782.

White, A. F. and Blum, A. E. (1995). "The effect of climate on weathering rates in watersheds", *Geochim. Cosmochim. Acta*, **59** (in press).

White, A. F. and Peterson, M. L. (1990). "Role of reactive- surface-area characterization in geochemical kinetic models", in *Chemical Modelling of aqueous Systems*, Vol. II (Eds. D. C. Melchior and R. L. Bassett, *American Chemical Society Symposium Series*, **416**, pp. 461−475.

White, A. F., Blum, A. E., Bullen, T. D., Peterson, M. L., Schulz, M. S. and Hardin, J. W. (1992). "A three million year weathering record for a soil chronosequence developed in grantitic alluvium, Merced, California, USA", in *Proceedings of 7th International Symposium on Water Rock Interaction* (Eds. Y. K. Kharaka and A. S. Maest, pp. 607−610, Park City, Utah.

White, A. F., Blum, A. E., Schulz, M. S., Peterson, M. and Hardin, H. W. (1995). "Chemical weathering in a soil chronosequence: solid state weathering rates", *Geochim. Cosmochim. Acta*, (in review).

Williams, M. W. and Melack, J. M. (1991). "Solute chemistry of snowmelt and runoff in an alpine basin, Sierra Nevada", *Water Resour. Res.*, **27**, 1575−1588.

Williams, M. W., Melack, J. M. and Everson, D. A. (1993). "Export of major ionic solutes from two sub-alpine watersheds before and after fire", *Trans. Am. Geophys. Union*, **74**, 258.

Yuretich, R. F. and Batchelder, G. L. (1986). "Hydrochemical cycling and chemical denudation in the Fort River Watershed, Central Massachusetts: an appraisal of mass-balance approaches", *Water Resour. Res.*, **24**, 105−114.

Yuretich, R., Knapp, E. and Irvine, V. (1993). "Chemical denudation and weathering mechanisms in central Massachusetts, USA", *Chem. Geol.*, **107**, 345−347.

Zeman, L. J. (1978). "Mass balance model for calculation of ionic input loads in atmospheric fallout and discharge from a mountainous basin", *Hydrol. Sci. Bull.*, **23**, 103−117.

5

Methods for Modelling Solute Movement in Structured Soils

ADRIAN ARMSTRONG[1]

ADAS Soil & Water Research Centre, Trumpington, UK

TOM ADDISCOTT

IACR-Rothamsted, Harpenden, Hertfordshire, UK

and

PETER LEEDS-HARRISON

Silsoe College, Cranfield University, Silsoe, Bedford, UK

5.1 INTRODUCTION: THE PHYSICS OF WATER AND SOLUTE MOVEMENT IN CRACKING SOILS

Water and solute movement play a vital part in the provision of suitable conditions for plant growth and in the replenishment and quality of groundwater supplies. In clay soils, the effective management of water requires the mechanisms of water movement to be understood. In order to provide adequate explanations of the observations of water and solute behaviour made in the field, the challenge for soil physicists and modellers (not always one and the same thing) has been to describe adequately the flow phenomena in structured soils that contain bimodal pore structures with a network of macropores separating aggregated soil that form regions of micropores.

As a basis for such work, a fundamental framework of relevant physics is required. Our understanding of the physics of soil water flow processes has developed largely from experimental studies of simple uniform porous materials ("sand physics") and has led to the widespread acceptance of Darcy's law for saturated flow and the Richards equation for

---

[1] Current address: ADAS Land Research Centre, ADAS Gleadthorpe, Meden Vale, Mansfield, Nottinghamshire NG20 9PF, UK.

Solute Modelling in Catchment Systems. Edited by Stephen T. Trudgill

unsaturated flow, in which it is assumed, *a priori*, that water flow in *all* situations is under the influence of an energy gradient (see, for example, Childs, 1969; Hillel, 1980; Marshall and Holmes, 1988). The conservation of mass forms the cornerstone of this approach to the description of water flow (indeed Richards' equation is derived from Darcy's empirical equation and the equation for the conservation of mass). However, water flow in cracked soils is not adequately described by conventional methods which assume a uniform porous medium.

In cracking clay soils, water flow may at times be dominated by flow in large pores which arise due to swelling and shrinking of the pods or biopores formed by soil fauna and the roots systems of plants, usually called macropores. Although rapid flow of both water and solutes through large pores was described as early as 1882 by Lawes *et al.* (1882), the importance of these pores to water flow in clay soils has been recently identified by many workers, in particular by the work of Bouma in the Netherlands and Beven and Germann working in the UK and North America. Bouma and Dekker (1978), Bouma (1981) and Beven and Germann (1982) have provided us with a framework for our studies of these soils. More recently, work in North America and Europe has given considerable attention to preferential flow processes in macropores in respect of water and its associated solutes draining to groundwater and surface water resources (e.g. Gish *et al.*, 1991; Chen and Wagenet, 1992a and 1992b). The same problems have also been addressed in a parallel fashion by hydrogeologists concerned with water movement in fractured strata. In all these models, the essential conceptual model is that the flow of water and solutes can be partitioned into two distinct but interacting components, normally termed macro- and micropores (Germann and Beven, 1985; Gerke and van Genuchten, 1993).

In this chapter we discuss possible ways of describing water and solute movement in the saturated and unsaturated zone in cracking clay soils. The chapter cannot in a short space provide a comprehensive review of all possible models. Instead, by giving a detailed description of three models, it illustrates the various approaches that can be employed to describe flow phenomena in soils where preferential flow may occur, and allows other modelling approaches to be put into context. Three models have been chosen which offer three different conceptualisations of the processes operating in such soils.

It is, of course, possible to ignore the presence of cracks and the two flow domains within a soil and to predict system behaviour using large-scale models which include the effects of the small-scale processes within the large-scale, bulk, parameters. Such a procedure may give acceptable numerical predictions, but does not further the understanding of the processes occurring at the small scale. For example, the transfer function models developed by Jury *et al.* (1986) and used by White *et al.* (1986) may describe the behaviour of a macroporous soil in an empirical sense, and so generate empirically useful results, but they do not require any detailed mechanistic assumptions either for their measurement or physical understanding. Nevertheless, the interpretation even of these empirical models may be improved by the recognition of the two flow components where this is appropriate (Sposito *et al.*,1986). Although this paper concentrates on physically based models, such empirical models may be useful in practical situations and should not be overlooked.

5.1.1 Water flow in cracking clays

Cracking clays have a structure to the pore system such that they have a few large pores often about 1000 μm in width and many micropores often about 1 μm in size. This distribution of

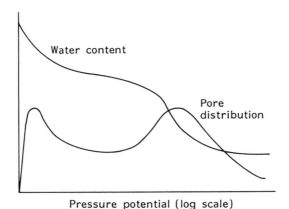

FIGURE 5.1   A water release characteristic for a structured clay soil showing a bimodal pore size distribution

pores is often referred to as being bimodal, (Figure 5.1), although some workers have argued that a multi-modal system is more appropriate (Steenhuis *et al.*, 1990). For instance, Luxmore *et al.* (1990) have described soils as having pores in three size ranges, macro, meso and micro, as shown in Figure 5.2, where mesopores lie in the size range between micropores and macropores (10—1000 μm). Beven and Germann (1982) review the problem of nomenclature. Nevertheless, we shall retain the simple macropore — micropore division that is now in common usage, although we acknowledge that this is itself a significant simplification of a complex physical situation (Gerke and van Genuchten, 1993). Because of their large size, macropores drain easily and in many clay soils the drainable porosity is equated with the macro porosity. In contrast the water held in micropores of the soil is tightly held and these pores provide a high resistance to flow.

Because of the bimodal nature of the pore spacing in cracking clays the resistance to the flow of water through the bulk of the soil is dominated by the large cracks and fissures. Depending on the relative saturation of the bulk soil in the profile the effect of cracks may be to provide a

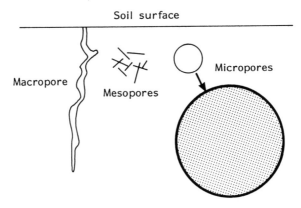

FIGURE 5.2   Classification of pores into three class sizes as suggested by Luxmore *et al.* (1990)

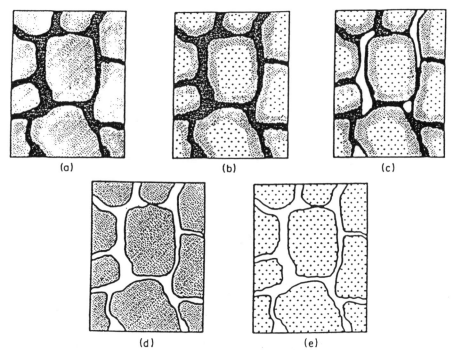

FIGURE 5.3 Various modes of saturation possible in aggregated soil: (a) macropore and micropore regions both fully saturated, (b) macropore region saturated but micropore region unsaturated, (c) macropore region partly saturated but micropore region unsaturated, (d) macropore region unsaturated but micropore region saturated, (e) both macropore and micropore regions unsaturated. (Reproduced from *Journal of Soil Science* (1991) 41, 667 by permission of Blackwell Scientific Publications)

low-resistance path for percolating water, as is the case at or near saturation, or they may provide a high-resistance path, as is the case when the soil is unsaturated with the macropores empty.

Youngs and Leeds-Harrison (1990) proposed five possible modes of saturation for consideration in terms of the transport behaviour in aggregated soils (see Figure 5.3). These are:

1. Where the macropore and micropore regions are both fully saturated. Here the hydraulic conductance of the soil is determined by the macropores. Such a situation occurs below the water table in a soil having a stable structure.
2. Where the macropore region is saturated but the micropore region is unsaturated. In this situation, which may occur where intense rainfall infiltrates rapidly into the macropore region, the aggregates (the micropore region of the soil) are wetted by the macropore water. Trapped air in this situation may lead to rapid transmission of water through the profile and drainage from the base of the profile, while aggregates higher in the profile may remain unsaturated.
3. Where the macropore region is partially saturated and the micropore region is unsaturated. This is a variant of case 2 which more usually occurs during heavy rainfall in these soils. Trapped air is less likely to be a problem, the rate of water movement down the profile

depends on the rate that water arrives at the soil surface and the rate at which the micropore region can take it up.

4. Where the macropore region is air filled but the micropore region is saturated. This situation may occur where a saturated structured soil profile is drained rapidly, which can sometimes be the case with mole-cum-tile drainage systems. Water movement in the profile occurs at the menisci at aggregate contact points.
5. Where the micropore and macropore regions are both unsaturated, as may be the case where a structured soil profile drains with no further addition of water. Transfer of water from layer to layer in the soil is very slow, and as in case 4, can occur only at contact points between aggregates.

A similar description of the two domains is also given by, *inter alia*, Chen and Wagenet (1992a).

In order to predict water movement, the energy (or potential) gradients must be identified in both the micropore and macropore regions. Usually, we express these gradients in terms of the energy per unit weight per unit length in the flow direction. Its units in the SI system are $J N^{-1} m^{-1}$ = m head of the working fluid. The energy per unit weight is here expressed as the hydraulic head of water which cannot be a discontinuous function with distance within continuous water films in the soil. In principle, therefore, a description of water movement in cracking soils should be possible by linking water potentials and potential gradients throughout the soil. It is then possible to build, at least in mathematical−conceptual terms, a complete description of water and solute movement in soils from two basic equations: the Richards equation for water movement and the convection−dispersion equation for solutes.

Richards (1931) gives a full description of the water movement in terms of the Cartesian coordinate system applied to the soil mass under consideration. The equation that bears his name is normally stated for one-dimensional vertical flow (i.e. assuming lateral horizontal flow is negligible) as

$$\frac{\partial \theta}{\partial t} = \frac{\partial}{\partial Z}\left[K(\theta)\,\frac{\partial H}{\partial Z}\right] \tag{1}$$

where

$\partial\theta/\partial t$	= rate of change of water content of the soil
$K(\theta)$	= hydraulic conductivity of the soil which is taken as a function of the water content, θ
H	= hydraulic gradient (i.e. $H = h - Z$, where h is pressure head)
Z	= vertical Cartesian coordinate (e.g. Wagenet, 1990)

Analytical solutions to this equation are only possible for some simple boundary condition problems and usually where simple empirical hydraulic conductivity functions are available, for instance that of Gardner (1958). Generally difficulties arise because of the non-linear nature of the hydraulic conductivity function, which must be overcome by the use of numerical solutions to this equation. Several models have been published which describe soil water flow using such an approach, for instance SWATRE (Belmans *et al.*, 1983) and LEACHN (Hutson and Wagenet, 1991).

Added complications arise in most clay soils due to the swelling and shrinking nature of these soils. The water content of a swelling clay depends on the overburden pressure of the soil above the point considered as well as on the soil water pressure. This leads to an extra potential term which is the overburden potential. As water flows into a swelling material, it expands and the soil particles move. Therefore Darcy's law must be stated in terms of the relative velocity of the water to the soil particles rather than to a fixed datum, as would be the case in a rigid soil (Towner, 1987). Further complications for modelling studies are that the degree of macroporosity will also change through time, as the peds shrink and swell in response to the movement of water in and out of them. Although the physics of the process is well understood, field studies of the process are not common (Bronswijk, 1991).

In water flow studies, but not necessarily in solute flow studies, attention has been concerned with macropores as the route for rapidly percolating and draining water in clay soils. Although macropores represent only a small fraction of the total porosity of clay soils, and have a small surface area, because of the power relationship between laminar flow in a pore and its diameter, they act as the major conduits for water. Childs (1969) shows that for a unit hydraulic gradient, the rate of flow, q, in a cylindrical tube of radius R (sometimes called Poiseuille's law) is

$$q = \left(\frac{\rho g}{8\eta}\right) \pi R^4 \qquad (2a)$$

where

ρ = density of water
η = dynamic viscosity of water
g = acceleration due to gravity

For linear cracks of width D, the relationship is

$$q = \left(\frac{\rho g}{12\eta}\right) D^3 \qquad (2b)$$

When carrying flowing water macropores may be either completely filled of partially filled by water flowing as a surface film down the pore. If a volume of soil contains cracks with porosity e_c and the cracks are partially saturated to degree S_c, then the total flow through the cracks per unit area, again for a unit hydraulic gradient, is given by

$$q = \left(\frac{\rho g}{12\eta}\right) D^2 e_c S_c^v \qquad (2c)$$

where the additional parameter, v, introduced by Germann and Beven (1981), represents the effects of path tortuosity. Large macropores are thus capable of transmitting large amounts of water, and can easily dominate the transport properties of any block of soil containing them (Towner, 1987; Chen and Wagenet, 1992a).

The ease with which macropores can transmit water is illustrated by Figure 5.4, which plots the rate of discharge, q, from eqn (2c) as a function of crack width, D, for a variety of values of percentage saturation and for a crack porosity of 5%. Discharges are shown as mm hr^{-1} fluxes. It is clear that soils can transmit water at a rate far in excess of that normally delivered by rainfall in the United Kingdom. The fact that they do not do so indefinitely in a field situation is a reflection of the fact that naturally cracks decrease in size with depth, that they do not have a free drainage outlet, and so fill up with water, and that the impact of rain is inevitably to cause some sealing of the surface either by particle movement or by ped swelling. Nevertheless, the fact remains that well-developed crack systems in soils have the potential to transmit large volumes of water rapidly.

Critical for the consideration of macropore flow is the process whereby water starts to flow down the macropores. Water arriving at the soil surface as rainfall or irrigation may meet either a crack or an aggregate. The proportion of cracks at the soil surface is small and few, if any, cracks extend vertically from the soil surface to depth in the profile. As a result most rainfall encounters the micropore space at, or very close to, the soil surface and infiltrates into this matrix when it is unsaturated. Even when clay aggregates are saturated they still have the capability to take up some additional water by swelling. The rate of uptake into the micropore space decreases with time during each rainfall event. At some point in time, depending on the rate at which the water is arriving at the soil surface, the water cannot continue to enter the fine pores and excess water flows across the surface of the aggregates (Leeds-Harrison and Jarvis, 1986). In a cracking soil this excess water will then enter a macropore and flow downwards through the profile (see Figure 5.5). During this downward flow the water may infiltrate into as yet unwetted aggregates. If water continues to arrive at the soil surface at this excess rate, then it may be transmitted deep into the profile. In some cases where the soil is particularly dry and the storm event is of short duration the fine pore matrix may be capable of taking up all the water that infiltrates in this way within the top few centimetres of the profile. On cessation of the storm event, water may redistribute within the micropore region. We note that, at this time, water initially at a high potential is adsorbed into the micropore region, where its potential is considerably lower. Water in this state is easily retained by the soil and is relatively immobile. Usually only plant roots in contact with that part of the micropore system can extract this water unless the aggregates (which define individual micropore regions) reach a soil water pressure of zero.

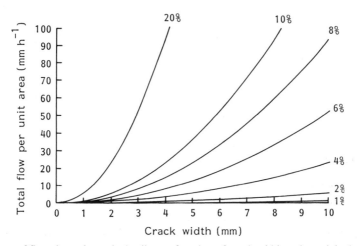

FIGURE 5.4   Rate of flow through cracked soils as a function of crack width under unit hydraulic gradient. Assuming a crack porosity of 5%, values are shown for successive values of crack saturation

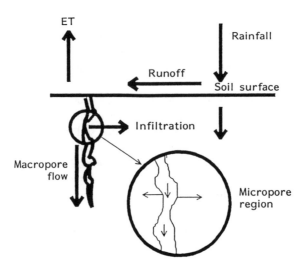

FIGURE 5.5   Water flow routes in a macroporous soil. Runoff into a macropore can wet the profile by
         lateral infiltration into the micropore region (after Leeds-Harrison and Jarvis, 1986)

5.1.2   Solute movement

Solute moves in response to two processes: movement by the moving water and the diffusion of
solute within static water in response to concentration gradients. In principle, dissolved solutes
move down the concentration gradient, so that the tendency is always to equalise the
concentrations. However, because water is also moving within a soil in response to potential
gradients, solutes also move with the water in which they are dissolved.

   In a continuous permeable medium, the movement of a chemical is given by the
convection−dispersion equation (CDE), which describes the combined effects of these two
processes. For one- dimensional vertical flow this is given by (Hutson and Wagenet, 1991)

$$\frac{\partial(\theta c)}{\partial t} + \frac{\partial(\rho_s s)}{\partial t} = \frac{\partial}{\partial z}\left[\theta D(\theta,q)\frac{\partial c}{\partial z} - qc\right] \tag{3}$$

in which

c	= chemical concentration in the liquid phase
s	= chemical concentration in the sorbed phase
ρ_s	= soil bulk density
$D(\theta,q)$	= effective dispersion coefficient
q	= water flux density

The solution of this equation, coupled to that of the Richards equation, can offer a complete
description of water and solute movement in soils, provided that the soil can be considered as a
continuous system. The LEACH family of models, for example, offers such a tool (Hutson and
Wagenet, 1991). However, extra terms must be added to this equation to describe the behaviour
of non-conserved solutes (such as nitrate and pesticides) and to model the uptake of the solute
by the plant and gains or losses of the solute from or to the soil (see for example, Wagenet,

1990; Hutson and Wagenet, 1991). The application of these equations to cracking soils encounters a further complication as the bimodal nature of the soil means that the boundary conditions for the micropore region are difficult to define with any certainty and the conductance of the soil exhibits extreme heterogeneity. The equation has mainly been applied to isotropic soils and is essentially a "sand" physics equation.

When considering the movement of some chemicals of interest, particularly pesticides, it is also necessary to consider the sorption of the chemical on the soil. The sorption distribution coefficient, K_d, controls the relationship between the concentration in a soil, c, and the concentrations in the soil water, s. Normally the sorbed and solution phases are related by a sorption isotherm, the Freundlich equation (Green and Karickhoff, 1990):

$$s = K_f c^N \tag{4a}$$

For many solutes, it is possible to assume that the exponent N is approximately equal to one, so the equation becomes

$$s = K_d c \tag{4b}$$

This equation is the one generally used for studies of pesticide movement, and considerable importance is attached to the value of the sorption coefficient, K_d, which is a property for each solute and describes the propensity of the solute to remain affixed to the soil. Mobile, and hence potentially leachable, pesticides have low K_d values, whereas highly sorbed, immobile pesticides have much higher K_d values. Considerable experimental effort has thus been expended in ascertaining the correct values for this property. It is, however, also affected by soil properties, most notably by the soil organic carbon (OC) content, and thus appears to vary with depth in the soil profile. To give a value independent of soil type, K_{OC} can be estimated from the melting point and solubility of the pesticide.

Applying the convection–diffusion equation to structured soils also encounters the difficulty that the vast majority of water, and hence also of solute movement, occurs close to the surface of the peds. In the same way that water can diffuse only slowly into peds, so solutes generally can also only move slowly. In conventional soil physical models ("sand physics"), equilibration of solute concentrations is assumed to be sufficiently fast compared to the transport velocities for it to be ignored. However, for structured soils this is no longer the case. Contact times between the rapidly moving water across the ped surfaces are short and the rates of solute movement within the peds are slow. For this reason, it is possible for the water moving through macropores to retain solute concentrations close to those they had when they first started to move down the profile.

5.1.3 The solute mass balance

The conservation of mass is fundamental to all descriptions of water and solute movement. In very simple models mass balance provides a useful way of determining the amount of water and solute that goes to drainage and that which is stored within the soil. Thresholds for drainage (or the maximum storage) have to be set, and the imprecise but often used concept of field capacity is usually evoked. In its simplest form the equation is written as

$$\text{Input} - \text{output} = \Delta\text{store} \tag{5a}$$

Such an approach treats the soil as a black box for which no detail of the processes involved are described and for which flow is implied rather than explicitly stated. Such a simple mass balance approach is restricted to non-reactive solutes (such as chloride or bromide, which are for this reason often used as tracers), but for many solutes of practical interest (notably nitrate and pesticides), it is also necessary to describe the transformations occurring within the soil profile, which may remove or add mass to the store:

$$\text{Input} - \text{output} \pm \text{transformations} = \Delta\text{store} \tag{5b}$$

Such a simplified description is most commonly applied to vertical profiles, in which case the terms in the equation are expressed as depths of water. To improve on such a description the soil can be partitioned into layers, so one layer may drain into the layer below it when it reaches field capacity. This has the main advantage that different thresholds (field capacities) can be set for each layer. Such models are often described as "capacitance" and several examples can be found in the literature (Burns, 1974; van der Molen, 1973).

To cope with preferential flow in cracks and fissures, models have to introduce bypass arrangements for water arriving at the soil surface. One method is to transfer a proportion of the water (determined by some means, often empirically) arriving at the soil surface directly to the bottom of the soil profile. Alternatively, a proportion of the incoming water can be transferred to some layer within the profile, so bypassing higher layers. The level of empiricism that has to be introduced into such models means that the need for "calibration" restricts their portability from one situation to another.

5.1.4 Data requirements for solute models

Undoubtedly the simplest data requirement is for the capacitance type of model and for this reason such models are still widely used. Estimates of evapotranspiration and measurement of rainfall inputs on a daily or even monthly basis may be all that is required. Any problems with these models lie in the empiricism that is often included with them when they are modified to cope with water flow processes in the soil.

Evapotranspiration is in fact a component that is frequently troublesome for all soil water and solute movement models. By the combined processes of evaporation from the surface and transpiration by the growing plant, water is removed from the profile. In response to the potential gradients that are set up, both water and solutes can move upwards. However, estimation of the evaporative demand is frequently difficult, and may consume a considerable proportion of the modelling effort involved. In addition, within the soil, the abstraction of water by roots must also be distributed within the soil mass, often on the basis of root habit. The upward redistribution of water from layer to layer cannot be handled by such models without introducing some description of water movement — often an empirical generalisation. In general, however, the estimation of the evaporative component is not a major integral part of solute models and many rely on the use of externally supplied evapotranspiration estimates (such as that provided in the United Kingdom by MORECS; see Thompson *et al.*, 1981).

Rather than take an empirical approach to modelling processes, approaches based on the physics of water flow offer more portability from one situation to another. Data inputs are independently measurable parameters such as hydraulic conductivity and drainable porosity and the water retention curve. However, herein lies one of the drawbacks in physically based models, as these parameters are notoriously difficult to estimate. In particular, the unsaturated hydraulic conductivity function, which varies spatially as well as with water content, is especially difficult to estimate for clay soils. Other physical parameters such as the swelling characteristic for the soil and the sorptivity of the matrix are not parameters that are measured on a routine basis. Procedures for estimation of the properties of two-domain soils are still being developed (e.g. Ankeny *et al.*, 1991: Grant *et al.*, 1991; Chen *et al.*, 1993; Messing and Jarvis, 1993).

This has led to many physical models being matched to field data in an empirical manner such that parameters are chosen to fit observed data rather than being independently measured. It can be argued that where this is the case simpler empirical models of the mass balance or transfer function type may be preferable. These have the main advantage that their construction and execution is usually very swift and they usually tend to be robust in the sense that little instability in iteration is observed, whereas the physics-based differential equations solved numerically can exhibit instability in iterative sequences which are difficult to resolve, particularly where water table fluctuation has to be taken into account.

5.2 MODELS OVERVIEW

Many models have been developed to describe the flow of water and solute in soils, and it is neither possible nor profitable to undertake a complete review of all the models. In practice, then, we concentrate on, and describe in some detail, three models which offer different approaches to the problems of modelling water solute movement. These are SLIM, MACRO and CRACK. In particular these three models offer three different ways of conceptualising the soil water and solute system. Of these, SLIM is applicable to predicting solute behaviour over whole years, and is thus compared with the results from a monitored site at Swavesey, Cambridgeshire (Harris and Parish, 1992). MACRO and CRACK are capable of predicting the short-term behaviour of solutes, and thus these complete and relatively developed models were applied to the issues of modelling solute movement as recorded at the Brimstone Farm site (Cannell *et al.*, 1984; Harris *et al.*, 1984, 1993; Goss *et al.*, 1993). Models of the complexity of MACRO and CRACK can also be run for many years although these then place significant demands on computer time. By contrast, simple models like SLIM can be embedded within other applications, such as GIS-based modelling exercises, in which the model may be run at many sites for many years.

The SLIM (Addiscott and Whitmore, 1991) model offers a simple model with parameters that can be easily estimated, at the cost of a major simplification of the processes of soil water movement. By contrast, more complex models deal with the transport of water in macroporous soils by regarding them as double-porosity materials in which the larger scale macrostructure surrounding the micropore regions in aggregates constitutes a porous material that is superimposed on the smaller scale microstructure. In each region transport of water can be described by equations governing the continuum physics at the appropriate scale (the Richards equation in the micropore region for instance). Source and sink terms, which may be physics based or entirely empirical, then have to be applied to allow the transport of water between the two porosities (see, for instance, Barenblatt *et al.*, 1960). The MACRO model of Jarvis (1991) has used just such an approach, in which the Richards equation is used to solve the movement of water in the soil matrix and simple gravity flow is assumed in the macropore region, with the addition of soil drainage from the saturated zone calculated using seepage potential theory.

However, there is no overall consensus between modellers as to the most efficient way of describing macropore flow. Thus J. L. Hutson and R. J. Wagenet (personal communication) have also attempted to model bypass flow in the model LEACHN by envisaging a profile of several columns, each having a different hydraulic conductivity and providing for water transport between the columns using source and sink terms derived from consideration of the physics of lateral flow, while Gerke and van Genuchten (1993) have presented a model in which the movement of water and solutes in both domains is predicted using standard soil physical (Richards' equation) methods, and Germann and Beven (1985) have discussed the use of kinematic wave theory to predict flow in macropores. Workman and Skaggs (1990) present a scheme in which the movement of water in macropores is described by Poiseuille's law (eqn(2)) and the lateral infiltration by a modified version of Darcy's law determined by a "characteristic" distance between pores, and this volume added an extra term to the solution of the Richards' equation for micropore water movement.

Although it uses the same representation of macropore flow (essentially a drainage equation), the CRACK model (Jarvis and Leeds-Harrison, 1987; Jarvis, 1989) uses a different physical basis for the description of water movement into soil aggregates, that based on infiltration theory, so giving an explicit description of soil water movement in cracked soils. A similar use of infiltration theory has also been described by Davidson (1985) and by Beven and Clark (1986).

Other models that approach the same set of problems include: Armstrong (1983), Armstrong and Arrowsmith (1986), Bronswijk (1988a, 1988b), Chen and Wagenet, (1992a, 1992b), Corwin *et al.* (1991), Edwards *et al.* (1979), Gerke and van Genuchten (1993), Germann and Beven (1981 and 1985), Hoogmoed and Bouma (1980), Steenhuis *et al.* (1990) and Workman and Skaggs (1990), although not all of these consider both water and solute movement.

5.3 SIMPLIFIED SOLUTE MOVEMENT MODELLING: SLIM

Many models have attempted to model solute, and in particular nitrate movement from soils. They range, for example, from the simple leaching equation of Burns (1975) to the complete description of water, heat and solute movement of the SOILN model (Johnsson *et al.*, 1987; Bergström and Jarvis, 1991). However, the majority of these models considers the soil as an homogeneous medium in which the concepts of the Richards' equation and the convection–dispersion equation apply, although they may use one of many simplifications to model them.

Of the many models that attempt to predict solute movement those models deriving from SL3 model of Addiscott (1977) are particularly appropriate to clay soils because of the conceptual subdivision of the soil water into mobile and immobile phases. Although this subdivision is *not* identical to the subdivision into macro- and micropore flow, it is in many ways similar, and so this model offers the prospect of being able to model solute leaching from macroporous soils. Although the pore system may be divided into mobile and immobile water, this does not necessarily mean that a model can simulate preferential flow. In general, the essential requirement is that different solute concentrations can exist in the two pore systems. SLIM does in fact allow for these differences, and so is an acceptable model for cracking soils. The SLIM model has been widely tested and found to offer an acceptable method for estimating the amount of nitrate leached from a profile during the winter months (Addiscott and Bland, 1988; Addiscott *et al.*, 1991; Lord and Bland,1991), although it has not been used widely on clay soils.

SLIM is essentially a management-oriented tool, its main objective being the agricultural management of nitrogen reserves in the soil. It thus uses simplified input parameters that might possibly be available in an advisory situation, and is also suitable for integration with distributed models. In the work reported here we have used the SLIM model of Addiscott and Whitmore (1991) in the extended version, SACFARM, which includes predictions of mineralization and crop uptake described by Addiscott and Whitmore (1987), so that changes in soil mineral nitrogen and losses of nitrate by leaching can be computed on a daily basis. The model has simple parameter requirements and should be applicable to a range of soil types.

The water balance component of the SLIM model divides the soil into layers and the water within each layer is divided into mobile and immobile categories, w_m and w_r which are estimated from the soil moisture characteristic as follows:

$$w_m \geq 0 \tag{6a}$$

except in the top layer where

$$d(\theta_s - \theta_{0.33}) \geq w_m \geq 0 \tag{6b}$$

or for immobile water:

$$d(\theta_{0.33} - 0.5\theta_{15}) \geq w_r \geq 0 \tag{6c}$$

where θ_s, $\theta_{0.33}$ and θ_{15} are the volumetric water contents at saturation, 0.33 and 15 bar respectively and d is the layer thickness, usually 50 mm.

The downward movement of water is controlled by a rate parameter, α, which is the proportion of the mobile water in a layer which moves down to the layer below each day. The value of α, which is estimated from a regression on the percentage clay in the profile, acts as a surrogate for the hydraulic conductivity of the soil (Addiscott and Whitmore, 1991) and has some of the properties of a velocity (D. M. Cooper, Institute of Hydrology, personal communication). Upward movement of water and solute is possible in both phases in response to evaporation from the soil surface. Rapid movement of water to the drains can take place if the top layer becomes saturated, as water reaching the surface in excess of its water-holding capacity thus becomes macropore flow. Nitrate in the top layer is leached via this mechanism but there is assumed to be no interaction with soil layers further down the profile. A version of the model that specifically includes this interaction with macropores has been developed (Hall, 1993).

Solute movement between layers is described within the model in the following way. Water and solute entering a layer are added to the current mobile water and solute, W_m and S_m, without displacing them. The top soil layer may reach saturation, and any water in excess of saturation does not pass through the soil matrix but "bypasses" it and appears at the base of the profile. This "bypass" flow is not the same as flow in the mobile category of water; it constitutes, when it happens, a third category of water in the soil, and it can cause rapid loss of solute applied to the surface of the soil. It is assumed that restricting water entry to the top layer prevents oversaturation in lower layers.

The leaching section of the model then operates in the five stages for each layer:

Stage 1. Incoming water and solute is added to W_m and S_m, and full mixing is assumed to occur.
Stage 2. A proportion (α) of the augmented water in the layer is designated to move to the next layer down.

Stage 3. Half the water designated to move to the next layer down does so, carrying with it the solute it contains. The losses from this layer become the inputs to the next layer down.

Stage 4. Solute and water move between the mobile and immobile categories in the layer. Solute may move by convection or diffusion.

Stage 5. The other half of the water designated to move down to the next layer does so, carrying with it the solute it contains.

The model is structured such that it performs stage 3 for all layers, then stage 4 for all layers and then stage 5 for all layers. As there will always be some mobile water in the soil, unless α is 1.0 or evaporation exceeds rainfall, stages 3, 4 and 5 are performed whether or not any rain falls. The model does not gave an explicit time step but rain and evaporation are presented on a daily basis. This can be changed by using a system that "chops" inputs of rain into smaller aliquots.

In stage 4, convective movement of solute between W_m and W_r is considered before diffusion. If W_r is less that the upper limit shown in equation (1c), water moves from W_m to W_r to make good the deficiency, carrying solute with it in proportion. The limits to lateral movement imposed by diffusion can be described by equalising concentrations between W_m and W_r only partially, using a "hold-back" factor, β, to limit the degree of equalisation. The reallocation of solute is thus given by

$$S_m \text{ (after)} = \beta S_m(\text{before}) + (1-\beta)S_t W_m/(W_m + W_r) \tag{7a}$$

$$S_r \text{ (after)} = \beta S_r(\text{before}) + (1-\beta)S_t W_r/(W_m + W_r) \tag{7b}$$

in which "before" and "after" relate to the pseudo-diffusive transfer and where $S_t = S_m + S_r$. The structure of the subsoil often differs appreciably from that of the topsoil, necessitating separate values of, β, β_t and β_s for topsoil and subsoil respectively.

The nitrate amount within the soil can also be affected by the process of mineralisation and crop uptake. Net mineralisation is computed using a linear relationship with time (Addiscott, 1983), in which the amount of nitrogen, N_t, released in time (t) is given by

$$N_t = k_m t \tag{8}$$

where k_m, the rate constant, is supplied as k_{20} for a fixed temperature of 20°C and adjusted for temperature effects:

$$k_m = k_{20} \exp \langle -B[(T - 273)^{-1} - 293^{-1}] \rangle \tag{9}$$

in which k_{20} is the rate constant at 20°C obtained from incubation experiments. The rate constant is assumed to decline with diminishing volumetric moisture content (θ) only when the soil moisture tension exceeds 0.33 bar (Stanford and Epstein, 1974). The base rate of mineralization is assumed to be uniform within the plough layer, but beneath the plough layer mineralization is assumed to show a sharp exponential decrease with depth.

The model also includes a crop model which simulates dry matter production, nitrogen uptake and root development of winter wheat. Degree-days of soil temperature in the topsoil is the driving variable behind the crop development. The root distribution is used to determine the fraction of water and solute in each layer that can be taken up by the crop. This fraction decreases exponentially with depth. Details are given in Whitmore and Addiscott (1987).

The results of applying this model are illustrated to the Tipple's subcatchment of the Swavesey site (Harris and Parish, 1992). This catchment consists of 56 ha of arable land with cracking clay soils of the Drayton and related soil series (Clayden and Hollis, 1984). The model was run for each of the six fields in the catchment and their patterns of nitrate leaching were then compared to the patterns of nitrate concentrations in the ditch draining from the catchment (Figure 5.6). It can be seen that the model has predicted the patterns of nitrate leaching only moderately well, and in particular fails to predict the "flush" of high nitrate concentrations that has been observed at the start of the drainage season at this and many other sites (Rose *et al.*, 1991).

5.4 MACRO

The MACRO model of Jarvis (1991) attempts to offer a complete model for the movement of water and solutes in a soil. Currently, the model can be used to simulate transport of non-reactive solutes or reactive pesticides. It is intended to link the model to the SOIL-N model of nitrogen transformations (Johnsson *et al.*, 1987) to enable simulation of nitrate leaching in structured soils (N. J. Jarvis, personal communication). The model includes the explicit modelling of the interaction of the growing crop, both in terms of the evaporative demand of water and the uptake of the solute by the crop. However, in this review, the soil−water−solute component of the model is the focus of interest.

The MACRO model subdivides the soil and its water into two components — the macro- and micropore systems. In order to represent the influence of macroporosity, MACRO subdivides each soil layer into two domains, each with their own state variables of degree of saturation, conductivity and flux. Conceptually, then, MACRO, considers the parallel systems of the macropore and micropore. It makes no attempt, however, to describe the soil structure explicitly, but merely accounts for the total porosity involved in the macropores.

Within the micropore regime, the water and solute movement is described using the classical Richards' and CDE systems. MACRO can be run in this mode alone, and so reduces to the standard solutions equivalent to that offered by other models such as the LEACH models described by Hutson and Wagenet (1991).

Within the macropore region, it is assumed that water flows under a unit hydraulic gradient using Poiseuille's law, eqn (2). Water transfer between the macropore and micropore regions is also assumed to occur at a rate given by

$$S_{\mathrm{w}} = \beta K_{\mathrm{b}} \left(\frac{\theta_{\mathrm{b}} - \theta_{\mathrm{mi}}}{\theta_{\mathrm{b}} - \theta_{\mathrm{r}}} \right) \theta_{\mathrm{ma}} \tag{10}$$

where θ is the moisture content, and the subscript ma refers to the macropores, mi to the micropores, b to the boundary and r to the residual condition. The major problem with this relationship is that the mass-transfer coefficient, β, cannot be directly measured, and so must be estimated, normally by fitting the model to an observed data set.

Solute modelling in MACRO follows the standard CDE approach, except that it includes an additional term which describes the interaction between the two phases of flow. This term is given by

$$U_{\mathrm{e}} = \alpha D_0 (c_{\mathrm{ma}} - c_{\mathrm{mi}}) \theta_{\mathrm{ma}} + (S_{\mathrm{w}} c') \tag{11}$$

where D_0 is the diffusion coefficient in free water, c are the solute concentrations and ma and mi refer to macro- and micropore components respectively, as before. The concentration c' depends on the direction of the mass transfer of water, S_{w}, defined previously, referring to the

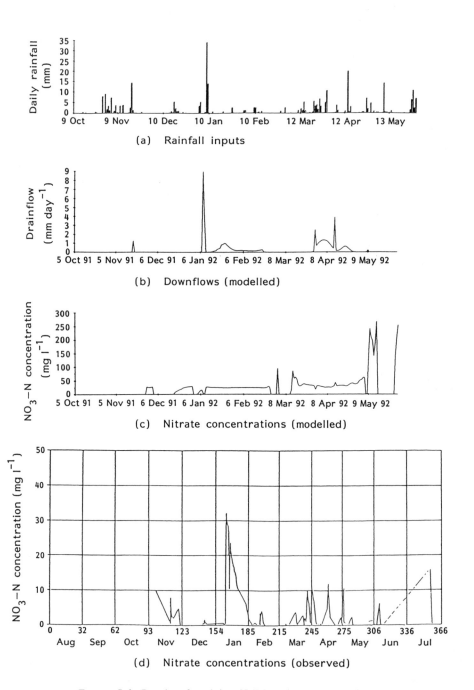

FIGURE 5.6 Results of applying SLIM to the Swavesey site

concentration in the source area. The mass-transfer coefficient, α, is however, another parameter that cannot be directly measured. (Readers should note that although Jarvis uses the parameter name α here, it is *not* the same as the parameter used by Addiscott in his SLIM model.)

The problem of the two critical rate parameters, α and β, controlling the rate of interaction between the solutes and water contents respectively in the two phases, illustrates a common problem with models, that they can define, and hence require values for, parameters that are conceptually well defined but for which no measurement procedures currently exist (see also, for example, Gerke and van Genuchten, 1993, who also comment on the importance of the interaction term between the two domains). Because, at present, these parameters must be estimated by fitting observations to model predictions, this places severe restrictions on the use of the model. The problem is, however, not restricted to the MACRO model, and similar remarks can be made about the α parameter of the SLIM model. The 3.1 version of MACRO, to be released in spring 1994, will contain physically based, although still approximate, descriptions of water and solute exchange between the pore systems (N. J. Jarvis, personal communication).

The outputs from the model can be illustrated by its application to data from the Brimstone Farm site (Harris *et al.*, 1984: Goss *et al.*, 1993). The soil of this site is a cracking clay soil of the Denchworth series. In order to fit the model, it was necessary to set the interaction parameters, α and β, to a low value, 0.05, typical for clay soils, and to give the soil matrix a very low conductivity. In this way the majority of the water and solute movement is concentrated in the macropore component.

Results for a 14 day period in December 1989 are shown in Figure 5.7. In it, the predicted values for drainflow, water table position and nitrate concentrations in the drainwater are plotted and compared with the observed performance reported by Rose *et al.* (1991). It can be seen that the model offers a fairly good fit to the observed data, apart from failing to reproduce the initially high concentrations of nitrate. The model predictions in Figure 5.7 were derived by assuming that nitrate could be treated as a non-reactive solute. Strictly speaking this is incorrect, but it is considered that for a short period in December nitrogen transformations (mineralisation, denitrification and crop uptake) are sufficiently small that they can be safely ignored. However, the same assumption could not be made if the model were to be run over a longer period. It is thought this lack of fit is due to the unavailability of accurately measured initial conditions. The nitrate concentrations in the profile at the start of the simulation were in fact inferred from general information, and were not site-specific. However, the amounts of solute involved are also very small, as the observed very high concentrations are associated with very low (almost negligible) flows. The pattern of dilution and recovery through each rainfall event is, however, reproduced very well, and this is considered to be excellent confirmation that the model adequately reproduces the behaviour of this site.

5.5 CRACK

An alternative conceptualisation of the soil water process is included in the CRACK model described by Jarvis and Leeds-Harrison (1987) and Jarvis (1989). This was developed as an hydrologic model for structured cracking clay soils to provide simulations for single rainfall events, but has been subsequently modified to model multiple wetting and drying events over a complete annual cycle. The hydrologic processes described in the model are illustrated

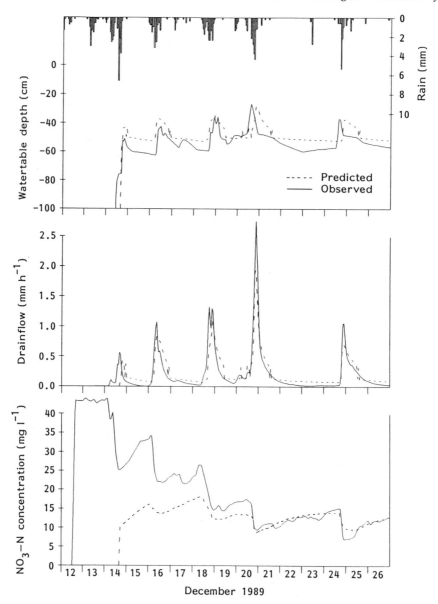

FIGURE 5.7   Results of MACRO applied to the Brimstone Farm site

schematically in Figure 5.8 (after Leeds- Harrison and Jarvis, 1986, and comparable to the schema provided by Germann and Beven 1985). Water reaching the soil surface may infiltrate, but when the rainfall intensity (1) exceeds the infiltration capacity of the peds at the soil surface (2), it begins to enter the cracks (3). The amount of water flowing in the cracks (5) may be reduced by sorption losses to peds within the soil (4). If a water table in the cracks rises above rain depth, drain outflow occurs (6).

CRACK is conceptually different from previous models in that it is built around a model of the entry of water into soil peds. The macroporosity of the soil is thus explicitly described by parameters which define the mean size of the peds in each layer. Rates of water movement and solute exchanges are described by infiltration theory, rather than the conventional use of the Richards equation. It thus includes the important limitation that it assumes that downwards water movement through the soil matrix is sufficiently small that it can be ignored. It does not, therefore, like MACRO, default to the Richards' solution for a uniform soil if invoked with no macroporosity. It is thus a model that deals explicitly and exclusively with cracking clay soils in which these assumptions are likely to be valid. However, because it attempts to describe the interaction of water and solutes in physical terms, it has no parameters that cannot at least in principle be measured explicitly. In this respect it offers a significant improvement over the current version of the MACRO model.

In CRACK, water arriving at the soil surface is partitioned into that entering the soil matrix and that excess which flows into cracks, using the first term of Philip's (1957) infiltration equation, in which the sorption capacity, I_p is given by

$$I_p = \frac{S}{2t^{0.5}} \tag{12}$$

in which S is the ped sorptivity and t is time since the start of rainfall (see also Grant et al., 1991). The time to ponding, t_p for rainfall rate, R, is then defined by

$$t_p = \left(\frac{S}{2R}\right)^2 \tag{13}$$

The sorptivity, S, is itself a linear function of the soil water content between field capacity, θ_f, and wilting point, θ_w:

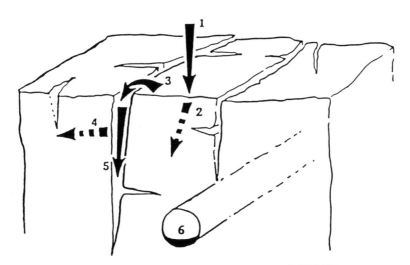

FIGURE 5.8 Water movement components of CRACK

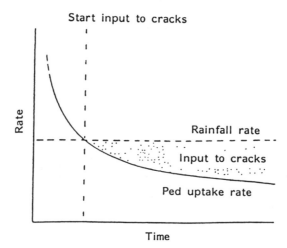

FIGURE 5.9   Partitioning of rainfall into flow of water into cracks and uptake by aggregates (after Jarvis, 1989, Fig. 3)

$$S = S_w \left[1 - \left(\frac{\theta - \theta_w}{\theta_f - \theta_w} \right) \right]$$ (14)

where S_w is the sorptivity at the wilting point.

Water flow in cracks is initiated when the uptake rate into peds at the soil surface, as defined by eqn (12), is exceeded by the rainfall intensity. Rainfall intensity is thus an important variable in the CRACK model, and must be supplied either as a combination of rainfall amounts and durations (for simulations with time-steps of a day) or by using hourly rainfall data. It is a characteristic of models of this degree of complexity that rainfall data with a time resolution of the order of 1 hour are required.

As soon as crack water flow is started, peds take up water not only through the soil surface but also through internally wetted surfaces. The amount of wetted surface area increases with time, so that the normal decay in infiltration rate below a wetted surface is compensated for by the generation of new wetted surfaces. The same infiltration equations, (9) to (11) are used for all subsurface layers, after accounting for the fractional aggregate wetted surface area, a, in each layer. For layers below the water table, a is 1, but above the water table a is estimated from the field studies of Bouma and Dekker (1978) to be 0.05. Water infiltrating into peds in any given layer is assumed to be stored entirely within that layer. If the layer reaches field capacity, any excess is transmitted to the layer below. If the bottom layer reaches its field capacity moisture content, any further excess is assumed to be lost as deep seepage.

In the crack model, soil structural status in the model is dependent on the moisture status of the soil. The volume of cracks is a function of the difference between the initial moisture content in each layer and the "field capacity" of moisture content (Figure 5.9). It is assumed that the vertical movement is small compared to the swelling and shrinking in the horizontal dimension.

The height of the water table in the cracks is controlled by the balance between the input at the soil surface and the loss due to ped water uptake within the soil and drain outflow, calculated using the seepage analysis described by Youngs (1980). The rate of rise or fall is also a function of the crack volume in the layer in which the water table is located. If the water table rises to the surface, any further rainfall input is immediately lost as surface runoff. Water thus moves from the soil surface into both the aggregates and the macropores, is lost from the macropores by seepage and by drainage, but can be removed from the aggregates only by root extraction.

Solute may be transported in the water flowing in the macropores and in the model these act as sources and sinks for solute in the aggregates. The velocity of water flow is small in the aggregates, and in addition swelling may mean that diffusion is the dominant transport mechanism for solute movement. To describe the movement of solutes within the aggregates, CRACK considers them to consist of layers of equal volume, and Fick's law is used to describe the movement of solute between the layers. Based on Addiscott's (1982) analysis of diffusion within cubic aggregates, whose mean size is given by the crack spacing, the rate of diffusion, q_s, between adjacent segments is given by

$$q_s = D_0 \, \theta_p \, k * A_s \, \frac{dc}{dx} \tag{15}$$

in which

D_0 = diffusion coefficient in free water
$k*$ = impedance factor
A_s = mean cross-sectional area per unit soil volume
dc/dx = solute gradient
θ_p = volumetric water content of the aggregates

Diffusive exchange of solutes is assumed to occur between the outermost segments of the aggregates and the water-filled cracks, using eqn (15) modified to take account of the degree of saturation of the cracks. The rate of exchange is thus proportional to the fraction contact area. In the model as currently implemented, this is the only mode of solute loss from the soil. In developments currently under way, other processes that affect the total quantity of solutes (denitrification and mineralisation of nitrate and degradation and adsorption of pesticides) are being included in the model.

The modelling of interchange of solute between the aggregates and the water in the macropores defines the concentrations of water in the macropores, which is then moved by mass transport. Solute in the drainage water thus originates from within the profile and reaches the drain by flow in the cracks. Mass balances for solute and water between the layers in the soil allow the mass flow of solute in the drainage water to be found, and thus its concentration.

Results are given in which the CRACK model is used to predict the movement of solutes from the Brimstone Farm experiment, using the same test period as used for the previous results obtained from the MACRO model (Figure 5.10). The results are obtained by assuming that nitrate can, for a short period in December, be treated as a conserved solute. The parameters used to derive the results were, as far as was possible, physically measured in the field. Mean ped size (and hence also mean crack spacing) in the topsoil (0–20 cm) was measured as 10 cm, but was larger, mean size 20 cm, in the layer beneath (20–40 cm). Ped sorptivity at the wilting point, S_w, was measured as 12.0 mm $t^{-0.5}$. The stable, minimum macropore volume was 5% in the topsoil and 2% in the subsoil, the hydraulic conductivity of the macropores was 40 mm h^{-1}. Data were available for a period during which drain discharge was recorded at half-hourly

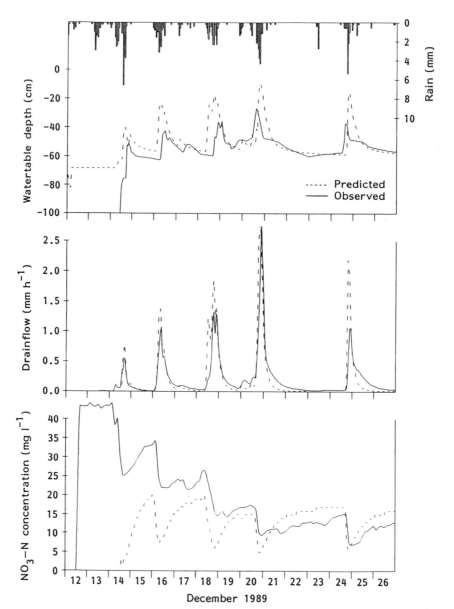

FIGURE 5.10   Results of applying CRACK to the Brimstone Farm site

increments, water table position (within the macropores) recorded at hourly increments and
water samples had been taken from the drainflow at a minimum of every 3 hours. The detailed
field data (Rose *et al.*, 1991) thus providing an ideal validation check for the model.

The results show a very good fit for both the hydrology and the pattern of solute movement.
Both the drain discharge and water table levels in the macropores are reproduced very well by
the model, so there is no doubt that the model has reproduced the hydrology of the site very

well. The simultaneous prediction of several components of the model while using field measured and not fitted parameters, indicates that the model is indeed appropriate to the field situation. It is often possible to get a model to fit the behaviour of one variable by adjusting parameters, but it is rarely possible to do so with multiple variables. The results also predict the nitrate concentrations in the drain discharges with a fair degree of accuracy. As with the predictions from MACRO, the model fails to predict the initially very high solute concentrations, and this is again due to the lack of field data for the initial conditions.

The results of CRACK and MACRO are very similar, both in their behaviour and in the goodness of fit. The main difference between them is that the critical parameters, those governing the interactions between micro- and macropore movement are derived from field observations in CRACK, but are derived by fitting in MACRO.

The value of such a model is also illustrated by its use to evaluate the impact of varying cultivation treatments on solute leaching patterns. Leeds-Harrison *et al.* (1992) have shown that the model predicts greater mass removal of solute from the soil profile where the soil structure is fine compared to a coarse structured soil. Such trends are observed in field studies, for instance by Czapar and Kanwar (1991) and Harris *et al.* (1984). It is noted, however, that the opposite phenomenon can arise when the solute is taken up at the soil surface during intense rainfall and transported rapidly to the drains in coarse structured soils, as observed by Isensee and Sadeghi (1993). In their studies a well-structured no-till treatment transported pesticide rapidly to the drains under intense rainfall events in contrast to a conventionally ploughed soil. This aspect of solute transport where the solute is not diffused through the soil has not been modelled by CRACK at the present time, but both CRACK and MACRO should be able to predict this phenomenon by a suitable choice of input parameters.

5.6 DISCUSSION

Three approaches to the modelling of water and solute movement in cracking clay soils have been demonstrated, based on conceptualisations of the soil and water domains. Thus SLIM uses a simple split into mobile and immobile water which imitates to some extent the division into macro- and micropore flow. MACRO, on the other hand, uses soil physics to describe the flow of water in parallel but interacting systems representing the two domains and CRACK develops a conceptualisation of water movement in cracked soils in which the dominant processes are the movement of water through the macropores and its infiltration into discrete and unconnected soil peds. These three models can all describe the behaviour of a cracking soil with varying degrees of success, but no one model can be said to be definitive for all situations. Each model has its strengths and its weaknesses, and the choice of any model for a specific investigation requires the matching of the model to the application.

Indeed, successful modelling requires three major steps: the formulation of the aims of the modelling exercise, the conceptualisation of the processes operating and the derivation of the necessary field input and validation parameters. Successful models are thus more than simply those models that reproduce a given set of validation data well, but which also reflect the structure of the physical system that they attempt to represent. Where a model is based on a detailed study of the physics of the phenomena under study, then a successful modelling exercise can learn from the lack of fit between model predictions and observed behaviour. The results presented in this paper have shown that accurate modelling of solute behaviour is possible using models that are designed specifically for such soils by the inclusion of the cracks in the conceptual scheme. However, conventional models that ignore the cracking behaviour of soils perform much less well. It is thus important that where cracks are an important part of the soil, models are used that take note of this fact.

In general, where the model is developed entirely from a combination of physical theory and independently measured parameters, there is then no question of "adjusting" or "fitting" the model to reproduce system behaviour; the model can be in some sense verified (Beven, 1987). Where, however, the models fail to reproduce the observations, three possible sources can be identified:

1. The inadequacies of the conceptual representation of the model, notably the omission of critical processes.
2. The failure to record the observed data sufficiently accurately. Errors in the observation data are frequently ignored in modelling studies, but can in fact lead to major problems if models attempt to reproduce spurious data. This problem is a major issue for empirically derived models, such as response function models, which derive their parameters from observational data.
3. The failure to measure the parameters of the model accurately. This is a particular problem for models that can easily postulate parameters for which there are no techniques for physical observation. Typical among these parameters are the "tortuosity" parameter for flow in macropores or the interaction term between macro- and micropore components.

The resolution of these problems depends on the ability of the modeller to identify the correct sources of the error and the robustness of the modelling procedure. Where the model uses, for example, least-squares fitting to "calibrate" the critical parameters, then it is often very difficult to identify the sources of error. Only with independently measured parameters is it possible to verify, or more properly falsify (Popper, 1935), the model. However, it is almost invariably the case that the best soil physical model cannot, even in the laboratory, produce an exact fit to observational data. Indeed, most models fail to reproduce one of the main features of observational data, which is its inherent variability (Armstrong *et al.*, 1993), with even the best data fitting the theoretically expected single value only in a statistical sense. Verification of a model in a formal sense is thus never possible, as no model will exactly fit a whole sequence of observations. In the end, the process of comparing models to observations generates a degree of confidence in the model, what might be called statistical verification (Harvey, 1969).

All the models and results mentioned so far in this chapter have been developed in the context of predicting the behaviour of a single, representative, soil profile. Major problems arise, however, when attempting to apply the results to wider contexts. Even models as complex as CRACK and MACRO make no allowance for the fact that soils are laterally inhomogeneous. The vertical inhomogeneity and the bimodal nature of cracking clay soils is described by the parameters of the soil, but there is no attempt to take the lateral variation of these parameters.

Extending the results beyond a single profile to a catchment scale thus presents a challenge that is as yet unresolved. It is possible to devote the enormous amounts of computer resource required to reproduce models such as CRACK or MACRO on multiple sites, and then to use lateral transfer models to represent the third dimension, using a scheme such as that used, for example, in the SHE package (Abbott *et al.*, 1986a, 1986b). However, in this procedure, there is both redundancy, if replicated parallel sites behave in very similar ways for much of the time, and problems with parameterisation, and in particular the problem of deriving meaningful parameters that describe the procedure of areas over which the intrinsic physical parameters may be expected to vary in some way (Beven, 1989). In macroporous soils, the parameterisation problem is particularly acute, as the representative elementary volume (REV) that describes the soil may be of the order of metres, and thus larger than conventional measurement techniques, and yet smaller than convenient units for model development (Beven and Clark, 1986).

As a result of these considerations, at different scales different conceptualisations may be more appropriate (Beven, 1987). The challenge is thus to use the detailed physical understanding of soil profile processes derived from the detailed physically based models to inform and calibrate much more general models which can be applied on the wider scale. This is one of the subjects of our current research.

ACKNOWLEDGEMENTS

The authors are grateful to Andrew Matthews and Andrew Portwood for their work in producing many of the results in this paper, to colleagues, notably Steve Rose, of ADAS SWRC for providing the data from the Swavesey experimental site and to Graham Harris of ADAS SWRC and John Catt and Roger Howse of ICAR for making data from the Brimstone Farm site available.

The authors are also grateful to Nick Jarvis of the University of Uppsala, for his comments on the manuscript, for the many helpful discussions on the development of these models and for providing a copy of his MACRO model. The (mis)uses to which it was put, however, remain the authors' responsibility!

Financial support for this work from the UK Ministry of Agriculture, Fisheries and Food (MAFF) and the AFRC/NERC joint initiative on solute transport in soils and rocks is gratefully acknowledged.

The models mentioned in this paper are available as follows:

SLIM from Dr T. Addiscott, ICAR-Rothamsted, Harpenden, Herts, AL5 2JQ;

CRACK from Dr P. B. Leeds-Harrison, School of Agriculture, Food and Environment, Cranfield University, Silsoe Campus, Silsoe, Bedford, MK45 4DT, or Dr A. C. Armstrong, ADAS Land Research Centre, ADAS Gleadthorpe, Meden Vale, Mansfield, Notts NG20 9PF;

MACRO from Prof. N. J. Jarvis, Department of Soil Science, SLU, Box 7072, 75007 Uppsala, Sweden.

REFERENCES

Abbott, M. B, Bathurst, J. C., Cunge, J. A., O'Connell, P. E. and Rasmussen, J. (1986a). "An introduction to the European Hydrological System — Système Hydrologique Européen. SHE, 1. History and philosophy of a physically-based modelling system", *J. Hydrol.* **87**. 45–59.

Abbott, M. B, Bathurst, J. C., Cunge, J. A., O'Connell, P. E. and Rasmussen, J. (1986b). "An introduction to the European Hydrological System — Système Hydrologique Européen. SHE, 2. Structure of a physically-based distributed modelling system", *J. Hydrol.* **87**. 61–77.

Addiscott, T. M. (1977). "A simple computer model for leaching in structured soils", *J. Soil Sci.*, **28**, 554–563.

Addiscott, T. M. (1982). "Simulating diffusion within soil aggregates: a simple model for cubic and other regularly shaped aggregates", *J. Soil Sci.*, **33**, 37–45.

Addiscott, T. M. (1983). "Kinetics and temperature relationships of mineralisation and nitrification in Rothamsted soil with differing histories". *J. Soil Sci.*, **34**, 343–353.

Addiscott, T. M. and Bland, G. J. (1988). "Nitrate leaching models and soil heterogeneity", in *Nitrogen Efficiency in Agricultural Soils* (Eds D. S. Jenkinson and K. A. Smith), pp. 394−408, Elsevier Applied Sciences, Barking.

Addiscott, T. M. and Whitmore, A. P. (1987). "Computer simulation of changes in soil mineral nitrogen and crop nitrogen during autumn, winter and spring", *J. Agric. Sci., Cambridge*, **109**, 141−157.

Addiscott, T. M. and Whitmore, A. P. (1991), "Simulation of solute leaching in soils of differing permeabilities", *Soil Use and Management*, **7**, 94−102.

Addiscott, T. M., Bailey, N. J., Bland, G. J. and Whitmore, A. P. (1991). "Simulation of nitrogen in soil and winter wheat crops: a management model that makes best use of limited information", *Fertilizer Res.*, **27**, 305−312.

Ankeny, M. D., Ahmed, M., Kaspar, T. C. and Horton, R. (1991). "Simple field method for determining unsaturated hydraulic conductivity", *Soil Sci. Soc. of Am. J.*, **55**, 467−470.

Armstrong, A. C. (1983). "A heuristic model of soil water regimes in clay soils in the presence of mole drainage", *Agricultural Water Management*, **6**, 191−201.

Armstrong, A. C. and Arrowsmith, R. (1986). "Field evidence for a bi-porous soil water regime in clay soils", *Agricultural Water Management*, **11**, 117−125.

Armstrong, A. C., Castle, D. A. and Matthews, A. M. (1993). "Drainage models to predict soil water regimes in drained soils", *Transactions, Workshop on Subsurface Drainage Simulation Models, 15th International Congress on Irrigation and Drainage (ICID)*, The Hague, The Netherlands, 1993, pp.95−105.

Barenblatt, G. E., Zheltov, I. P. and Kochina, I. N. (1960). "Basic concepts in the theory of seepage of homogeneous liquids in fissured rocks", *Journal of Applied Mathematical Methods USSR*, **24**, 1286−1303.

Belmans, C., Wesseling, J. G. and Feddes, R. A. (1983). "Simulation model of the water balance of a cropped soil: SWATRE", *J. Hydrol.*, **63**, 271−286.

Bergström, L. and Jarvis, N. J. (1991). "Prediction of nitrate leaching losses from arable land under different fertilization intensities using the SOIL−SOILN models", *Soil Use and Management*, **7**, 79−85.

Beven K. J. (1987). "Towards a new paradigm in hydrology", *in Water for the Future: Hydrology in Perspective*, IAHS Publication 164 pp, 393−403.

Beven K. J. (1989). "Changing ideas in hydrology — the case of physically-based models", *J. Hydrol.* **105**, 157−172.

Beven K. J. and Clark, R. T. (1986).'On the variation of infiltration into a homogenous soil matrix containing a population of macropores", *Water Resour. Res.*, **22**, 383−388.

Beven, K. J. and Germann, P. F. (1981). "Water flow in soil macropores. II. A combined flow model", *J. Soil Sci.*, **32**, 15−29.

Beven, K. J. and Germann, P. F. (1982). 'Macropores and water flow in soils", *Water Resour. Res.*, **18**, 311−325.

Bouma, J. (1981). "Soil morphology and preferential flow along macropores". *Agric. Water Management*, **3**, 235−250.

Bouma, J. and Dekker, L. W. (1978). "A case study on the infiltration into dry clay soil. I. Morphological observations", *Geoderma*, **20**, 27−40.

Bronswijk, J. J. B. (1988a). "Modelling of water balance, cracking and subsidence of clay soils", *J. Hydrol.*, **97**, 199−212.

Bronswijk, J. J. B. (1988b). "Effect of swelling and shrinkage on the calculation of water balance and water transport in clay soils", *Agric. Water Management*, **14**, 185−193.

Bronswijk, J. J. B. (1991). "Relation between vertical soil movements and water-content changes in cracking clays". *Soil Sci. Soc. of Am. J.*, **55**, 1220−1226.

Burns, I. G. (1974). "A model for predicting the redistribution of salts applied to fallow soils after excess rainfall or evaporation", *J. Soil Sci.*, **25**, 165−178.

Burns, I. G. (1975). "An equation to predict the leaching of surface-applied nitrate", *J. Agric. Sci. Cambridge*, **85**, 443−454.

Cannell, R. Q., Gross, M. J., Harris, G. L., Jarvis, M. G., Douglas, J. T., Howse, K. R. and Le Grice, S. (1984). "A study of mole drainage with simplified cultivation for autumn-sown crops on a clay soil. 1. Background, experiment and site details drainage systems, measurements of drainflow, and summary of results, 1978−80", *J. Agric. Sci, Cambridge*, **102**, 539−559.

Chen, C. and Wagenet, R. J. (1992a). "Simulation of water and chemicals in macropore soils. Part 1. Representation of the equivalent macropore influence and its effect on soil water flow", *J. Hydrol.*, **130**, 105−126.

Chen, C. and Wagenet, R. J. (1992b). "Simulation of water and chemicals in macropore soils. Part 2. Application of linear filter theory", *J. Hydrol.*, **130**, 127−149.

Chen, C., Thomas, D. M., Green, R. E. and Wagenet, R. J. (1993). "Two-domain estimation of hydraulic properties in macropore soils", *Soil Sci. Soc. of Am. J.*, **47**, 680−686.

Childs, E. C. (1969). *An Introduction to the Physical Basis of Soil Water Phenomena*, Wiley-Interscience, London, 493 pp.

Clayden, B. and Hollis, J. M. (1984). *Criteria for Differentiating Soil Series*, Technical Monograph No. 17, Soil Survey of England and Wales, Harpenden.

Corwin, D. L., Waggoner, B. L. and Rhoades, J. D. (1991). "A functional model of solute transport that accounts for bypass", *J. Environ. Quality*, **20**, 647−658.

Czapar, G. F. and Kanwar, R. S. (1991). "Field measurement of preferential flow using subsurface drainage tiles", in *Preferential flow*, (Eds. T. J. Gish, and A. Shirmohammadi), Proceedings, ASAE National Symposium, Chicago, December 1991, pp. 122−128.

Davidson, M. R. (1985). "Numerical calculation of saturated−unsaturated infiltration in a cracked soil", *Water Resour. Res.*, **21**, 709−714.

Edwards, W. M., van der Ploeg, R. R. and Ehlers, W. (1979). "A numerical study of the effects of noncapillary-sized pores upon infiltration", *Soil Sci. Soc. of Am. J.*, **43**, 851−856.

Gardner, W. R. (1958). "Some steady state solutions to the unsaturated flow equation with application to evaporation from a water table", *Soil Sci.*, **85**, 228−232.

Gerke, H. H. and van Genuchten, M. T. (1993). "A dual-porosity model for simulating the preferential movement of water and solutes in structured porous media", *Water Resour. Res.*, **29**, 305−319.

Germann, P. F. and Beven, K. (1981). "Water flow in soil macropores. III. A statistical approach", *J. Soil Sci.*, **32**, 31−39.

Germann, P. F. and Beven, K. (1985). "Kinematic wave approximation to infiltration into soils with sorbing macropores", *Water Resour. Res.*, **21**, 990−996.

Gish, T. J., Shirmohammadi, A. and Helling, C. S. (1991). "Modelling preferential movement of agricultural chemicals", in *Preferential flow*, (Eds. T. J. Gish, and A. Shirmohammadi), Proceedings, ASEA National Symposium, Chicago, December 1991, pp. 214−222.

Goss, M. J., Howse, K. R., Lane, P. W., Christian, D. G. and Harris, G. L. (1993). "Losses of nitrate-nitrogen in water draining from under autumn-sown crops established by direct drilling or mouldboard ploughing", *J. Soil Sci.*, **44**, 35−48.

Grant, S. A., Jabro, J. D., Fritton, D. D. and Baker, D. E. (1991). "A stochastic model of infiltration which simulates 'macropore' soil water flow", *Water Resour. Res.*, **27**, 1439−1446.

Green, R. E. and Karickhoff, S. W. (1990). "Sorption estimates for modelling", in *Pesticides in the Soil Environment: Processes, Impacts, and Modelling*, (Ed. H. H. Cheng), pp. 80−101, Soil Science Society of America, Madison, Wisconsin.

Hall, D.G.M. (1993). "An amended functional leaching model applicable to structured soils. I. model description", *J. Soil Sci.*, **44**, 579−88.

Harris, G. L. and Parish, T. (1992). "Influence of farm management and drainage on leaching of nitrate from former floodlands in a lowland clay catchment", in *Lowland Floodplain Rivers: Geomorphological Perspectives*, (Eds. P. A. Carling and G. E. Petts), pp. 203−216, John Wiley, Chichester.

Harris, G. L., Goss, M. J., Dowdell, R. J., Howse, K. R. and Morgan, P. (1984)."A study of mole drainage with simplified cultivation for autumn sown crops on a clay soil. 2. Soil water regimes, water balances and nutrient loss in drain water", *J. Agric. Sci. Cambridge*, **102**, 561−581.

Harris, G. L., Howse, K. R. and Pepper, T. J. (1993). "Effects of moling and cultivation of soil-water and runoff from a drained clay soil", *Agric. Water Management*, **23**, 161−180.

Harvey, D. (1969). *Explanation in Geography*, Arnold, London, 521 pp.

Hillel, D. (1980). *Fundamentals of Soil Physics*, Academic Press, New York, 413 pp.

Hoogmoed, W. D. and Bouma, J. (1980). "A simulation model for predicting infiltration into cracked soil", *Soil Sci. Soc. of Am. Proc.*, **44**, 458−461.

Hutson, J. L. and Wagenet, R. J. (1991) "Simulation nitrogen dynamics in soils using a deterministic model", *Soil Use and Management*, **7**, 74−78.

Isensee, A. R. and Sadeghi, A. M. (1993). "Impact of tillage practice on runoff and pesticide transport", *J. Soil and Water Conservation*, **48**, 523−527.

Jarvis, N. J. (1989) *CRACK: a model of water and solute movement in cracking clay soils*, Report 159, Swedish University of Agricultural Sciences, Department of Soil Sciences, Uppsala, 38 pp.

Jarvis, N. J. (1991). *MACRO — a model of water movement and solute transport in macroporous soils*, Swedish University of Agricultural Sciences, Department of Soil Science, Reports and Dissertations No. 9, Uppsala, Sweden, 58 pp.

Jarvis, N. J. and Leeds-Harrison, P. B. (1987). "Modelling water movement in drained clay soil. I. Description of the model, sample output and sensitivity analysis", *J. Soil Sci.*, **38**, 499−509.

Johnsson, H., Bergström, L., Jansson, P.-E. and Paustian, K. (1987). "Simulated nitrogen dynamics and losses in a layered agricultural soil", *Agriculture, Ecosystems and Environment*, **18**, 333−356.

Jury W. A., Sposito, G. and White, R. E. (1986). "A transfer model of solute transport through soil. 1. Fundamental concepts", *Water Resour. Res.*, **22**, 243−247.

Lawes, J. B., Gilbert, J. H. and Warrington, R., (1982). *On the Amount and Composition of the Rain and Drainage Water Collected at Rothamsted*, Williams, Clowes & Sons, London.

Leeds-Harrison, P. B. and Jarvis, N. J. (1986). "Drainage modelling in heavy clay soils", in *Proceedings of the International Seminar on Land Drainage, Helsinki, July 9−11*, 1986. (Eds. J. Saavalainen and P. Vakkilainen), University of Helsinki, pp. 198−221.

Leeds-Harrison, P. B., Vivian, B. J. and Chamen, W. C. T. (1992). "Tillage effects in drained clay soils", ASAE Paper 92-2648.

Lord, E. and Bland, G. (1991). "Leaching of spring-applied fertilizer nitrogen: measurement and simulation", *Soil Use and Management*, **7**, 110−114.

Luxmore, R. J., Jardine, R. H., Wilson, G. V., Jones, J. R. and Zelazny L. W. (1990). "Physical and chemical controls of preferred path flow through a forested hillslope", *Geoderma*, **46**, 139−154.

Marshall, T. J. and Holmes, J. W. (1988). *Soil Physics*, 2nd edition, Cambridge University Press, Cambridge, 374 pp.

Messing, I. and Jarvis, N. J., (1993). "Temporal variation in the hydraulic conductivity of a tilled clay soil as measured by tension infiltrometers", *J. Soil Sci.* **44**, 11−24.

Philip, J. R. (1957). "The theory of infiltration. 4. Sorptivity and algebraic infiltration equations", *Soil Sci*, **84**, 257−264.

Popper, K. R. (1935). *Logik der Forschung*, English translation: *The Logic of Scientific Discovery*, Hutchinson, London, 1959, and subsequent revisions.

Richards, L. A. (1931). "Capillary conduction of liquids through porous mediums", *Physics*, **1**, 318−333.

Rose, S. C., Harris, G. L., Armstrong, A. C., Williams, J. R., Howse, K. R. and Tranter, N. (1991). "The leaching of agrochemicals under different agricultural land uses and its effect on water quality", *IAHS Publ.*, **203**, 249−257.

Sposito, G., White, R. E., Darrah, P. R. and Jury, W. A. (1986). "A transfer function model of solute transport through soil. 3. The convection−dispersion equation", *Water Resour. Res.*, **22**, 255−262.

Stanford, G. and Epstein, E. (1974). "Nitrogen mineralisation — water relationships in soils', *Soil Sci. Soc. of Am. Proc.*, **38**, 103−107.

Steenhuis, T. S., Parlange, J.-Y. and Andreini, M. S. (1990). "A numerical model of preferential solute movement in structured soils", *Geoderma*, **46**, 193−208.

Thompson, N., Barrie, I. A. and Ayles, M. (1981). "The Meteorological Office rainfall and evaporation calculation systems: MORECS", Hydrological Memorandum No. 45, The Meteorological Office, Bracknell, 69 pp.

Towner, G. D. (1987). "Formulae for calculating water flow in macro-pores in soil", *Int. Agrophysics*, **3**, 5−15.

van der Molen, W.H. (1973). "Salt balance and leaching requirement", in *Drainage Principles and Applications*, Volume II: *Theories of Field Drainage and Watershed Runoff*, pp. 59−100, International Institute for Land Reclamation and Improvement, Publication 16, Wageningen, The Netherlands.

Wagenet, R. J. (1990). "Quantitative prediction of the leaching of organic and inorganic solutes in soil",

Phil. Trans. R. Soc. Lond. B, **329**, 321−330.

White, R. E., Dyson, J. S., Haigh, R. A., Jury, W. A. and Sposito, G. (1986). "A transfer function model of solute transport through soils. 2. Illustrative applications", *Water Resour. Res.*, **22**, 248−254.

Whitmore, A. P. and Addiscott, T. M. (1987). "A function for describing N uptake dry matter production and rooting by wheat crops", *Plant and Soil*, **101**, 51−60.

Workman, S. R. and Skaggs, R. W. (1990). "PREFLO: a water management model capable of simulating preferential flow", *Trans. Am. Soc. of Agric. Engrs*, **33**, 1939−1948.

Youngs, E. G. (1980). "The analysis of groundwater seepage in heterogeneous aquifers", *Hydrol. Sci. Bull.*, **25**, 149−165.

Youngs, E. G. and Leeds-Harrison, P. B. (1990). "Aspects of transport processes in aggregated soils", *J. Soil Sci.*, **41**, 665−675.

SECTION III

ECOSYSTEM PROCESSES IN SOLUTE MODELLING

6

Predicting Nitrate Concentrations in Small Catchment Streams

M. J. WHELAN,[1] M. J. KIRKBY

School of Geography, University of Leeds, UK

and

T. P. BURT

School of Geography, University of Oxford, UK

6.1 INTRODUCTION: PATHWAYS AND PROCESSES

Average stream water nitrate (NO_3^-) concentrations in many catchments have increased markedly in recent years (e.g. Burt *et al.*, 1988). Concern over the consequences of these trends for aquatic ecosystems, drinking water quality and agricultural production has prompted an increase in research activity oriented towards identifying the pathways by which nitrogen (N) may enter streams (cf. Addiscott *et al.*, 1991; Burt *et al.*, 1993). Several pathways exist. These include direct wet and dry deposition from the atmosphere, N in overland flow (often as ammonium or organic material associated with sediment losses in soil erosion) and point sources such as sewage outfalls or from accidental spillage from silage or slurry tanks on farms. By far the most important pathway for N losses from agricultural catchments, however, is the diffuse leaching of nitrate from soils.

Two groups of processes control nitrate leaching in soils, both of which are affected by the management of the soil, by the soil type and by the effects of the environment on conditions in the soil. The first group encompasses those processes which control the availability of nitrate in the soil (no nitrate — no nitrate leaching loss). The second includes those processes which affect the solute transport mechanism, i.e. soil and hillslope hydrological processes (no transport — no leaching, regardless of soil nitrate concentrations). This chapter describes a process-oriented approach to modelling nitrate leaching losses from small catchments which attempts to take both of these groups of processes into account.

The model has the acronym MONITOR (model of nitrogen turnover and runoff) and is described in detail by Whelan (1993). It is deterministic (for a given set of inputs it will give the

---

[1] Current Address: Department of Biological Sciences, University of Exeter, Exeter, EX4 4PS, UK

Solute Modelling in Catchment Systems. Edited by Stephen T. Trudgill
© 1995 John Wiley & Sons Ltd

same output), runs on a fixed daily time-step and has individual fields or a small catchment as the basic spatial unit. It requires standard daily meteorological data (rainfall, air temperature, humidity deficit, radiation balance and wind velocity) as the basic driving variables, along with some parameters that describe the soil characteristics and the area, topography and layout of the fields within the catchment.

6.2 ROOT ZONE HYDROLOGY

A successful estimation of the gains, losses and stores of moisture in the root zone (i.e. interception of rainfall by vegetation, throughfall, evaporation of intercepted water from the canopy, evapotranspiration from the soil and soil drainage) is an essential prerequisite to modelling root zone solute budgets. In order to make these estimates it is first necessary to characterise the soil in terms of its hydraulic properties. This is most commonly done by estimating the saturated hydraulic conductivity and by deriving a moisture retention curve for different horizons. As the soil matric potential ψ decreases water is drawn out of progressively smaller pores, causing a decrease in the volumetric moisture content (θ). From this relationship it is possible to estimate the proportion of the total pore space occupied by different ranges of pore size.

In MONITOR the total pore volume within the root zone is divided into four fractions (*vbigpor*, *xstore*, *ystore* and *resid*) covering different pore size ranges (Figure 6.1). The moisture content of each fraction, expressed as a mm depth equivalent, is accounted for on a daily basis. The residual pores (*resid*) represent those very small pores which remain water filled at the nominal permanent wilting point (PWP) of $\psi = -1500$ kPa. Water is tightly held within these pores and is effectively unavailable to plant uptake and transpiration. The *vbigpor* fraction refers to large cracks and macropores, i.e. those pores which remain air filled at a matric potential selected to represent the "field capacity", e.g. at $\psi = -10$ kPa, and down which water may drain rapidly under some conditions in "bypass" flow. Water in pores with sizes in the range between the boundaries of these two fractions is sometimes referred to as "plant available" moisture. In MONITOR this range is further subdivided between larger and smaller pores (*xstore* and *ystore* respectively). *Xstore* is always exploited first by the crop because water is held within it at lower tensions (i.e. less negative pore water pressures). When *xstore* is water filled the soil presents no appreciable water stress to the crop so that actual evapotranspiration (*aet*) = potential evapotranspiration (*pet*). *Ystore* refers to pores with sizes ranging between those of *resid* and *xstore* which are exploited by the crop only when *vbigpor* and *xstore* are empty and which, when exploited, cause a soil moisture related resistance response in the crop such that *aet* < *pet*. The lower the value of *ystore*, the harder it is for plants to imbibe soil water which is progressively held at increasingly higher tensions (i.e. more negative pore water pressures) in increasingly smaller pores until the point is reached (at the PWP) when they are unable to obtain enough water to maintain cell turgor and consequently wilt. The threshold point at which *aet* begins to fall below *pet* is uncertain and will vary depending on the atmospheric evaporative demand and the distribution of moisture (and, therefore, $K(\theta)$) around the roots (e.g. Denmead and Shaw, 1962). Arbitrarily it may be defined as the split of all the pores in the plant available moisture range between *ystore* and *xstore*, in a ratio of 3:2 respectively. Progressive drying from saturation will result in the emptying of the pores in each of these classes in the order *vbigpor* → *xstore* → *ystore*; *ystore* will always take the value *ystoresat* until *xstore* = 0. Similarly, *xstore* will always equal *xstoresat* unless

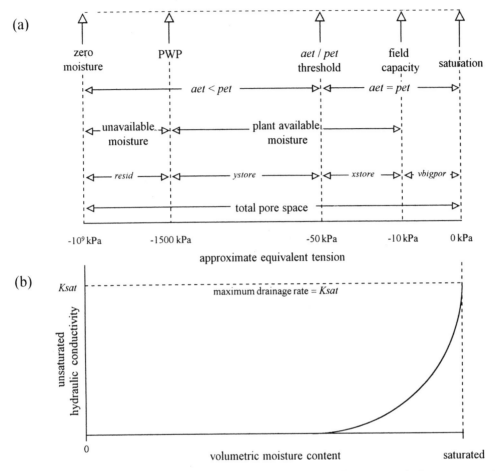

FIGURE 6.1 Schematic conceptualisation of (a) the division of total soil pore space in the root zone and (b) the dependence of unsaturated hydraulic conductivity on water content

$vbigpore = 0$, although there may be an exception to this latter condition if "bypass" (macropore) flow occurs.

Total daily rainfall is adjusted for interception losses depending on the time of year and the stage of crop development, and the soil moisture store is updated by canopy throughfall. Evapotranspiration losses from the soil are estimated using a modification of the meteorological office rainfall and evaporation calculation system (MORECS) described by Thompson *et al.* (1981). This model is based on the Penman–Monteith equation (Penman, 1948; Monteith, 1965) and accounts for the effects of crop leaf development on transpiration and net radiation, the effects of crop height on the aerodynamic resistance presented to water vapour flux from the surface to the atmosphere and the effects of soil moisture content on the resistance presented by the crop and bare soil surface to moisture exchange with the atmosphere. Values for leaf area index and crop height are assumed to change linearly between predefined marker values for sowing date, crop emergence date, time of maximum canopy development, harvest date for arable crops (which also represents the point at which grassland species begin to decline) and date at which grassland decline ceases. Marker dates and values may be altered according to typical cropping practice in the area under consideration. The principal deviation of MONITOR from the MORECS scheme is in its conceptualisation of soil water dynamics.

It is generally held that root zone transport processes occur predominantly in the vertical direction and can, therefore, be adequately described in one dimension. If we assume that, after infiltration, the matric potential gradient in the wetted part of a homogeneous soil approaches zero (i.e. uniform moisture content in the profile) then, from Darcy's law, for one-dimensional vertical flow the hydraulic gradient can be approximated by the gravity component of the total potential gradient (i.e. -1), provided that no shallow water table exists. The vertical drainage flux, q (i.e. the volume discharged per unit cross-sectional area perpendicular to the direction of flow, per unit time), is therefore equal to the unsaturated hydraulic conductivity, i.e.

$$q = K(\psi) \tag{1}$$

This is called the steady-state gravity flow equation (e.g. Jury *et al.*, 1991). It has been shown to be a reasonable approximation to the flow regime of high-frequency irrigation systems where water storage changes are minimal (Rawlins, 1974), although, in reality, most field soils will rarely, if ever, have uniform moisture content and potential, and therefore will rarely have steady-state water flow (e.g. Jury *et al.*, 1991). However, in considering the drainage flux from the bottom of the profile this may not be important since matric potential gradients are generally lower than at the surface. In addition, drainage generally takes place only when the soil profile is reasonably wet. If the gradient in matric potential is high then, in general, the very low unsaturated hydraulic conductivity associated with the dry part of the soil will limit the flux. The fact that most soils are non-homogeneous, contain horizons with different hydraulic characteristics, have non-uniform root density and root hydraulic resistance and may have a capillary fringe associated with a shallow water table (e.g. Hillel *et al.*, 1976; Cassel and Nielsen 1986) are assumed to be relatively unimportant for the estimation of solute and water flux through the lower boundary of the root zone at a daily time-step and at the spatial scale of a field or small catchment. The unsaturated hydraulic conductivity is approximated by a functional relationship between $K(\theta)$ and the moisture content (van Genuchten, 1980) based on the moisture-retention characteristics of the soil in question. Once the soil moisture content has been adjusted for daily drainage loss, it can be used to drive the N turnover model.

6.3 N SUPPLY: THE INTERNAL SOIL N CYCLE

The focus for modelling the availability of nitrate to leaching loss is the internal soil N cycle (Figure 6.2). Most N in soils is organic and its dynamics are intricately associated with the decomposition of soil organic matter (SOM). Organic N may be broken down by soil microorganisms to release inorganic N in the form of ammonium (NH_4^+), which itself may be transformed by microbially mediated oxidation to nitrite (NO_2^-), most of which is usually further oxidised to nitrate. N accumulation in soils is principally via atmospheric deposition of mineral N, the fixation of atmospheric gaseous N (N_2) by certain soil microorganisms and, in agricultural systems, by the addition of fertiliser (which might be organic, i.e. manure, or inorganic, i.e. nitrate and/or ammonium). Both nitrate and ammonium may be taken up by plants and also by soil microorganisms (here referred to as immobilisation) which use inorganic N to synthesise biomass proteins. Similarly, both nitrate and ammonium may be lost to the

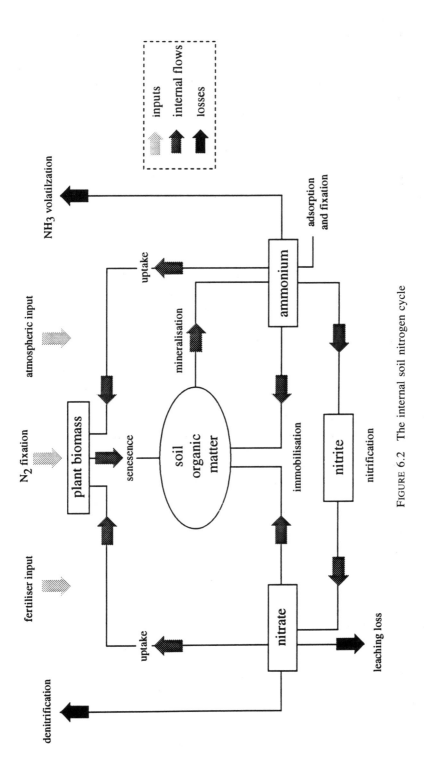

FIGURE 6.2 The internal soil nitrogen cycle

atmosphere by different processes, respectively termed denitrification and ammonia volatilisation. Ammonium, being positively charged, is attracted to the predominantly negatively charged soil solid phase and, in addition, because of its size and shape, can become lodged within the lattice structure of clay minerals (adsorption and fixation respectively). Nitrate, on the other hand, has a negative charge and is, for the most part, repelled by the solid phase, which makes it much more mobile should the soil solution move under drainage.

6.3.1 Mineralisation and immobilisation of N

In MONITOR the root zone is compartmentalised into a number of conceptual layers (*nlayers*) of equal depth. The uppermost layer (*ca.* 20 cm) represents the "topsoil" of agricultural soils with higher organic matter content, higher mineral nutrients and permeability and lower bulk density. Since, in general, most organic matter and, therefore, N is found in the topsoil, simulated N supply by the soil N cycle is restricted to this layer. Within it N may be biologically and chemically transformed (on a mass of substrate or product per unit mass of dry soil basis) and may enter and exit in various forms via the upper and lower boundaries. N transformations, except for "sink" processes such as plant uptake, denitrification and leaching, are not accounted for explicitly in the layers below the topsoil. Although restricting the simulation of decomposition and N mineralisation to the top layer is a gross simplification of the distribution of these processes (cf. Cassman and Munns, 1980) it serves as a useful first approximation.

Because almost all of the soil N transformation processes are mediated by the soil microflora and fauna the key to understanding the controls on the balance and rates of these processes lies in understanding the soil biota. The soil contains a huge variety of organisms, all of which interact with one another and probably play some part in the N cycle. Since the behaviour of each group of organisms is complex and difficult to model the soil microflora are often bundled together into a conceptual "black box" called the microbial biomass. Laboratory and field investigations give insight into the general response of the microbial biomass to different levels of substrate availability and environmental conditions. These general response relationships are used as the basis for the prediction of nitrate supply in MONITOR.

In the case of SOM decomposition and the mineralisation and immobilisation of inorganic N the amount of energy (organic carbon compounds) available to the microbial biomass is assumed to be the factor that most limits the metabolic activity of the heterotrophic organisms involved (e.g. Jenkinson and Rayner, 1977). Soil organic carbon (C) dynamics are, therefore, modelled in parallel with N.

Because SOM is a composite of a number of different compounds, each behaving differently, it is fractionated into conceptual pools of different availabilities to the decomposers according to the scheme of van Veen and Paul (1981) and Veen *et al.* (1985) (Figure 6.3). The pools marked "residue" represent fresh plant material (dead roots, leaf litter and reincorporated cereal straw) and the rest represent fractions of the native SOM or humus. Despite the fact that SOM decomposition involves a complex series of enzymically mediated reactions, overall rates can be described, successfully in many cases, by relatively straightforward kinetics (Paul and Clark, 1989). Decomposition and N mineralisation − immobilisation turnover (MIT) is assumed to approximate first-order kinetics (e.g. Jenkinson and Rayner, 1977; van Veen and Paul, 1981; Wagenet and Hutson, 1987): the rate of decomposition is proportional to the concentration of the substrate under consideration. This assumption is largely based on observations of weight loss or by monitoring the evolution of $^{14}CO_2$ from ^{14}C labelled material.

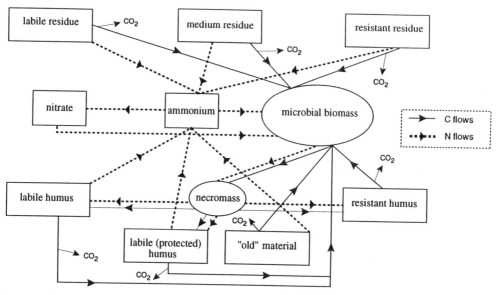

FIGURE 6.3   A schematic representation of the soil organic C and N decomposition component in MONITOR (based on the scheme of van Veen *et al.*, 1985)

Thus, assuming an optimal soil environment for microbial activity, the daily degradation of C substrate in each pool may be described by

$$\frac{dC}{dt}(p,t) = k(p) \cdot C(p,t) \tag{2}$$

$$C(p,t) = C(p,t-1) - \frac{dC}{dt}(p,t) \tag{3}$$

where $C(p,t)$ and $C(p,t-1)$ are the substrate concentrations remaining in pool p at time t and $t-1$ respectively (days) and where $k(p)$ is the decomposition rate constant for pool p (day^{-1}). The decomposition characteristics of each pool are described by individual rate constants. The portion of decomposed C from each pool used for microbial biosynthesis is represented with an efficiency factor *eff(p)* (0 to 1). Thus, provided mineral N is not limiting, the microbial biomass accrues C at a rate

$$totCin(t) = \sum_{p=1}^{7} eff(p) \frac{dC}{dt}(p,t) \tag{4}$$

where *totCin(t)* is the sum of the assimilated C from each of the seven pools, per day. Daily CO_2 evolution *(totCO2(t))* may therefore be described by

$$totCO2(t) = \sum_{p=1}^{7} (1-eff(p)) \frac{dC}{dt}(p,t) \tag{5}$$

which may be used to monitor the overall activity of the biomass over time by comparison with observations of CO_2 evolution.

Similar decomposition schemes have been used by, for example, Parton *et al.* (1987), Jenkinson and Rayner (1977) and Nordmeyer and Richter (1985). An interesting feature of the scheme used in MONITOR is its explicit emphasis on the importance of physical protection of a part of the native organic matter. This is believed to be important in phenomena such as the mineralisation flush, which is often observed following soil disturbance (e.g. ploughing, freeze − thaw or severe drying and rewetting). In such situations some of the material that is biologically "palatable" but physically protected by the soil matrix becomes exposed to the soil biota and may be decomposed, causing an increase in mineralisation.

Ammonification may be described using eqns (2) and (3) if $C(p,t)$ and $C(p,t\text{-}1)$ are replaced by $N(p,t)$ and $N(p,t\text{-}1)$, which refer respectively to concentrations of organic N in each pool at time t and $t\text{-}1$. The total N mineralised is thus

$$totNmin(t) = \sum_{p=1}^{7} \frac{dN}{dt}(p,t) \qquad (6)$$

which is instantaneously added to the ammonium pool ($NH(t)$). All organic N is assumed to be mineralised before it can be assimilated by the heterotrophic microbial biomass.

On any day, the amount and availability of C and the C/N ratio of each pool determines the net N MIT after allowing for suboptimal soil temperature and moisture content constraints, efficiency in C assimilation from each pool and the production of CO_2. The addition of fresh plant residues with large amounts of available C and with relatively low concentrations of available N (high C/N ratios) will result in the net immobilisation of soil inorganic N in the microbial biomass. Material with low C/N ratios will generally cause net mineralisation. In the decomposition of complex, composite substrates, the rate and direction of MIT will often change as decomposition progresses.

Total biosynthetic N demand (*imdemt*) is determined by the availability of C and the assumed C/N ratio of the microbial biomass (*CNbug*) according to

$$imdemt = totCin(t)/CNbug \qquad (7)$$

Ammonium is assimilated preferentially over nitrate in accordance with current consensus in the literature (e.g. Stanier *et al.*, 1977). If the concentration of soil ammonium is capable of satisfying this demand, i.e. if $NH(t) > imdemt$, then no immobilisation demand is made on the soil nitrate and nitrite pools. However, if $NH(t) < imdemt$ then all the soil ammonium pool will be immobilised and the remainder of *imdemt* will be met by nitrate and nitrite. If *imdemt* exceeds the total soil mineral N then all the mineral N is immobilised and overall microbial biosynthesis becomes N limited. In this case

$$totCinl(t) = (NH(t) + NO3(t) + NO2(t)) \cdot CNbug \qquad (8)$$

where *totCinl(t)* is the C assimilated by the biomass if N is limiting. It is assumed that the difference between *totCin(t)* and *totCinl(t)* consists of C already catabolised by enzymic hydrolysis and can, therefore, simply supplement the labile humus pool. This is in accordance with the idea that excess available C is immobilised as polysaccharides when N is limiting.

6.3.2 Environmental influences

As well as being governed by the quantity and quality of available substrate, MIT is strongly influenced by environmental factors, in particular temperature, pH, soil moisture content and aeration. There are several different ways in which the effects of changes in these variables manifest themselves, including change in population dynamics, change in metabolic rates (and/ or metabolic pathways, e.g. aerobic to anaerobic), change in the types of organism operating, change in enzyme synthesis and activity and change in abiotic chemical kinetics. Stress factors and their manifestations seldom act independently of one another and it is usually difficult to interpret and isolate the interactions between them. The broad effect of environmental variables, with only an implicit reference to actual mechanisms, is expressed in MONITOR, using reduction factors $(0-1)$ which depress the process rate below some defined optimum should unfavourable conditions prevail.

It is assumed that decomposition rates increase with temperature up to some optimum, above which rates are inhibited (e.g. Cassman and Munns, 1980). Various workers have established optimal soil water potentials for collective microbial activity and N mineralisation which are generally between -10 and -50 kPa, corroborating early suggestions (e.g. Greaves and Carter, 1920) that the optimum was around "field capacity". If high moisture contents are sustained, zones of anaerobiosis may develop once available O_2 is consumed in respiration. Although many organisms (principally bacteria) may be able to make the switch from aerobic to anaerobic respiration, provided that there are suitable alternative electron acceptors available, anaerobiosis inhibits the activity of organisms such as fungi and actinomycetes (Vinten and Smith, 1993), resulting in a lower rate of decomposition under anaerobic conditions than under aerobic conditions (e.g. Tenny and Waksman, 1930; Reddy *et al.*, 1980). The proportion of the optimum rate at saturation is uncertain and probably ranges from between 0.4 (Reddy *et al.*, 1980) and 0.6 (McConnaughey and Bouldin, 1985). At low moisture contents microbial activity is inhibited because of restrictions in bacterial mobility and in the transport of nutrients, as well as the direct influence of induced internal turgor stresses in the microbes on overall metabolism. Implicitly it is assumed that fungi, by virtue of their hyphae, can access available soil moisture at low potentials more readily than bacteria, which become restricted by thinning water films (Paul and Clark, 1989; Brown, 1990).

Microbial death rates are assumed to be related to substrate availability and moisture stresses (cf. Smith, 1979). Dead biomass is distributed between the native SOM pools in accordance with the assumed cellular composition of the biomass.

6.3.3 Nitrification, denitrification and plant uptake

Nitrification, denitrification and plant uptake are all accounted for explicitly in MONITOR. Under moderate climatic conditions nitrification is generally considered to be faster than ammonification in many agricultural soils, i.e. ammonification is the rate-limiting step in the production of nitrate. Nevertheless, explicit account is taken of the nitrification process (rather than simply assuming all N mineralised is nitrified) since nitrification reactions are mediated by a narrower range of soil organisms that tend to be more sensitive to soil environmental conditions than the ammonifiers. This can lead to ammonium accumulation. First-order reaction kinetics are assumed and both steps in the oxidation of ammonium to nitrate ($NH_4^+ \rightarrow NO_2^- \rightarrow$ and $NO_2^- \rightarrow NO_3^-$) are modelled explicitly. pH, soil temperature and soil moisture status are assumed to affect the metabolic rate of the nitrifying organisms. Nitrifiers respond to changes in temperature and soil moisture content in a similar way to the general response of oxidative ammonifiers, although functions representing the response of ammonium and nitrite oxidising bacteria to the soil environment are defined separately so as to account for known differences under some conditions. Soil nitrifiers are particularly sensitive to pH change (e.g. Paul and Clark, 1989) and this is accounted for explicitly in MONITOR.

The mean rate of plant uptake of either nitrate or ammonium, U_x (mg N cm^{-1} root length day^{-1}) may be described by the hyperbolic function

$$U_x = \frac{U_{max} \cdot C_x}{K_m + C_x} \tag{9}$$

where C_x is the concentration of the ion under consideration (mg N kg^{-1}), U_{max} is the maximum uptake rate (mg N cm root length^{-1} day^{-1}) and K_m is the half-saturation constant (mg N kg^{-1}). In order to estimate the daily uptake rate on a unit mass of dry soil basis U_x must be multiplied by the root length per unit mass, *rden* (cm kg^{-1}). The depth distribution of root length density over time is described by a crude empirical function using the same marker dates defined for functions describing leaf area and crop height development. It is assumed that root densities diminish exponentially with depth, which constrains N uptake from the subsoil layers. The effects of very low soil water contents on mineral N uptake (due to a combination of physiological stress, mineral metabolism and restrictions in the movement of solute from the bulk soil to the roots) have been included using a drought reduction factor, which reduces the rate of N uptake when soil moisture deficits are high. In the absence of universal relationships on the effect of other nutrients on N uptake, it is assumed that they are not limiting.

In some soils large losses of N are believed to occur via denitrification (the dissimilatory reduction of nitrate and nitrite to nitrous oxide (N_2O) and nitrogen gas (N_2) by a range of facultative anaerobic microorganisms). When the partial pressure of O_2 is low, nitrate and nitrite serve as alternative electron acceptors in the oxidation of reduced carbon compounds. Since it is principally an anaerobic process, denitrification is strongly controlled by soil oxygen concentration (which, in turn, is affected by soil moisture content). Nitrate and nitrite availability, temperature, pH and the presence of a readily oxidised electron source, such as organic carbon or reduced sulphur compounds, are also important factors (e.g. Reddy and Patrick, 1983; Burton and Beauchamp, 1985).

In accordance with observed patterns, hyperbolic kinetics (see eqn (9)) are used to describe the rate of nitrate loss with nitrate concentration in MONITOR, provided the degree of anaerobiosis is not limiting. Reduction factors are used to account for the influence of temperature and soil moisture content (a surrogate for the degree of anaerobiosis). Since most denitrifiers are facultative they can metabolise organic matter under both aerobic and anaerobic conditions and probably make the metabolic switch to using nitrate or nitrite as the terminal electron acceptor when O_2 concentrations fall below a threshold level. Since they are usually already present in the soil when conditions become suitable for denitrification, their response can be quite rapid once the available oxygen in the soil has been utilised by respiration. This is exacerbated at higher temperatures since the solubility of oxygen is reduced and microbial respiration is enhanced (e.g. Grant, 1991).

It is now fairly widely accepted that in well-aggregated soils perennial anaerobic or semi-anaerobic microsites exist (even under low soil moisture contents) in which some denitrification probably takes place (e.g. Arah and Smith, 1989). Denitrification usually remains low below a threshold moisture content but above this threshold rates may increase dramatically, even with relatively small changes in water content following rainfall or flooding (Sextone *et al.*, 1985). The denitrification reduction factor *Drf* is described using

$$Drf = podry + (1 - podry) \exp[-(100 - Se)/c] \tag{10}$$

where *podry* is the proportion of the optimum rate at very low soil moisture contents (0–1), *Se* is the percentage saturation and *c* is a dimensionless curve parameter describing the rate of decrease in *Drf* at lower soil moisture contents.

The availability of an organic substrate is often ranked closely after the availability of oxygen in terms of its limiting effects on denitrification (e.g. Burton and Beauchamp, 1985). Ideally, denitrification kinetics should be examined within the same framework as aerobic respiration (Grant, 1991), including all the factors that enhance decomposition rates such as freeze–thaw, wetting and drying and litterfall events (Groffman and Tiedje, 1991). Because of the high organic matter content, the topsoil is assumed to be the primary site for denitrification in MONITOR. Conveniently, Burford and Bremner (1975) observed that denitrification capacity was linearly related to mineralisable organic carbon in a study of 17 different soils. They also estimated that approximately 1 mg of C was required to reduce 1 mg of $NO_3^- - N$ to N_2O and N_2, although some variation was observed depending on the ratio of N_2O to N_2 evolved. C availability limits may, therefore, be approximated fairly easily, subject to the constraints of temperature and soil moisture, using eqn (4). Since it is assumed that the same organisms act as denitrifiers and decomposers, the carbon available for denitrification (*dncdem*) is approximately equal to *totCin(t)*. Provided the demand for nitrate reduction (calculated from nitrate availability and water content) does not exceed *dncdem* then carbon is assumed to be non-limiting. If this demand is greater than *dncdem* then it may be assumed that *dncdem* is the limiting factor. An exponential decrease in the concentration of organic matter with increasing depth is assumed and subsoil denitrification is reduced accordingly.

6.4 LEACHING

There are two broad approaches to the deterministic modelling of leaching processes in soils (e.g. Hall, 1993). One (e.g. Wagenet and Hutson, 1987) is to use the classical equations describing the continuum physics of porous media (Richards' equation for water transport and the advection dispersion equation (ADE) for solute movement). The other is to use functional models, which usually involve dividing the soil up into a number of conceptual horizontal compartments and describing the processes within the interactions between each compartment (e.g. Burns, 1974; Addiscott, 1977). The latter approach, whilst less rigorous, is often simpler and easier to apply to field soils. In any case both approaches may require calibration of parameters using empirical data.

The occurrence of structural aggregates in many soils contributes to the exhibition of bimodality in the frequency distribution of the pore sizes, with a large number of pores with small diameters (mostly within aggregates) and a large number with big diameters (the inter-aggregate cavities). In these soils the larger pores act as the principal avenues for water and solute losses and are consequently labelled the "mobile" pore fraction. The small intra-aggregate pores are termed the "immobile" or retained fraction. In MONITOR, as a first approximation to the division of total pore space between these fractions, water held within *vbigpor* and *xstore* is considered "mobile" while that held within *ystore* and *resid* is considered "immobile" (Figure 6.1). Such a division of the pore volume has been accounted for in modelling solute movement using both the continuum physics approach (e.g. van Genuchten and Wierenga, 1976) and the conceptual "layer" approach (e.g. Addiscott, 1977; Addiscott and Whitmore, 1991).

As in Addiscott's (1977) model the vertical transfer of nitrate (Fv, mg N m^{-2} layer^{-1} day^{-1}) between layers, in MONITOR, is assumed to occur with the water flux itself (by advection) in the mobile pore fraction:

$$Fv = C_{\mathrm{m}} \cdot q \qquad (11)$$

where C_{m} is the mass concentration in mg N l^{-1} (mg N mm water^{-1} m^{-2}) of nitrate N in the mobile fraction of each layer and q is as previously defined.

Between drainage events solute in the remaining mobile pore fraction may partially equilibrate with that in neighbouring pores in the immobile fraction by diffusion along concentration gradients. The equilibrium concentration (*eqcon*, mg^{-1}) in each layer can be written as

$$eqcon = \frac{M_{\mathrm{m}} + M_{\mathrm{i}}}{(xstore + ystore + 0.5\ resid)/nlayers} \qquad (12)$$

where M_{m} and M_{i} are the mass of nitrate N (mg N m^{-2} layer^{-1}) in the mobile and immobile fractions of each layer respectively and where only half of the residual pore volume is included because of anion exclusion (due to electrostatic repulsion) from the smallest pores (after Addiscott and Whitmore, 1987). Solute will move by diffusion from regions of high concentration to regions of low concentration. If the concentration of nitrate in the mobile fraction exceeds that in the immobile fraction (C_{i}) then the rate of change in the concentration of nitrate in the immobile fraction, *eqrate*, may be expressed by

$$eqrate = \alpha\ (eqcon - C_{\mathrm{i}}) \qquad (13)$$

where α is an equilibration "retardation" factor $(0-1)$ related to the rate of diffusion between the two fractions. When $\alpha = 1$ complete concentration equilibration between the solute in each fraction is approached and when $\alpha = 0$ then the interaction between the two phases becomes negligible. C_{i} increases by *eqrate* and M_{m} is adjusted by mass balance. If $C_{\mathrm{i}} > C_{\mathrm{m}}$ then the flux will occur in the opposite direction.

Explicit account may be taken of hydrodynamic dispersion (i.e. the spreading of a solute peak as it travels through tortuous pathways of varying lengths and at varying velocities in the soil) by introducing an additional conceptual parameter β $(0-1)$ to hold back that part of the solute in the mobile fraction most affected by viscous drag (i.e. in the smaller pores). β is included as a product on the right-hand side of eqn (11). In situations where $q < (xstore/nlayers)$ only a fraction of the mobile solution in any layer will be displaced by q. The rest will remain behind. This will introduce an unwanted artificial (numerical) dispersion and may be accounted for using the parameter *nd* in eqn (11), which becomes

$$Fv = \beta \cdot C_{\mathrm{m}} \cdot (q + nd) \qquad (14)$$

where $nd = (xstore/nlayers) - q$ if $(xstore/nlayers) > q$ and where $nd = 0$ if not.

This model is able to reproduce a wide range of steady-state breakthrough curve forms, i.e. curves generated by plotting the concentration of solute in water draining from the base of a column of soil after a pulse of solute is applied to the top with time or drainage volume (Figure 6.4).

6.5 LATERAL RUNOFF

An essential prerequisite to the simulation of N dynamics at a catchment scale is the satisfactory approximation of the catchment water balance. Since the principal conveyor of nitrate from the catchment is the stream, this essentially means predicting the stream hydrograph which, in small catchments (e.g. < 1 km^{-2}), can be reasonably approximated by the areally averaged hillslope hydrograph.

Lateral flow on hillslopes flow may be generated at permeability discontinuities or at a water table if present. If the direction of flow is assumed to be along the line of maximum slope (i.e. water movement is induced predominantly by gradients in gravity potential) then the flow lines will be orthogonal to the contours resulting in a convergence of flow in hollow areas and at slope feet and a divergence of flow in others (e.g. Anderson and Burt, 1978). Such a topographic concentration of moisture has long been recognised as having an important influence on catchment hydrological response (Hewlett, 1961), manifested in the variable source area (VSA) model. Lateral discharge in MONITOR is described using a formulation of TOPMODEL (Beven and Kirkby, 1979), a semi-lumped hydrological forecasting model which accounts for the influence of topography and antecedent moisture content on the catchment hydrological response. A temporal saturation deficit (D, mm) is defined for the zone below the root zone, adjusted by simple moisture accounting of input (drainage from the root zone store) and output (lateral runoff,j), where j(mm) at any point on the hillslope may be defined by

$$j = \frac{K_0 \tan S}{a} \exp\left(-\frac{D}{m}\right) \tag{15}$$

where K_0 is the saturated hydraulic conductivity, S is the local ground surface slope, a is the area drained per unit contour length and m is a curve parameter related to the soil permeability. The higher the deficit the lower the runoff response to a unit of rainfall input, so that wet winter catchments respond readily to moderate storms whereas dry summer catchments do not. A fundamental assumption of TOPMODEL is that, under steady state, water input and the consequent runoff produced are spatially uniform over the whole catchment, resulting in a spatially uniform adjustment of D. Thus small catchments may be modelled in a lumped fashion, provided suitable parameter values are selected. The latter may be achieved by calibration with a limited set of data.

6.6 NITRATE IN LATERAL FLOW

Modelling the behaviour of solutes in catchment drainage water in a process-oriented fashion is not easy, principally because of uncertainties in the inherently variable subsurface properties and water content (including the position of the water table). Precise internal mechanisms remain a mystery in most catchments and usually the only information available about how the

(a)

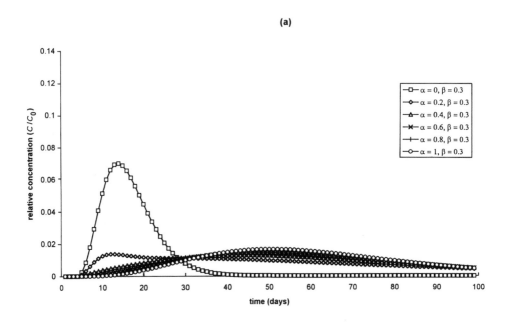

(b)

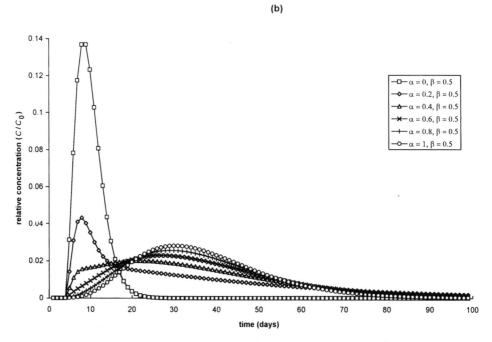

FIGURE 6.4 A range of steady-state breakthrough curves produced using the MONITOR leaching scheme for a solute pulse scenario in a five-layer soil (corrected for numerical dispersion). (a) $\beta = 0.3$, (b) $\beta = 0.5$, (c) $\beta = 0.7$. In all curves the rate of flux = 4 mm day^{-1}

(c)

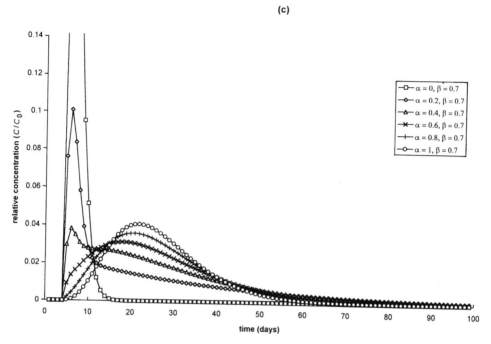

FIGURE 6.4   continued

system operates is derived from observations of the outflow. It is inevitable, therefore, that models of catchment hydrochemistry must be, to a reasonably large extent, empirically based.

On a seasonal basis a pronounced annual cycle may be observed in the concentrations of nitrate in stream water in many catchments (e.g. Figure 6.5). Highest concentrations generally occur in winter when catchment soils wet up and soil nitrate becomes mobile. In spring and early summer both the average concentration and the day-to-day variation in concentration generally decline. This probably reflects a "decoupling" of the relatively well mixed base flow from rain storm events as the root zone dries out and becomes less conducive to drainage.

In MONITOR the subsurface moisture deficit D is used to describe the proportion of nitrate leached from the root zone which is transported relatively rapidly to the stream (*pquick*) and that which is assumed to travel more slowly, mixing with the baseflow ($1 - pquick$). Assuming lateral flow does not interact with the root zone then

$$pquick = \exp\left(-\frac{D}{p}\right) \tag{16}$$

where p is a parameter describing the rate of decrease in *pquick* with increase in D. Thus the higher the subsurface antecedent moisture content, the more rapidly root zone-leached nitrate may be transported to the stream. The remainder supplements a well-mixed baseflow reservoir and the final concentration in the stream is assumed to be a mixture of the two pathways. Near-stream (riparian zone) and in-stream transformations of nitrate, which are known to be very significant in some catchments (e.g. Haycock, 1991), are not accounted for in MONITOR.

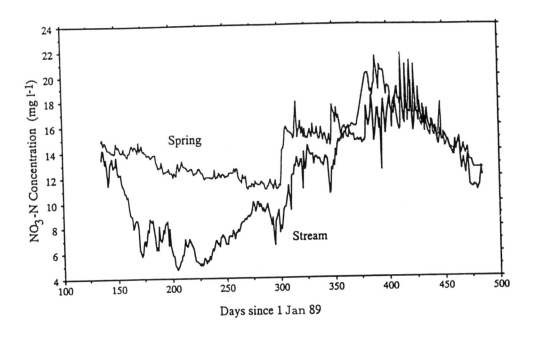

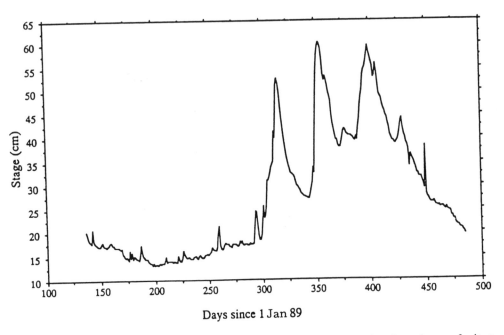

FIGURE 6.5 Concurrent stream hydrograph and chemograph showing the dependence of nitrate concentration on discharge (from Burt and Haycock, 1993)

6.7 ROOT ZONE VALIDATION

Predictions of root zone moisture content, produced by MONITOR, were compared with total profile moisture content to 90 cm depth, estimated using a neutron probe by Rushton (1993), in a winter wheat field at the University of Leeds field station near Tadcaster, North Yorkshire (Figure 6.6). The total profile moisture content is reproduced by the model reasonably well, especially when allowing variability and uncertainties in the measured values.

Nitrate concentrations in 50 (10 g) soil samples taken from the top 10 cm of this soil were measured every month for a year. Figure 6.7 shows a comparison between the model predictions for top soil nitrate $-$ N concentration and the mean of the measured concentrations. Predicted concentrations follow the measured patterns quite well. Both follow an annual pattern which is not unusual for temperate agricultural soils. Both are lowest at the beginning of the calendar year when winter sink processes (such as denitrification and leaching) have been operating for some time and when plant uptake commences. Apart from the spikes introduced by the addition of fertiliser in spring, peak soil nitrate concentrations generally occur in late summer when uptake has effectively ceased but when soil mineralisation is still active and winter sink processes have not been significant. In general, the measured concentrations are overestimated by the model. This may, in part, reflect a slight difference in the measured and modelled soil volume, in that the model simulates concentrations in the top 20 cm whereas the measurements were made in the top 10 cm where nitrate might easily be leached to lower levels of the topsoil by small rain storms and could, therefore, be a misleading guide to the nitrate concentration in the topsoil as a whole. The underestimation of concentrations immediately after the application of mineral N fertiliser could also be a reflection of differences between the soil being measured and the part of the soil being modelled. Fertiliser applications in the model are assumed to equilibrate instantaneously with the nitrate in the top 20 cm. In reality fertiliser N may remain in the top few centimetres for some time after application and, therefore, be included in soil samples.

Measured and modelled soil nitrate concentrations decrease in early September immediately after harvest because of the reincorporation of the cereal stems. This material has a lot of readily available C compounds and a high C/N ratio. Its decomposition results in the immobilisation of mineral N from the surrounding soil in the microbial biomass and a flush of CO_2 output from the soil (Figure 6.8). Increasing the assumed C input in the model resulted in an increase in the immobilisation of soil mineral N.

6.8 SMALL CATCHMENT VALIDATION

Validation of the small catchment component of the model was performed using data collected from a V-notch weir in the Slapton Wood catchment, South Devon. The nutrient budget and hydrology of this catchment have been monitored since 1970 (see Burt *et al.*, 1988) and a large archive of data on stream water nitrate concentrations and steam discharge exists.

Calibration of the lumped model was performed using data covering the period 11 April 1990 to 10 April 1991. Predicted runoff (mm) was compared with the observed hydrograph, expressed as a depth equivalent over the area of the gauged catchment (Figure 6.9). Parameter values were "optimised" by trial and error so as to minimise the mean daily absolute percentage error. Subsequently several continuous records of stream discharge were examined and compared with MONITOR predictions for the same periods. The model was able to reproduce

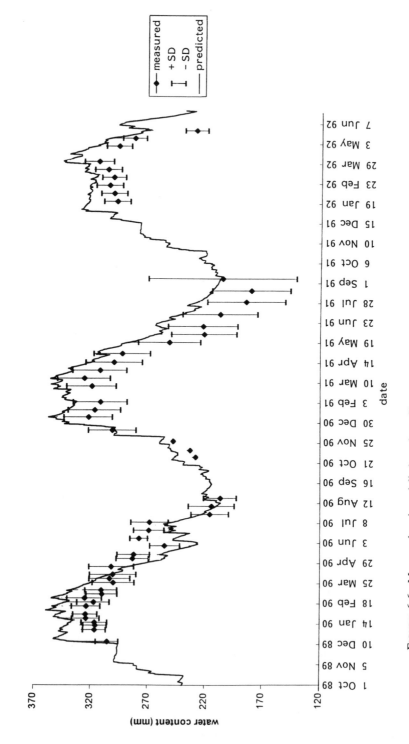

FIGURE 6.6 Measured and modelled total profile water content (1989–92) for an arable field at the Tadcaster site. Error bars indicate plus and minus one standard deviation

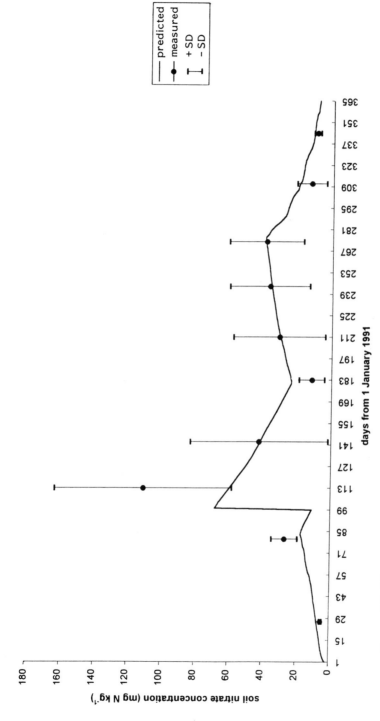

FIGURE 6.7 Measured and modelled soil nitrate concentration over one year at the Tadcaster site. Error bars denote plus and minus one standard deviation. Measured concentrations are arithmetic means

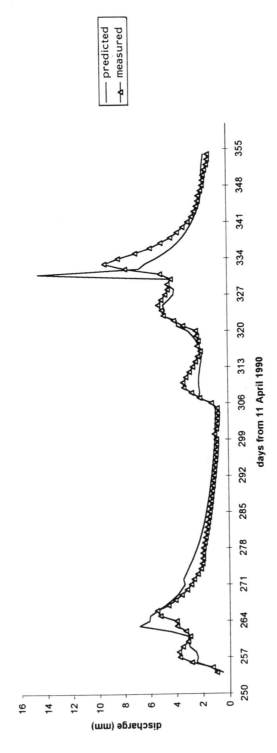

FIGURE 6.8 Measured and optimised lumped model stream discharge for the Slapton Wood catchment for the winter of 1990–1. Both measured and modelled flow was relatively low during the early part of the record used for calibration

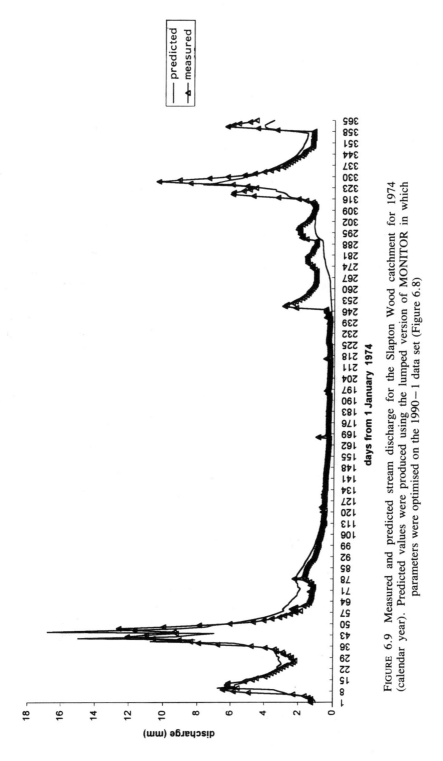

FIGURE 6.9 Measured and predicted stream discharge for the Slapton Wood catchment for 1974 (calendar year). Predicted values were produced using the lumped version of MONITOR in which parameters were optimised on the 1990—1 data set (Figure 6.8)

the majority of stormflow peaks over considerable periods of time reasonably well (an example is shown for 1974 in Figure 6.10) as well as the annual catchment hydrological balance over a 20 year period from 1970 to 1990 (Table 6.1). The main systematic discrepancy between the measured and modelled runoff occurred in the autumn rewetting period. At this time the root zone is still fairly dry and predicted unsaturated hydraulic conductivity and the resultant root zone drainage are consequently low. Model predictions were significantly better for October water years when initial soil moisture deficits were assumed to be low, thereby minimising root zone buffering of streamflow. In fact, the lumped model does not take into account the significant hydrological contribution made by near-stream source areas which respond much more rapidly to rewetting than the bulk of the catchment; nor does it account for direct channel precipitation (usually small), nor overland flow processes (which have been observed in limited areas of the catchment; e.g. Heathwaite *et al.*, 1990).

Figure 6.11 shows the results of a comparison between measured and modelled mean monthly nitrate concentrations in the Slapton Wood stream over the 20 year record. Both plots clearly show the expected pronounced annual cycle discussed above. The model predicts the broad trends reasonably well, although it misses significant features such as the peak in the winter of 1976−7 which has been attributed to the very dry summer of 1976 (Burt *et al.*, 1988). This is probably because large topsoil nitrate leaching losses, which were predicted for 1976−7 did not cross the lower boundary of the root zone and were therefore not incorporated in the throughflow model until the following winter, which suggests that the leaching component of the model may be in error for this catchment. Nitrate concentrations in the following winter were significantly overestimated.

The match in the first half of the record is not as good as in the latter half. This was not surprising because the modelled results were actually produced by assuming a constant land-use scenario of a grassland system with constant annual fertiliser applications at estimated mean catchment input levels for 1985. In fact, the land use of the catchment has been changing with an increase in the area of temporary grassland and arable crops and a reduction in the area of permanent pasture (Johnes and O'Sullivan, 1989).

A comparison between measured and modelled annual nitrate−N load is shown in Figure 6.12. Both predicted and measured loads are higher in the latter half of the record compared with the first. Whilst concentrations did increase in the latter half of the record, the increase in load has probably more to do with the increase in discharge in these later years. In the early part of the record, predicted mean monthly concentrations were consistently higher than those in the measured data set. Overall losses are underpredicted by the model in 1971, 1973 and 1975, however, because low measured streamflows were underpredicted in these years (Table 6.1). This emphasises the importance of successfully modelling discharge in order to model total load (Burt *et al.*, 1988). The increase in the concentration of nitrate in the later years is probably due to the generally wetter winters of the 1980s and stresses the strong role climate may play in influencing catchment N budgets and the importance of this type of modelling approach which attempts to take climate influences into account.

6.9 CONCLUSIONS

Describing process rates using mathematical models is, at best, an approximation and must be recognised as such. Nevertheless, the broad direction and magnitude of individual processes may be reasonably described, even by deterministic models, despite an inability to cope with

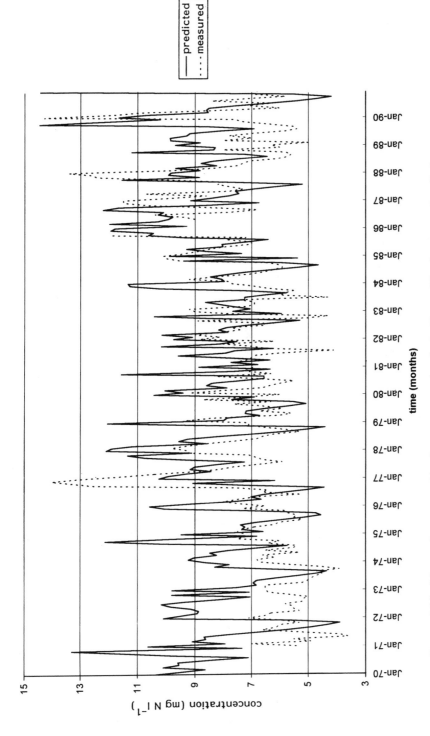

FIGURE 6.10 Measured mean monthly nitrate concentrations in the Slapton Wood stream and predictions from the lumped version of MONITOR, assuming grassland with 1985 fertiliser inputs, from 1970 to 1990

TABLE 6.1 Measured and predicted annual water balance for the Slapton Wood catchment between 1970 and 1990. All values are in mm. No overland flow was predicted by the model

Year	Rainfall	pet	aet	rzd	int loss	dmc	pro (1)	pro (2)	mro (3)	pro (4)	mro (4,5)
1970	1110.3	682.63	435	571.86	106.8	3.39	552.89	620.47	—	—	—
1971	698.6	800.27	469.03	197.18	67.96	35.56	244.49	337.24	272.89	437	540
1972	1170.4	709.22	468.21	608.3	102.5	8.56	583.32	575.48	575.76	497	333
1973	733	786.67	475.33	180.1	87.79	10.22	180.04	199.85	232.95	489	436
1974	1248	783.33	483.33	645.98	116.6	−2.09	620.6	663.84	679.76	554	484
1975	736.7	881.26	428.5	252.05	80.17	24.03	259.26	293.9	326.16	442	538
1976	841.6	960.92	340.98	406.25	76.72	−17.64	373.65	426.62	448.34	178	168
1977	1057.9	569.45	400.71	548.03	109.1	−0.1	514.87	551.47	—	783	689
1978	968.2	681.08	398.47	457.19	95.92	−16.62	442.58	469.97	572.05	572	569
1979	1163.6	757.01	481.92	570.11	104.1	−7.47	527.26	535	681.98	622	580
1980	1116.8	666.63	430.34	588.31	105.9	7.75	561.96	593.1	667.3	629	557
1981	1165.3	774.7	493.64	535.73	113.1	−22.82	482.13	507.88	—	652	601
1982	1159.8	736.15	486.08	471.17	109.3	6.7	541.09	597.28	—	655	607
1983	910	676.91	396.76	420.57	95.27	2.6	386.35	415.19	—	630	654
1984	1118.3	848.71	412.86	615.75	86.82	−2.87	581.97	603.66	—	495	488
1985	1024.6	702.96	454.46	463.7	99.83	−6.61	423.64	495.59	—	764	630
1986	1221.6	638.84	439.75	658.21	108.9	−14.76	623.89	656.53	—	642	656
1987	954.7	723.64	460.11	394.11	90.72	−9.75	369.12	409.9	—	603	531
1988	1122.1	718.9	474.51	571.78	97.22	21.41	560.01	631.1	—	756	581
1989	1025.2	825.23	394.47	548.08	82.71	0.06	512.63	520.37	—	458	493
1990	910.7	707.38	400.02	416.65	86.37	−7.66	432.52	457.47	—	635	574
Mean	1021.78	744.38	439.26	481.96	96.37	0.57	464.97	502.95	490.24	575	536

pet is potential evapotranspiration.
aet is actual evapotranspiration.
rzd is root zone drainage.
int loss is interception loss.
dmc is change in moisture content of the root zone.
mro is measured runoff (streamflow).
pro is predicted catchment runoff (streamflow).

(1) For the calendar year from lumped model.
(2) For the calendar year from field matrix configured for the Carness subcatchment.
(3) For the calendar year calculated from stream hydrograph data.
(4) For the water year starting 1 October.
(5) After Burt et al. (1988).

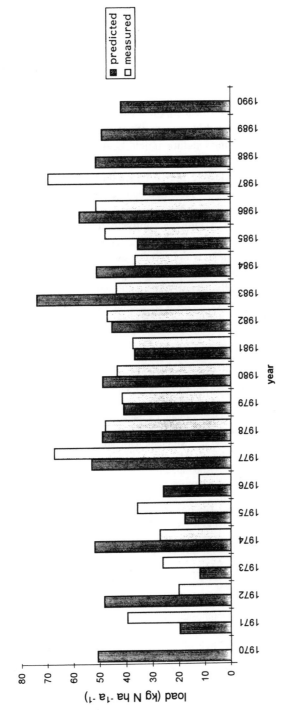

FIGURE 6.11 Measured and predicted total annual losses of nitrate − N from the Slapton Wood catchment (1970 − 90). Predicted losses produced from the lumped version of MONITOR, assuming grassland with 1985 fertiliser inputs

the inherent variability in catchment processes. Models provide a unique way of synthesising and exploring the complexity of natural systems. An important by-product of modelling exercises is the identification of inadequacies in empirical work. At the same time, simulation modelling allows the exploration of the complex system in question, indicating directions for future research with the ultimate aim of a better understanding of that system (Kirkby *et al.*, 1993, p. 149).

The idea of representing stream water as a mixture of different conceptual reservoirs of different concentrations has also been used in models of catchment hydrochemistry developed within the context of better understanding catchment acidification (e.g. Christophersen *et al.*, 1982). Models of this type have been able to reproduce daily and seasonal variations in stream water quality reasonably successfully despite the fact that they do not attempt to reproduce internal catchment processes and are seldom validated against detailed internal catchment data (Christophersen *et al.*, 1993).

Although the validity and utility of a lumped approach to modelling catchment solute dynamics is well recognised, future developments should attempt to focus on the distribution of processes occurring within the catchment which contribute to the patterns observed at the outlet. There is certainly much room for improvement in the fit obtained between model output and observations. This will undoubtedly require further development of the concepts upon which the model is based and a better estimation of model parameters. In addition, further investigation into the potential of the model as an indicator of the effects of climate and land-use change is required.

REFERENCES

Addiscott, T. M. (1977). "A simple computer model for leaching in structured soils", *J. Soil Sci.*, **28**, 554−563.

Addiscott, T. M. and Whitmore, A. P. (1987). "Computer simulation of changes in soil mineral nitrogen and cross nitrogen during autumn, winter and spring", *J. Agric. Sci.*, Cambridge, **109**, 141−157.

Addiscott, T. M. and Whitmore, A. P. (1991). "Simulation of solute leaching in soils of different permeabilities", *Soil Use and Management*, 7, 94−102.

Addiscott, T. M., Whitmore, A. P. and Powlson, D. S. (1991). *Farming, Fertilizers and the Nitrate Problem*, CAB International, Wallingford, Oxon.

Anderson, M. G. and Burt, T. P. (1978). "The role of topography in controlling throughflow generation", *Earth Surface Processes and Landforms*, 3, 331−334.

Arah, J. R. M. and Smith, K. A. (1989). "Steady-state denitrification in aggregated soils: a mathematical model", *J. Soil Sci.*, **40**, 139−149.

Beven, K. and Kirkby, M. J. (1979). "A physically-based, variable contributing area model of basin hydrology", *Hydrol. Sci. Bull.*, **42**, 43−69.

Brown, A. D. (1990). *Microbial Water Stress Physiology: Principles and Perspectives*, John Wiley, Chichester.

Burford, J. R. and Bremner, J. M. (1975). "Relationships between the denitrification capacities of soils and total water soluble and readily decomposable soil organic matter", *Soil Biol. Biochem.*, **7**, 389−394.

Burns, I.G. (1974). "A model for predicting the redistribution of salts applied to fallow soils after excess rainfall or evaporation", *J. Soil Sci.*, **25**, 165−178.

Burt, T. P., Arkell, B. P., Trudgill, S. T. and Walling, D. E. (1988). "Stream nitrate levels in a small catchment in south west England over a period of 15 years (1970−1985)", *Hydrol. Process*, 2, 267−284.

Burtt, T. P. and Haycock, N. E. (1993). "Controlling losses of nitrate by changing land use", in *Nitrate: Processes, Patterns and Management* (Eds. T. P. Burt, A. L. Heathwaite and S. T. Trudgill), pp. 341−367, John Wiley, Chichester.

Burt, T. P., Heathwaite, A. L. and Trudgill, S. T. (1993). "Catchment sensitivity to landuse controls", in *Landscape Sensitivity* (Eds. D. S. G. Thomas and R. J. Allison), pp. 229–240, John Wiley, Chichester.

Burton, D. L. and Beauchamp, E. G. (1985). "Denitrification rate relationships with soil parameters in the field", *Commun. in Soil Sci. Plant Annal.*, **16**, 539–549.

Cassel, D. K. and Nielsen, D. R. 1986). "Field capacity and available water capacity", in *Methods of Soil Analysis, Part 1. Physical and Mineralogical Methods*, (Ed. A. Klute), Agronomy Monograph No. 9, pp. 901–926.

Cassman, K. G. and Munns, D. N. (1980). "Nitrogen mineralisation as affected by soil moisture, temperature and depth", *Soil Sci. Soc. of Am. J.*, **44**, 1233–1237.

Christophersen, N., Neal, C. and Hooper, R. P. (1993). "Modelling the hydrochemistry of catchments: a challenge for the scientific method", *J. Hydrol.*, **152**, 1–12.

Denmead, O. T. and Shaw, R. T. (1962). "Availability of soil water to plants as altered by soil moisture content and meteorological conclusions", *Agron. J.*, **54**, 385–390.

Grant, R. F. (1991). "A technique for estimating denitrification retae at different soil temperatures, water contents, and nitrate concentrations", *Soil Sci.*, **152**, 41–51.

Greaves, J. E. and Carter, E. G. (1920). "Influence of Moisture on the Bacterial Activity of Soil", *Soil Science*, **10**, 361–387.

Groffman, P. M. and Tiedje, J. M. (1991). "Denitrification in north temperate forest soils. I. Spatial and temporal patterns at the landscape and seasonal scales", *Soil Biol. Biochem.*, **21**, 613–620.

Hall, D. G. M. (1993). "An amended functional leaching model applicable to structured soils. I. Model description", *J. Soil Sci.*, **44**, 579–588.

Heathwaite, A. L., Burt, T. P. and Trudgill, S. T. (1990). "The effect of agricultural land use on nitrogen, phosphorus and suspended sediment delivery to streams in a small catchment in south west England", in *Vegetation and Erosion*, (Ed. J. B. Thornes), pp. 161–179, John Wiley, Chichester.

Hewlett, J. D. (1961). 'Watershed management", in *Report for 1961 Southeastern Forest Experiment Station*, US Forest Service, Ashville, North Carolina.

Hillel, D., Talpaz, H. and van Keulen, H. (1976). "A macroscopic scale model of water uptake by non-uniform root system and of water and salt movement within the soil profile", *Soil Sci.*, **122**, 242–255.

Jenkinson, D. S. and Rayner, J. H. (1977). "The turnover of soil organic matter in some of the Rothamsted classical experiments", *Soil Sci.*, **123**, 298–305.

Johnes, P. J. and O'Sullivan, P. E. (1989). "The Natural History of Slapton Ley Nature Reserve XVIII. Nitrogen and phosphorus losses from the catchment — an export coefficient approach", *Field Studies*, **7**, 285–309.

Jury, W. A., Gardner, W. R. and Gardner, W. H. (1991). *Soil Physics*, John Wiley, New York.

Kirkby, M. J., Naden, P. S., Burt, T. P. and Butcher, D. P. (1993). *Computer Simulation in Physical Geography*, 2nd edition, John Wiley, Chichester.

McConnaughey and Bouldin (1985). "Transient microsite models of denitrification. III. Comparison of experimental and model results", *Soil Sci. Soc. of Am. J.*, **49**, 895–901.

Monteith, J.L. (1965). 'Evaporation and environment', *Proc. Symp. on Environmental Biology*, **19**, 205–234.

Nordmeyer, H. and Richter, J. (1985). "Incubation experiments on nitrogen mineralisation in loess and sandy soils", *Plant and Soil*, **83**, 433–445.

Parton, W. J., Schimel, D. S., Cole, C. V. and Ojima, D. S. (1987). "Analysis of factors controlling soil organic matter levels in Great Plains grasslands", *Soil Sci. Soc. of Am. J.*, **51**, 1173–1179.

Paul, E. A. and Clark, F. E. (1989). *Soil Microbiology and Biochemistry*, Academic Press.

Penman, H. L. (1948). "Natural evaporation from open water, bare soil and grass", *Proc. R. Soc. .*, **193**, 120–145.

Rawlins, S.L. (1974). "Principles of managing high frequency irrigation", *Soil Sci. Soc. of Am. Proc.*, **37**, 626–629.

Reddy, K. R. and Patrick, J. R. (1983). "Effects of aeration on reactivity and mobility of soil constituents", in *Chemical Mobility and Reactivity in Soil Systems* (Eds. Nelson *et al.*), Soil Science Society of America, Madison, Wisconsin.

Reddy, K. R., Khaleel, R. and Overcash, M. R. (1980). "Carbon transformations in the land areas receiving organic wastes in relation to non-point source pollution: a conceptual model", *J. Environ. Qual.* 9, 434–442.

Rushton, K. M. (1993). "A field study on nitrate leaching and hydrology of farm woodlands planted on lowland arable soil", Unpublished PhD thesis, University of Leeds, UK.

Sextone, A. J., Parkin, T. B. and Tiedje, J. M. (1985). "Temporal response of soil denitrification rates and irrigation", *Soil Sci. Soc. of Am. J.*, **49**, 99–103.

Smith, O. L. (1979). "An analytical model of the decomposition of soil organic matter", *Soil Biol. Biochem.*, **11**, 585–606.

Stanier, R. Y., Adelberg, E. A. and Ingraham, J. L. (1977). *General Microbiology*, Macmillan Press, London.

Tenny, F. G. and Waksman, S. A. (1930). "Composition of organic materials and decomposition in the soil: V. Decomposition of various chemical constituents in plant materials under anaerobic conditions", *Soil Sci.*, **30**, 143–160.

Thompson, N., Barrie, I. A. and Ayles, M. (1981). *The Meteorological Office Rainfall and Evaporation Calculation System: MORECS (July 1981)*, Meteorological Office Memorandum No. 45.

van Genuchten, M. Th. (1980). "A closed-form equation for predicting the hydraulic conductivity of unsaturated soils", *Soil Sci. Soc. of Am. J.*, **44**, 892–898.

van Genuchten, M. Th. and Wierenga, P. J. (1976). "Mass transfer studies in sorbing porous media. I. Analytical solutions", *Soil. Sci. Soc. of Am. J.*, **40**, 473–480.

van Veen, J. A. and Paul, E. A. (1981). "Organic carbon dynamics in grassland soils. Background information and computer simulation", *Can. J. Soil Sci.*, **61**, 185–201.

van Veen, J. A., Ladd, J. N. and Amato, M. (1985). "Turnover of carbon and nitrogen through the microbial biomass in a sandy loam and a clay soil incubated with ^{14}C glucose and ^{15}N $(NH_4)_2SO_4$ under different moisture regimes", *Soil Biol. Biochem.*, **17**, 747–756.

Vinten, A. J. A. and Smith, K. A. (1993). "Nitrogen cycling in agricultural soils", in *Nitrate: Processes, Patterns and Management* (Eds. T. P. Burt, A. L. Heathwaite and S. T. Trudgill), pp. 39–73, John Wiley, Chichester.

Wagenet, R. J. and Hutson, J. L. (1987). *LEACHM: Leaching Estimation and Chemistry Model*, CONTINUUM Series, Water Resources Institute, USA.

Whelan, M. J. (1993). "Numerical modelling of small catchment nitrogen dynamics with particular reference to the Slapton Wood catchment, South Devon", Unpublished PhD thesis, University of Leeds, UK.

7

Interaction of Ecosystem Processes and Weathering Processes

MICHAEL ANTHONY VELBEL

Department of Geological Sciences, Michigan State University, USA

7.1 INTRODUCTION

The purpose of this chapter is to explore the conceptual treatment of elemental exchange between botanical and inorganic compartments of small forested catchments, and the consequences of that treatment for geochemical mass balance of small catchments. The geochemical mass balance approach is used by geochemists (to estimate rates of inorganic processes, e.g. weathering of minerals in soils) and by ecologists (as one component of quantifying elemental additions to, losses from and internal cycling within terrestrial ecosystems). Geochemical studies commonly ignore biological effects; ecological studies commonly oversimplify the representations of inorganic geochemical processes. However, for the few ecosystems that have been studied in the requisite detail, the two approaches gave surprisingly consistent quantitative results.

How important are ecosystem processes in geochemical mass balances of catchments? Can they be neglected? Should they be treated as an input, or perhaps as a governing factor? This paper quantitatively investigates the consequences of different explicit assumptions regarding biomass behaviour for weathering rates calculated using mass balance methods. The geochemical mass balance method was modified by Velbel (1985a, 1986) and Taylor and Velbel (1991) to include botanical exchange terms, and applied to forested catchments of the southern Appalachian Mountains (USA). This chapter (a) reviews the previous work on biomass-inclusive mass balance calculations in the southern Appalachian Mountains, (b) examines the sensitivity of previous biomass-inclusive results to variations in the assumed biomass composition and (c) extends the biomass-inclusive mass balance approach to other forested catchments in the southern and central Appalachian Mountains for which biomass effects on elemental budgets have not been previously incorporated into geochemical mass balance calculations of weathering rates.

7.2 BACKGROUND

Silicate mineral weathering reactions are fundamental processes in the reaction of acid precipitation with susceptible landscapes (Cosby *et al.*, 1985a, 1985b; Reuss *et al.*, 1987;

Solute Modelling in Catchment Systems. Edited by Stephen T. Trudgill
© 1995 John Wiley & Sons Ltd

Binkley *et al.*, 1989; Turner *et al.*, 1990; Kauffmann *et al.*, 1991) and in the contributions of continental crustal weathering to global geochemical cycles (Berner and Berner, 1987). The rates at which weathering reactions proceed are crucial to the consumption (neutralisation) of environmental acidity and to the consumption (transfer from atmosphere to oceans) of atmospheric carbon dioxide. Geochemical mass balance (input − output budgeting) is commonly used to calculate weathering rates (Garrels and Mackenzie, 1967; Cleaves *et al.*, 1970; Clayton, 1979; Velbel, 1985a, 1992; Taylor and Velbel, 1991, and references therein), especially in small catchments. Small catchments are hydrologically and topographically well-defined drainage basins, for which inputs, outputs and internal properties (e.g. rock type, soil type, vegetation type) can be characterized with relative ease. Small catchments are thus both easily studied in terms of hydrologic and geochemical dynamics, and are a fundamental, landscape-scale unit appropriate for studying regional (bio)geochemical processes at natural space and time-scales (as opposed to simulations of such processes in the laboratory).

Many researchers employ the "balance sheet" approach, a system of simultaneous linear equations with constant coefficients. Weathering rates are calculated by solving the system of equations that represent the steady-state input − output behaviour of the catchment (Plummer and Back, 1980, eqn 3):

$$\sum_{j=1}^{\phi} \alpha_j \beta_{c,j} = \Delta m_c, \qquad c = 1, \ldots, n \tag{1}$$

where phi (ϕ) is the number of (mineral) phases j whose transformations account for the change of mass of species c within the system, Δm_c is the total (known) change in mass of species c across the boundaries of the system (output minus input), $\beta_{c,j}$ is the stoichiometric coefficient of species c released to solution by the weathering of phase j and α_j is the weathering rate of phase j. Where multiple mass balance equations are stoichiometrically linked to one another through one or more phase(s) j containing more than one species c, a system of n equations results. If the number of mass balance equations (n, each for a different species c) equals the number of unknowns (ϕ, the number of phases j whose transformations account for changes of mass of c within the system), eqn (1) can be solved for mineral weathering rates (α_j) by conventional techniques of linear algebra. The "balance sheet" method of Garrels and Mackenzie (1967), which has also been employed by numerous other workers (see Taylor and Velbel, 1991) to calculate individual mineral weathering rates, is merely a long-hand, pencil-and-paper solution to the system of mass balance equations (1). The relationship between this use of mass balance equations (1) and that Plummer and Back (1980) is discussed by Velbel (1986).

This system of linear equations represents the steady-state input − output behaviour of the modelled system. Most workers also assume that the steady-state assumption for the entire system means that botanical factors can be ignored, in other words, that there is no *net* elemental transfer between biomass and inorganic compartments of the system. There are certainly types of catchments (e.g. alpine catchments, above tree-line, which lack forest biomass) where element transfers involving biomass may be justifiably ignored, because there is no significant biomass (Reynolds and Johnson, 1972; Drever and Hurcomb, 1986; Giovanoli *et al.*, 1988; Frogner, 1990; Turner *et al.*, 1990; Stauffer, 1990; Mast *et al.*, 1990). However, in many forested catchments, even small changes in the abundant biomass might consume or release significant amounts of elements. The assumption of steady-state input − output behaviour of the system is not equivalent to stating that there is no net change in individual compartments (e.g. biomass; see Taylor and Velbel, 1991). The widely invoked assumption that biomass is at "steady state" (usually interpreted to mean that total biomass is constant) is both (a) mathematically a second, explicit assumption, unrelated to the assumption of overall steady state and (b) also generally erroneous. One potential consequence is systematic errors in weathering rates calculated by mass balance.

Taylor and Velbel (1991) constructed geochemical mass balances for seven small forested catchments of the US Department of Agriculture (Forest Service) Coweeta Hydrologic Laboratory in the Nantahala Mountains of the southern Blue Ridge Appalachians, North Carolina, USA. All seven catchments are "control" catchments, which have been free of any intentional anthropogenic disturbances since 1924. Four of the catchments (2,18, 34, 36) are underlain by the Tallulah Falls Formation; the other three (14, 27, 32) are underlain by the Coweeta Group. The two lithostratigraphic units on which the soils and saprolites of the study area are developed both consist of high-grade (amphibolite facies) metasedimentary schists and gneisses. The Tallulah Falls Formation was metamorphosed from sedimentary protoliths of slightly lower compositional maturity than the Coweeta Group; consequently, the rocks of the Tallulah Falls Formation have a higher abundance of weatherable parent minerals than the rocks of the Coweeta Group. The higher "weatherability" of the Tallulah Falls Formation relative to the Coweeta Group accounts for the fact the Tallulah Falls Formation catchments have (a) higher long-term average cation fluxes (especially Na and K; see Taylor and Velbel, 1991, and Table 7.1), (b) higher calculated rates of mineral weathering (Table 7.1) and (c) higher long-term average stream water pHs and dissolved silica and cation concentrations than Coweeta Group catchments (Velbel, 1985b). The geology, hydrology, topography, climate and land-use history of the study area are described in more detail by Velbel (1985a)and Swank and Crossley (1988).

Taylor and Velbel (1991) constructed geochemical mass balances for the seven Coweeta control catchments by using a system of linear equations formalized as eqn (1). Mineralogical and petrographic (micromorphological) data determine stoichiometries of mineral weathering reaction. By evaluating and reinterpreting published hydrological and hydrogeochemical data and combining them with the known stoichiometries of the mineral weathering reactions, Velbel (Velbel, 1985a, 1985b; Taylor and Velbel, 1991) found that four transformations influence dissolved elemental budgets in the catchment. The four reactions are the weathering of three major weatherable rock-forming minerals:

1. Biotite mica to vermiculite:

$$K_{0.85}Na_{0.02}(Mg_{1.2}Fe^{II}_{1.3}Al_{0.45})^{VI}(Al_{1.2}Si_{2.8})^{IV}O_{10}(OH)_2 + 0.19\ O_2 + 0.078\ H^+ + 0.31\ H_2O$$
$$+\ 0.016\ Ca^{2+} + 0.04\ Na^+ + 0.35\ Al(OH)_2{}^+{}_{(aq)} + 0.3\ Fe(OH)_2{}^+{}_{(aq)} \Rightarrow$$
$$K_{0.25}Na_{0.06}Ca_{0.16}(Mg_{1.1}Fe^{II}_{0.5}Fe^{III}_{1.1})^{VI}(Al_{1.2}Si_{2.8})^{IV}O_{10}(OH)_2 \times 0.133\ Al_6(OH)_{15} + 0.6\ K^+$$
$$+\ 0.1\ Mg^{2+}$$

2. Almandine garnet to gibbsite, goethite and solutes:

$$Ca_{0.2}Mg_{0.5}Mn^{II}_{0.2}Fe^{II}_{2.1}Al_2Si_3O_{12} + 0.625\ O_2 + 2.5\ H^+ + 8.35\ H_2O \Rightarrow$$
$$2FeOOH_{(s)} + 0.1\ Fe(OH)_2{}^+{}_{(aq)} + Al(OH)_{3(s)} + Al(OH)_2{}^+{}_{(aq)}$$
$$+\ 3\ H_4SiO_{4(aq)} + 0.5\ Mg^{2+} + 0.2\ Ca^{2+} + 0.2\ MnO_{2(s)}$$

3. Plagioclase feldspar to kaolinite, gibbsite and solutes:

TABLE 7.1   Mineral weathering rates from catchment balances of North Carolina (Coweeta) catchments with alternate biomass

			Biomass			
Catchment	Mineral	Without biomass[a]	Coweeta[a]	Hubbard Brook	UK deciduous	Walker Branch
2	Plagioclase	535	567	553	545	540
	Garnet	164	240	179	207	235
	Biotite	124	432	302	293	195
14	Plagioclase	250	257	254	252	251
	Carnet	176	195	180	187	194
	Biotite	97	175	144	141	116
18	Plagioclase	439	460	451	445	442
	Garnet	192	242	202	220	238
	Biotite	126	325	241	235	172
27	Plagioclase	232	230	231	232	232
	Garnet	201	197	200	199	197
	Biotite	104	85	93	93	100
32	Plagioclase	351	353	352	352	351
	Garnet	297	293	288	290	293
	Biotite	123	150	139	138	129
34	Plagioclase	483	500	493	489	486
	Garnet	263	304	271	286	301
	Biotite	131	294	225	221	169
36	Plagioclase	641	662	653	648	644
	Garnet	285	334	294	312	330
	Biotite	141	338	255	250	187

Units: moles of reaction (as written in text) per hectare per year.
[a] From Taylor and Velbel (1991).

$$Ca_xNa_{(1-x)}Al_{(1+x)}Si_{(3-x)}O_8 + [4.5 - 3.5x + 2.5y(1+x)]\, H_2O + (1+x)\, H^+ \Rightarrow$$
$$x\, Ca^{2+} + (1-x)\, Na^+ + [(2.2x) + y(1+x)]\, H_4SiO_{4(aq)} + (1+x)y\, Al(OH)_3$$
$$+ 0.5(1+x)(1-y)\, Al_2Si_2O_5(OH)_4$$

(where x is the mole fraction anorthite in the plagioclase; $x = 0.25$ for rocks of the Coweeta Group and $x = 0.32$ for rocks of the Tallulah Falls Formation (Velbel, 1985a) and y is the (unknown) fraction of Al released by feldspar weathering which goes into gibbsite).

4. Uptake of mineral nutrients by the forest biota.

The rates of these four transformations are the four unknowns in eqn (1); therefore, mass balance equations for four elements are needed in order to solve the system of equations in the desired number of unknowns. The stoichiometries of these reactions are known for minerals and can be approximated for biomass (using the data of Day and Monk, 1977; Boring *et al.*,1981; Taylor and Velbel, 1991; see Table 7.2), and net fluxes for Na, Mg, K and Ca are known (from Swank and Waide, 1988; Taylor and Velbel, 1991; their Table 2), so the system

TABLE 7.2 Molar ratios of major rock-derived cations in temperate forest biomass (Mg = 1)

Catchment	Na	Mg	K	Ca	Reference
Coweeta	0.143	1.0	2.68	2.57	Taylor and Velbel (1991)
Hubbard Brook	0.2	1.0	4.22	6.41	Likens *et al.* (1977)
UK deciduous	n.r.	1.0	2.63	4.29	Cole and Rapp (1981)
Walker Branch	n.r.	1.0	1.0	4.0	Katz (1989)

n.r. = not reported.

of equations can be solved. Rates of primary mineral weathering and mineral nutrient supply to the terrestrial biota were calculated by this method (Taylor and Velbel, 1991).

Taylor and Velbel (1991) demonstrated the effects on calculated weathering rates of including and ignoring botanical terms by comparing the results of two different sets of geochemical mass balance calculations of weathering rates for the Coweeta control catchments (using different sets of assumptions but identical input data). The results of the comparative mass balance calculations for seven undisturbed catchments at Coweeta (Taylor and Velbel, 1991) are included in Table 7.1. The "Coweeta biomass" column shows the results of solving a system of four equations (steady-state mass balance equations for K, Na, Ca and Mg) in four unknowns (weathering rates of almandine garnet, oligoclase-andesine plagioclase feldspar and biotite mica, and exchange with forest biomass). The "without biomass" column shows mineral weathering rates calculated without the botanical exchange term. Weathering rates calculated using the two different assumptions differ by as much as a factor of four (Taylor and Velbel, 1991). The difference is most pronounced for minerals that contain major nutrient elements and for rates calculated from mass balances of those nutrient elements. These results, although provocative, are only applicable to the specific catchments studied. In this paper, similar comparisons of calculated mass balances with and without biomass terms permit examination of the interactions between geochemical and ecological terms in other Appalachian catchments.

7.3 METHODS

The present study examines only rates from studies in which weathering rates of multiple silicate minerals were determined simultaneously. Ideally, there would be numerous solute mass balance studies of forested catchments which report weathering rates of multiple silicate minerals. However, many mass balance studies calculate only "lumped" weathering rates (e.g. total release by weathering of cations or silica, in moles or eq ha^{-1} yr^{-1}), without apportioning the fluxes among individual minerals. In other instances, there are obvious errors and/or inconsistencies in the published mass balances (e.g. negative weathering rates for rock-forming silicates, implying formation of minerals like feldspar during weathering, a physically unreasonable result), rendering suspect the previously published fluxes and/or stoichiometries. Additionally, many mass balance studies that calculate weathering rates for individual silicate minerals involve catchments with ecosystems other than forests (e.g. bog, alpine). For this study, only forested catchments in the southeastern United States were examined. There are many previous geochemical mass balance studies in this geographic region; many of these have neglected possible biomass exchange of mineral-derived elements. Forested catchments outside this region, and other ecosystem types, will be the subjects of future research.

Following Taylor and Velbel (1991), the present study examines variations in calculated weathering rates which result from different mathematical assumptions regarding biomass effects which are demonstrated by comparing the results of two different sets of geochemical mass balance calculations of weathering rates for the catchments examined. In most cases examined here, the original calculations assume *both* steady state for the overall system *and* no net exchange of elements between biomass and the rest of the system (Taylor and Velbel, 1991). Although more mathematically restrictive and less likely to represent the behaviour of natural botanical systems, the introduction of these two assumptions simplifies the mass balance calculation by eliminating the necessity of introducing botanical coefficients into the equations.

In the present study, only the overall input – output behaviour of the system is assumed to be at steady state, and mathematical coefficients are introduced into the equations to allow elemental exchange with biomass (Taylor and Velbel, 1991). Data on the chemical composition of annual increments and/or net primary production (NPP) for ecosystems relevant to the study catchments in the southern and central Appalachians (northern hemisphere temperate hardwood forests) were retrieved from the ecological literature, to estimate the appropriate abundance ratios of major rock(weathering)-derived elements (K, Ca, Mg,Na) in biomass NPP in the individual study areas (Table 7.2). It is through the use of these ratios that biological effects are included in geochemical mass balance calculations (Velbel, 1985a, 1986; Taylor and Velbel, 1991).

In order to eliminate the possibility that the choice of a different stoichiometry for a mineral common to the two approaches is the cause of different results, this study uses the same stoichiometric coefficients for mineral weathering reactions as those of the original studies. In addition, to prevent different input parameters from dominating differences between previous results and the present study, the same (measured) net elemental fluxes (output minus input) are used.

7.4 SENSITIVITY OF CALCULATED WEATHERING RATES TO VARIATIONS IN ASSUMED BIOMASS COMPOSITION: NORTH CAROLINA REVISITED

Taylor and Velbel (1991) calculated weathering rates for seven control catchments in the Coweeta Hydrologic Laboratory, using elemental exchange ratios for biomass from a detailed study of one of the seven catchments. Because the biomass composition they used is derived from the forests of the study area itself, the rates so calculated are the best possible estimate, and inferences based on these calculations are likely to be quite reliable. However, to test how sensitive the rates would be to error or variation in the biomass composition, weathering rates for the seven Coweeta control catchments were recalculated using the three other biomass compositions compiled for this study.

The results are shown in Table 7.1. Although the numerical values of the calculated rates vary widely, two major points emerge:

1. The magnitude of the variation introduced into the estimates of weathering rates by changing the cation composition of the modelled forest biomass follows the pattern of sensitivity originally described by Taylor and Velbel, (1991). Rates of plagioclase weathering are not significantly affected by including or excluding biomass terms (Taylor and Velbel, 1991) or by varying the biomass composition used (Table 7.1), because the weathering rate of plagioclase is determined almost exclusively by the mass balance equation for Na. Sodium is

not a major plant nutrient, so, even when invoked, the botanical exchange coefficient is very small and has little effect on the Na mass balance equation. Rates of garnet weathering are invariably higher in the calculations which included botanical exchange than in the commonly used method which ignores biomass (Taylor and Velbel, 1991), regardless of the specific biomass composition used (Table 7.1). Garnet weathering rates are calculated from the mass balances of Ca and Mg, both of which are important plant nutrients. Biotite mica weathering rates are also invariably higher in the calculations which included botanical exchange than in the commonly used method which ignores biomass (Taylor and Velbel, 1991), again regardless of the specific biomass composition used (Table 7.1). Furthermore, the biotite weathering rates show the largest variability with changing biomass composition (Table 7.1) — in other words, of all the minerals investigated at Coweeta, the weathering rate of biotite is the most sensitive to variability in assumed biomass composition. The biotite weathering rate is calculated primarily from the mass balance equation for K, a major plant nutrient. The sensitivity of mineral weathering rates to biomass composition is most pronounced for minerals that contain major nutrient elements and for rates calculated from mass balances of those nutrient elements (Taylor and Velbel, 1991). The one exception to the described patterns (catchment 27) is discussed in more detail below.

2. The observed patterns of change in the calculated weathering rate with varying biomass composition are the same, regardless of the specific biomass composition used. The absolute value of the calculated rate may depend upon the specific composition used to calculate it. However, the rate calculated with any specific biomass included always bears the same relationship to the rates calculated without biomass, regardless of the biomass composition used. If the biomass-included rate using one biomass is higher than the rate calculated without biomass, the use of any other biomass also yields a rate higher than the rate calculated without biomass.

7.5 CHANGES IN CALCULATED WEATHERING RATES UPON INCORPORATION OF BIOMASS TERMS IN CATCHMENTS IN MARYLAND AND VIRGINIA

7.5.1 Maryland Piedmont: Pond Branch, Maryland

The geochemical mass balance of the Pond Branch (Maryland) catchment was studied by Cleaves *et al.* (1970). Pond Branch is situated in the Piedmont Province, is almost entirely forested and is underlain by the Lower Pelitic Schist Member of the Wissahickon Schist. The schist consists of quartz, plagioclase, muscovite, biotite and staurolite, with minor garnet, kyanite, apatite, tourmaline, zircon and chlorite. Mass balances for three elements (Na, Mg and SiO_2) and certain specific assumptions regarding the relative proportions were used to solve for the rates of four weathering reactions:

1. Plagioclase to kaolinite:

$$Ca_{0.22}Na_{0.78}Al_{1.22}Si_{2.78}O_8 + 0.61\ H_2O + 1.22\ H_2CO_3 \Rightarrow$$
$$0.22\ Ca^{2+} + 0.78\ Na^+ + 1.56\ SiO_{2(aq)} + 1.22\ HCO_3^- + 0.6\ Al_2Si_2O_5(OH)_4$$

2. Kaolinite to gibbsite:

$$Al_2Si_2O_5(OH)_4 + H_2O \Rightarrow 2\ Al(OH)_3 + 2\ SiO_{2(aq)}$$

3. Biotite to vermiculite:

$$3[K_2(Mg_3Fe_3)Al_2Si_6O_{20}(OH)_4] + 8\ H_2O + 12\ H_2CO_3 \Rightarrow$$
$$2[(Mg_3Fe_3)Al_3Si_5O_{20}(OH)_8 \times 8\ H_2O + 6\ K^+ + 3\ Mg^{2+} + 3\ Fe^{2+}$$
$$+ 8\ SiO_{2(aq)} + 12\ HCO_3{}^-$$

4. Vermiculite to kaolinite:

$$2[(Mg_3Fe_3)Al_3Si_5O_{20}(OH)_8 \times 8\ H_2O + 30\ H_2CO_3 \Rightarrow$$
$$3\ Al_2Si_2O_5(OH)_4 + 6\ Mg^{2+} + 6\ Fe^{2+} + 4\ SiO_{2(aq)} + 29\ H_2O + 30\ HCO_3{}^-$$

7.5.2 Maryland Blue Ridge

Katz *et al.* (1985) and Katz (1989) examined geochemical mass balances in several catchments in the Catoctin Mountains of Maryland, all of which are underlain by greenstone metabasalt of the Catoctin Formation. The Catoctin Formation consists of chlorite, albite, epidote and actinolite, with minor quartz and calcite.

7.5.2.1 *Hauver Branch Maryland*

Katz *et al.* (1985) studied the Hauver Branch catchment, which hosts a mature deciduous forest cover. Mass balance for four elements (Na, Ca, Mg and SiO_2) were used to solve for the rates of weathering reactions:

1. Albite to kaolinite:

$$NaAlSi_3O_8 + H^+ + 9/2\ H_2O \rightarrow Na^+ + 1/2\ Al_2Si_2O_5(OH)_4 + 2\ H_4SiO_4$$

2. Calcite dissolution:

$$CaCO_3 + H^+ \rightarrow Ca^{2+} + HCO_3{}^-$$

3. Chlorite to kaolinite:

$$Mg_5Al_2Si_3O_{10}(OH)_8 + 10\ H^+ \rightarrow 5\ Mg^{2+} + H_4SiO_4 + Al_2Si_2O_5(OH)_4 + 5\ H_2O$$

4. Actinolite dissolution:

$$Ca_2(Mg_3Fe_2)Si_8O_{22}(OH)_2 + 14\ H^+ + 8\ H_2O \rightarrow 2\ Ca^{2+} + 3\ Mg^{2+} + 2\ Fe^{2+} + 8\ H_4SiO_4$$

Fluxes were corrected for anthropogenic addition of deicing salts.

7.5.2.2 *Owens Creek, Maryland*

Katz (1989) examined the Owens Creek catchment, which is located several kilometres north of Hauver Branch. Bedrock geology (Catoctin Formation metabasalt and metarhyolite) and vegetation (hickory, poplar, birch and oak) in the Owens Creek catchment are similar to Hauver Branch. Mineral weathering reaction stoichiometries used to constrain the Owens Creek mass balance are identical to those listed above for Hauver Branch. In addition to correcting fluxes for deicing salts, Katz (1989) introduced a mass balance equation for K and included in the system of equations coefficients for botanical exchange of K, Ca and Mg in molar proportions derived from the forest ecosystem in Walker Branch, Tennessee, several hundred kilometres to the southwest. The biomass terms were invoked primarily to account for the abundance of K observed in the catchment efflux; note that none of the major weatherable minerals typical of the Catoctin metabasalt releases K upon weathering. Therefore, the observed excess of K in output relative to input must be derived from some non-geological source, such as decay of biomass.

7.5.3 Virginia Blue Ridge: Mill Run, Virginia

The Mill Run catchment, located in the Blue Ridge Mountains of Virginia, was studied by Afifi and Bricker (1983).Bedrock consists of the Massanutten Sandstone (Silurian), actually a low-grade metasandstone comprising predominantly quartz, with minor albitic plagioclase, orthoclase, chlorite and amphibole. Complete forest cover consists of second-growth mixed oak−hickory with scattered stands of conifers.

The original Afifi and Bricker (1983) mass balance invoked the following weathering reactions for the weatherable silicate minerals of the Massanutten Sandstone:

1. Albite to kaolinite:

$$NaAlSi_3O_8 + H^+ + 9/2\ H_2O \rightarrow Na^+ + 1/2\ Al_2Si_2O_5(OH)_4 + 2\ H_4SiO_4$$

2. Orthoclase to kaolinite:

$$KAlSi_3O_8 + H^+ + 9/2\ H_2O \rightarrow K^+ + 1/2\ Al_2Si_2O_5(OH)_4 + 2\ H_4SiO_4$$

3. Chlorite to kaolinite:

$$Mg_5Al_2Si_3O_{10}(OH)_8 + 10\ H^+ \rightarrow 5\ Mg^{2+} + H_4SiO_4 + Al_2Si_2O_5(OH)_4 + 5\ H_2O$$

4. Amphibole to kaolinite and goethite:
$$NaCa_2(Mg_3Fe_2)(Al_3Si_5)O_{22}(OH)_2 + 11\ H^+ + 3/2\ H_2O \rightarrow$$
$$Na^+ + 2\ Ca^{2+} + 3\ Mg^{2+} + 3/2\ Al_2Si_2O_5(OH)_4 + 2\ FeOOH + 2\ H_4SiO_4$$

Each of the catchment studies listed above produced silicate weathering rates estimated from geochemical mass balance. The rates reported by the original researchers or recalculated for this study are shown in Table 7.3. Also shown in Table 7.3 are the weathering rates recalculated in this study using the original stoichiometries and fluxes, with the addition of stoichiometric coefficients for biomass.

7.6 DISCUSSION

Cleaves *et al.* (1970) were among the first to address the impact of biomass on mass balance calculations of mineral weathering rates. They observed that, when mass balance for Na, Mg and silica was achieved, the weathering reactions produced more K and Ca than was observed in the catchment output. From this they concluded that K and Ca were taken up by an unmeasured sink within the catchment and that the unmeasured sink was the forest. Likens *et al.* (1977) showed that failure to account for botanical uptake of elements resulted in estimated weathering rates which underestimated the actual rates by a factor of two. Despite these demonstrations, only a relatively small number of mass balance studies (Paces 1983; Schnoor and Stumm, 1985; Velbel, 1985a, 1992; Clayton, 1988; Sverdrup, 1988; Sverdrup and Warfvinge, 1988; Katz, 1989; Taylor and Velbel, 1991; Probst *et al.* 1992) have incorporated botanical uptake and/or release terms into their mass balance equations. However, most mass

TABLE 7.3 Mineral weathering rates from catchment balances of Maryland and New York Catchments with alternate biomass

Catchment	Mineral	Original published rate	Biomass			
			Coweeta	Hubbard Brook	UK deciduous	Walker Branch
Pond, Branch, Maryland	Plagioclase	148	148	147	146	146
	Biotite	32	8	7	6	5
	Vermiculite	12	6	5	6	7
	Kaolinite	30	37	44	45	50
Hauver Branch, Maryland	Albite	316	315	315	316	316
	Calcite	249	235	227	226	192
	Chlorite	115	113	114	113	112
	Actionilite	23	24	24	23	24
Owen and Hunting Creek, Maryland[a]	Albite	45	44	44	45	45
	Calcite	479	511	504	502	469
	Chlorite	11	10	10	10	8
	Actionolite	22	26	26	26	26
Mill Run, Virginia[b]	Albite	73	66	60	58	13
	Orthoclase	121	137	139	140	151
	Chlorite	40	37	33	32	10
	Amphibole	17	25	31	33	77

Units: moles of reaction (as written in text) per hectare per year.
[a] Original rate calculated from mass balance including biomass terms. For all others, the original published rate was determined from mass balances that excluded consideration of botanical factors.
[b] Original rate in micromoles litre^{-1} converted using measured discharge.

balance studies in forested catchments, and some stream acidification models, leave out botanical exchange terms entirely.

The widespread neglect of botanical factors in geochemical estimates of weathering and soil formation rates stems from misunderstanding of the nature of "steady state" as it applies to the biological compartment of terrestrial biogeochemical cycles (Taylor and Velbel, 1991). At the landscape scale, "steady state" in forest biomass exists in the form of the "shifting-mosaic steady state" of Bormann and Likens (1981), in which a steady-state landscape consists of a geographic patchwork of different ecosystems, each at a different stage of succession and nutrient demand. The landscape-scale distribution of these successionally diverse patches reaches a steady-state distribution if the landscape as a whole is free from large-scale regional perturbations. On geologic time-scales and global spatial-scales, the local and temporary fluctuations in elemental storage and cycling by the shifting mosaic of ecosystem stages average out spatially and temporally. However, on the time-scales (months to decades) and the spatial-scales (small forested catchments) of interest in the present discussion, ignoring botanical exchange effects can cause large errors in weathering and soil-formation rates calculated from elemental budgets (Taylor and Velbel, 1991).

Taylor and Velbel (1991) proposed that the ratios of weathering rates calculated with and without including expressions for botanical exchange are related to the state of the catchment forest ecosystems. Table 7.1 illustrated this. For most Coweeta control catchments, mineral weathering rates are greater when calculated with biomass terms in the mass balance than when calculated without such terms (the ratio of "biotic/abiotic" rates exceeds unity). If the botanical compartment is actively taking up nutrients, the measured net cation (e.g. K, Ca, Mg) efflux (by streams) is not the only output term; biomass is also an output from the mineral compartment (Sollins *et al.*, 1980). Because both rates are calculated using the same net solute fluxes, this behaviour implies that (a) more weathering must take place to produce the observed flux when biomass is included than when biomass is excluded and (b) there must be an intra-catchment sink to account for the excess solutes that are released by the greater amount of weathering but that do not show up in the solute effluxes. The most likely sink is the catchment forest biomass itself.

The role of net uptake (aggradation) of forest biomass is discussed in detail by Taylor and Velbel (1991). Not all of the mineral nutrient elements in the growing biomass come directly from the mineral compartments. A given year's net primary production includes a large proportion of mineral nutrients recycled from decay of the previous year's litter. Much of the nutrient content of "new" tissue (e.g. leaves, sapwood) consists of nutrients translocated from other parts of the tree (Whittaker *et al.*, 1979; Tanner, 1985: Mahendrappa *et al.*, 1986; Monk and Day, 1988). again indicating that the nutrients in new tissue are only partly derived from the minerals in the soil. Taken together, these observations indicate that some 50−90% of the annual uptake from the soil litter pool is recycled; only the remaining 10−50% are "new" additions, and these can be added by both atmospheric input (Swank and Henderson, 1976) and weathering (Taylor and Velbel, 1991). It is the weathering portion of this "new" input that is estimated by the mass balance calculations. The biomass "uptake" calculated by geochemical mass balance represents only the annual increment of nutrient "withdrawal" from the mineral compartment, and is therefore considerably less than the actual net primary production measured by the forest ecologists.

Mineral components of the forest soil are included in the calculations, through the stoichiometric coefficients of the mineral weathering reactions. Organically mediated non-biologic soil materials such as calcium oxalate accumulations can be regarded as internal components of the biological nutrient cycle, which do not act as sinks for soil-derived nutrients, but as "regulators" of botanically recycled nutrients. Graustein *et al.* (1977) and Cromack *et al.* (1979) observed that fungal hyphae in contact with mineral fragments are devoid of calcium oxalate crystals, and concluded that the oxalate has been removed from the weathering site by forming soluble complexes with Fe and Al derived from the weathering of the mineral fragments. Any Ca released by the weathering of the mineral fragment was similarly mobilized, because there is no free, uncomplexed oxalate with which it may precipitate. In contrast, fungal hyphae in the spaces between soil grains possess calcium oxalate crystals. The Ca in these crystals is probably not the same Ca that was released by weathering of nearby mineral fragments: it is more likely to be Ca "stripped" from the ambient soil solution percolating through the intergranular pore spaces.

The "ambient" Ca may be partially derived from mineral weathering, but must also include significant proportions of atmospherically derived Ca and/or recycled Ca released by the decay of plant litter on the forest floor. If the botanical cycle of Ca is similar to that of chemically analogous Sr, the forest biota, litter and soil may be dominated by externally (atmospherically) derived Ca, either primary or recycled. Graustein (Graustein, 1981, 1989; Graustein and Armstrong, 1983) showed that Sr in forest biomass was derived largely from atmospheric input and that plant Sr is largely botanically recycled Sr of atmospheric origin. Furthermore,

Graustein found that groundwaters and streamwaters of his study area bore no imprint of atmospherically derived Sr, implying that the atmospherically derived nutrient is effectively stripped from incoming solutions and recycled botanically. Leakage of Sr from the botanical cycle is minimal, relative to the release of bedrock Sr to groundwaters below the soil. In other words, the contribution to the groundwaters by weathering of bedrock minerals must be much larger than the contribution to the groundwater by leakage of atmospherically derived nutrients from the botanical cycle. If the botanical cycle of Ca is as tight as that of Sr, this may imply that the ecological function of forest floor oxalate may be more to regulate the botanical recycling ("tightness") of recycled Ca, rather than to extract Ca from soil minerals. Externally derived and recycled Ca are stored in fungal mats of calcium oxalate; the mineral-derived Ca, like the Sr with "bedrock" isotopic signature (Graustein, 1981, 1989; Graustein and Armstrong, 1983), is released from soils minerals, and much or most of it passes out of the soil and into the groundwaters and streamwaters.

Fungally regulated Ca is most likely either primary or recycled atmospheric Ca. If it is atmospheric (Swank and Henderson, 1976), it is implicitly included in the Δm_c term of eqn (1); if recycled, it is merely storage, derived from the decay of the plant tissue with known composition. Growth of this plant tissue by removal of nutrients from the dissolved compartment is explicitly incorporated into the mass balance equations. Botanical recycling is an internal component of the biomass term in the dissolved mineral-nutrient budget and need not be considered separately, Non-biological forest soil components (e.g. forest floor mats of fungal calcium oxalate, etc.) are a regulatory device in the botanical nutrient cycles, and are not sinks or reservoirs of elements derived from mineral weathering.

Degradation of biomass can add elements to the efflux (thereby returning more elements to the solutions than are taken up into new growth; see Whittaker *et al.*, 1979; Sollins *et al.*, 1980: Jordan, 1982), which would be an internal source (Katz, 1989). In such cases, mineral weathering rates are slower when calculated with biomass terms in the mass balance than when calculated without such terms (the ratio of "biotic/abiotic" rates is less than unity). This implies that (a) less weathering must take place to produce the observed flux when biomass is included than when biomass is excluded and (b) there must be an intra-catchment source to supply the additional solutes that must be added to the amount supplied by weathering to produce the solute effluxes. The most likely source is the catchment forest biomass itself. This behaviour is known from managed landscapes (Katz, 1989) and from a few undisturbed forested catchments (Taylor and Velbel, 1991).

Interpreted in this light, the ratio of "biotic/abiotic" weathering rates can distinguish catchments in which forest biomass was aggrading over the period of solute sampling from those with declining biomass (Taylor and Velbel, 1991). This hypothesis is borne out by the known state of the forest ecosystems in the seven Coweeta control catchments. Most Coweeta control catchments exhibit increases in calculated weathering rates when biomass exchange coefficients are introduced (indicating net uptake by biomass; see Taylor and Velbel, 1991). However, one catchment (catchment 27) exhibits slight decreases, suggesting net loss of biomass, adding nutrients of biological origin to the stream efflux (Taylor and Velbel, 1991). Catchment 27 is the only control catchment on Coweeta Group rocks which was partially defoliated by fall cankerworm during the 1970s (Swank and Douglass, 1977; Swank and Crossley, 1988). Taylor and Velbel (1991) conclude that the "anomalous" ratios for mineral weathering rates in catchment 27 calculated with and without biomass exchange terms are due to decay of defoliation products. Catchment 36 (on rocks of the Tallulah Falls Formation) also experienced fall cankerworm infestation over part of the same period, but the infestation on catchment 36 may not have been extensive enough to cause extensive biomass decay or otherwise detectably perturb major element budgets. The possibility that the effects of fall cankerworm infestation were significant on catchment 27, and present but less significant on catchment 36, is supported by two observations; infestation was of shorter duration on catchment 36 than on catchment 27, and perturbations of budgets of solutes which are extremely sensitive to biomass disturbance (e.g. nitrate nitrogen) were significantly smaller on catchment 36 than on catchment 27 (Swank and Waide, 1988, Tables 4.6 and 4.11). Thus, the results of Taylor and Velbel's (1991) mass balance calculations with biomass terms included are consistent with what is known about the actual state of the forest ecosystems in the modelled catchments.

Table 7.1 of this study shows that, although the magnitude of the weathering rates calculated when biomass is included can vary widely, the ratio of biomass-inclusive/biomass-exclusive rates exhibits the same behaviour described by Taylor and Velbel (1991) using only local biomass composition. For all catchments except catchment 27, rates calculated including biomass are greater than rates calculated without biomass, implying that the forest biomass is a net internal sink for major rock-derived cations. Only for the catchment with degrading biomass (catchment 27) does the incorporation of botanical terms decrease the calculated weathering rate, implying that the biomass is acting as an additional source of major cations to the solute effluxes. This conclusion is robust in the face of any possible uncertainty regarding the biomass composition; the same relationships hold regardless of the specific biomass composition used to make the calculation.

The results from catchments in the North Carolina Blue Ridge suggest that catchment mass balances can distinguish catchments in which forest biomass was aggrading over the period of solute samples (Coweeta catchments 2, 14, 18, 34 and 36) from those with declining biomass (Coweeta catchment 27). By this reasoning, Pond Branch and Hauver Branch, Maryland, were catchment systems with declining forest biomass (Table 7.3), at least during their respective periods of study. Biomass in the Owen Creek, Maryland, catchment was also declining; the original mass balance of Katz (1989) invokes biomass decay to achieve potassium balance, and the mass balance is insensitive to varying biomass composition (Table 7.3). Thus, in contrast to the pattern of aggrading biomass in the North Carolina catchments, geochemical mass balances suggest a pattern of general loss of forest biomass in all catchments previously studied in the state of Maryland. In order to explain observed solute fluxes, weathering rates for orthoclase (potassium feldspar) in the Mill Run, Virginia, catchment must be higher when biomass is accounted for (Table 7.3). This suggests that the forest in the Mill Run catchment was most likely an aggrading ecosystem; this is consistant with its obvious second-growth character.

7.7 SUMMARY AND CONCLUSIONS

Ecosystem processes are important in geochemical mass balances of catchments. The uptake or release of common rock-derived elements (K, Ca, Mg) during aggradation or degradation of forest biomass can have noticeable effects on catchment solute mass balances, and on weathering rates derived from these mass balances. The magnitude of "error" introduced into the calculated weathering rates by omitting biomass from the mass balance depends largely on the botanical significance of each specific element. Weathering rates vary the least for minerals whose rate is calculated from the mass balance for botanically unimportant elements (e.g. sodic plagioclase feldspar, whose weathering rate is determined from the catchment mass balance for Na; see Tables 7.1 and 7.3). The most variable and uncertain weathering rates are those for minerals that supply botanically important elements. Weathering rates for biotite mica (whose weathering rate is determined from the catchment mass balance for K) vary by a factor of six or more, depending on whether biomass is accounted for in the mass balance or not (Tables 7.1 and 7.3), and can vary by more than a factor of two with different choices of biomass composition (Table 7.1). Thus, the determination of weathering rates of individual minerals in catchment ecosystems can be very sensitive to the precise manner in which botanical factors are accounted for in constructing the mass balance.

The choice of how to treat ecosystems (as an input, or perhaps as a governing factor) depends in part on the purpose of the specific study, in part on how much is already known about a specific system and in part on the resources available. Where known in great detail (from ecological nutrient cycle studies; e.g. Coweeta, Hubbard Brook), ecosystem processes might be treated as a known set of inputs to catchment solute models. However, where site-specific information is lacking, it may be informative to use reasonable estimates of biomass composition (as was done in this study) and to treat the magnitude of elemental cycling as an unknown, to be freely solved for along with the weathering rates. Sensitivity of the calculated rates to variation in the selected botanical compositions can be done with selected specific examples (as in this study) or more rigorously (statistically, by, for instance, creating biomass compositions from random numbers between specified limits).

Ecosystem processes cannot be neglected in catchment solute mass balances. Catchment mass balance studies can be undertaken for a variety of reasons. The goal of a particular study may be to estimate the rate of replenishment of soil nutrient pools by primary-mineral weathering (a common ecological application of catchment mass balances), the rate of supply of "base" cations in soil cation exchange complexes and/or soil and surface waters (a common need in studies of environmental acidification) or the rate of denudation of the landscape (a common geological application). All such studies commonly (although not invariably) invoke a "lumped-parameter" modelling approach, in which key parameters (in this case, bulk "weathering" rates at which specific cations are supplied to the rest of the system by "weathering") are assigned a single value intended to represent the behaviour of the entire catchment. However, such approaches are of limited predictive value. Rates of elemental release from different mineral sources (e.g. Ca release by weathering of calcite versus calcic silicates) have very different functional dependences on various environmental parameters (solution pH, soil temperature, hydraulic flow rates; see Drever and Hurcomb, 1986). It is therefore not enough to know the "bulk" rate of, for instance, Ca release from the rocks and soils of a catchment. One must apportion the weathering contribution of Ca among the various contributing mineral species, as is done by solving systems of simultaneous linear equations for the weathering rates of individual minerals. Furthermore, quantifying the contributions of individual mineral species cannot be accomplished in the absence of botanical information. The weathering rate of biotite in Coweeta catchment 2 is nearly four times higher when biomass is accounted for than the rate estimated from solute fluxes alone. Thus, biotite weathering is four times as effective at releasing K than might otherwise be thought. Factor-of-four variations like this may not be trivial when estimating sustainable silvicultural harvest intervals.

In order to predict how solute budgets will respond to environmental stress (removal of forest products, acid deposition, global warming), one must know what mineral species (individually or in combination) are responsible for the "bulk" release rates. Regardless of the potential application, physically meaningful apportionment of elements among realistic suites of sources and sinks is essential to a proper understanding of the system. Both biotic and abiotic compartments of the catchment ecosystem must be included.

ACKNOWLEDGEMENTS

The author is grateful to O. B. Bricker, J. I. Drever, L. O. Hedin, A. B. Taylor, N. L. Romero, K. Rice, J. F. Dowd, B. P. Boudreau, E. T. Cleaves, D. S. Brandt and W. T. Swank for valuable criticisms, comments and encouragement at all stages of the work discussed here.

REFERENCES

Afifi. A. A. and Bricker, O. P. (1983). "Weathering rates, water chemistry and denudation rates in drainage basins of different bedrock types: I — sandstone and shale", in *Dissolved Loads of Rivers and Surface Water Quantity/Quality Relationships*, IAHS Publication 141, pp. 193–203.

Berner, E. K. and Berner, R. A. (1987). *The Global Water Cycle: Geochemistry and Environment*, Prentice-Hall, Englewood Cliffs, New Jersey.

Binkley, D., Driscoll, C. T., Allen, H. L., Schoeneberger, P. and McAvoy, D. (1989). *Acidic Deposition and Forest Soils: Context and Case Studies in the Southeastern United States*, Springer Ecological Studies Series No. 59, 119 pp.

Boring, L. R., Monk, C. D. and Swank, W. T. (1981). "Early regeneration of a clear-cut southern Appalachian forest", *Ecology*, **62**, 1244–1253.

Bormann, F. H. and Likens, G. E. (1981). *Pattern and Process in a Forested Ecosystem* (second corrected printing), Springer-Verlag, New York, 253 pp.

Clayton, J. L. (1979) "Nutrient supply to soil by rock weathering", in *Impact of Intensive Harvesting on Forest Nutrient Cycling*, pp. 75–96, State University of New York, College of Environmental Science and Forestry, Syracuse, New York.

Clayton, J.L. (1988). "Some observations on the stoichiometry of feldspar hydrolysis in granitic soil", *J. Environ. Qual.*, **17**, 153–157.

Cleaves, E. T., Godfrey, A. E. and Bricker, O. P. (1970). "Geochemical balance of a small watershed and its geomorphic implications", *Geol. Soc. Am. Bull.*, **18**, 3015–3032.

Cole, D. W. and Rapp, M. (1981). "Elemental cycling in forest ecosystems", in *Dynamic Principles of Forest Ecosystems* (Ed. D.E. Reichle), pp. 341–409, Cambridge University Press, London and New York.

Cosby, B. J., Hornberger, G. M., Galloway, J. N. and Wright, R. F. (1985a). "Modelling the effects of acid deposition: assessment of a lumped-parameter model of soil water and stream water chemistry", *Water Resour. Res.*, **21**, 51–63.

Cosby, B. J., Hornberger, G. M., Galloway, J. N. and Wright, R. F. (1985b). "Freshwater acidification from atmospheric deposition of sulfuric acid: a quantitative model", *Environ. Sci. Technol.*, **19**, 1144–1149.

Cromack, K. Jr, Sollins, P., Graustein, W. C., Speidel, K., Todd, A. W., Spycher, G., Li, C. Y. and Todd, R. L. (1979). "Calcium oxalate accumulation and soil weathering in mats of hypogeous fungus *Hysterangium crassum*", Soil Biol. Biochem., **11**, 463–468.

Day, F. P. and Monk, C. D. (1977). "Seasonal nutrient dynamics in the vegetation on a southern Appalachian watershed". *Am. J. Botany*, **64**, 1126–1139.

Drever, J. I. and Hurcomb, D. R. (1986). "Neutralization of atmospheric acidity by chemical weathering in an alpine drainage basin in the North Cascade Mountains", *Geology*, **14**, 221–224.

Frogner, T. (1990). "The effect of acid deposition on cation fluxes in artificially acidified catchments in western Norway", *Geochim. Cosmochim. Acta*, **54**, 769–780.

Garrels, R. M. and Mackenzie, F. T. (1967). "Origin of the chemical compositions of some springs and lakes", in *Equilibrium Concepts in Natural Water Systems* (Ed. W. Stumm), American Chemical Society Advances in Chemistry Series No. 67, pp. 222–242.

Giovanoli, R., Schnoor, J. L., Sigg, L., Stumm, W. and Zobrist, J. (1988). "Chemical weathering of crystalline rocks in the catchment area of acidic Ticino lakes, Switzerland", *Clays, Clay Minerals*, **36**, 521−529.

Graustein, W. C. (1981). "The effect of forest vegetation on solute acquisition and chemical weathering: a study of the Tesuque watersheds near Santa Fe, New Mexico", Unpublished PhD dissertation, Yale University, New Haven, Connecticut, 645 pp.

Graustein, W. C. (1989). "$^{87}Sr/^{86}Sr$ ratios measure the sources and flow of strontium in terrestrial ecosystems", in *Stable Isotopes in Ecological Research* (Eds. P. W. Rundel, J. R. Ehleringer and K. A. Nagy), pp. 491−512, Springer-Verlag, New York.

Graustein, W. C. and Armstrong, R. L. (1983). "The use of strontium-87/strontium-86 ratios to measure atmospheric transport into forested watersheds", *Science*, **229**, 289−292.

Graustein, W. C., Cromack, K. Jr and Sollins, P. (1977). "Calcium oxalate: occurrence in soils and effect on nutrient and geochemical cycles", *Science*, **198**, 1252−1254.

Jordan, C. F. (1982). "The nutrient balance of an Amazonian rain forest', *Ecology*, **63**, 647-654.

Katz, B. G. (1989). "Influence of mineral weathering reactions on the chemical composition of soil water, springs, and groundwater, Catoctin Mountains, Maryland", *Hydrol. Process.*, **3**, 185−202.

Katz, B. G., Bricker, O. P. and Kennedy, M. M. (1985). "Geochemical mass-balance relationships for selected ions in precipitation and stream water, Catoctin Mountains, Maryland", *Am. J. Sci.*, **285**, 931−962.

Kaufmann, P. R., Herlihy, A. T., Mitch, M. E., Messer, J. J. and Overton, W. S. (1991). "Stream chemistry in the eastern United States, 1. Synoptic survey design, acid-base status, and regional patterns", *Water Resour. Res.*, **27**, 611−627.

Likens, G. E., Bormann, F. H., Pierce, R. S., Eaton, J. S. and Johnson, N. M. (1977). *Biogeochemistry of a Forested Ecosystem*, Springer-Verlag, New York, 146 pp.

Mahendrappa, M. K., Foster, N. W., Weetman, G. F. and Krause, H. H. (1986). "Nutrient cycling and availability in forest soils", *Can. J. Soil Sci.*, **66**, 547−572.

Mast, M. A., Drever, J. I. and Baron, J. (1990). "Chemical weathering in the Loch Vale watershed, Rocky Mountain National Park, Colorado", *Water Resour. Res.*, **26**, 2971−2978.

Monk, C. D. and Day, F. P. Jr (1988). "Biomass, primary production, and selected nutrient budgets for an undisturbed watershed", in *Forest Hydrology and Ecology at Coweeta* (Eds. W.T. Swank and D.A. Crossley Jr), pp. 151−159, Springer-Verlag, New York.

Paces, T. (1983). "Rate constants of dissolution derived from the measurements of mass balance in hydrological catchments", *Geochim. Cosmochim. Acta*, **47**, 1855−1862.

Plummer, L. N. and Back, W. (1980). "The mass balance approach: application to interpreting the chemical evolution of hydrologic systems", *Am. J. Sci.*, **280**, 130−142.

Probst, A., Viville, D., Fritz, B., Ambroise, B. and Dambrine, E. (1992). "Hydrogeochemical budgets of a small forested granitic catchment exposed to acid deposition: the Strengbach catchment case study (Vosges Massif, France)", *Water, Air and Soil Pollut.*, **62**, 337−347.

Reuss, J. O., Cosby, B. J. and Wright, R. F. (1987). "Chemical processes governing soil and water acidification", *Nature*, **329**, 27−32.

Reynolds, R. C. and Johnson, N. M. (1972). "Chemical weathering in the temperate glacial environment of the Northern Cascade Mountains", *Geochim. Cosmochim. Acta*, **36**, 537−554.

Schnoor, J. L. and Stumm, W. (1985). "Acidification and aquatic and terrestrial systems", in *Chemical Processes in Lakes* (Ed. W. Stumm), pp. 311−338, John Wiley, Chichester.

Sollins, P., Grier, C. C., McCorison, F. M., Cromack, K. Jr, Fogel, R. and Fredriksen, R. L. (1980). "The internal element cycles of an old-growth Douglas-fir ecosystem in western Oregon", *Ecol. Monogr.*, **50**, 261−285.

Stauffer, R. E. (1990). "Granite weathering and the sensitivity of alpine lakes to acid deposition", *Limnol. Oceanogr.*, **35**, 1112−1134.

Sverdrup, H. (1988). "Calculation of critical loads for acid deposition", *Vatten*, **44**, 231−236.

Sverdrup, H. and Warfvinge, P. (1988). "Weathering of primary silicate minerals in the natural soil environment in relation to a chemical weathering model", *Water, Air and Soil Pollut.*, **38**, 387−408.

Swank, W. T. and Crossley, D. A. (1988). "Introduction and site description", in *Forest Hydrology and Ecology at Coweeta* (Eds. W. T. Swank and D. A. Crossley Jr), pp. 3−16, Springer-Verlag, New York.

Swank, W. T. and Douglass, J. E. (1977). "Nutrient budgets for undisturbed forest ecosystems in the mountains of North Carolina", in *Watershed Research in Eastern North America* (Ed. D.L. Correll), pp. 343–364, Smithsonian Institution.

Swank, W. T. and Henderson, G. S. (1976). "Atmospheric inputs of some cations and anions to forest ecosystems in North Carolina and Tennessee", *Water Resour. Res.*, **12**, 541–546.

Swank, W. T. and Waide, J. B. (1988). "Characterization of baseline precipitation and stream chemistry and nutrient budgets for control watersheds", in *Forest Hydrology and Ecology at Coweeta* (Eds. W. T. Swank and D. A. Crossley Jr), pp. 57–79, Springer-Verlag, New York.

Tanner, E. V. J. (1985). "Jamaican montane forests: nutrient capital and cost of growth", *J. Ecol.*, **73**, 553–568.

Taylor, A. B. and Velbel, M. A. (1991). "Geochemical mass balance and weathering rates in forested watersheds of the southern Blue Ridge. II. Effects of botanical uptake terms", *Geoderma*, **51**, 29–50.

Turner, R. S., Cook, R. B., Van Miegrot, H., Johnson, D. W., Elwood, J. W., Bricker, O. P., Lindberg, S. E. and Hornberger, G. M. (1990). "Watershed and lake processes affecting surface water acid–base chemistry", in *Acidic Deposition: State of Science and Technology*, NAPAP Report 10, Washington, D.C., National Acid Precipitation Assessment Program, 167 pp.

Velbel, M. A. (1985a). "Geochemical mass balances and weathering rates in forested watersheds of the southern Blue Ridge", *Am. J. Sci.*, **285**, 904–930.

Velbel, M. A. (1985b). "Hydrogeochemical constraints on mass balances in forested watersheds of the southern Appalachians", in *The Chemistry of Weathering* (Ed. J.I. Drever), pp. 231–247, D. Reidel, Dordrecht.

Velbel, M. A. (1986). "The mathematical basis for determining rates of geochemical and geomorphic processes in small forested watersheds by mass balance: examples and implications", in *Rates of Chemical Weathering of Rocks and Minerals* (Eds. S.M.Colman and D.P. Dethier), pp. 439–451, Academic Press, Orlando.

Velbel, M. A. (1992). "Geochemical mass balances and weathering rates in forested watersheds of the southern Blue Ridge, III. Cation budgets in an amphibolite watershed, and weathering rates of plagioclase and hornblende", *Am. J. Sci.*, **292**, 58–78.

Whittaker, R. H., Likens, G. E., Bormann, F. H., Eaton, J. S. and Siccama, T. G. (1979). "The Hubbard Brook ecosystem study: forest nutrient cycling and element behaviour", *Ecology*, **60**, 203–220.

8

Forest Manipulation and Solute Production: Modelling the Nitrogen Response to Clearcutting

B. REYNOLDS

Institute of Terrestrial Ecology, Bangor Research Unit, University of Wales Bangor, UK

W. H. ROBERTSON

Department of Statistics and Modelling Science, University of Strathclyde, Glasgow, UK

M. HORNUNG

Institute of Terrestrial Ecology, Merlewood Research Station, Grange-over-Sands, UK

and

P. A. STEVENS

Institute of Terrestrial Ecology, Bangor Research Unit, University of Wales Bangor, UK

8.1 INTRODUCTION

Over the last 25 years there have been many studies of the impact of forest growth and management on the solute chemistry of drainage waters. These studies have involved monitoring of undisturbed systems, of standard forestry practice and of experimental manipulations. The majority of the manipulative studies have involved assessment of the impact of harvesting regimes/practices. Perhaps the most famous manipulative study is the experimental clearcutting carried out in the early 1970s at Hubbard Brook in New Hampshire (Likens *et al.*, 1977: Bormann and Likens, 1979). The results of this study gave rise to concerns about the impact of clearcutting on water quality and site fertility and spawned a

Solute Modelling in Catchment Systems. Edited by Stephen T. Trudgill
© 1995 John Wiley & Sons Ltd

generation of studies on the impacts of harvesting in many countries, including the United Kingdom.

Clear cutting represents a dramatic perturbation of the ecosystem which influences the rate and/or direction of all the major ecosystem processes with the consequent impacts on water and element fluxes into and out of catchment ecosystems and between pools within the system. Removal of the forest canopy changes atmosphere−canopy interactions, with a reduction in atmospheric inputs of solutes, gases and particles, and also leads to a reduction in interception losses of moisture. Plant uptake is much reduced, thus eliminating the main sink for nutrient solutes in soil solution. Evapotranspiration is reduced with a consequent increase in drainage fluxes. The climate at ground level is changed with increases in both the average and range of temperature and moisture conditions, which will influence the rates of biological processes such as decomposition, mineralisation and nitrification. There is a sudden large input of litter from felling debris and the death of fine roots.

The impact of these changes on the fluxes and concentrations of solutes varies with site and soil properties but a number of trends are common to the results from a range of sites. There is often an increase in nitrate, potassium and phosphorus concentrations in the fluxes in soil solution and drainage waters but a reduction of sulphate, chloride and sodium (Likens *et al.*, 1970; Knighton and Stiegler, 1981; Hornbeck *et al.*, 1986). The changes in outputs of base cations or aluminium show considerable variations between sites, reflecting site properties but also the interactions with the changes in anion outputs (Johnson *et al.*, 1988; Lawrence *et al.*, 1987). However, the directions of the changes in the concentrations and fluxes of these ions can be predicted, given information on site and soil properties.

The direction of the response of a given forest system to manipulations, in terms of the concentrations and fluxes of solutes, can be predicted given information on the atmospheric inputs, soil and site properties, but the magnitude and timing of the response is still uncertain. These aspects of the response are the net result of the interaction of a complex of physical and biologically mediated processes. The investigation of these interactions and improved predictions of response are being explored through the development of dynamic models. In most cases the model development has been linked to field and plot-based studies which have investigated whole system responses and specific processes and provided tests for model outputs.

This chapter considers the linking of system, process and modelling studies in a series of investigations in the United Kingdom. The series of catchment and plot-based studies have investigated the response of plantation forest systems to clearcutting and to variations in the intensity of harvesting. From the results of these studies we have chosen to examine the behaviour of nitrogen because it demonstrates the role of the major ecosystem processes, and the interactions of these processes, in controlling the response of the system. Furthermore, the results have been used in the development and testing of a dynamic model of nitrate production and leaching following forest harvesting.

8.2 EFFECTS OF FOREST HARVESTING ON SITE NITROGEN DYNAMICS IN THE UNITED KINGDOM

Since the early 1980s a number of plot and catchment-scale experiments have been undertaken in the United Kingdom to determine the effects of forest harvesting on stream water quality and long-term nutrient status (Reynolds and Stevens, 1993a). These were initiated because the combination of site type and forest management practice found in the United Kingdom was not represented in the wealth of studies from overseas and because large areas of plantation forestry were approaching harvestable age.

In the British uplands, commercial forestry relies almost entirely upon even-aged stands of exotic conifers, predominantly Sitka spruce (*Picea sitchensis* (Bong.) Carr.). The crop is usually harvested at between 40 and 50 years of age when entire stands of trees are clearcut. A few individuals or groups of trees may be left if these have particular amenity value. Conventional harvesting (CH) practice involves removal of the bole, leaving felling debris or "brash" (branches, needles and stems < 7 cm in diameter) randomly spread across the site. On wetter sites the brash may be distributed in parallel rows (windrows). These provide routes for timber extraction machinery, thereby protecting the surface from rut formation and erosion. The brash remaining on site provides a potential source of nutrients to the subsequent forest crop which is normally planted within two years of felling. The presence of brash also makes conditions difficult for trees to establish, so gaps left by felled trees have to be filled in subsequent years. However, for established seedlings, brash can provide protection from weed competition and drought.

An alternative to conventional harvesting that has found favour at various times but has not been widely practised is whole-tree harvesting (WTH). This technique involves removal of all above- ground tree components from the site and provides clear ground and easier, more successful replanting. Loss of nutrients from the site through removal of the nutrient-rich needles and branches is substantial, however.

The harvesting of conifers in the British uplands was not expected to produce large quantities of nitrate in runoff waters (Heal *et al.*, 1982). It was thought that the combination of large C/N ratios in the crop residues, low temperatures and wet clay-rich soils would limit decomposition, mineralisation and nitrification processes and that microbial immobilisation would further reduce nitrate leaching losses. Results from the British studies have now shown that these predictions were generally inaccurate (Stevens and Reynolds 1993a), and nitrate losses similar to those described in the United States have been observed following clearcutting of British forests.

8.2.1 Stream water response

At Kershope Forest in northern England (Table 8.1), where 2 ha plots were 100% clearcut, mean nitrate N concentrations in outlet drains were 2 mg N l^{-1} before felling, but rose to around 3−6 mg N l^{-1} within one year of felling, followed by a decline to levels similar to those in the unfelled reference catchment in the fourth year after felling (Adamson *et al.*, 1987; Adamson and Hornung, 1990). Annual losses of nitrate N in the two years after felling were between 35 and 40 kg N ha^{-1} yr^{-1} compared with less than 10 kg N ha^{-1} yr^{-1} before felling. This increase was the result of the higher nitrate concentrations and an increase from 50 to 68% in the proportion of rainfall leaving the site as runoff (Anderson *et al.*, 1990).

At Beddgelert Forest in North Wales (Table 8.1), where stream chemistry has been monitored since 1982, one 4.7 ha stream catchment (D3) has remained as an unfelled reference, whilst 62 and 28% of two other catchments (D2, 1.4 ha, and D4, 6.1 ha) were harvested respectively in September and June 1984 (Stevens and Hornung, 1988). As the catchments were only partially clearcut, changes in response to felling were difficult to

TABLE 8.1 A summary of the forest harvesting experiments described in this chapter

Site	Soils/drift/geology	Crop and age at harvesting	Exptl. design	Measurements	Source references
Kershope forest, Northern England	Peaty gley over glacial boulder clay	Sitka spruce, 35 years old	2 ha plots bounded by drainage ditches	Soil and stream water chemistry, hydrology	Adamson et al. (1987); Adamson and Hornung (1990); Anderson et al. (1990)
Beddgelert forest, North Wales	Podzols on lower Palaeozoic slates and shales	Sitka spruce, 50 years old	Plot studies +4 small catchments	Soil and stream water chemistry, nutrient cycling studies	Stevens and Hornung (1988); (1990); Stevens et al. (1989)
Plynlimon forest, Mid Wales	Podzolic and peaty gley soils on lower Palaeozoic mudstones, shales and grits	Sitka and Norway spruce, 38–40 years old	Large and small catchments + plot studies	Soil and stream water chemistry, hydology, sediments	Neal et al. (1992); Reynolds et al. (1992)
Kielder forest, Northern England	Pety gleys over glacial boulder clay	Sitka spruce, 40–42 years old	Plot studies	Soil water chemistry, nutrient cycling studies	Titus (1985); Titus and Malcolm (1991); (1992)

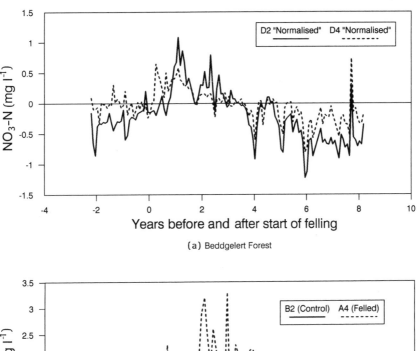

(a) Beddgelert Forest

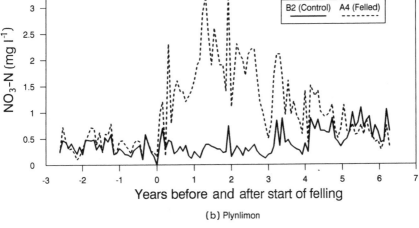

(b) Plynlimon

FIGURE 8.1 (a) Nitrate concentrations in streams D2 (62% felled) and D4 (28% felled) at Beddgelert in north Wales, normalised with respect to the reference stream D3; (b) nitrate concentrations in streams draining felled (A4) and unfelled reference catchments at Plynlimon in mid Wales

disentangle from variations due to seasonality and storm events. For clarity, therefore, results from the two streams which drain clearfelled areas (D2 and D4) have been "normalised" with respect to the reference stream D3. This simply means that the concentrations from D2 and D4 were subtracted from D3 values for each sampling occasion before plotting as a time series in Figure 8.1(a). This plot shows that nitrate concentrations in streams draining from the partially felled catchments increased within six months of felling from less than 1.0 mg N l^{-1} to around 1.5 mg N l^{-1}, but concentrations then dropped to control levels within 3 years of felling. No data are available from this site on stream water nitrate loads (i.e. kg N ha^{-1} yr^{-1}) due to the absence of a continuous streamflow record, but C horizon soil water nitrate concentrations up to 11 mg N l^{-1} and equivalent nitrate flux approaching 70 kg N ha^{-1} yr^{-1} were observed in the second year after felling in the conventionally felled plots at the site (Stevens and Hornung, 1988).

In the 335 ha Hore catchment at Plynlimon in mid Wales (Table 8.1), nitrate concentrations started to increase within one year of felling from pre-felling and reference values of 0.2−0.5 mg N l^{-1} up to 0.7−1.0 mg N l^{-1} (Neal et al.,1992). Felling of the Hore catchment took 4 years to complete and elevated nitrate concentrations were sustained throughout this period (Neal et al., 1992). Nitrate concentrations in a small 6 ha subcatchment (A4) of the Hore, which was 100% clearcut, increased to a maximum of nearly 3.5 mg N l^{-1} about one year after the start of felling, subsequently decreasing to concentrations similar to those in an unfelled reference stream (B2) after about 5 years (Figure 8.1(b)), (Stevens and Reynolds, 1993b). The differences between the subcatchment and the Hore demonstrate the extent to which dilution by runoff from unfelled forest and moorland can ameliorate the effects of harvesting.

Concentrations of ammonium N in the streams at Plynlimon and Beddgelert were negligible, perhaps because of nitrification in the predominantly freely drained soils or active oxidation to nitrate in the streams. At Kershope, however, the soils are predominantly poorly drained peaty gleys and ammonium N was present at mean concentrations of 0.2 mg N l^{-1} in the reference stream and 0.9 mg N l^{-1} in the streams draining harvested plots during the year after felling (Adamson et al., 1987).

8.2.2   The soil response

The pulse of nitrate observed in streams following felling has been widely investigated in temperate forest systems (Likens et al., 1977; Vitousek et al., 1979). Within the context of British plantation forests the most detailed work on soil nitrogen dynamics after felling has probably been undertaken at Beddgelert Forest in north Wales (Stevens and Hornung, 1988, 1990; Emmett and Quarmby, 1991; Emmett et al., 1991a, 1991b) and Kielder forest in northern England (Malcolm and Titus, 1983; Titus, 1985; Titus and Malcolm, 1991, 1992). These two studies provide an insight into the nitrogen response for two very different soil types and contrasting harvesting techniques. At Beddgelert, the effects of conventional harvesting (CH) were compared with those of whole-tree harvesting (WTH) on a relatively freely drained peaty podzol. At Kielder Forest conventional harvesting was undertaken on a poorly drained peaty gley soil.

At Beddgelert, nitrate N accounted for approximately 95% of the inorganic N in mineral horizon soil waters both before and after harvesting (Stevens and Hornung, 1990). Following harvesting with both CH and WTH treatments, nitrate N concentrations in mineral horizons increased substantially above those observed in the standing forest to reach a maximum of between 6 and 7 mg N l^{-1} 14 months after felling (Figure 8.2) (Stevens and Hornung, 1990). This compares with maximum concentrations of approximately 2 mg N l^{-1} prior to felling. Nitrate N concentrations subsequently declined in both treatments, although the decrease was more rapid in the WTH plots. After five years, nitrate N was virtually absent in WTH mineral horizon soil waters.

In the surface organic (L and O) horizons, there was rather more ammonium N prior to felling, with nitrate N accounting for 57 and 70% of the inorganic N in the L and O horizons

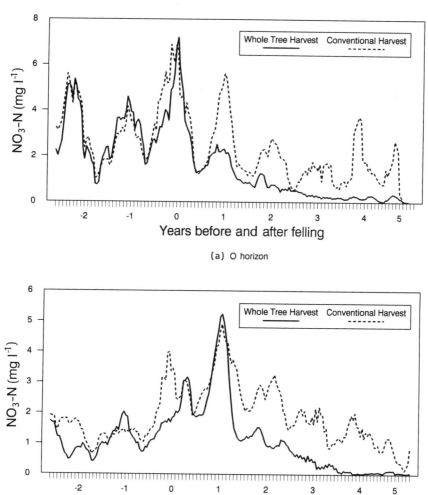

(a) O horizon

(b) B horizon

FIGURE 8.2   Nitrate concentrations in soil waters beneath conventionally and whole-tree harvested Sitka spruce at Beddgelert Forest, north Wales: (a) O horizon and (b) B horizon

respectively (Stevens and Hornung, 1990). Following harvesting the proportion of nitrate N increased in the WTH plots to 80% in the L and 83% in the O horizon. There was little change in the relative amounts of nitrate N and ammonium N in the L horizon of the CF plots, but in the O horizon, the proportion of nitrate N increased to 81%. Following harvesting in the WTH plots, there was a rapid fall in inorganic N concentrations in the surface horizons which was followed over the next three years by a prolonged and steady decline to virtually zero (Figure 8.2). In contrast, inorganic N concentrations and fluxes changed very little in the L horizon of the CH plots, although a decline was evident in the O horizon (Stevens and Hornung, 1990) (Figure 8.2).

8.2.3 Processes controlling nitrate leaching after clearcutting

Disruption of the nutrient cycle by removing trees would, of itself, make more inorganic nitrogen available for leaching, irrespective of the effects of soil nitrogen transformations. The relative contributions from these processes can be judged by comparing uptake and leaching losses from the site. Nitrogen uptake by the standing crop at Beddgelert was estimated as 50 kg N ha^{-1} yr^{-1} (Stevens *et al.*, 1989), comparable with a mean of 47.4 kg N ha^{-1} yr^{-1} for 17 temperate forests quoted by Cole and Rapp (1981). In the second year after felling, the loss of inorganic N from below the C horizon of the CH plots was determined as 70 kg N ha^{-1} yr^{-1} (Stevens and Hornung, 1988), indicating that some "internal" production of inorganic nitrogen occurs after felling. Before harvesting, inorganic N losses from below the C horizon were typically between 9 and 19 kg N ha^{-1} yr^{-1}.

A number of processes can contribute to the internal production of inorganic N and nitrate after felling. While root uptake of N is terminated, mineralisation of organic matter continues in the soil and inorganic N becomes available in the soil solution. Furthermore, decomposition and mineralisation of the large biomass of fine tree roots may be initially very important (Natural Environment Research Council, 1991). Removal of the tree as a nutrient "sink" also eliminates a major competitor for available inorganic N. This may allow the soil microbial population to multiply and immobilise some of the available nitrogen in the microbial biomass. Eventually this nitrogen may be mineralised and nitrified. At Beddgelert, nitrification was active both prior to and after felling, so that ammonium N was readily converted to nitrate N (Emmett, 1989).

Effects of brash cover

An obvious feature of conventional harvesting is the production of large quantities of felling debris or brash. The presence or absence of this material has an important influence on the nitrogen dynamics of the site either directly through its effect on the chemistry of percolating rainwater or indirectly by its influence on microclimate and re-vegetation.

At Beddgelert approximately two thirds of the inorganic N deposited in bulk precipitation was immobilised by the brash during the first two years after felling (Stevens and Hornung, 1988). The brash therefore acted as a net sink for inorganic N and was not directly the source of the additional leached N. Brash may have been a source of dissolved organic N which, after mineralisation and nitrification, could have contributed to the observed nitrate pulse.

Lysimeter experiments at Beddgelert Forest indicated that the presence of brash probably induced microclimatic conditions favourable to organic matter mineralisation and nitrification (Emmett *et al.*, 1991a). Similarly, in Sweden, maintenance of a more constant, higher moisture content beneath the brash was identified as a particularly important microclimatic factor, leading to increased rates of nitrogen mineralisation and nitrogen leaching beneath brash piles (Rosén and Lundmark-Thelin, 1987). In areas without brash cover, microclimatic conditions are also more favourable for nitrogen transformations compared to the pre-felling situation (Emmett, 1989). Since nitrate is a very mobile anion leaching takes place, unless there is denitrification or uptake by vegetation.

Ground vegetation is generally absent under standing Sitka spruce plantations. However, after cutting, the WTH plots at Beddgelert were colonised very rapidly by acid grassland species such as *Agrostis* spp., *Deschampsia flexuosa* and on wetter ground, *Juncus effusus*

(Fahey *et al.*, 1991a). Colonisation rates were much faster in the WTH plots compared to the CH plots, where vegetation cover was significantly negatively correlated with brash cover (r^2 = 0.965, P = 0.001), indicating that the presence of brash inhibited re- vegetation. In addition, nutrient uptake and immobilisation by the re-established vegetation was greater in the WTH plots (Fahey *et al.*, 1991a). It was concluded, therefore, that re-vegetation was a major factor in reducing the duration of the nitrate pulse in the WTH compared to the CH plots (Stevens and Hornung, 1990).

This conclusion was substantiated by studies using lysimeters containing surface forest floor and peaty organic horizons at Beddgelert. In these, the presence of less than 50% cover of *Agrostis capillaris* L. growing on soil organic horizon material without brash reduced nitrate N leaching losses to 19% of the control and to 10% of that from lysimeters in which the soil was covered with brash (Emmett *et al.*, 1991b). The presence of Sitka spruce seedlings also reduced the nitrate leaching compared to the control and brash only treatments. In the Swedish study quoted above, root uptake by vegetation was much reduced under brash piles, compared with clear areas (Rosén and Lundmark-Thelin, 1987).

Effects of soil type

The studies of nitrogen transformations at Beddgelert were conducted on the dominant, freely drained podzol soils in which nitrification was active. These were also the dominant soils at Plynlimon, which were actively nitrifying prior to felling (B. Emmett personal communication, 1993). Both sites contain areas of less well-drained peaty gley soils and evidence from Kielder Forest (Table 8.1) suggests that these may have responded very differently to felling.

The poorly drained, peaty gley soils at Kielder did not nitrify freely, and much of the runoff occurred laterally below the forest floor, especially after felling when the water table rose. Inorganic N concentrations in forest floor lysimeter leachates were significantly higher beneath brash bands than in open areas (Titus, 1985; Malcolm and Titus, 1983) but nitrate N concentrations were actually lower beneath the brash than in the open. In contrast, amounts of ammonium N were significantly larger under the brash than in the open being an order of magnitude greater than the nitrate N concentrations. Nitrate N and ammonium N continued to leach for six years after felling, during which time the site remained virtually free of vegetation except for re-stocked Sitka spruce (Titus and Malcolm, 1992).

The comparatively low concentrations of nitrate at Kielder may reflect the fact that samples were collected beneath the forest floor only, where nitrate concentrations are not expected to be large. Also, nitrification in soils is not restricted to surface organic horizons, so it is not possible to say whether increased nitrate concentrations occurred at depth in the Kielder soil.

Prior to felling at Plynlimon, ammonium N and nitrate N concentrations in the surface organic-rich O horizon of the peaty gley soils were greater than those observed in the same horizon of the podzols (Figures 8.3(a) and (b)). Following felling, there was a large and immediate increase in ammonium N in the podzol O horizon, but a less marked change in the gley. Nitrate concentrations increased in the O horizon of both soil types, but the response lagged behind that of ammonium, with the largest nitrate peak occurring approximately one year after felling. The relative timing of the ammonium and nitrate responses suggests that some nitrification occurred, although this may have been periodically inhibited in the gley by waterlogging and anaerobic conditions resulting from the rise in water table following felling (Reynolds *et al.*, 1992). No increases in nitrate in concentrations were observed in the deeper

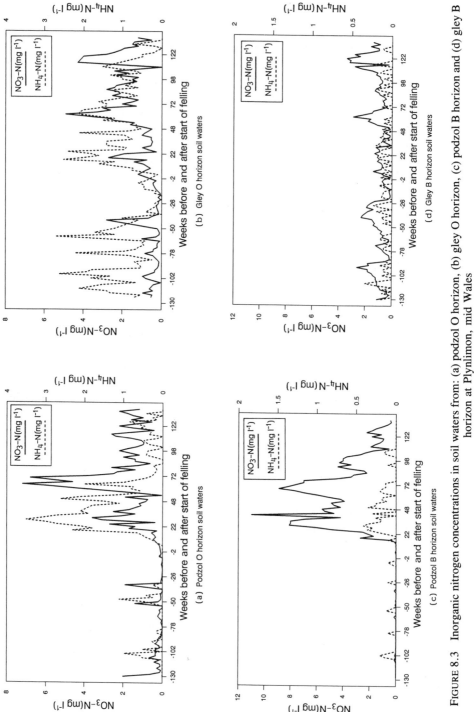

FIGURE 8.3 Inorganic nitrogen concentrations in soil waters from: (a) podzol O horizon, (b) gley O horizon, (c) podzol B horizon and (d) gley B horizon at Plynlimon, mid Wales

horizons of the Plynlimon gley in response to felling (Figure 8.3(d)). In contrast there was a large flush of nitrate in the deeper horizons of the podzol (Figure 8.3(c)). These results and those from Kielder reflect observations at Kershope Forest, also on peaty gley soils, where substantial increases in nitrate in drainage water from the surface LF horizon (J. K. Adamson, personal communication, 1994) occurred after felling. A common feature of all three sites on peaty gleys was the slow re-establishment of natural vegetation after felling.

8.2.4 Synthesis for modelling

The discussion to this point has illustrated how manipulation of the forest canopy can have major repercussions for ecosystem processes. In addition, the response of these processes will be determined by factors such as soil type and harvesting practice. The challenge to the modeller is to represent this complexity whilst at the same time producing a model that has a limited number of identifiable parameters and general applicability.

The most obvious effect on nitrogen dynamics of forest harvesting is the pulse of nitrate leaching seen in the stream water following clearfelling. A model has therefore been constructed which aims to reproduce this behaviour, with a minimum number of input parameters in order to allow its application at a number of different sites with different properties.

8.3 MODELLING EFFECTS OF FOREST HARVESTING ON NITRATE LEACHING

Two main factors control forest N dynamics: hydrology-driven N transport through the soil and biologically mediated N transformations. By representing these two processes in the model the major determinants of the nitrogen cycle, in particular the amount of nitrate leaching to streams, are included, and by using measurable, physically meaningful parameters to define these processes the model should be transportable between those sites where sufficient data are available.

8.3.1 Hydrology

The process of water movement through the soil has two main effects on nitrogen leaching: (a) N is transported through the soil in solution and (b) soil moisture status affects rates of N transformations such as mineralisation and denitrification. A multitude of hydrological models exist, at all levels of time and spatial resolution. Complex and accurate models based on well-established hydrodynamic principles, which can predict soil water fluxes at hourly intervals or less, may be used if the intensive input data and numerous soil parameters which they require are available (Jansson and Halldin, 1979). Similarly, models of transpiration and evaporation processes have evolved from the Penman equation (Calder, 1990) and, operating at the stomatal level, may be equally accurate. For our purposes — data at daily or weekly resolution, rather than hours or years — a less intensive approach is necessary and desirable.

Components of the hydrological model are:

- Interception
- Transpiration
- Soil water flux

as shown in Figure 8.4(a) and described in turn below.

Interception

In afforested upland areas of the United Kingdom interception losses may be considerable — typically twice those due to transpiration and up to 30% of precipitation. Calder (1990) suggests a stochastic model for daily interception (I) of rainfall by a forest canopy, which is an appropriate level of time resolution for our purposes and which has the form

$$I = \gamma(1 - e^{-\delta P})$$

where P is daily precipitation in mm and δ, γ are empirically derived parameters which depend upon canopy architecture.

Transpiration

The second component of Calder's evapotranspiration model is the daily transpiration rate T, given by

$$T = \beta E_t (1 - \omega)$$

where β is the "transpiration fraction" i.e. the ratio of actual evapotranspiration (AET) to potential evapotranspiration (PET), E_t is a climate-based Penman evapotranspiration estimate and ω is the fraction of the day for which the canopy is wet, estimated by

$$\omega = \begin{cases} 0.045\, P, & P < 22 \text{ mm} \\ 1, & P > 22 \text{ mm} \end{cases}$$

(following Calder and Newson, 1979).

For a mechanistic model we have not used the transpiration fraction β: instead the soil water status is allowed to limit AET (see below). Penman E_t is estimated using a modification of a method due to Linacre (1977), requiring only temperature data and the latitude and altitude of the site.

Soil water flux

The model used to predict the flow of water through the soil is an extension of the so-called "bucket model" approach, first presented by Terkeltoub and Babcock (1971). Such models are reasonably accurate in the longer term (of the order of weeks) for freely drained, non-structured soils in which neither ponding nor bypass flow occur. Short-term accuracy is here sacrificed to reduce the data requirements; soil texture and daily values for rainfall and PET are sufficient.

In the original scheme a fallow soil was modelled. Any excess of precipitation over potential evaporation, on a daily basis, was assumed to enter the first of several soil layers. If the water in this layer exceeded the prescribed field capacity then the excess percolated to the next layer down, and the process repeated down to the freely draining bottom layer. If potential evaporation was greater than rainfall then evaporation was assumed to occur from the top layer down to its minimum water-holding capacity, and then capillary rise invoked to transfer water upwards as necessary. This scheme has been extended here to modify the hydrology under a

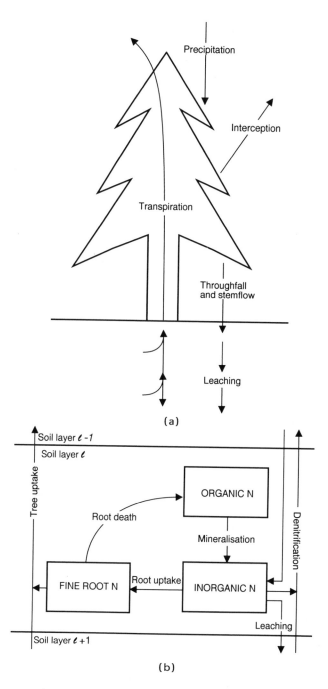

FIGURE 8.4   Components of (a) the forest hydrological cycle and (b) the nitrogen cycle used in the forest harvesting model

forest canopy: throughfall and stemflow (obtained from interception as above) replace rainfall as the input to the soil, and transpiration demand replaces potential evaporation. It is assumed here, following Greenwood *et al.* (1974), that the trees take up water from the top soil layer until the lower water content limit is reached, then from the next layer and so on as necessary. This is not physically realistic since it assumes that availability of soil water to roots is independent of soil water content, but is here preferred to fitting arbitrary uptake−water content curves for each soil layer.

8.3.2 N transformations

Models of nitrogen cycling and nitrogen transformations are many and varied, and depend largely on their desired use: the level of detail may be anywhere from representing decomposition as exponential decay to including protozoan grazing of bacteria. In this study, we have attempted to isolate the qualitatively important components of the system, given a time-scale of years rather than centuries or hours. Decomposition of coarser roots (> 2 mm), for example, is not considered: according to Fahey *et al.* (1991b) around 6 kg N ha^{-1} were immobilised by roots > 5 mm in the first four years after the clearfelling of Beddgelert Forest, and the same amount released in the following six years; in comparison, around 62 kg N ha^{-1} was released from foliage in the first four years. We therefore consider only fine roots and foliage in the N transformation model, and represent each subsoil horizon as follows (Figure 8.4(b)).

An additional organic N pool in the surface layer represents needle litter. Fine root death is assumed to be first order, i.e. directly proportional to the amount of live fine roots present, and the fine root pool in each horizon assumed constant. Denitrification is taken to be constant as an initial approximation. Tree uptake is also assumed constant, with the total uptake given by (net N immobilised in tree) + (litterfall). This is divided between the soil layers in the same proportions as the fine roots. Dry deposition and canopy leaching/uptake are accounted for implicitly as far as possible by using measured throughfall N as input to the soil.

The organic N pool represents dead fine roots undergoing decomposition (and leaf litter at the surface). This process is based on a model by Bosatta and Staaf (1982), modified to allow for a continuous litter input, described below (Figure 8.5).

By making the simplyfing assumptions that

(a) microbial C/N ratio f is fixed,
(b) microbial production is proportional to substrate C content,
(c) microbial carbon use efficiency e is fixed and
(d) microbial population is in steady state, i.e. mortality = production,

it follows that

$$\frac{dC}{dt} = -\frac{1-e}{e}\,pC = -kC$$

$$\frac{dN}{dt} = \frac{e}{1-e}\frac{k}{f}C - \frac{k}{1-e}N$$

where C represents total litter carbon, N total litter nitrogen and p the rate of microbial production, and hence

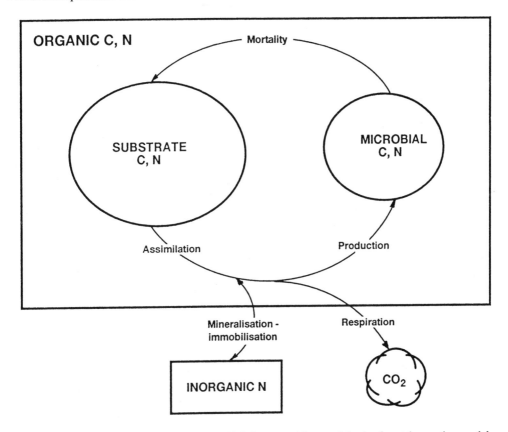

FIGURE 8.5 Conceptual outline for microbial decomposition used in the forest harvesting model

$$C(t) = C_0 \exp(-kt)$$

$$N(t) = \frac{C_0}{f} \exp(-kt) + \left(\frac{N_0 - C_0}{f}\right) \exp\left(\frac{-k}{1-e}t\right)$$

Examination of these equations shows that

$C(t)/N(t) > f/e$ implies $dN/dt > 0$, i.e. there is a net immobilisation of N at time t

$C(t)/N(t) < f/e$ implies $dN/dt < 0$, i.e. net mineralisation

In other words, f/e corresponds to the critical C/N ratio of the litter, $(C/N)_c$, and the model behaviour is in line with the common observation that $(C/N)_c$ determines whether litter immobilises or mineralises N. The model can also be shown numerically to depend on f and e only through their ratio and not individually.

Further, Edmonds (1987) suggests an empirical relation between decay rate k and critical C/N:

$$k = 0.49 \times \exp[-0.01(C/N)_c]$$

derived from a wide range of decomposition studies in the literature. Therefore, for a given mass loss rate k an estimate of f/e can be obtained, and hence given initial C and N contents, C_0, N_0, the carbon and nitrogen dynamics of the system can be determined.

8.3.3 Application

The model described here has been applied to Beddgelert Forest, using the data from the experiment detailed above in Section 8.2. In particular, we require the model to reproduce the increase in nitrate leaching following conventional harvesting, in order to further investigate the effects of different harvesting practices, timing of felling, etc.

For simplicity all mineral nitrogen is assumed to be present as nitrate, a reasonable assumption at Beddgelert (see Section 8.2 above). The dependence of decomposition on soil temperature and moisture is assumed to follow the common forms found in the literature: a Q_{10} relation and linear dependence on saturation percentage respectively.

As we concentrate on the time around felling, the assumption of constant annual nitrogen uptake is reasonable since the trees will be mature and undergoing steady-state growth at this time. At the time of clearfelling, a switch in uptake is applied: the constant value for the forest is replaced with a Richards curve model to account for the regrowth of ground vegetation (including regenerating trees) following removal of the forest canopy. The parameters for this curve are obtained by fitting to measured data (Fahey *et al.*, 1991).

Values of the interception parameters γ and δ were obtained by Calder (1990) for Sitka spruce plantations at sites in mid Wales and across Scotland, where δ was found to be 0.099 mm^{-1} at all sites and γ has a mean value of 7 mm (range $6.1-7.6$).

Literature values were used for those parameters not measured at Beddgelert, e.g. rates of fine root turnover. For the period outside the intensive monitoring, daily meteorological data were not available so mean monthly values were applied.

An example input parameter file is given in the Appendix below.

8.3.4 Results

Hydrology

Figure 8.6 shows the results from the hydrological component of the model, with measured results for comparison. These data are from an unfelled control block at Beddgelert. Note that the quantities which are physically measured are precipitation, transpiration and runoff (i.e. percolation below the C horizon): "measured" interception is calculated as the difference between precipitation and throughfall, and transpiration is calculated as throughfall minus runoff. The daily output data from the model have been combined to correspond to the monthly or fortnightly measurements.

Interception and transpiration are generally represented reasonably well by the model, apart from periods of large fluctuations in rainfall when canopy storage of rainwater will have a relatively large effect. Due to the large volume of rainfall at Beddgelert (annual average 2600 mm) these quantities are small in comparison with runoff, and consequently measured and predicted runoff are in good agreement, particularly when the simple model structure and minimal input requirements are taken into consideration.

Decomposition

Figure 8.7(a) shows the results from the decomposition submodel for the fine root data given in the Appendix. The plot gives the relative amount of N remaining in a single cohort of litter as it decays with time, and demonstrates the initial period of immobilisation during which the amount of N in the litter increases, followed by rapid mineralisation.

Figure 8.7(b) shows the total amount of root litter in the soil, determined by assuming continuous litter input (steady state) and integrating over time. Input is ceased after ten years to simulate the effects of felling. We see that the amount of root litter reaches a steady state after around ten years and that felling causes a sudden fall in the size of the pool as mineralisation losses continue without any balancing inputs. Note that the decay curve is not exponential in shape, due to the immobilisation phase in the top figure: during year 10, for example, the net loss of organic N is the result of a small amount of immobilisation by the most recent litter and a larger amount of mineralisation of older litter. The rate of loss of N from litter after felling is, therefore, slower than the rate of mass loss, given by the exponential decay constant k.

N leaching

Combining these two submodels allows us to predict the effect on nitrogen leaching of clear felling. To conform to the steady-state assumption, the model is run for a number of years with long-term meteorological data as input, until the N pools have reached equilibrium values. The

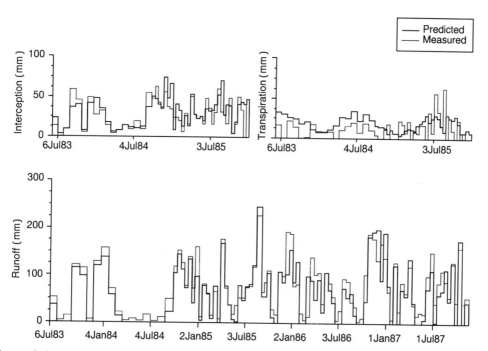

FIGURE 8.6   Model simulations and observations of interception, transpiration and runoff at Beddgelert Forest

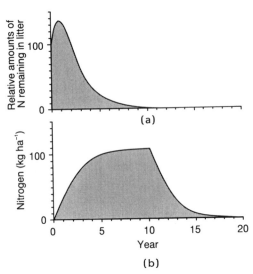

FIGURE 8.7 Results from the decomposition submodel showing (a) the relative amount of nitrogen remaining in a single cohort of litter and its decay over time and (b) the total amount of root litter in the soil

intensive meteorological data are then applied, for a short period before felling and a number of years following. Model predictions and measured data are shown for nitrogen leaching from the C horizon in Figure 8.8 for a conventionally harvested plot at Beddgelert which was felled in August 1983. Again, the daily model output has been aggregated to fortnightly, to allow comparison with the measured data.

The two sets of results show good qualitative agreement, in that the timing of the peak N leaching periods coincide, and there is a broad quantitative agreement in the amount leached over each six-month period, seen by comparing the area under each curve, but the magnitude of each peak is not very well predicted. In the latter years of the simulation, 1985–7, the total amount of N leaching is consistently underestimated, which suggests that N uptake by ground vegetation following felling has been overestimated. Apart from this factor, however, and considering the minimal site-specific parameter input requirements, the model performs satisfactorily and does indeed reproduce the effects of clearfelling on nitrogen leaching in this case.

As an example of the possible use of the model, we can investigate the effect which climate has on the timing and magnitude of the nitrate pulse following clearfelling, by hypothetically felling the model forest one year later. Model results for this simulation are shown in Figure 8.9.

Comparison with the previous figure shows that the timing of the peak N leaching is similar in both cases: an immediate release of nitrate immediately after felling, and a further pulse one year later. The shape of the pulses is, however, quite different, and this appears to be due to the dry summer in 1984 (see the runoff plot in Figure 8.6). In the case of the 1983 felling this leads up to a buildup of nitrate in the soil as mineralisation continues (though at a reduced rate due to the low soil moisture). Nitrate is not leached from the soil as there is no runoff for a number of months; when runoff does occur it then removes this accumulated nitrate and results in large fluxes over a short time. By contrast, the felling in 1984 takes place in a more representative year for precipitation and the nitrate leaching pulse is lower and broader. Note that the area under each curve for the first 2 years following felling is approximately equal, indicating that only the timing and concentration of N leaching is affected and not the total amount over time.

8.4 CONCLUSIONS

The dynamics of nitrogen in harvested forest plantations are determined by the interaction of a number of complex biotic and abiotic processes. The main biotic processes include the removal of tree nitrogen uptake at felling and an increase in rates of decomposition, mineralisation and nitrification. However, the magnitude of these responses is largely determined by soil type, site climate and harvesting intensity. The former directly influences the environment under which nitrogen transformations occur as nitrification is limited in waterlogged, anaerobic soils. Under these conditions, denitrification may be important in determining site nitrogen losses. Harvesting intensity exerts an important control on the rate of re-vegetation, which in turn limits leaching losses from the site.

The removal of the canopy has a major influence on abiotic processes by reducing atmospheric inputs through diminished scavenging of gaseous, particulate and dissolved nitrogen compounds and by decreasing interception losses. The latter results in an increased water flux following harvesting. Site microclimate is also changed, and may become more favourable for microbial nitrogen transformations, although the direction and magnitude of this response is strongly influenced by the presence or absence of felling debris.

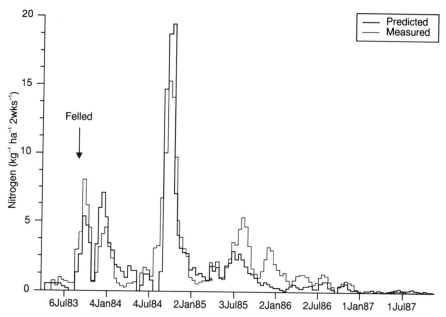

FIGURE 8.8 Modelled and observed nitrate leaching following conventional harvesting of a 50 year old stand of Sitka spruce at Beddgelert Forest

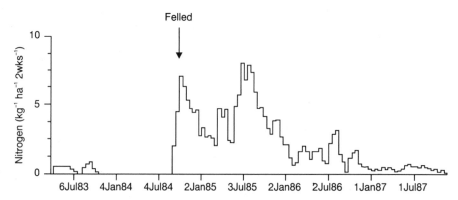

FIGURE 8.9   Effect of changing year of felling on predicted nitrate leaching from a 50 year old stand of Sitka spruce at Beddgelert Forest

Despite the complexities, a dynamic model has been constructed which satisfactorily simulates nitrate leaching losses following conventional harvesting of a Sitka spruce plantation in north Wales. The model is constructed around a minimum of site-specific input parameter requirements and aims to capture the processes identified as qualitatively most important. Preliminary predictive use of the model has indicated that changing the timing of felling can alter the shape and size of the nitrate leaching pulse due to interactions between microbial processes, nitrate leaching and climate. However, further testing of the model is required. In particular, it may also be desirable to link the model to a forest growth model to allow simulation of one or more complete forest rotations. A link to a soil chemical model could also be developed so that the effects of the nitrate pulse on other components of the soil solution can be evaluated.

REFERENCES

Adamson, J. K., Hornung, M., Pyatt, D. G. and Anderson, A. R. (1987). "Changes in solute chemistry of drainage waters following the clearfelling of a Sitka spruce plantation", *Forestry*, **60**, 165 – 177.

Adamson, J. K. and Hornung, M. (1990). "The effect of clearfelling a Sitka spruce (*Picea sitchensis*) plantation on solute concentrations in drainage water", *J. Hydrol.*, **116**, 287-297.

Anderson, A. R., Pyatt, D. G. and Stannard, J. P. (1990). "The effects of clearfelling of a Sitka spruce stand on the water balance of a peaty gley soil at Kershope Forest, Cumbria", *Forestry*, **63**, 51 – 71.

Bormann, F. H. and Likens, G. E. (1979). *Pattern and Process in a Forested Ecosystem*, Springer-Verlag, New York.

Bosatta, E. and Staaf, H. (1982). "The control of nitrogen turn-over in forest litter", *Oikos*, **39**, 143 – 151.

Calder, I. R. (1990). *Evaporation in the Uplands*, John Wiley, Chichester.

Calder, I. R. and Newson, M. D. (1979). "Land use and upland water resources in Britain — a strategic look", *Water Resour. Bull.*, **16**, 1628 – 1639.

Cole, D. W. and Rapp, M. (1981). "Element cycling in forest ecosystems", in *Dynamic Properties of Forest Ecosystems* (Ed. D. E. Reichle). pp. 341 – 410, IBP Vol. 23, Cambridge University Press, Cambridge.

Edmonds, R. L. (1987). "Decomposition rates and nutrient dynamics in small-diameter woody litter in four forested ecosystems in Washington, USA", *Can. J. Forest Res.*, **17**, 499 – 509.

Emmett, B. A. (1989). "The effects of harvesting on soil nitrogen transformations in a Sitka spruce (*Picea sitchensis* (Bong.) Carr.) plantation at Beddgelert Forest (N. Wales)", PhD thesis, University of Exeter (unpublished).

Emmett, B. A. and Quarmby, C. (1991). "The effect of harvesting intensity on the fate of applied ^{15}N-ammonium to the organic horizons of a coniferous forest in N. Wales", *Biogeochemistry*, **15**, 47 – 63.

Emmett, B. A., Anderson, J. M. and Hornung, M. (1991a). "The controls on dissolved nitrogen losses following two intensities of harvesting in a Sitka spruce forest (N. Wales)", *Forest Ecology and Management*, **41**, 65 – 80.

Emmett, B. A., Anderson, J. M. and Hornung, M. (1991b). "Nitrogen sinks following two intensities of harvesting in a Sitka spruce forest (N. Wales) and the effect on the establishment of the next crop", *Forest Ecology and Management*, **41**, 81 – 93.

Fahey, T. J., Hill, M. O., Stevens, P. A., Hornung, M. and Rowland, A. P. (1991a). "Nutrient accumulation in vegetation following conventional and whole-tree harvest of Sitka spruce plantations in North Wales", *Forestry*, **64**, 271 – 288.

Fahey, T. J., Stevens, P. A., Hornung, M. and Rowland, P. (1991b). "Decomposition and nutrient release from logging residue following conventional harvest of Sitka spruce in North Wales, UK", *Forestry*, **64**, 289 – 301.

Greenwood, D. J., Wood, J. T. and Cleave, T. J. (1974). "A dynamic model for the effects of soil and weather conditions on nitrogen response", *J. Agric. Sci., Cambridge*, **82**, 455 – 467.

Heal, O. W., Swift, M. J. and Anderson, J. M. (1982). "Nitrogen cycling in United Kingdom forests: the relevance of basic ecological research", *Phil. Trans. R. Soc. Lond*, **B296**, 427 – 444.

Hornbeck, J. W., Martin, C. W., Pierce, R. S., Bormann, F. H., Likens, G. E. and Eaton, J. S. (1986). "Clearcutting northern hardwoods: effects on hydrologic and nutrient ion budgets", *Forest Sci.*, **32**, 667 – 686.

Jansson, P.-E. and Halldin, S. (1979). "Model for annual water and energy flow in a layered soil", in *Comparison of Forest Water and Energy Exchange Models*, (Ed. S. Halldin), pp. 145 – 163, International Society for Ecological Modelling, Copenhagen.

Johnson, D. W., Kelly, J. M., Swank, W. T., Cole, D. W., Van Migroet, H., Hornbeck, J. W., Pierce, R. S. and Van Lear, D. (1988). "The effects of leaching and whole-tree harvesting on cation budgets of several forests", *J. Environ. Qual.*, **17**, 418 – 424.

Knighton, M. D. and Steigler, J. H. (1981). "Phosphorus release following clearcutting of a black spruce fen and a black spruce bog", *Proceedings of the Sixth International Peat Congress 1980*, pp. 677 – 683.

Lawrence, G. B., Fuller, R. D. and Driscoll, C. T. (1987). "Release of aluminum following whole-tree harvesting at the Hubbard Brook Experimental Forest, New Hampshire", *J. Environ. Qual.*, **16**, 383 – 390.

Likens, G. E., Bormann F. H., Johnson, N. M., Fisher, D. W. and Pierce, R. S. (1970). "Effects of forest cutting and herbicide treatment on nutrient budgets in the Hubbard Brook watershed-ecosystem", *Ecological Monographs*, **40**, 23 – 47.

Likens, G. E., Bormann, F. H., Pierce, R. S., Eaton, J. S. and Johnson, N. M. (1977). *Biogeochemistry of a Forested Ecosystem*, Springer-Verlag, New York.

Linacre, E. T. (1977). "A simple formula for estimating evaporation rates in various climates using temperature alone", *Agric. Meteorology*, **18**, 409 – 424.

Malcolm, D. C. and Titus, B. D. (1983). "Decomposing litter as a source of nutrients for second-rotation stands of Sitka spruce established on peaty gley soils", in *IUFRO Symposium on Forest Site and Continuous Productivity* (Eds. R. Ballard and S. P. Gessel), pp. 138 – 145, USDA Forest Service General Technical Report PNW-163, Portland, Oregon.

Natural Environment Research Council (1991). *Effects of atmospheric pollutants on forests and crops*, Report on the National Power/PowerGen Special Topic.

Neal, C., Smith, C. J. and Hill, S. (1992). *Forestry impact on upland water quality*, Project Report 114/5/W to NRA Bristol; also Institute of Hydrology Report No. 119.

Reynolds, B., Stevens, P. A., Adamson, J. K., Hughes, S. and Roberts, J. D. (1992). "Effects of clearfelling on stream and soil water aluminium chemistry in three UK forests", *Environ. Pollut.*, **77**, 157 – 165.

Rosén, K. and Lundmark-Thelin, A. (1987). "Increased nitrogen leaching under piles of slash — a consequence of modern harvesting techniques", *Scand. J. Forestry Res.*, **2**, 21–29.

Stevens, P. A. and Hornung, M. (1988). "Nitrate leaching from a felled Sitka spruce plantation in Beddgelert Forest, North Wales", *Soil Use and Management*, **4**, 3–9.

Stevens, P. A. and Hornung, M. 1990. "Effect of harvest intensity and ground flora establishment on inorganic-N leaching from a Sitka spruce plantation in north Wales, UK", *Biogeochemistry*, **10**, 53–65.

Stevens, P. A. and Reynolds, B. (1993a). "A review of water quality implications of conifer harvesting in the UK. 1. Literature review and recommendations for research", NRA R&D Note 156, National Rivers Authority, Bristol.

Stevens, P. A. and Reynolds, B. (1993b). "A review of the water quality implications of conifer harvesting in the UK 2. Unpublished results from ITE clearfelling studies and management options", NRA R&D Note 159, National Rivers Authority, Bristol.

Stevens, P. A., Hornung, M. and Hughes, S. (1989). "Solute concentrations, fluxes and major nutrient cycles in a mature Sitka spruce plantation in Beddgelert Forest, North Wales", *Forest Ecology and Management*, **27**, 1–20.

Terkeltoub, R. W. and Babcock, K. L. (1971). "A simple method for predicting salt movement through soil", *Soil Sci.*, **111**, 182–187.

Titus, B. D. (1985). "Nutrient dynamics of Sitka spruce stands after clearfelling on peaty gley soils", PhD thesis, University of Edinburgh.

Titus, B. D. and Malcolm, D. C. (1991). "Nutrient changes in peaty gley soils after clearfelling of Sitka spruce stands", *Forestry*, **64**, 251–270.

Titus, B. D. and Malcolm, D. C. (1992). "Nutrient leaching from the litter layer after clearfelling of Sitka spruce stands on peaty gley soils", *Forestry*, **65**, 389–416.

Vitousek, P. M., Gosz, J. R., Grier, C. G., Melillo, J. M., Reiners, W. A. and Todd, R. L. (1979). "Nitrate losses from disturbed ecosystems", *Science*, **204**, 469–474.

APPENDIX: SAMPLE INPUT DATA FILE

Length of simulation:

75	FIRST__TRIP	Start of simulation (tripno)
305	LAST__TRIP	End of simulation (tripno)
87	FELLTRIP	Time of felling (tripno)

Constant rate parameters (kg N ha^{-1} yr^{-1}):

25.	DEPOSN	Total atmospheric N deposition
9.	ACCUM	Net N mineralisation by tree accumulation
25.	LITTERFALL__N	N in litterfall
*	DENITR(*)	Dentrification

First-order rate parameters (yr^{-1}):

0.6	K__ROOT__DECOMP	Decomposition rate coefficient, fine root litter
1.48	K__ROOT__DEATH	Death rate (first order) for roots
0.52	K__NEEDLE__DECOMP	Decomposition rate coefficient (needle litter)

Initial values (kg N ha^{-1}):

105.	TOTAL__LITTER__N	Over all horizons
42.	TOTAL__ROOT__N	Over all horizons
*	INORG__N(*)	Initial inorganic N
50.	NEEDLE__N	Needle litter pool (in L layer)

Root proportions in each layer:

*	ROOT__FRAC(*)	

Ground vegetation regrowth after felling:

2.3	T-HALF	(yr) Parameters for Richards
1.6	NU	Curve for vegetation uptake
160.	VEG__ACCUM	(kg ha^{-1}) Expected final vegetation N

Decomposition temperature dependence (Q10 relation):

2.0	Q10	Usual figure
8.0	T__REF	(degrees C) — Mean annual temperature

Needle litter decomposition constants:

0.45	NEEDLE__C0	Initial C fraction in fresh needles
0.011	NEEDLE__N0	Initial N fraction in fresh needles
0.50	NEEDLE__E	Bug C use efficiency
12.5	NEEDLE__F	Bug C/N (assumed constant)

Fine root litter decomposition constants:

0.45	FR__C0	
0.010	FR__N0	
0.50	FR__E	As for needle litter
20.0	FR__F	

Soil moisture limits (mm water):

*	FC(*)	Field capacity
*	WILT(*)	Wilting point

New parameters post__felling:

11.	DEPOSN
0.6	K__ROOT__DECOMP
1.48	K__ROOT__DEATH
0.52	K__NEEDLE__DECOMP

(* indicates parameters to be defined for each soil horizon.)

SECTION IV

HYDROLOGICAL AND HYDROCHEMICAL PROCESSES IN SOLUTE MODELLING

9

Soil Water Isotopic Residence Time Modelling

P. V. UNNIKRISHNA

Department of Civil and Environmental Engineering, Utah State University, USA

J. J. MCDONNELL

College of Environmental Science and Forestry, State University of New York, USA

and

M. K. STEWART

Institute of Geological and Nuclear Sciences Ltd, Lower Hutt, New Zealand

9.1 INTRODUCTION

While this book is concerned with the transport of solutes in natural systems, this chapter provides a view of the hydrological system through the use of naturally occurring stable isotope tracers (i.e. deuterium and oxygen-18). Chapter 10 outlines the isotope technique as it is applied typically to stream hydrograph separations and detection of runoff sources. Isotope tracers, however, may be quite useful in providing additional information to scientists interested in solute transport, solutional and non-solutional processes and landscape evolution. This additional insight is gained by addressing the issue of water residence time in the catchment. Obviously, residence times, if known, may allow one to constrain other models of solutional weathering, downslope geochemical evolution of pore waters and slope geomorphological development. The advantage of isotopes over other types of tracers (dyes, solutes, etc.) is that they are added naturally to the system during rain events and, more importantly, that they *are* the water molecule. As a result, stable isotope tracers are not subject to some of the non-conservative interpretive problems associated with these other tracers; they are truly conservative. Only mixing and fractionation (as outlined in Chapter 10) may alter their composition.

Many studies have been conducted to study flow generation mechanisms and proportions of pre-event and event waters in streamflow using isotope tracers (Sklash *et al.*, 1976; Sklash and Farvolden, 1979; Rodhe, 1981; McDonnell *et al.*, 1991; Harris *et al.*, 1995). Relatively fewer

Solute Modelling in Catchment Systems. Edited by Stephen T. Trudgill
© 1995 John Wiley & Sons Ltd

studies, however, have focused on the residence time modelling of the streamflow using the isotope technique. Several reasons may be responsible for the paucity of studies in this area: the large number of samples required and resulting expense, the long-term nature of such studies (often beyond the PhD time-scale) and the difficulty of some of the numerical and statistical techniques. In this chapter, four case studies are reviewed, focusing on event-based residence time modelling and longer term intra-storm residence time modelling approaches in different geomorphological and hydrological settings. Before outlining these studies, some basic theoretical concepts will be presented, outlining the various model approaches and model assumptions.

9.2 GENERAL DEFINITIONS — DISTRIBUTED MODELS

The continuity equation for flow in the unsaturated zone is given by

$$\frac{\partial \theta}{\partial t} + \nabla \cdot \mathbf{q} = 0 \tag{1}$$

where $\theta = nS_w$ is the moisture content, n is the porosity, S_w is the water saturation and $\mathbf{q}$ is the moisture flux vector. Let the total potential $\Phi = p_w/\gamma + z$, where z is the elevation head, p_w is the pressure of water in the void space and γ is the specific weight of water. The negative value of the capillary pressure head is also called suction or tension (Bear, 1979) and is denoted by $\psi = -p_w/\gamma$, $\psi > 0$. The equation of motion (Darcy's law) is given by

$$\mathbf{q} = -K(\theta) \cdot \nabla \Phi \tag{2}$$

where $K(\theta)$ is the unsaturated hydraulic conductivity. Combining eqns (1) and (2) and considering one-dimensional vertical flow only, we have

$$\frac{\partial \theta}{\partial t} = \frac{\partial}{\partial z}\left[K(\theta)\left(-\frac{\partial \Psi}{\partial z} + 1\right)\right] \tag{3}$$

This is Richards' equation for moisture flow through the unsaturated zone. The presence of the gravity term in the motion equation makes it difficult to arrive at exact solutions for the continuity equation by analytical methods. Another form of this equation is obtained by defining moisture diffusivity $D'(\theta) = -K(\theta)(d\psi/d\theta)$. Substituting in eqn (3) we obtain

$$\frac{\partial \theta}{\partial t} = \frac{\partial}{\partial z}\left[D'(\theta)\frac{\partial \theta}{\partial z} + K(\theta)\right] \tag{4}$$

The functions $D'(\theta)$ and $K(\theta)$ appearing in the continuity equations above are assumed to be known functions of the water content. All the partial differential equations of motion and continuity presented above are non-linear equations with hydraulic conductivity and moisture diffusivity being functions of variables θ and ψ. Without the assumption of $D'(\theta)$ and $K(\psi)$ being single-valued functions, they are subject to hysteresis and antecedent conditions become important. Such an assumption is tenable in hysteresis-prone soils only if everywhere in the flow domain there is either drainage or imbibition (Bear, 1979).

The transport of an ideal, conservative tracer is described by the convective—dispersive equation

$$\frac{\partial C}{\partial t} = D\frac{\partial^2 C}{\partial x^2} - V\frac{\partial C}{\partial x} \tag{5}$$

where C is the concentration of the tracer, D is the longitudinal dispersion coefficient and V is the mean flow velocity.

Equations (3) and (5) are solved simultaneously with the appropriate initial and boundary conditions (Bear, 1979) to obtain a spatial and temporal description of the velocity and concentration fields in the flow domain (i.e. a distributed model of the system). Although such a modelling scheme is physically based and is able to accommodate the varying field properties such as porosity and permeability in a distributed fashion, there is need for detailed field data and special numerical schemes to handle problems arising from non-linearity.

In most cases detailed field data characterizing spatially varying hydraulic conductivity, dispersion coefficient and porosity and the soil moisture characteristic are not available. A detailed knowledge of the spatial and temporal moisture and concentration fluxes is also not necessary for most practical purposes. Alternative non-distributed modelling approaches using conservative tracers such as environmental stable isotopes can be used with advantage to yield important information on soil water residence times, flowpaths and the composition of pre-event and event waters in the streamflow. These are often referred to as black box models.

Yurtsever (1983) outlines the main objectives of a mathematical model for tracer study in hydrological investigations as:

1. Simulation of the temporal and spatial variations of the tracer observed in a hydrological system and, consequently, derivation of quantitative estimates related to the dynamics of the system.
2. Prediction of future tracer output from the system, either for a known (actual tracer input) or for a hypothetical tracer input, relevant, for example, to problems of water pollution.

As a prelude to the discussion of various modelling approaches, a few definitions will be presented, following Maloszewski and Zuber (1982, 1993).

Resident concentration (C_R) is the mass of solute (Δm) per unit volume (ΔV) in the system at any given instant:

$$C_R(t) = \frac{\Delta m(t)}{\Delta t} \tag{6}$$

Flux concentration (C_F) is the ratio of solute flux ($\Delta m/\Delta t$) to the volumetric fluid flux ($\Delta V/\Delta t$) passing through a given cross-section:

$$C_F(t) = \frac{\Delta m(t)/\Delta t}{\Delta V/\Delta t} = \frac{\Delta m(t)}{Q\,\Delta t} \tag{7}$$

C_R is not equal to C_F whenever flow lines of different velocities have different solute contents or when there is a gradient of concentration along the flow lines. Kreft and Zuber (1978) discuss solutions for eqn (5) for different initial and boundary conditions for both infinite and semi-infinite media for the case of a conservative tracer. They stress the need to make a distinction between C_R and C_F in dispersive systems and show that the theory of age–distribution functions applies only if the tracer is injected and measured in flux.

The *mean age* (t_0) of water leaving the system or the *mean transit time* of water or the *turnover time* of water is defined as the ratio of the volume of mobile water (V_m) in the system to the volumetric flow rate (Q):

$$t_0 = \frac{V_m}{Q} \tag{8}$$

For unidirectional flow systems which can be approximated by piston flow or dispersion model, $t_0 = x/v$, where x is the length of the systems and v is the mean transit velocity of water. The relationship could also be applied along a chosen flow line.

The *mean transit time* of a conservative tracer (t_t) or the *age of tracer leaving the system* is given by

$$t_t = \frac{\int_0^\infty t C_1(t) \, dt}{\int_0^\infty C_1(t) \, dt} \tag{9}$$

where $C_1(t)$ is the concentration of tracer resulting from instantaneous injection at $t = 0$. The tracer age t_t is equal to the water age t_0 only if an ideal conservative tracer is both injected and measured in flux, i.e. if the tracer is injected and measured proportionally to volumetric flow rates contributed to the total flow by individual flowpaths. Environmental tracers are introduced into groundwater systems through precipitation and are, in principle, injected proportionally to the volumetric flow rates by nature itself. Consequently, the tracer age may be taken to be equal to the water age. It is assumed, however, that there are no stagnant zones accessible to the tracer. The presence of such zones would induce matrix diffusion, resulting in the conservative tracer age being much greater than the water age. The tracer and water ages would then be related by

$$R_p = \frac{t_t}{t_0} = \frac{V_p + V_f}{V_f} \simeq \frac{n_p + n_f}{n_f} \tag{10}$$

where R_p is the retardation factor resulting from matrix diffusion and n_p and n_f are the matrix and fissure porosities respectively. V_p and V_f are the stagnant water volumes in the micropores of the soil matrix and the mobile water volumes in the fissures respectively.

The *weighting function* $g(t)$ or the *system response function* or the *exit age distribution function* of a tracer describes the residence time distribution of the tracer particles at the outlet:

$$g(t) = C_1(t)Q/M \tag{11}$$

where M is the injected mass or activity of the tracer.

9.3 CONVOLUTION INTEGRAL

The input–output relationship based on the conservation of mass in a hydrological system is given by the following convolution integral:

$$Q_o(t)C_o(t) = \int_0^\infty Q_i(t - \tau)C_i(t - \tau)g(\tau, t) \, d\tau \tag{12}$$

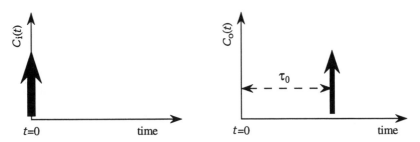

FIGURE 9.1   Schematic of the piston flow model (modified from Yurtsever, 1983)

where $C_o(t)$ and $C_i(t)$ are output and input tracer concentrations and $Q_o(t)$ and $Q_i(t)$ are outflow and inflow rates, τ is the residence time and $g(\tau, t)$ is the system response function for the tracer. For steady-state flow and response, the above equation becomes

$$C_o(t) = \int_0^\infty C_i(t - \tau)g(\tau)\,d\tau$$

(13)

There are two approaches to modelling with this method: the lumped parameter models and the solution of the inverse problem or system response identification. Yurtsever (1983) gives a detailed account of the commonly used models.

9.3.1   Lumped parameter models

In the first approach, lumped parameter models using assumed weighting functions are used with the input field data in eqn (13) to give the best estimates of the output concentrations. The parameters of the final weighting function give a qualitative insight into the flow mechanism prevailing and a quantitative estimate of the distribution of residence times of the tracer within the system. The latter is obtained from the parameters of the weighting function.

Piston flow model (PFM)

The piston flow for stable isotope tracers (Figure 9.1) represents the case where dispersion is low. The input tracer pattern travels without distortion to the output, because there is little mixing between successive parcels of water or adjacent flow lines. The weighting function is the δ function at time $t = \tau$ (the mean residence time):

$$g(t) = \delta(t - \tau)$$

(14)

The tracer input − output relationship for a piston flow model is therefore

$$C_o(t) = C_i(t - \tau)$$

(15)

Exponential model

The exponential model (Figure 9.2) was first introduced in hydrology by Eriksson (1963). This is one of the popular models used for the interpretation of environmental isotope data. The exponential weighting function is of the form

$$g(t) = \frac{1}{\tau} e^{-t/\tau} \tag{16}$$

where τ is the mean residence time. This model was derived for unconfined aquifers under the assumption of exponential distribution of transit times between the recharge area and the discharge site with no mixing of tracer between the flow lines (Maloszewski and Zuber, 1993). The exponential model of transit times also corresponds to the situation of decreasing permeability with aquifer depth. It is mathematically identical to the model of ideal mixing and thus applicable when incoming rainfall mixes immediately with all water in the soil reservoir (Stewart and McDonnell, 1991).

Environmental stable isotopes in precipitation often exhibit a nearly sinusoidal variation over annual cycles (Maloszewski *et al.*, 1983; Stichler and Herrmann, 1983) and even weekly cycles (Stewart and McDonnell, 1991). The input can therefore be represented by a sine function of appropriate amplitude and frequency, and the mean residence time and delay can be determined from the decrease in amplitude between input and output, as follows:

$$\tau = \omega^{-1}[(A/B)^2 - 1]^{1/2} \tag{17}$$

$$\cos\Phi = (\omega^2\tau^2 + 1)^{-1/2} = B/A \tag{18}$$

where A and B are the amplitudes of the input and output delta sine curves, ω the angular frequency of variation and Φ the phase lag.

Dispersion model

The solutions for the dispersion equation (eqn (5)) have been summarized by Kreft and Zuber (1978). The C_{IFF} solution corresponds to the case of injection and sampling proportional to the volumetric flow rates, which is often realized in nature. The corresponding weighting function is given by Maloszewski and Zuber (1982) as

$$g(t) = \left(\frac{4\pi D_p t}{\tau}\right)^{-1/2} t^{-1} \exp\left[-\left(\frac{1-t}{\tau}\right)^2 \left(\frac{\tau}{4D_p t}\right)\right] \tag{19}$$

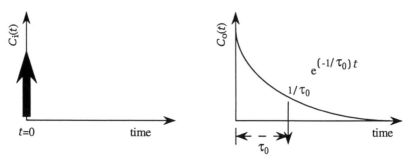

FIGURE 9.2 Schematic of the completely mixed reservoir or 'exponential' model (modified from Yurtsever, 1983)

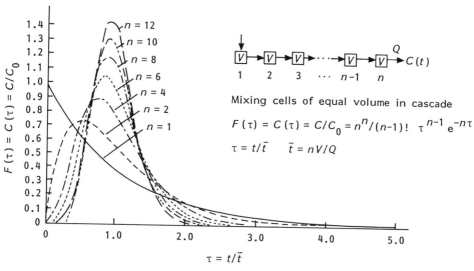

FIGURE 9.3 System showing cascade of equal volume mixing compartments, based on dimensionless concentration distribution curves at outflow (modified from Yurtsever, 1983)

where τ and D_p $(= D/vx)$ are the model parameters (mean residence time and dispersion parameter respectively, and D is the dispersion coefficient, v the mean transit velocity of water in the system and x the length of the lines of flow).

Again, the sine input function can be used in the convolution integral for stable isotopes. The following expression is obtained for the residence time:

$$\tau = \omega^{-1}\left(\frac{-\ln f}{D/vx}\right)^{1/2} \tag{20}$$

where $f = B/A$, the ratio of the output and input sine amplitudes.

The dispersion model represents an intermediate case in the extreme cases of no mixing with the piston flow model and complete mixing with the exponential model. Yurtsever (1983) gives the following analytical solution for the conceptual model of dispersion in a cascade of uniform mixing compartments of equal volume (Figure 9.3):

$$^{n}C_0 = \frac{C_0\left(\frac{Q}{V}\cdot t\right)^{n-1}\cdot e^{-Q/V\cdot t}}{(n-1)!} \tag{21}$$

where $^{n}C_0$ is the tracer concentration of the outflow of the nth compartment in the cascade, C_0 is the initial concentration of the first compartment, n is the number of compartments in the cascade, Q is the flow rate through the system, and V is the volume of compartments (constant for each compartment in the cascade). Equation (21) has the form of the gamma probability density function and the normalized response-time curve is given by

$$C(\tau) = \frac{C}{C_0} = \frac{n^n \tau^{(n-1)} e^{-n\tau}}{(n-1)!} \qquad (22)$$

where

$$\tau = \frac{t}{\bar{t}} \qquad (23)$$

and $\bar{t} = nV/Q$ and C_0 is the initial average concentration of tracer if uniformly distributed through the system. $C(\tau)$ values are shown in Figure 9.3. In this model, when $n = 1$ the response corresponds to the case of the exponential model and it can be proved that when $n = \infty$ the response will be that of a piston flow model.

Lumped parameter dispersive models with two parameters giving information about the mean transit time and the hydrodynamic dispersion coefficient of the system are most suitable for situations where velocity field of flow is essentially unidirectional. Examples are applications to infiltration in the unsaturated zone, lysimeter studies, and to confined aquifers where longitudinal dispersion is the governing feature of the system.

Combinations of g(t)

The different system response functions described above may be combined by assuming that they act simultaneously or in sequence during the model selection process. The most popular model in this group is the exponential−piston flow model (EPM), which is given by

$$g(t) = \left(\frac{\tau}{\eta}\right)^{-1} \exp\left(\frac{-\eta t}{\tau} + \eta - 1\right) \qquad (24)$$

for $t \geq \tau(1 - \eta^{-1})$ and $g(t) = 0$ otherwise. The exponential−piston flow model is supposed to represent an aquifer with an exponential distribution of transit times of flow paths in the unconfined (recharge) area, which changes into piston flow on its way (in the confined part). In the unsaturated flow, it could be applied in situations where the initial movement is in a well-mixed zone followed by a pulse-like movement due to change in soil properties.

In theory, any number of system response functions can be combined in order to better reproduce the observed output. But this also involves the estimation of more parameters. Maloszewski and Zuber (1993) contend that even with an output record of a period of a few years, a unique calibration of a model with more than two parameters seems to be impossible. With this limitation, the commonly used models are the exponential model (EM), the dispersion model (DM) and the combined exponential−piston flow model (EPM).

The convolution integral approach can also be used in unsteady flow conditions. This would involve more complications by having more variables in the form of $Q(t)$ and $V(t)$. Also, there would be no justification for adopting this technique if the overall improvement in results is small.

9.3.2 System identification approach

The second solution method in the convolution integral approach involves the determination of the system response function using input and output concentrations as in an identification or an inverse problem solving scheme. Eriksson (1963) used this method to interpret tritium data available for the Ottawa River basin in Canada. Details of the method are also given by Yurtsever (1983).

For a steady-state flow condition in a given hydrological system, conservative tracer outflow concentrations at selected discrete time intervals can be related to respective tracer input concentrations by the following linear system of equations:

$$(C_o)_1 = (C_i)_1 h_1$$

$$(C_o)_2 = (C_i)_1 h + (C_i)_2 h_1 \tag{25}$$

$$(C_o)_n = (C_i)_1 h_n + (C_i)_2 h_{n-1} + \ldots + (C_i)_{n-1} h_2 + (C_i)_n h_1$$

where

$(C_o)_1, (C_o)_2, \ldots, (C_o)_n$ = tracer outflow concentrations for respective time intervals of $\Delta t = 1$, $2, \ldots, n$

$h_1, h_2, \ldots, h_n$ = fractions of the inflowing water that leaves the system at respective $\Delta t = 1, 2$, $\ldots, n$

The system of n simultaneous equations are solved for the n unknowns $h_1, h_2, \ldots, h_n$, which will define the fractions according to the time they spent in the system or the transit time distribution function of the system.

9.4 MIXING CELLS

9.4.1 Solution by quadratic programming

In order to identify and quantify multiple recharge sources, Adar *et al.* (1988) present a method which utilizes information about groundwater chemistry and environmental isotopes. The motivation for such an approach is that it is often difficult to estimate spatial and temporal distributions of recharge rates, especially in arid and semi-arid zones. The solution of the highly non-linear Richards equation is infeasible for basin-wide recharge studies. Moreover, the method of solving the "inverse problem" by treating recharge as an unknown parameter in a numerical aquifer flow model is not possible in many cases due to non-availability of data. Groundwater chemistry and environmental isotope data, on the other hand, are easily collected. This method is proposed for situations where very little is known about the hydraulic and transport characteristics of the aquifer. A mass balance equation expressing conservation of water, isotopes and dissolved chemicals is written for each mixing cell. These equations are solved simultaneously by quadratic programming for unknown rates of recharge into the various cells. Individual dissolved constituents are assumed to be conservative. The validity of this assumption for a constituent may be tested with a chemical equilibrium model (e.g. WATEQF) and those that do not pass the test may be ignored or assigned a suitably small weight.

The aquifer is divided into N discrete cells, within each of which complete mixing is assumed to take place. Flow into any given cell, n, can be derived from I_n different sources. Similarly, discharge from cell n can occur through J_n different routes. Q_{ni} represents the average volumetric flow rate into n from the ith source during discrete time Δt, and Q_{nj} represents average volumetric flow rate out of n through the jth discharge avenue during time Δt (Figure

9.4). If H_n is the average change in hydraulic head within cell n during Δt, the water balance for cell n is as follows:

$$\sum_{i=1}^{I_n} Q_{ni} - \sum_{j=1}^{J_n} Q_{nj} = S_n \frac{\Delta H_n}{\Delta t} \tag{26}$$

where S_n is the storage capacity of the cell. If the cell does not contain a water table, $S_n = V_n S_{sn}$, where V_n is the saturated volume of n and S_{sn} is its specific storage.

The mass balance for the conservative chemical species in the nth cell is given by

$$\sum_{i=1}^{I_n} C_{nik} Q_{ni} - C_{nk} \sum_{j=1}^{J_n} Q_{nj} = V_n \Phi_n \frac{\Delta C_{nk}}{\Delta t} \tag{27}$$

where C_{nik} is the average concentration over Δt of the kth isotope or chemical constituent in the water entering into cell n at rate Q_{ni}, C_{nk} is the average concentration of the kth species inside n, ΔC_{nk} is the change in the concentration of k in n during Δt and Φ_n is the porosity of n. If water is flowing from cell n at rate Q_{nj} into cell m at rate Q_{mi}, then

$$Q_{mi} = Q_{nj} \tag{28}$$

$$C_{mik} = C_{nk} \tag{29}$$

With real data, the mass balance eqns (26) and (27) will not be closed due to errors in both the model and the data. An error term is therefore included to obtain

$$\sum_{i=1}^{I_n} Q_{ni} - \sum_{j=1}^{J_n} Q_{nj} - S_n \frac{\Delta H_n}{\Delta t} = \epsilon_n \tag{30}$$

and

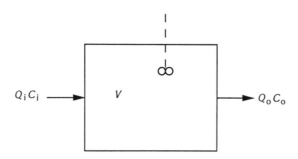

FIGURE 9.4   The well-mixed cell, with volume V, inflow volume Q_i, outflow volume Q_o, inflow tracer concentration C_i and outflow tracer concentration C_o

$$\sum_{i=1}^{I_n} C_{nik} Q_{ni} - C_{nk} \sum_{j=1}^{J_n} Q_{nj} - V_n \Phi_n \frac{\Delta C_{nk}}{\Delta t} = \epsilon_{nk} \tag{31}$$

for every n and k. ϵ_n and ϵ_{nk} represent the mass balance errors associated with the water and the K species in the nth cell respectively. Designating the known terms by Y_{ni}, Y_{nj} and the unknown terms by X_{ni} and X_{nj} for inflows and outflows, the following equations are obtained in matrix form:

$$[C_n][X_n] - [D_n] = [E_n] \tag{32}$$

where $[X_n]$ is a vector of unknowns including X_{ni} and X_{nj}, $[C_n]$ is a matrix having $K + 1$ rows (K being the number of species) and as many columns as there are unknown X terms, $[D_n]$ is a vector of known terms of length $K + 1$ and $[E_n]$ is an unknown vector of errors having similar length. Unknowns X_{ni} and X_{nj} are estimated by minimizing the weighted sum of squared error terms:

$$J = \sum_{n=1}^{N} ([E_n^T][W][E_n]) \tag{33}$$

subject to the cell interface constraints (eqns (28) and (29)) and non-negativity constraints $X_{ni} > 0$ and $X_{nj} > 0$ for all n, i and j. In eqn (33), T designates transpose and $[W]$ is a $(K+1) \times (K+1)$ matrix of weights to account for any prior knowledge about the nature of errors ϵ_n and ϵ_{nk}.

Substitution of eqn (32) into eqn (33) gives

$$J = \sum_{n=1}^{N} ([C_n][X_n] - [D_n])^T[W]([C_n][X_n] - [D_n]) \tag{34}$$

Minimization of eqn (34) subject to eqns (28) and (29) is performed by quadratic programming. Assumption of steady-state conditions eliminates the storage terms $S_n \Delta H / \Delta t$ and $V_n \Phi_n \Delta C_{nk} / \Delta t$.

9.4.2 Solution by linear regression

Gieske and de Vries (1990) suggest an alternative solution method to the above problem using the singular value decomposition (SVD) algorithm, which also provides information about the existence of a unique solution. This approach makes it possible to deal with ill-conditioned problems and also provides the variances and covariances of the calculated flow parameters.

This approach involves arranging all the relations of eqns (26) and (27) in the following matrix form:

$$[D][x] - [b] = [e] \tag{35}$$

where [D] is called the design matrix, [x] contains the unknown flow components, [b] is the vector of known terms and [e] is the error vector. The following expression for weighted sum of squared error terms is to be minimized:

$$F = ([\mathbf{D}][\mathbf{x}] - [\mathbf{b}])^{\mathrm{T}}[\mathbf{W}]([\mathbf{D}][\mathbf{x}] - [\mathbf{b}]) \tag{36}$$

where T indicates matrix transpose and the diagonal matrix [W] contains the normalization terms (Wagner and Gorelick, 1986) and/or weighting terms for individual tracers (Adar *et al.*, 1988). The random errors [e] are assumed to be independent and normally distributed with zero mean and associated variance. Also, the standard deviation of the random error is assumed proportional to the tracer concentration (Wagner and Gorelick, 1986). The diagonal elements w_{ii} of [W] are taken as

$$w_{ii} = \frac{1}{(c_i)^2} \tag{37}$$

where c_i are the tracer concentrations of a representative flow component, e.g. the downstream outflow from the aquifer. With these assumptions the least-squares estimate becomes a maximum likelihood estimate and eqn (37) can be simplified to

$$\chi^2 = \|\mathbf{Dx} - \mathbf{b}\|^2 \tag{38}$$

where the vertical lines indicate the Euclidean vector norm and χ^2 the chi-square function. The SVD algorithm is used to minimize eqn (38). A description of how this is achieved is presented in Press *et al.* (1986). Using the SVD algorithm, [D] is decomposed into

$$[\mathbf{D}] = [\mathbf{U}][\mathbf{S}][\mathbf{V}^{\mathrm{T}}] \tag{39}$$

where [U] is a column orthogonal matrix, [V] is a square orthogonal matrix and [S] is a diagonal matrix with singular values s_{jj}. The solution [x] is obtained from

$$[\mathbf{x}] = [\mathbf{V}]\mathrm{diag}(1/s_{jj})([\mathbf{U}^{\mathrm{T}}][\mathbf{b}]) \tag{40}$$

The flow system is designed such that all flow components yield positive values. The linear regression approach has the advantage that it also provides the variances of the flow components.

9.5 ADAPTIVE OR VARIABLE PARAMETER MODELLING

Reflecting the fact that hydrological systems are often highly non-steady, especially on short time-scales, Turner *et al.* (1987) and Turner and Macpherson (1990) used variable parameter models to describe catchment water and tracer flows. These authors determined the average transit (or mean residence) times for the systems as time-varying quantities. Kalman filtering (Bryson and Ho, 1975) and smoothing techniques are used for recursive estimation of the non-stationary parameters. The smoothing algorithm used was based on the technique by Shumway and Stoffer (1982) which maximizes, for linear systems, the joint likelihood of estimates of model parameters and system noise statistics. Where simple linear models are not adequate, rainfall−streamflow relations are plotted to obtain functional forms for models. Non-linear models used numerical differentiation for linearization or extended Kalman filter formulations.

To estimate time lags between rainfall and streamflow, two simultaneous observation equations were used:

$$F_t = \sum_{i=0}^{n} f_{t-i} + v_{t1} \qquad (41)$$

$$F_t \delta_{f,t} = \sum_{i=0}^{n} \delta_{r,t-i} f_{t-i} + v_{t2} \qquad (42)$$

where F_t = the flow on day t, f_{t-i} = the amount of flow on day t coming from rain within the ith lagged period before t, $\delta_{f,t}$ = the isotopic composition of streamflow on day t, $\delta_{f,t-i}$ = the volume-weighted average δ of rain during the ith lagged period before day t, n is the maximum lag and v_{t1} and v_{t2} are non-serially correlated random variables with zero mean and covariance matrix R.

The f_{t-i} are non-stationary parameters of the system to be estimated. The time-varying lag $T(t)$ is obtained from

$$T(t) = \frac{\Sigma \tau_{t-i} f_{t-i}}{\Sigma f_{t-i}} \qquad (43)$$

where τ_{t-i} is the lag from day t to the middle of the ith lag period before t. The error variance of the estimate of T can be calculated from the product of the vector of τ_{t-i} and the error covariance matrix of the f_{t-i}.

9.6 MODEL APPLICATIONS

9.6.1 The Maimai catchment: long-term residence time modelling

Mean soil water residence times were modelled by Stewart and McDonnell (1991) at the Maimai catchment in New Zealand. The catchment is steep with shallow soil and is highly responsive to rainfall. Mean annual rainfall is 2600 mm, with 1550 mm of runoff from 1950 mm of net rainfall. The slopes are short (<30 m) and steep (mean $34°$), with local relief of the order of $100-150$ m. The soil is underlain by a firmly compacted and weathered conglomerate. The mean soil depth is 0.6 m with a range of $0.2-1.8$ m.

Samples were collected for deuterium analysis from rainfall, soil water, groundwater and stream baseflow. Rainfall samples were collected sequentially within the catchment at 2.5, 5 and then successive 9.2 mm increments during each storm event. (The soil water content and flow measurements are described by McDonnell, 1990.) The rainfall was collected in a sequential sampler containing a series of 0.3, 0.6 and successive 1.0 litre bottles. Specially designed glass valves and an air outlet tube ensured that each bottle was filled before water flowed into the next, and also that there was no mixing between successive rain samples. Soil and groundwater were sampled using 40 mm diameter porous cup suction lysimeters which were inserted at 200, 400 or 800 mm depths in different slope positions within the M8 catchment. A suction of 60 kPa was established within the tube. Soil water or groundwater was drawn in through the porous cup and recovered by a hand pump into a collection bottle. Weekly samples were taken from September to December 1987. Maximum-rise piezometers

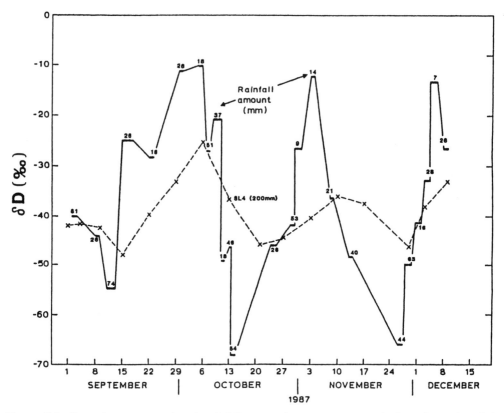

FIGURE 9.5   Deuterium concentrations in rainfall averaged over storm events during a 3 month period. The amount of rainfall is given for each event in millimetres. Soil water deuterium concentrations from SL4 are also shown (modified from Stewart and McDonnell, 1991)

constructed using polystyrene floats in 20 mm OD Perspex tubes indicated whether sample extracted was from the unsaturated or saturated zone.

All water samples were analysed for deuterium. A plot of rainfall deuterium values weighted by rainfall amounts and summed over a week showed three sinusoidal cycles over a period from September to December 1987 (Figure 9.5). All deuterium signatures from the suction lysimeters were smoothed compared to those of rainfall. However, the soil water at shallow levels and upstream locations showed considerable δD responses to the rainfall input as compared to that closer to the stream. In the shallow unsaturated zone (SL4, Figure 9.5), suction lysimeters showed considerable isotopic responses (12–15‰) compared to precipitation (36‰).

Three approaches were adopted to obtain the soil water residence times in order to gain an understanding of water flow within the soil matrix. In the first approach, the sinusoidal nature of the rainfall deuterium input was used to evaluate soil water residence times assuming a steady-state well-mixed flow model. Short-term variations in the amount of rainfall or soil water flow were not considered. The second approach allowed the model parameter describing the new water fraction or mean residence time to vary on a weekly or longer time basis. Again, the well-mixed or exponential model was used. The third approach attempted to reproduce measured outputs using different lumped parameter models. In each case, the input function was transformed into the output by convolution with the appropriate transit time distribution or the system response function.

With the first approach, mean residence times calculated ranged from 12 days to greater than 100 days with the lag times relatively constant at one week. The shallow depth suction lysimeters (SL4) were most responsive with a mean residence time of 12−15 days. The suction lysimeters at 400 mm depth (SL5) showed much less variation of δD and had mean residence times of 50−70 days. Finally, the deep soil water samplers located at depths of 800 mm (SL8) showed the least variation with mean residence times around 100 days or more.

The second approach assumed simple mixing between the incoming rainfall and the pre-event soil water to give the final soil water. The fraction of the rain water in the mixture (x) was obtained from

$$\delta_{SL} = (1 - x)\delta_{SL0} + x\delta_R \tag{44}$$

or

$$x = \frac{\delta_{SL} - \delta_{SL0}}{\delta_R - \delta_{SL0}} \tag{45}$$

where δ_R, δ_{SL0} and δ_{SL} represent the δD of rainfall during the week, that of the previous suction sample and of the final soil water respectively. The varying fraction of the rain water for each time step made the model unsteady. The mean residence times (τ) for this case is calculated from

$$\tau = \frac{-1}{\ln(1 - x)} \tag{46}$$

Best results were obtained for SL4 (Figure 9.5) with a mean residence time of 13 days and average fraction of new water of 42%. Other locations had lesser δD variations and resulted in larger errors.

In the second approach, the Kalman filtering technique was also applied which allows variation in model parameters. The parameter estimates were smoothed by including the effect of the preceding weeks in an exponentially decreasing manner (Minchin and Troughton, 1980; Turner *et al.*, 1987). The mean residence time for SL4 was 15 days, with the fraction of rain water as 34%. This compares well with the results obtained by the unsteady mixing model.

The third approach validated the selection of a model by applying the chi-square test to compare the closeness of the model output to the deuterium content of the soil water. The chi-square statistic is given by

$$\chi^2 = \sum_{i=0}^{n} \left(\frac{|X_i - M_i|}{\epsilon_i}\right)^2 \tag{47}$$

where X_i are the simulated values, M_i the measured values and ϵ_i the measurement error in δD (+1‰). The models that performed well were the exponential model (EM) and the dispersion models (DM1 and DM2). Here, DM1 is the C_{IFF} solution to the dispersion equation given by eqn (5). In this solution the tracer is assumed to be introduced and sampled in proportion to the flux in a semi-infinite medium. Since the DM1 solution produced unreasonable values for this study, a combination of the C_{IFF} and C_{IRR} solution was used. The C_{IRR} part is for the tracer being injected and sampled instantaneously in water volumes. The DM2 weighting function is given by

$$ g(t) = (16\pi D_p)^{-1/2} \left[\left(\frac{t}{\tau} \right)^{1/2} + \left(\frac{\tau}{t} \right)^{1/2} \right] t^{-1} \exp \left[- \left(1 - \frac{t}{\tau} \right)^2 \left(\frac{\tau}{4D_p t} \right) \right] \tag{48} $$

where the parameters τ and D_p have the same significance as in eqn (19). The piston flow model and combinations with the exponential and dispersion models were found to be unsatisfactory. The dispersion−piston flow, while being flexible, was not found to be much better than the dispersion model.

In the third approach, the input deuterium signature was semi-weighted (weighted by rainfall up to 40 mm) and fully-weighted to study the model response. It was observed that results became progressively worse with weighting for all model types tried. This indicated that the rainfall bypassed the soil matrix.

Overall, the mean residence times obtained by the three approaches were comparable. In the third approach, the EM and dispersion models DM1 and DM2 gave good results. The fact that DM2 was better than the DM1 model indicated the importance of capillary flow as opposed to hydrodynamic dispersion in the water flow in the unsaturated SLA soil matrix.

9.6.2 The Maimai catchment: event-based residence time modelling

Application of the above modelling approach to individual storm events allows the residence time distribution of the event water to be determined (Stewart and Rowe, 1994). This allows a more accurate separation of the event and pre-event waters in storm hydrographs because the appropriate weighted average value of the event water can be determined for each point in time (thus solving the problem of what value to use for the tracer concentration of the event water if it varies strongly, as it usually does; see McDonnell *et al.*, 1990; Turner and Macpherson, 1990). The rainfall rate can be included as part of the weighting. The parameters optimized are the fraction of event water in the runoff and the flow model parameters (mean residence time and dispersion parameter for the dispersion model). Optimization is carried out by fitting the model-generated curve to the experimental data (i.e. minimizing the chi-square statistic given by eqn (47)).

The technique was applied to an experiment carried out at Maimai catchment on 18−21 January 1994. The stream, as well as the largest and smallest soil water flows from a sequence of 60 troughs along the base of a hillslope (i.e. along the stream), were sampled at hourly intervals during a storm event and subsequent recession. The oxygen-18 concentrations and results of the modelling for the stream are shown in Figure 9.6. According to the analysis, 14% of the water discharged in the stream was event water, the remainder being pre-event water with $\delta^{18}O$ of $-5.3‰$. The event water residence time distribution had a peak at two hours, a mean of five hours and two-thirds of the water had residence times in the range 0−5 hours. These observations indicate the power of the method for elucidating water sources of runoff.

Analysis of flow from troughs along the slope base showed that flow varied greatly from trough to trough. Results from low flow troughs showed that 60% of the flow was event water,

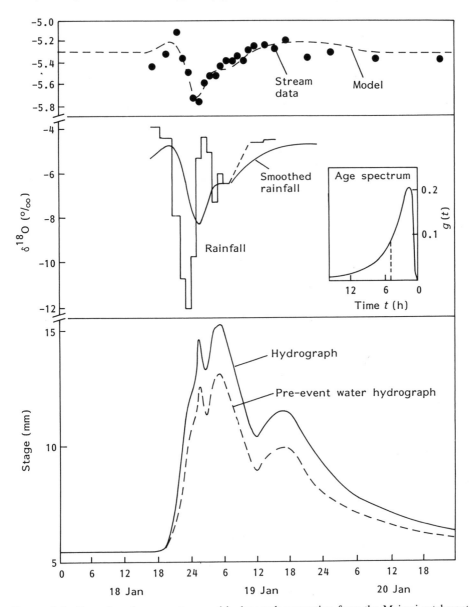

FIGURE 9.6 Event-based age spectrum and hydrograph separation from the Maimai catchment

but this water had relatively long residence times. The distribution had a peak at $4-5$ hours, a mean of 28 hours and 50% of the water discharged had residence times less than 12 hours. Flow commenced about four hours after the onset of rain when sufficient water had infiltrated the soil.

High-flow troughs showed a quite different pattern. The water was initially considerably more negative ($-6.2‰$) than that of the stream or T6, suggesting a source from groundwater

with much longer residence times (the annual mean $\delta^{18}O$ value for the catchment rainfall is
$-6.5\%o$). After 8 hours of rainfall, the $\delta^{18}O$ changed to $-5.0\%o$ and remained there for 16
hours before gradually climbing back towards the groundwater value ($-6\%o$) during 36 hours.
Different assumptions are needed to model this flow; i.e. groundwater supplies a near-constant
flow that gradually has soil water added to it after rain has been falling for some hours. Later,
groundwater again became dominant as the soil water flow decreased.

9.6.3 The Salmon River catchment: long-term modelling

Turner *et al.* (1987) conducted a study in the Salmon catchment located in southwest Western
Australia with the objectives of: (a) using stable isotope data to trace the path of rainfall events
through the catchment to the outflow discharge during winter 1985; (b) determining the
contribution of selected individual rainfall events to the total streamflow using conventional
isotope hydrograph separation; (c) using the stable isotope data to test a conclusion based on the
chloride concentration in streamflow that the groundwater does not contribute significantly to
runoff; and (d) analysing the complete seasonal response of the streamflow isotopic
composition with filtering techniques to define the time-scale for discharge of rainfall events to
streamflow.

The Salmon catchment is broadly rectangular in shape with a catchment area of 82 ha and
total basin relief of 112 m. The basin length is approximately 1500 m. Soils are composed of
gravels, pale sands and lateritic duricrust on the uplands through to red and yellow earths on the
lowlands. Vegetation consists of a relatively dense, high eucalypt forest with shrub
understorey. The local climate is characterized by cool, wet winters and warm to hot summers.
Average annual rainfall is 1220 mm with 80% occurring between May and October. Class A
pan evaporation is 1600 mm per year. The stream is ephemeral and flows from the end of May
to early December.

The deep groundwater system was monitored by six bores. The potentiometric data records
of the bores located in the upper portion of the catchment showed a cyclic seasonal trend
superimposed upon longer term trends. A transect of shallow bores installed over a distance of
150 m in the headwater region of the catchment (Figure 9.7) revealed seasonal response in
hydraulic head and isotopic composition of the shallow groundwater. The total rainfall in 1985
was 1070 mm. Cores through the unsaturated zone into the saturated zone were taken at WBS 3
in December 1982 and December − January 1983 − 4 to characterize the isotopic composition of
recharge water in the unsaturated zone. Rainfall samples were analysed for δD, $\delta^{18}O$ and
chloride over a 21 month period from April 1984 to the end of 1985 at WBS 3. During the
period of streamflow in 1985, the rainfall collection frequency was increased to approximately
one week. The exact timing of a one or two day rainfall in the total exposure time was kept track
of by the tipping-bucket data. Streamflow was recorded continuously at the catchment outlet
and daily, and stage height initiated stream water samples were taken automatically at the
gauging station.

For the 21 month period, the amount-weighted mean isotopic composition of rainfall was
$\delta^{18}O = -4.71\%o$ and $\delta D = -19.5\%o$. During the period of streamflow from 25 May 1985 to
30 November 1985, amount-weighted mean isotopic composition of rainfall was $\delta^{18}O = -4.37\%o$ and $\delta D = -17.9\%o$. The meteoric water line (MWL) for the catchment was $\delta D = 6.59\delta^{18}O + 9.6$. The MWL from rainfall at Salmon for the period of streamflow was $\delta D = 6.90\delta^{18}O + 10.4$. Interestingly, no seasonal trends were found in the rainfall delta values. This

would indicate a stronger influence of local meteorological conditions than a seasonal effect induced by a variable such as temperature. Stable isotope profiles of the unsaturated zone showed remarkable uniformity of isotopic composition, indicating a good mix of the recharge water with the soil water during infiltration. Deep groundwater had mean $\delta^{18}O = -4.99\%_o$ and mean $\delta D = -21\%_o$. The deep groundwater isotopic composition was found to be uniform, indicating good mixing. Based on bore studies on 6 August 1985, it was found that discharge from shallow groundwater into streamflow was significant. The $\delta^{18}O$ and δD responses of the first streamflow were found to be identical to that of the initiating rainfall. Streamflow isotopic composition responded more strongly to individual rainfall events early in the season. This response was damped out with time due to isotopic mixing and dilution of individual rainfall events with increasing storage of shallow groundwater. Response in streamflow isotopic composition was initiated by rainfall events of 10 mm day^{-1} or greater. Towards the end of the streamflow period, direct evaporation from the stream resulted in isotopic enrichment. A two-component hydrograph separation of three individual rainfall events showed that between 60 and 95% of the streamflow hydrograph was pre-event water.

No reasonably stationary linear model was found relating the rainfall to streamflow. In non-stationary models where streamflow is a linear function of lagged rainfall values, the coefficients at all lags tended to rise sharply during periods of high flow, while stationary

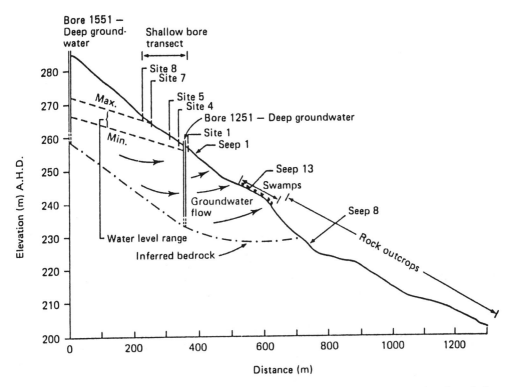

FIGURE 9.7 Hillslope section monitored at the Salmon catchment showing cross-section through deep groundwater bores (modified after Turner *et al.*, 1987)

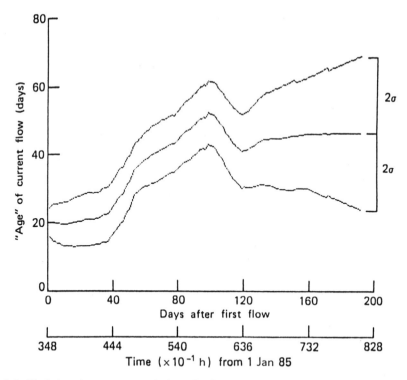

FIGURE 9.8   Variations in average transit times for flow in the Salmon catchment (modified after Turner *et al.*, 1987)

models of this type showed low R^2 values. An attempt to use an extended Kalman filter to estimate a system in which flow was proportional to the level of a hypothetical storage variable was also not successful. Finally, a non-linear model of the following form was adopted:

$$F_t = a \exp\left(b \sum_{i=0}^{100} c^i r_{t-i}\right) \tag{49}$$

where a, b and c are coefficients and F_t and r_{t-i} are as given in eqns (41) and (42). Three models were tried for streamflow dating, i.e. using $\delta^{18}O$ and δD individually and together. All three gave good fits to streamflow isotope data. For this analysis, rainfall delta values were aggregated by using moving ten-day volume-weighted averages. The procedure was justified based on a high autocorrelation of all system variables over periods less than ten days, the fact that rainfall collection is already a form of aggregation and the insensitivity of results to the precise value used. The value of n used was 7, indicating the use of rainfall delta values from up to 70 days before. When both of the isotopes were used simultaneously in eqns (41) and (42), the close relationship between δD and $\delta^{18}O$ induced near-singular matrices in the Kalman filtering algorithm when R^2 was high and these results were considered unreliable. Figure 9.8 shows the variation in average estimated delay between rainfall and streamflow during the period of observations with confidence limits of 2 standard deviations derived from the estimation error of the covariance matrix of the flow components f_{t-i}. Instability in the form of negative f_{t-i} appeared in only 19 of the 1330 days (190 days with 7 components) and was not a serious problem.

The Kalman filtering scheme showed that flow was non-linearly related to rainfall history. The time lag between rainfall and streamflow ranged from 20 to about 50 days over the period of observation.

9.6.4 The Susannah Brook catchment: short-term modelling

In another study, Turner and Macpherson (1990) used deuterium and the hydrogeochemical data nitrate and chloride of rainfall, streamflow, deep and shallow groundwaters to study streamflow generation mechanisms and nutrient discharge in the Susannah Brook catchment located approximately 25 km east of metropolitan Perth in Western Australia. They also estimate a time-dependent or non-stationary transit time using an extension of the Kalman filtering technique used in Turner *et al.* (1987). Isotopic responses in streamflow to depleted rainfall input are identified on three time-scales. The role of shallow groundwater in streamflow generation was shown to be in marked contrast to that in the previous study by Turner *et al.* (1987).

Susannah Brook is a third-order stream draining an area of 27 km². The catchment is 35% forested and 40% pasture, with the remaining land for minor residential and farming purposes. The basement rock is a deeply weathered Archaean igneous complex resulting in a deep lateritic profile of up to 30 m thickness. Surface elevation ranges from 200 m at the gauging station to 300 m at the eastern catchment boundary. Gradients close to the gauging station are about 10% and level out further above to about 4%. The Susannah Brook catchment lies between the 950 and 1000 mm annual rainfall isohyet. Annual rainfall in the catchment for 1987 and 1988 was 820 mm and 957 mm respectively. Total Class A pan evaporation measured at Perth Airport 15 km away was 1950 and 2180 mm respectively. About 90% of the annual rainfall occurs between May and November. Streamflow in the catchment is ephemeral and occurs between early May and the end of November. The most significant hydrogeologic feature is the occurrence of a perched, ephemeral aquifer. This develops within the sloping lateritic duricrust during winter due to infiltration of rainfall through the permeable surficial sands, gravels and lateritic duricrust. The depth to the perched aquifer water table is around 2 m and that to the deep permanent groundwater is 12−15 m in the upper parts of the catchment.

A seasonal trend was seen for streamflow δD, which became enriched in the range of −5 to −15‰ during the low flow summer months. Isotopically depleted rainfall in May caused a shift towards isotopic depletion in streamflow in late June or July. Streamflow recovers in δD values towards the last three months of the flow season with the decrease in frequency and intensity of rainfall events. The shift to enrichment in δD is caused by seasonal enrichment in rainfall isotope values and also due to an isotopic recovery towards δD of deep groundwater, indicating the larger deep groundwater component in streamflow. Most of the isotopically depleted rainfall events induced a response of a similar nature in streamflow. Following the event, over a period that may last up to several weeks, streamflow δD returned to its former value. These are termed isotopic relaxation periods and may be easily determined by studying plots of streamflow δD recovery towards groundwater δD following a rainfall event.

A single event with total rainfall of 120 mm was studied in 1987. δD of streamflow showed responses for 1100 hours after the event. Lagged cross-correlation of rainfall and streamflow for isotopic signatures showed a lag of about 5 hours. Hydrograph separation was done using three procedures: (a) the amount-weighted mean δD of rainfall during the event was used, (b) the actual measured δD for each increment of rainfall was used with zero lag between occurrence of rainfall and its discharge from the catchment, and (c) the estimated lag of 5 hours and the measured δD values of rainfall were used. Under the first approach, the amount-weighted δD or rainfall was more negative than both streamflow and pre-event δD. The hydrograph separation calculations were stable. Under the second approach, however, the hydrograph separation breaks down when δD of rainfall becomes greater than δD of streamflow. There was a slight improvement in the third case, but even here the calculation breaks down when δD of rainfall becomes greater than δD of streamflow and pre-event water.

Finally, a lag analysis was done using Kalman filtering by determining the parameters given in eqns (44) and (45). Raw data were aggregated at one hour intervals to form consistent input and output sequences. No information regarding the flow pathway was incorporated into the calculations.

The results from hydrograph separation techniques and time-series analysis were coupled with hydrogeochemical measurements of chloride and nitrate. The contrasting behaviour of these two ions is due to their different sources. Chloride is naturally occurring and is derived from input of marine aerosols. Nitrate is applied as a fertiliser and the principal mechanism for discharge to streamflow is surface and, possibly, subsurface storm flow. In this basin, very low nitrate concentrations were seen in shallow and deep groundwaters, indicating that streamflow is the primary source of nitrate.

The response of deuterium and chloride in streamflow on time-scales of hours after the event highlighted the importance of surface and subsurface flow components. Hydrograph separation indicated that about 40% of the streamflow is the storm component. Isotopic relaxation periods after significant rainfall events were between 14 and 44 days depending on the timing and magnitude of the event.

The combined use of isotopic, hydrogeochemical and numerical analysis led to a better understanding of streamflow-generating mechanisms in the Susannah Brook catchment.

9.7 CONCLUSIONS

In this chapter we have described a number of modelling techniques for determining residence times of water in hydrological systems from isotopic measurements and case studies have been given of the few studies specifically applied to catchments. Many other isotopic studies have been made which use isotopes to investigate mechanisms of streamflow generation and hydrograph separation analysis, i.e. separation of event water from pre-event water, but these generally used the weighted mean tracer concentration of incoming rainfall to identify the new water, although there are usually large variations in the intensity and isotopic composition of the rainfall. These variations, along with time delays in new water reaching the stream, mean that the isotopic compositions of new water entering the stream will vary in time. Modelling in principle allows this variation to be accounted for.

Water entering the stream comes from a variety of sources (or paths through the catchment), each of which has an associated distribution of residence times. Together they produce the overall distribution of residence times of water in the stream. This distribution is expected to vary depending on streamflow stage, antecedent rainfall and other hydrologic characteristics.

The mean residence time is a weighted average of the residence times of input waters. Its variation in time (e.g. as derived by Turner and Macpherson, 1990) gives a gross measure of the changes in input waters during a rainfall event. However, of equal importance is the peak residence time, which is usually very different from the mean residence time because the residence time distribution of stream water is usually highly skewed (Stewart and McDonnell, 1991). Peak residence times reflect the influence of rapid-flow pathways through the catchment, such as preferred flow channels in soils, perched saturated layers, etc., and give important information on the quick-flow response of streams. Longer residence time waters are from soil matrix or deeper groundwaters, and support streamflow during stormflow and recessions.

Measurements and modelling on different time-scales reveal different storage characteristics of the systems. Longer term modelling at Maimai used samples collected at weekly intervals from soil matrix and the stream during baseflow conditions, hence related to sources of water with residence times of weeks to months. Measurements during storm events on samples collected at hourly intervals reveal the short-term storage characteristics — old water is assumed to have a constant composition and constitute a constant or variable fraction of the flow.

Variable parameter models have intuitive appeal and are a logical approach to interpreting the isotopic data in terms of flow-generation processes. However, application of the models needs to be made more accessible. Residence time distributions should be determined as well as mean residence times, because the latter alone are not very diagnostic of water flowpaths. More work is also needed to tune the filter to give optimum tracking of the parameters for the system investigated, so that the model output more closely approximates the system.

REFERENCES

Adar, E. M., Neuman, S. P. and Woolhiser, D. A. (1988). "Estimation of spatial recharge distribution using environmental isotopes and hydrochemical data: mathematical model and application to synthetic data", *J. Hydrol.*, **97**, 251–277.

Bear, J. (1979). *Hydraulics of Groundwater*, McGraw-Hill Publishing Company.

Bryson, A. E. and Ho, Y.-C. (1975). *Applied Optimal Control*, Wiley, New York.

Eriksson, E. (1963). "Atmospheric tritium as a tool for the study of certain hydrologic aspects of river basins", *Tellus*, **15**, 303–308.

Gieske, A. and De Vries, J. J. (1990). "Conceptual and computational aspects of the mixing cell method to determine groundwater recharge components", *J. Hydrol.*, **121**, 277–292.

Harris, D., McDonnell, J. J. and Rodhe, A. (1995). "Isotope hydrograph separation using a continuous open system mixing approach", *Water Resour. Res.*, **31**(1), 157–171.

Kreft, A. and Zuber, A. (1978). "On the physical meaning of the dispersion equation and its solutions for different initial and boundary conditions", *Chem. Engng Sci.*, **33**, 1471–1480.

McDonnell, J. (1990). "A rationale for old water discharge through macropores in a steep, humid catchment", *Water Resour. Res.*, **26**, 2821–2832.

McDonnell, J., Bonell, M., Stewart, M. K. and Pearce, A. J. (1990). "Deuterium variations in storm rainfall: implications for stream hydrograph separation", *Water Resour. Res.*, **26**, 455–458.

McDonnell, J., Stewart, M. K. and Owens, I. F. (1991). "Effect of catchment-scale subsurface mixing on stream isotopic response", *Water Resour. Res.*, **26**, 3065–3073.

Maloszewski, P. and Zuber, A. (1982). "Determining the turnover time of groundwater systems with the aid of environmental tracers. 1. Models and their applicability", *J. Hydrol.*, **57**, 207–231.

Maloszewski, P. and Zuber, A. (1993). "Principles and practice of calibration and validation of mathematical models for the interpretation of environmental tracer data in aquifers", *Adv. in Water Resour.*, **16**, 173–190.

Maloszewski, P., Rauert, W., Stichler, W. and Herrmann, A. (1983). "Application of flow models in an Alpine catchment area using tritium and deuterium data", *J. Hydrol.*, **66**, 319–330.

Minchin, P. E. H. and Troughton, J. H. (1980). "Quantitative interpretation of phloem translocation data", *Ann. Rev. Plant Physiol.*, **31**, 191–215.

Press, W. H., Flannery, B. P., Teukolsky, S. A. and Vetterling, W. T. (1986). *Numerical Recipes: The Art of Scientific Computing*, Cambridge University Press, Cambridge.

Rodhe, A. (1981). "Spring flood: meltwater or groundwater?", *Nord. Hydrol.*, **12**, 21–30.

Shumway, R. H. and Stoffer, D. S. (1982). "An approach to time series smoothing and forecasting using the EM algorithm", *J. Time Series Anal.*, **3**, 253–264.

Sklash, M. G. and Farvolden, R. N. (1979). "The role of groundwater in storm runoff", *J. Hydrol.*, **43**, 45–65.

Sklash, M. G., Farvolden, R. N. and Fritz, P. (1976). "A conceptual model of watershed response to rainfall, developed through the use of oxygen-18 as a natural tracer", *Can. J. Earth Sci.*, **13**, 271–283.

Stewart, M. K. and McDonnell, J. J. (1991). "Modeling base flow soil water residence times from deuterium concentrations", *Water Resour. Res.*, **27**(10), 2681–2693.

Stewart, M. K. and Rowe, L. K. (1994). "Water component analysis of runoff and soil water flows in small catchments", in *Tracer Modeling*, Proceedings of the Western Pacific Geophysics Meeting, p. 37.

Stichler, W. and Herrmann, A. (1983). "Application of environmental isotope techniques in water balance studies of small basins", in *New Approaches in Water Balance Computations*, Proceedings of the Hamburg Workshop, IAHS Publication No. 148, pp. 93–112.

Turner, J. V. and Macpherson, D. K. (1990). "Mechanisms affecting streamflow and stream water quality: an approach via stable isotope, hydrogeochemical and time series analysis", *Water Resour. Res.*, **26**(12), 3005–3019.

Turner, J. V., Macpherson, D. K. and Stokes, R. A. (1987). "The mechanisms of catchment flow processes using natural variations in deuterium and oxygen-18", *J. Hydrol.*, **94**, 143–162.

Wagner, B. J. and Gorelick, S. M. (1986). "A statistical methodology for estimating transport parameters: theory and applications to one-dimensional advective-dispersive systems", *Water Resour. Res.*, **22**(8), 1303–1315.

Yurtsever, Y. (1983). "Models for tracer data analysis", in *Guidebook on Nuclear Techniques in Hydrology*, pp. 381–402, International Atomic Energy Agency, Vienna.

10

Isotope Tracers of Water and Solute Sources in Catchments

CAROL KENDALL

United States Geological Survey, Menlo Park, California, USA

MICHAEL G. SKLASH

The Dragun Corporation, Farmington Hills, Michigan, USA

and

THOMAS D. BULLEN

United States Geological Survey, Menlo Park, California, USA

10.1 INTRODUCTION

The topic of streamflow generation has received considerable attention over the last two decades primarily in response to concern about "acid rain". Many sensitive, low-alkalinity streams in North America and Europe are already acidified. Still more streams that are not yet chronically acidic may undergo acid episodes in response to large rain storms or spring snowmelt. These acidic events can seriously damage local ecosystems. Future climate changes may exacerbate the situation by affecting biogeochemical controls on the transport of water, nutrients and other materials from land to freshwater ecosystems.

New awareness of the potential danger to water supplies posed by the use of agricultural chemicals has also focused attention on the nature of rainfall-runoff processes and the mobility of various solutes, especially nitrate and pesticides, in shallow systems. Dumping and spills of other potentially toxic materials are also of concern because these chemicals may eventually reach streams and other public water supplies. A better understanding of hydrologic flowpaths and solute sources is required to determine the potential impact of contaminants on water supplies, develop management practices to preserve water quality and devise remediation plans for sites that are already polluted.

Solute Modelling in Catchment Systems. Edited by Stephen T. Trudgill
Published 1995 by John Wiley & Sons Ltd

Of all the methods used to understand hydrologic processes in small catchments, applications of tracers — in particular isotope tracers — have been the most useful in providing new insights into hydrologic processes because they integrate small-scale variability to give an effective indication of catchment-scale processes. The main purpose of this chapter is to highlight recent research into the use of naturally occurring isotopes to track the movement of water and solutes through small catchments to streams. As we improve our understanding of how water and solutes are transported through catchments during high flow episodes, we can move closer to finding solutions to stream acidification and other water-quality problems.

10.1.1 Environmental isotopes as tracers in catchments

Environmental isotopes are naturally occurring (or, in some cases, anthropogenically produced) isotopes whose distributions in the hydrosphere can assist in the solution of hydrogeochemical problems. Typical uses of environmental isotopes in hydrology include:

- Identification of the mechanisms responsible for streamflow generation
- Characterization of the flowpaths that water follows from the time precipitation hits the ground to discharge at the stream
- Determination of the weathering reactions that mobilize solutes along those flowpaths
- Determination of the role of atmospheric deposition in controlling water chemistry
- Identification of the sources of solutes in contaminated systems
- Assessment of biologic cycling of nutrients within the ecosystem
- Testing flowpath and water-budget models developed using hydrometric data

Environmental isotopes can be used as tracers of waters and solutes in catchments because:

1. Waters that were recharged at different times, in different locations, or that followed different flowpaths are often isotopically distinct; in other words, they have distinctive "fingerprints".
2. Unlike most chemical tracers, environmental isotopes are relatively conservative in reactions with the catchment materials. This is especially true of oxygen and hydrogen isotopes in water. The waters mentioned above retain their distinctive fingerprints until they mix with other waters or undergo phase changes.
3. Solutes in the water that are derived from atmospheric sources are often isotopically distinct from solutes derived from geologic and biologic sources within the catchment.
4. Both biological cycling of solutes and water/rock reactions often change isotopic ratios in the *solutes* in predictable and recognisable directions. These interactions often can be reconstructed from the isotopic compositions.
5. If water from an isotopically distinctive source (e.g. rain with an unusual isotopic composition) is found along a flowpath, it must be hydrologically possible for that source to contribute to this flowpath, despite any hydraulic measurements or models to the contrary.

The applications of environmental isotopes as hydrologic tracers in low-temperature systems fall into two main categories:

- Tracers of the water itself (water isotope hydrology)
- Tracers of the solutes in the water (solute isotope biogeochemistry)

These classifications are by no means universal but they are useful conceptually and often eliminate confusion when comparing results using different tracers.

Water isotope hydrology

Water isotope hydrology addresses the application of the isotopes that form water molecules: the oxygen isotopes (oxygen-16, oxygen-17 and oxygen-18) and the hydrogen isotopes (protium, deuterium and tritium). These isotopes are ideal tracers of water sources and movement because they are constituents of water molecules, not something that is dissolved *in* the water like other tracers that are commonly used in hydrology (e.g. dissolved species such as chloride).

In most low-temperature environments, stable hydrogen and oxygen isotopes behave conservatively in the sense that as they move through a catchment, any interactions with oxygen and hydrogen in the organic and geologic materials in the catchment will have a negligible effect on the ratios of isotopes in the water molecule. Although tritium also exhibits insignificant reaction with geologic materials, it does change in concentration over time because it is radioactive and decays with a half-life of about 12.4 years. The main processes that dictate the oxygen and hydrogen isotopic compositions of waters in a catchment are: (a) phase changes that affect the water near the ground surface (evaporation, condensation, melting) and (b) simple mixing at or below the ground surface.

Oxygen and hydrogen isotopes are commonly used in catchment research for determining the contributions of "old" and "new" water to high flow (storm and snowmelt runoff) events in streams. "Old" water is defined as the water that existed in a catchment prior to a particular storm or snowmelt period. Old water includes groundwater, soil water and surface water. "New" water is either rainfall or snowmelt, and is defined as the water that triggers the particular storm or snowmelt runoff event. The application of oxygen and hydrogen isotopes to catchment studies has been summarised in a number of papers including Rodhe (1987), Sklash (1990) and others.

Solute isotope biogeochemistry

Solute isotope biogeochemistry addresses the application of isotopes of constituents that are dissolved in the water or are carried in the gas phase. Isotopes commonly used in solute isotope biogeochemistry research include the isotopes of: carbon, nitrogen and sulphur. Less commonly applied isotopes include: strontium, lead, uranium, radon, radium, lithium and boron.

Unlike the isotopes in the water molecules, the ratios of solute isotopes can be significantly altered by reaction with geological and/or biological materials as the water moves through the catchment. Although the literature contains numerous case studies involving the use of solutes (and sometimes solute isotopes) to trace *water* sources and flowpaths, such applications include an implicit assumption that these solutes are transported conservatively *with* the water. In a strict sense, *solute isotopes only trace solutes*. Solute isotopes also provide information on the reactions that are responsible for their presence in the water and the flowpaths implied by their presence.

Solute isotopes are not yet commonly used for determining weathering reactions and sources of solutes in catchment research. Although there has been extensive use of carbon, nitrogen and sulphur isotopes in studies of forest growth and of agricultural productivity, this chapter will focus only on applications of solute isotopes as tracers of solute *sources*. We will concentrate on applications of solute isotopes that apportion the solutes in a given catchment among weathering reactions, atmospheric deposition and biological release.

10.1.2 Isotope fundamentals

Isotopes are atoms of the same element (the same number of electrons and protons) that have different numbers of neutrons. The difference in the number of neutrons among the various isotopes of an element means that the isotopes have different masses. The superscript number to the left of the element designation denotes the number of protons plus neutrons in the isotope. For example, among the hydrogen isotopes, deuterium (denoted as D or ^{2}H) has one neutron and one proton; this is approximately twice the mass of protium (^{1}H), whereas tritium (^{3}H) has approximately three times the mass of protium.

The various isotopes of an element have slightly different chemical and physical properties because of their mass differences. For elements of low atomic numbers, these mass differences are large enough for many physical, chemical and biological processes or reactions to "fractionate" or change the relative proportions of various isotopes. Two different types of processes — equilibrium and kinetic isotope effects — cause isotope fractionation. As a consequence of fractionation processes, waters and solutes often develop unique isotopic compositions (ratios of heavy to light isotopes) that may be indicative of their source or of the processes that formed them.

Equilibrium isotope-exchange reactions involve the redistribution of isotopes of an element among various species or compounds. At equilibrium, the forward and backward reaction rates of any particular isotope are identical. This does not mean that the isotopic compositions of two compounds at equilibrium are identical, but only that the ratios of the different isotopes in each compound are constant. During equilibrium, the heavier isotope generally becomes *enriched* (preferentially accumulates) in the species or compound with the higher oxidation state. For example, sulphate is enriched in ^{34}S relative to sulphide; consequently, the sulphide is described as *depleted* in ^{34}S. During phase changes, the ratio of heavy to light isotopes in the molecules in the two phases changes. For example, as water vapour condenses (an equilibrium process), the heavier water isotopes (^{18}O and ^{2}H) become enriched in the liquid phase while the lighter isotopes (^{16}O and ^{1}H) tend towards the vapour phase.

Kinetic isotope fractionations occur in systems out of isotopic equilibrium where forward and backward reaction rates are not identical. The reactions may, in fact, be unidirectional if the reaction products become physically isolated from the reactants. Reaction rates depend on the ratios of the masses of the isotopes and their vibrational energies; as a general rule, bonds between the lighter isotopes are broken more easily than the stronger bonds between the heavy isotopes. Hence, the lighter isotopes react more readily and become concentrated in the products, and the residual reactants become enriched in the heavy isotopes.

Biological processes are generally unidirectional and are excellent examples of "kinetic" isotope reactions. Organisms preferentially use the lighter isotopic species because of the lower energy "costs", resulting in significant fractionations between the substrate (heavier) and the biologically mediated product (lighter). The magnitude of the fractionation depends on the reaction pathway utilised and the relative energies of the bonds being severed and formed by the reaction. In general, slower reaction steps show greater isotopic fractionation than faster steps because the organism has time to be more selective. Kinetic reactions can result in fractionations very different from, and typically larger than, the equivalent equilibrium reaction.

Many reactions can take place either under purely equilibrium conditions or can be affected by an additional kinetic isotope fractionation. For example, although evaporation can take place under purely equilibrium conditions (i.e. at 100% humidity when the air is still), more typically the products become partially isolated from the reactants (e.g. the resultant vapour is blown downwind). Under these conditions, the isotopic compositions of the water and vapour are affected by an additional kinetic isotope fractionation of variable magnitude.

The partitioning of stable isotopes between two substances A and B can be expressed by use of the isotopic fractionation factor α (alpha):

$$\alpha_{A-B} = \frac{R_A}{R_B}$$

where R is the ratio of the heavy to light isotope (e.g. $^2H/^1H$ or $^{18}O/^{16}O$). Values for α tend to be very close to 1. Kinetic fractionation factors are typically described in terms of enrichment or discrimination factors; these are defined in various ways by different researchers, using such symbols as β, ϵ or D.

Radioactive (unstable) isotopes have nuclei that spontaneously disintegrate over time to form other isotopes. The so-called stable isotopes have nuclei that do not appear to decay to other isotopes on geologic time-scales, but may themselves be produced by the decay of radioactive isotopes. For example, ^{14}C, a radioisotope of carbon, is produced in the atmosphere by the interaction of cosmic ray neutrons with stable ^{14}N. With a half-life of about 5730 years, ^{14}C decays back to ^{14}N by emission of a beta particle; the stable ^{14}N produced by radioactive decay is called "radiogenic" nitrogen.

The stable isotopic compositions of low-mass elements such as oxygen, hydrogen, carbon, nitrogen and sulphur are normally reported as delta (δ) values in parts per thousand (denoted as ‰ or permil, per mil or per mille) enrichments or depletions relative to a standard of known composition. The δ values are calculated by

$$\delta \text{ (in ‰)} = \left(\frac{R_{sample}}{R_{standard}} - 1 \right) \times 1000$$

where R is the ratio of the heavy to light isotope. For the elements sulphur, carbon, nitrogen and oxygen, the average terrestrial abundance ratio of the heavy to the light isotope ranges from 1:22 (sulphur) to 1:500 (oxygen); the ratio $^2H/^1H$ is 1:6410. A positive delta value means that the sample contains more of the heavy isotope than the standard; a negative delta value means that the sample contains less of the heavy isotope than the standard.

Stable oxygen and hydrogen isotopic ratios are normally reported relative to the SMOW standard (standard mean ocean water) (Craig, 1961b) or the equivalent V-SMOW (Vienna-SMOW) standard. Carbon, nitrogen and sulphur stable isotope ratios are reported relative to the V-PDB, AIR and CDT standards respectively, as defined later.

The delta values for these isotopes are determined in specialised laboratories using isotope ratio mass spectrometry. The analytical precisions are small relative to the ranges in delta values that occur in natural earth systems. Typical one standard deviation and analytical precisions for oxygen, carbon, nitrogen and sulphur isotopes are in the range of 0.05 to 0.2‰; typical precisions for hydrogen isotopes are poorer, from 0.2 to 2.0‰, because of the lower $^2H/^1H$ ratio.

10.2 WATER ISOTOPES

Oxygen and hydrogen isotopes can be used to determine the contributions of old and new water to a stream during periods of high runoff because the rain or snowmelt (new water) that triggers the runoff is often isotopically different from the water already in the catchment (old water). This section briefly explains why the old and new water components often have different isotopic compositions. For more detailed discussions of these and other environmental isotopes, the reader should consult reviews such as: Fritz and Fontes (1980, 1981), Gat and Gonfiantini (1981), Faure (1986), Hoefs (1987) and Coplen (1993).

10.2.1 Deuterium and oxygen-18

Craig (1961a) observed that the $\delta^{18}O$ and δ^2H (δD) values of precipitation that has not been evaporated are linearly related by

$$\delta D = 8\delta^{18}O + 10$$

This equation, known as the global meteoric water line (GMWL), is based on precipitation data from locations around the globe. The slope and intercept of the local meteoric water line (LMWL) for a catchment can be different from the GMWL. Water that has evaporated or mixed with evaporated water plots below the meteoric water line on a line with a slope of between 2 and 5 (Figure 10.1).

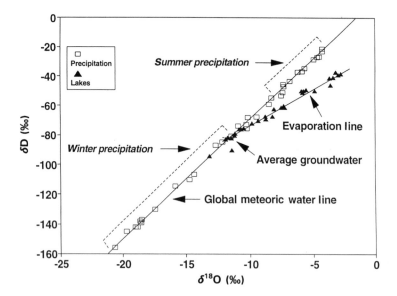

FIGURE 10.1 Isotopic compositions (δD versus $\delta^{18}O$) of precipitation and lake samples from northern Wisconsin. The precipitation samples define the local meteoric water line ($\delta D = 8.03\ \delta^{18}O + 10.95$), which is virtually identical to the global meteoric water line. The lake samples plot along an evaporation line with a slope of 5.2 that intersects the LMWL at the composition of average local groundwater (modified from Krabbenhoft *et al.*, 1994, with additional data)

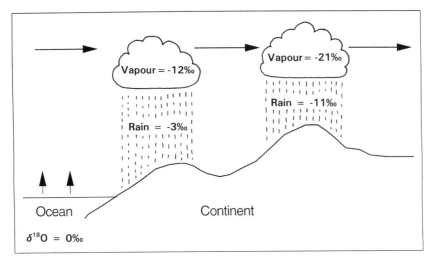

FIGURE 10.2 Origin of variations in the $\delta^{18}O$ of meteoric waters as moisture rains out of air masses moving across the continents

The two main factors that control the isotopic character of precipitation at a given location are the condensation temperature of the precipitation and the degree of rain-out of the air mass (the ratio of water vapour that has already condensed into precipitation to the initial amount of water vapour in the air mass). Progressive rain-out as clouds move across the continent causes successive rain storms to become increasingly lighter, as shown in Figure 10.2. In this example, non-equilibrium evaporation from the ocean of $\delta^{18}O = 0‰$ produces vapour of $-12‰$; equilibrium condensation of rain from this vapour results in water with a $\delta^{18}O = -3‰$ and residual vapour with a $\delta^{18}O = -21‰$.

At a given location, the seasonal variation in $\delta^{18}O$ and δ^2H values of precipitation and the weighted average annual $\delta^{18}O$ and δ^2H values of precipitation remain fairly constant from year to year. This happens because the annual range and sequence of climatic conditions (temperature, vapour source, direction of air mass movement, etc.) remain fairly constant from year to year. Figure 10.3 shows the roughly sinusoidal annual pattern of $\delta^{18}O$ values of individual storms at Panola Mountain, Georgia (USA). The cyclic nature is emphasised by calculating the volume-weighted average compositions for two-month intervals. The annual volume-weighted averages for the three years are very similar, with a total range of 0.7‰.

In general, rain in the summer is isotopically heavier than rain in the winter, as shown in data from Wisconsin (Figure 10.1) and Georgia (Figure 10.3). This change in average isotopic composition is principally caused by seasonal temperature differences, but is also affected by seasonal changes in moisture sources and storm tracks.

Shallow groundwater $\delta^{18}O$ and δ^2H values reflect the local average precipitation values but are modified to some extent by selective recharge and fractionation processes that may alter the $\delta^{18}O$ and δ^2H values of the precipitation before the water reaches the saturated zone (Gat and Tzur, 1967). Some of these processes include:

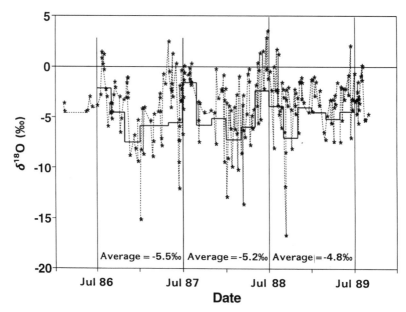

FIGURE 10.3   The $\delta^{18}O$ of storms at Panola Mountain, Georgia, shows a 20‰ range in composition. Superimposed on the inter-storm variability is the expected seasonal oscillation between heavy $\delta^{18}O$ values in the summer and lighter values in the winter caused by seasonal changes in temperature, storm track and rain-out. The volume-weighted $\delta^{18}O$ values for 2 month intervals are shown by heavy solid lines. The volume-weighted averages for the years 1987, 1988 and 1989 were each calculated from 1 July to 30 June

1. Evaporation of rain during infiltration
2. Selective recharge. Examples include recharge that occurs only in response to major storms whereas rain from smaller storms may be completely evaporated, and melting of the snowpack on frozen ground.
3. Interception of rain water (Kendall, 1993; DeWalle and Swistock, 1994) and snow (Claassen and Downey, 1995) by the tree canopy may result in significant evaporation.
4. Exchange of infiltrating water with atmospheric vapour (Kennedy *et al.*, 1986).
5. In the case of snow, various post-depositional processes, such as melting and subsequent infiltration of surface layers and evaporation, may alter the isotopic content of the snowpack, often leading to meltwater delta values that become progressively enriched (Stichler, 1987).

Once the rain or snowmelt passes into the saturated zone, the delta values of the subsurface water change only by mixing with waters that have different isotopic contents. The homogenising effects of recharge and dispersive processes produce groundwater with delta values that approach uniformity in time and space and that approximate a damped reflection of the precipitation over a period of years (Brinkmann *et al.*, 1963).

It is important to remember that although an individual storm may be large and isotopically very different from the old water in the catchment, the amount of precipitation that infiltrates will likely be small compared to the amount of old water in storage. Although there may be significant storm-to-storm and seasonal variations in precipitation $\delta^{18}O$ and δ^2H values (Figure 10.3), baseflow delta values remain relatively uniform in most streams in humid, temperate

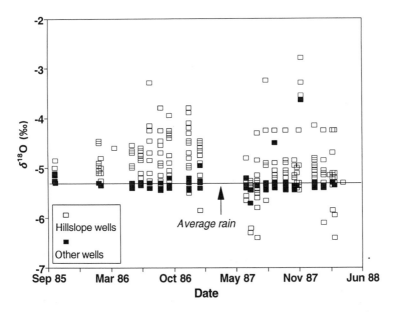

FIGURE 10.4 The $\delta^{18}O$ of groundwater from about 30 wells at Panola Mountain, a 41 ha watershed in Georgia. In general, shallow hillslope wells (open squares) show more temporal variability than the deeper and/or near-stream wells (solid squares) that approach the composition of average rain

areas. In small catchments, temporal and spatial variability in groundwater and baseflow delta values may reflect seasonal variability in precipitation delta values (Figure 10.4). However, these variations are less extreme and usually delayed relative to the temporal variations that occur in precipitation (Turner *et al.*, 1987: Sklash, 1990; McDonnell *et al.*, 1991). The chapter by Unnikrishna *et al.* in this volume (Chapter 9) describes how these variations can be utilized for water residence time calculations.

Superimposed on the seasonal cycles in precipitation delta values are storm-to-storm and intra-storm variations in the $\delta^{18}O$ and δ^2H values of precipitation. These variations may be as large as the seasonal variations (Figure 10.3). It is this potential difference in delta values between the relatively uniform old water and variable new water that permits isotope hydrologists to determine the contributions of old and new water to a stream during periods of high runoff (Sklash *et al.*, 1976).

10.2.2 Tritium

Tritium (^3H) is a radiogenic and radioactive isotope of hydrogen with a half-life of 12.43 years (IAEA, 1981). It is an excellent tracer for determining time scales for the mixing and flow of waters, and is ideally suited for studying processes that occur on a time-scale of less than 100 years. Tritium content is expressed in tritium units (TU) where 1 TU equals 1 ^3H in 10^{18} atoms of hydrogen. Prior to the advent of atmospheric testing of thermonuclear devices in 1952, the tritium content of precipitation was probably in the range of $2-8$ TU (Thatcher, 1962); this background concentration is produced by cosmic ray spallation. Since 1952, tritium produced by thermonuclear testing ("bomb tritium") has been the dominant source of tritium in

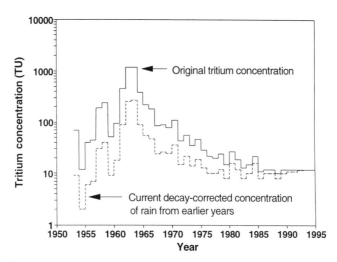

FIGURE 10.5 Tritium concentration in volume-weighted average annual rain in the southeastern United States. The solid lines show the original average tritium concentration of rain for each year. The dashed lines show the current (1995) composition of rain recharged from 1953 onwards. These values were calculated using a 12.43 yr half-life (data from Rose, 1993)

precipitation. A peak concentration of several thousand TU was recorded in precipitation in the northern hemisphere in 1963, the year that the Atmospheric Test Ban Treaty was signed. After 1963, the tritium levels in precipitation began to decline gradually because of radioactive decay and the cessation of atmospheric testing (Figure 10.5).

Tritium concentrations in precipitation are affected by latitude, distance from the ocean and season. Water vapour over the ocean has a low tritium concentration due to exchange with oceanic surface water, which has a low tritium content. With progressive movement of air masses across the continent, tritium concentrations in precipitation increase due to additions of tritium from the stratosphere and evapotranspiration. The seasonal effect is primarily caused by the breakup of the tropopause between 30 and 60°N latitude each spring, causing a leakage of stratospheric water vapour with a high tritium content into the troposphere. This leakage results in a north–south gradient in tritium concentrations in precipitation, with higher concentrations in the north and spring–summer high concentrations that are typically 2.5–6 times greater than the winter lows (Gat, 1980).

The simplest use of tritium is to check whether detectable concentrations are present in the water. Although pre-bomb atmospheric tritium concentrations are not well known, waters derived exclusively from precipitation before 1953 would have maximum tritium concentrations of 0.2–0.8 TU by the early 1990s (Plummer *et al.*, 1993). For waters with higher tritium contents, some fraction of the water must have been derived since 1953; thus, the tritium concentration can be a useful marker for recharge since the advent of nuclear testing.

Distinct "old" and "new" water ^3H values are required for storm and snowmelt runoff hydrograph separation. In small (first-order) catchments where the average residence time (the average time it takes for precipitation to enter the ground and travel to the stream) of the old water is on the order of months, the old and new water ^3H concentrations are not likely to be distinguishable. However, in some larger catchments with longer residence times, old and new waters may be distinctive as a result of the gradual decline in precipitation ^{3}H values since 1963 and the even more gradual decline in groundwater ^{3}H values (Figure 10.5). Figure 10.5 shows that the average tritium content of current precipitation (new water) is indistinguishable from the decay-corrected composition of precipitation that fell $10-15$ years ago (old water).

Tritium measurements are frequently used to calculate recharge rates, rates or directions of subsurface flow and residence times. For these purposes, the seasonal, yearly and spatial variations in the tritium content of precipitation must be accurately assessed. This is difficult to do because of the limited data available, especially before the 1960s. For a careful discussion of how to calculate the input concentration at a specific location, see Michel (1989) and Plummer *et al.* (1993). Several different approaches (e.g. piston-flow, reservoir, compartment and advective−dispersive models) to modelling tritium concentrations in groundwater are discussed by Plummer *et al.* (1993). The narrower topic of using environmental isotopes to determine residence time is discussed briefly below and in detail in the chapter by Unnikrishna *et al.* (Chapter 9).

10.3 DETERMINATION OF RUNOFF MECHANISMS

A major uncertainty in hydrologic and chemical modelling of watersheds has been the quantification of the contributions of water and solutes from various hydrologic pathways. Many hydrologic models can adequately reproduce the timing and discharge features of a hydrograph, but when chemical data are coupled to these physical models, they rarely reflect the streamwater chemistry accurately. It has therefore been concluded that the hydrologic pathways are incorrectly represented in the models (Mulholland *et al.*, 1990) or that the processes that are affecting the water chemistry are not well understood (McKnight and Bencala, 1990). Water isotope hydrology and solute isotope biogeochemistry can address these problems.

In their summary of runoff generation mechanisms, Pearce *et al.* (1986) suggested that there are three main hydrologic pathways that operate in humid, vegetated, headwater catchments: "partial-area overland flow", "saturation overland flow" and "subsurface flow". Although the exact names of these processes may vary from text to text, the following descriptions seem to be generally accepted:

1. "Partial-area overland flow" is overland flow that is generated on certain parts of the catchment after these areas become saturated from the surface downward by rainfall (or snowmelt) rates in excess of the soil infiltration rate. Subsequent rain (or snowmelt) on these areas runs off overland.
2. "Saturation overland flow" is overland flow that is generated near streams and in valley floors where surface saturation occurs because of a rising water table. Subsequent rain (or snowmelt) on these areas runs off overland.
3. "Subsurface flow" is the underground movement of water either rapidly as open-channel flow through macropores or slowly by Darcian flow through micropores (matrix flow).

The relative importance of these processes in a particular catchment to a large extent governs the temporal changes in the stream water chemistry during storm and snowmelt runoff events. Two typical scenarios where the dominant hydrologic pathway significantly affects subsequent stream water quality are:

1. Immediately after fertilization, a rain storm hits an agricultural catchment. How much fertilizer will reach the stream if some type of overland flow mechanism dominates storm runoff generation? How much will reach the stream if Darcian subsurface flow dominates storm runoff generation? How can the amount of fertilizer that reaches the stream be reduced in this catchment by changes in management practices, given the dominant pathway?

2. A small forested catchment in granitic terrain near a major industrial centre has received record snowfall during the winter. The snowpack is about to melt rapidly as very warm weather is predicted. How acidified will the stream become if some type of overland flow mechanism dominates storm runoff generation? How acidified will the stream become if Darcian subsurface flow dominates storm runoff generation? How can stream acidification be alleviated in this catchment, given the dominant pathway?

Of all the methods used to understand hydrologic processes in small catchments, applications of tracers — in particular isotope tracers — have been the most useful in terms of providing new insights into hydrologic processes because they integrate small-scale variability to give an effective indication of catchment-scale processes (McDonnell and Kendall, 1992; Buttle, 1994; Buttle, 1995). Internal watershed measurements, such as water level or groundwater composition, cannot be used without extrapolation or additional assumptions. Isotopes are "applied" at the watershed scale. In particular ^{18}O, ^{2}H and ^{3}H form parts of natural water molecules that fall year after year over a watershed as rain and snow, and are ideal tracers of water. This no-cost, long-term and widespread application of these tracers allows hydrologists to study runoff generation on scales ranging from macropores to hillslopes to first- and higher-order streams (Sklash, 1990).

10.3.1 Mixing models

Until the 1970s, the term "hydrograph separation" meant a graphical technique that had been used for decades in predicting runoff volumes and timing. For example, the graphical separation technique introduced by Hewlett and Hibbert (1967) is commonly applied to storm hydrographs (i.e. streamflow versus time graphs) from forested catchments to quantify "quick flow" and "delayed flow" contributions. According to this technique, the amounts of quick flow and delayed flow in stream water can be determined simply by considering the shape and timing of the discharge hydrograph.

Unfortunately, some hydrologists have come to equate the quick flow and delayed flow parts of separated hydrographs to specific precipitation-to-runoff conversion processes (flowpaths and/or runoff mechanisms) and to specific water sources. For example, quick flow has become associated with rapid precipitation-to-runoff conversion mechanisms such as partial-area overland flow, saturation overland flow and macropore flow; delayed flow has become associated with slow-moving subsurface Darcian flow. Hydrograph separations using chemical and isotopic tracers have shown that these concepts are often wrong (Fritz *et al.*, 1976; Sklash *et al.*, 1976).

Isotope hydrograph separation apportions storm and snowmelt hydrographs into contributing components based on the distinctive isotopic signatures carried by the old and new water. Hence, the method allows the calculation of the relative contributions of new precipitation and older groundwater to streamflow. A major limitation of the tracer method is that we cannot directly determine how the water reaches the stream (i.e. geographic source of the water) nor the actual runoff generation mechanism from knowledge of the temporal sources of water. In a very real sense, the isotope hydrograph separation technique is still a "black box" method that provides little direct information about what is actually going on in the subsurface. However, in combination with other types of information (e.g. estimates of saturated areas or chemical compositions of water along specific flowpaths), runoff mechanisms can sometimes be inferred.

10.3.2 Isotope hydrograph separation

The *basic principle* behind isotope hydrograph separation is that if the isotopic compositions of the sources of water contributing to streamflow during periods of high runoff are known and are different, then the relative amounts of each source can be determined. Since these studies began in the 1960s, the overwhelming conclusion is that old water is by far the dominant source of runoff in humid, temperate environments (Sklash *et al.*, 1976; Pearce *et al.*, 1986; Bishop, 1991).

Two-component mixing models

Isotope hydrograph separation normally involves a two-component mixing model for the stream. The model assumes that water in the stream at any time during storm or snowmelt runoff is a mixture of two components: new water and old water.

During baseflow conditions (the low flow conditions that occur between periods of storm and snowmelt runoff), the water in a stream is dominated by old water. The chemical and isotopic character of stream water at a given location during baseflow represents an integration of the old water discharges upstream. During storm and snowmelt runoff events, however, new water is added to the stream. If the old and new water components are chemically or isotopically different, the stream water becomes changed by the addition of the new water. The extent of this change is a function of the relative contributions by the old and new water components.

The contributions of old and new water in the stream at any time can be calculated by solving the mass balance equations for the water and tracer fluxes in the stream, provided that the stream, old water and new water tracer concentrations are known:

$$Q_s = Q_o + Q_n$$
$$C_s Q_s = C_o Q_o + C_n Q_n$$

where Q is discharge, C refers to tracer concentration of a component or the stream, and the subscripts s, o and n indicate the stream, the old water component and the new water component respectively. If stream samples are taken at a stream gauging station, the actual volumetric contributions of old and new water can be determined. If no discharge measurements are available, old and new water contributions can be expressed as percentages of total discharge.

Although the simple two-component mixing model approach to stream hydrograph separation does not directly identify the actual runoff generation mechanisms, the model can sometimes allow the hydrologist to evaluate the importance of a given conversion process in a catchment. For example, if a rapid conversion mechanism (partial-area overland flow, saturation overland flow or perhaps subsurface flow through macropores) is the dominant conversion process contributing to a storm runoff hydrograph, the isotopic content of the stream will generally reflect mostly new water in the stream. Conversely, if a slow conversion mechanism (Darcian subsurface flow) is dominant in producing the storm runoff, the isotopic content of the stream should indicate mostly old water in the stream. Evaluation of the flow processes can be enhanced by applying the mixing model to water collected from subsurface runoff collection pits, overland flow and macropores.

Sklash and Farvolden (1982) listed five assumptions that must hold for reliable hydrograph separations using environmental isotopes:

1. The old water component for an event can be characterised by a single isotopic value or variations in its isotopic content can be documented.
2. The new water component can be characterised by a single isotopic value or variations in its isotopic content can be documented.
3. The isotopic content of the old water component is significantly different from that of the new water component.
4. Vadose water contributions to the stream are negligible during the event or they must be accounted for (use an additional tracer if isotopically different from groundwater).
5. Surface water storage (channel storage, ponds, swamps, etc.) contributions to the stream are negligible during the runoff event.

The utility of the mixing equations for a given high runoff period is a function mainly of the magnitude of $(\delta_o - \delta_n)$ relative to the analytical error of the isotopic measurements and the extent to which the aforementioned assumptions are indeed valid (Pearce *et al.*, 1986). Clearly the relative amounts of new versus old water are affected by many environmental parameters such as: size of the catchment, soil thickness, ratio of rainfall rate to infiltration rate, steepness of the watershed slopes, vegetation, antecedent moisture conditions, permeability of the soil, amount of macropores and storage capacity of the catchment.

Figure 10.6 shows an example of a two-component hydrograph separation for four catchments at Sleepers River watershed, Vermont (Shanley *et al.*, submitted). During the snowmelt period, groundwater appeared to have an approximately constant $\delta^{18}O$ value of $-11.7 \pm 0.3\%$], whereas snowmelt collected in snow lysimeters ranged from -20 to -14%. The stream samples had $\delta^{18}O$ values intermediate between the snowmelt and groundwater. The percentage of new snowmelt in stream water during the melt period showed a range of values from 40% at W-9 (a 47 ha forested catchment) to 68% at W-2 (a 59 ha agricultural catchment).

Three-component mixing models

Recent investigations indicate that in some catchments the two-component mixing model is not appropriate. For example, data from some isotope hydrograph separation studies have shown that the stream $\delta^{18}O$ values fall outside of the mixing line defined by the two "end members" (DeWalle *et al.*, 1988). Other studies have shown that stream delta values fall off the mixing line when both $\delta^{18}O$ and δ^2H are used for fingerprinting (Kennedy *et al.*, 1986). If the stream water is *not* collinear with the two suspected end members, either the end members are inaccurate or there must be another component (probably soil water) that plots above or below the mixing line for the other components. In the example shown in Figure 10.7, the stream $\delta^{18}O$

and δD values clearly do *not* plot intermediate between the two presumed end members, new rain and pre-storm baseflow.

Although hydrograph separations are typically performed with only one isotope, both to save money and because of the high correlation coefficient between oxygen and hydrogen stable isotopes of meteoric waters ($r^2 > 0.95$; Gat, 1980), analysis of both isotopes can sometimes prove very beneficial. Figure 10.7 shows an example described by Kennedy *et al.* (1986). The figure shows that stream water samples at the Mattole River in California plot above the *mixing line* defined by the two supposed sources of the stream water, namely pre-storm baseflow and new rain. As is often the case, the mixing line is collinear with the LMWL. Thus, no mixture of waters from these two sources of water can produce the stream water. This means that either there is an additional source of water (above the line) which contributes significantly to streamflow or some process in the catchment has caused waters to shift in isotopic composition.

This third source of water is often soil water. Although no soil water samples were collected during the January 1972 storm, the composition of the last big storm to hit the area, which presumably was a major source of recharge to the soil zone, was analysed for δ¹⁸O and δD. The stream samples plot along a mixing line connecting the average composition of this earlier

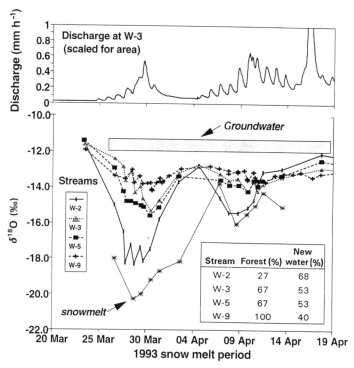

FIGURE 10.6   Early spring samples from four streams at Sleepers River Watershed, Vermont, have δ¹⁸O compositions intermediate between the compositions of snowmelt collected in pan lysimeters and groundwater. Diurnal fluctuations in discharge correlate with diurnal changes in δ¹⁸O, especially at W-2. W-2, a 59 ha agricultural basin, shows much greater contributions from snowmelt than the other three catchments. The three mixed agricultural/forested nested catchments — W-9 (47 ha), W-3 (837 ha) and W-5 (11.125 ha) — show higher contributions from new snowmelt at increased scale (modified from Shanley *et al.*, submitted)

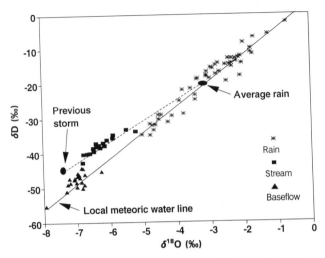

FIGURE 10.7   Stormflow samples from the Mattole River, California, 18–24 January 1972, have $\delta^{18}O$ and δD values that do not plot along the mixing line (solid line) between the two supposed sources: new rain and pre-storm baseflow. Therefore, there must be another source of water, possibly soil water derived from a previous storm, contributing significant amounts of water to stormflow. The stream samples plot along a mixing line (dashed line) between the composition of average new rain and the composition of previous rain (modified from Kennedy *et al.*, 1986)

storm and the average composition of January rain. Hence, the stream samples were probably mainly a mixture of soil water recharged during the previous storm and new rain water, with smaller amounts of baseflow.

If only $\delta^{18}O$ or δD had been analysed, the stream water would have plotted intermediate between the two end members (consider the projections of these water compositions on to the $\delta^{18}O$ and δD scales) and a two-component isotope hydrograph separation would have seemed feasible. Only when both isotopes were analysed did it become apparent that there was a mass balance problem. Hence, we recommend analysis of both isotopes on a subset of stream water and end member samples to ensure that an end member or process is not overlooked.

Sometimes it is obvious that the two-component model is inadequate: rain combined with glacial meltwater (Behrens *et al.*, 1979), for example. For rain on snow events where snowcores are used to assess the new water composition, a two- component model is also inadequate (Wallis *et al.*, 1979). However, if the catchment is snow-covered, overland flow is minimal, the surface area of the stream is small and several snowmelt lysimeters are used to catch sequential samples of new water derived from infiltration of both rain and melting snow, a two-component model may be sufficient. It is less obvious *a priori* when the isotopic difference between soil water and groundwater is sufficient to require a three-component model (Kennedy *et al.*, 1986; DeWalle *et al.*, 1988; Swistock *et al.*, 1989; Hinton *et al.*, 1994).

DeWalle *et al.* (1988) used a three-component model for the special case where the discharge of one component was known. Wels *et al.* (1991) used a three-component model for the case where the isotopic compositions of new overland and new subsurface flow were identical and the silica contents of new and old subsurface flow were identical. Ogunkoya and Jenkins (1993) used a model where one component was time-varying.

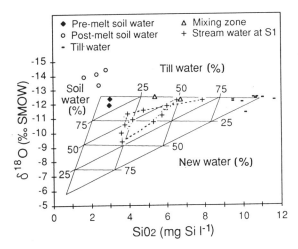

FIGURE 10.8 This diagram from Hinton *et al.* (1994) gives a good illustration of a three-component hydrograph separation; the three components are soil water ($\delta^{18}O = -12.5\%o$, $SiO_2 = 2.5$ mg Si l^{-1}), till water ($\delta^{18}O = -12.5\%o$, $SiO_2 = 11$ mg l^{-1}) and new water ($\delta^{18}O = -6\%o$, $SiO_2 = 0.5$ mg l^{-1}). Mixing lines on this diagram are linear. All stream samples plot within the triangle formed by the compositions of the three end members. (Reproduced by permission of the American Geophysical Union)

Hinton *et al.* (1994) listed five assumptions required for three-component models using two tracers:

1. Concentrations of each source must be distinct from the other two sources for one or both tracers.
2. The concentrations of the three sources cannot be collinear for the two tracers.
3. The concentration of each source must remain constant during the event.
4. There are only three sources contributing to discharge.
5. The tracers must mix conservatively.

Figure 10.8 shows an example of three-component hydrograph separation (Hinton *et al.*, 1994) where both $\delta^{18}O$ and SiO_2 are used to define three end members: new water (rain), soil water and till water. Changing percentages of contributions from the three end members in stream water during the storm can be seen on this plot.

10.3.3 Temporal and spatial variability in end members

During the last decade, the universal applicability of the simple two-component mixing model has been challenged (Kennedy *et al.*, 1986; Rodhe, 1987; Sklash, 1990; McDonnell *et al.*, 1991; Bishop, 1991; and Kendall and McDonnell, 1993). These challenges address the five assumptions governing the use of the two-component mixing model given in Sklash and Farvolden (1982) and listed above.

Firstly, the new water component is rarely constant in isotopic composition during events. Temporal variations in rain isotopic composition have been found to have an appreciable effect on hydrograph separations (McDonnell *et al.*, 1990). Secondly, investigators have concluded that in some catchments, soil water is an important contributor to storm runoff and is isotopically and/or chemically distinct from groundwater (Kennedy *et al.*, 1986; DeWalle *et al.*, 1988; Hooper *et al.*, 1990); in these cases, a three-component mixing model may be useful. Some recent studies suggest that in some catchments, the shallow soil and groundwaters are so heterogeneous in isotopic composition that isotope hydrograph separations are ill-advised (McDonnell *et al.*, 1991; Ogunkoya and Jenkins, 1991). Figure 10.4 shows an example of variable groundwater $\delta^{18}O$ values in a 40 ha catchment in Georgia.

None of these challenges to the simple two-component mixing model approach has caused any significant change in the basic conclusion of the vast majority of the isotope and chemical hydrograph studies to date, namely that *most stormflow in humid temperate environments is old water* (Bishop, 1991). There is however, a need to address the potential impact of natural isotopic variability on the use of isotopes as tracers of water sources. Most of these concerns can be addressed by applying more sophisticated two- and three-component mixing models that allow for variable isotopic signatures (e.g. Harris *et al.*, 1995).

The "new" water component

The isotopic composition of rain and snow can vary both spatially and temporally during storms. In general, rain becomes progressively more depleted in oxygen-18 and deuterium during an event because these isotopes preferentially rain-out early in the storm. Successive frontal and convective storms may have more complex isotopic variations. Intra-storm rainfall compositions have been observed to vary by as much as 90‰ in δD at the Maimai watershed in New Zealand (McDonnell *et al.*, 1990), 16‰ in $\delta^{18}O$ in Pennsylvania, USA (Pionke and DeWalle, 1992) and 15‰ in $\delta^{18}O$ in Georgia, USA (Kendall and Cappellato, submitted).

Snowmelt also varies in isotopic composition during the melt period and is commonly different from that of melted snowcores (Hooper and Shoemaker, 1986; Stichler, 1987). Part of the changes in snowmelt isotopic composition can be attributed to distinct isotopic layers in the snow, rain events on snow and isotopic fractionation during melting.

The variations in "new " water isotopic content are assessed by collecting sequential samples of rain or throughfall during rain storms and samples of meltwater from snow lysimeters during snowmelt. The method of integrating this information into the separation equations depends on the magnitude and rapidity of temporal variations in the "new" water isotopic content and the size of the catchment. If a varying "new" water value is used, one must account for the travel time to the stream and to the monitoring point. In very small catchments, travel time is probably negligible since most of the "new" water should run off from near-stream saturated areas, and reach the stream and monitoring point fairly rapidly.

In forested areas, rain intercepted by the tree canopy can have a different isotopic composition from rain in open areas (Saxena, 1986; Kendall, 1993; DeWalle and Swistock, 1994). On average, throughfall during individual storms at a site in Georgia was enriched by 0.5‰ in $\delta^{18}O$ and 3.0‰ in δD relative to rain; site-specific differences in canopy species, density and microclimate resulted in an average $\delta^{18}O$ range of 0.5‰ among 32 collectors over a 0.04 km^2 area for the same storm, with a maximum range of 1.2‰ (Kendall, 1993). Other sites appear to have little spatial variability (Swistock *et al.*, 1989). Because of this potential for spatial variability, putting out several large collectors and combining the throughfall for isotopic analysis is advisable (DeWalle *et al.*, 1988).

Different rainfall weighting methods can substantially effect estimates of new/old waters in storm runoff in basins with large contributions of new water (McDonnell *et al.*, 1990). Use of sequential rain values is probably the best choice in very responsive catchments or in catchments with high proportions of overland flow. When rain intensities are low and soils drain slowly, current rain may not infiltrate very rapidly and thus use of the cumulative approach (i.e. running average) is probably more realistic (Kendall and Gu, 1992).

In large watersheds and watersheds with significant elevation differences, both the "old" water and "new" water delta values may show large spatial variability; generally, the higher elevations will have more depleted delta values. Such large or steep watersheds would require many sampling stations to adequately assess the variability in old and new waters. Spatial variations in the $\delta^{18}O$ of up to 5‰ were observed for rain storms in a 2500 km^2, 35-station network in Alabama, USA (Kendall and McDonnell, 1993). Multiple collectors are especially critical for assessing temporal and spatial variability in both the timing and composition of snowmelt. Consequently, simple isotope hydrograph separation techniques are best suited for small first- and second-order catchments.

Preferential storage of different time fractions of rain in the shallow soil zone can be important because of the large amounts of rain potentially going into storage. For a storm that dropped 12 cm of rain over a 1-day period at Hydrohill, a 500 m^2 artificial catchment in China, 45% of the total rain went into storage (Kendall, 1993), with almost 100% of the first 2 cm of rain going into storage before runoff began.

The "old" groundwater component

Groundwater in catchments can frequently be heterogeneous in isotopic composition due to variable residence times (Figure 10.4). For example, groundwater at the 500 m^2 artificial Hydrohill catchment in China showed a 5‰ range in $\delta^{18}O$ after a storm (Kendall and Gu, 1992). Consequently, even if a large number of wells are sampled to characterise the potential variability in the groundwater composition, baseflow is probably a better integrator of the water that actually discharges into the stream than any average of groundwater compositions. Hence, it may be safer to use pre-storm baseflow, or interpolated intrastorm baseflow (Hooper and Shoemaker, 1986), or ephemeral springs as the old water end member. If groundwater is used for the end member, researchers need to check the effects of several possible choices for composition on their separations.

One major assumption in sampling baseflow to represent the "old" water component is that the source of groundwater flow to the stream during storms and snowmelt is the same as the source during baseflow conditions. It is possible that ephemeral springs remote from the stream or deeper groundwater flow systems may contribute differently during an event than between events, and if their isotopic signatures differ from that of baseflow, the assumed "old" water isotopic value may be incorrect. Although the occurrence of such situations could be tested by hydrometric monitoring and isotopic analyses of these features, quantification could be difficult.

Soil water

The role of soil water in stormflow generation in streams has been intensively studied using the isotope tracing technique (DeWalle and Swistock, 1989; Buttle and Sami, 1990). Most isotope studies have lumped soil water and groundwater contributions together, which may mask the importance of either component along. However, Kennedy *et al.* (1986) observed that as much as 50% of the stream water in the Mattole catchment was probably derived from rain from a previous storm stored in the soil zone (Figure 10.7). If there is a statistically significant difference between the isotopic compositions of soil and groundwater, then they need to be considered as separate components (McDonnell *et al.*, 1991).

The isotopic composition of soil water may be modified by a number of processes including direct evaporation, exchange with the atmosphere and mixing with infiltrating water (Gat and Tzur, 1967). Although transpiration removes large amounts of soil water, it is believed to be non-fractionating (Zimmermann *et al.*, 1967). Wenner *et al.* (1991) found that the variable composition of the percolating rain became largely homogenized at depths as shallow as 30 cm due to mixing with water already present in the soil. Buttle and Sami (1990) noted that waters extracted from clayey soils are more buffered by mixing with immobile water than are waters in sandy soils.

Intra-storm variability of soil water of up to 1.5‰ in $\delta^{18}O$ was observed during several storms at Panola Mountain, Georgia, and a 15‰ range in δD was seen during a single storm at Maimai, New Zealand (Kendall and McDonnell, 1993). Analysis of over 1000 water samples at Maimai (McDonnell *et al.*, 1991) showed a systematic trend in soil water composition in both downslope and downprofile directions. Multi-variate cluster analysis also revealed three distinct soil water groupings with respect to soil depth and catchment position, indicating that the soil water reservoir is poorly mixed on the time-scale of storms.

10.3.4 Need for other tracers

The underlying theme of the preceding section is that the isotopic composition of rain, throughfall, meltwater, soil water and groundwater is commonly variable in time and space. If such variability is significant at the catchment scale (i.e. if hillslope waters that are variable in composition actually reach the stream during the storm event) or if transit times are long and/or variable, then simple two- and three-component, constant composition, mixing models may not provide realistic interpretations of the system hydrology. A promising new approach is to develop models with variable end members (Harris *et al.*, 1995). Another possible solution is to include alternative, independent isotopic methods for determining the relative amounts of water flowing along different subsurface flowpaths. Possibilities include: radon-222 (Genereux and Hemond, 1990; Genereux *et al.*, 1993), beryllium-7 and sulphur-35 (Cooper *et al.*, 1991); carbon-13 (Kendall *et al.*, 1992), and strontium-87 (Bullen and Kendall, 1991, 1995). In particular, carbon isotopes, possibly combined with strontium isotopes, appear to be useful for distinguishing between deep and shallow flowpaths and, combined with oxygen isotopes, for distinguishing among water sources.

10.4   AGES AND RESIDENCE TIMES

Isotopes can be used for determining the ages and residence times of waters, gases, solutes and solid materials in catchments. A thorough discussion of the subject is beyond the scope of this chapter. Good reviews of age-dating methods include those by Davis and Bentley (1982), Faure (1986), Geyh and Schleicher (1990) and Plummer *et al.* (1993).

Two main types of environmental isotopes can be used for determining age: (a) those that mark the occurrence of a particular event and (b) those that are present world-wide as a result of processes continuing over extended time periods. Examples of this first "marker" category include isotopes that were produced and released into the atmosphere in high concentrations during nuclear testing (e.g. ^3H, ^{14}C and ^{36}Cl) or accidents such as Chernobyl (e.g. ^{137}Cs). Several isotopes that are continuously produced in the atmosphere have such short half-lives that they effectively label water derived from a particular storm or season (e.g. ^{35}S and ^7Be); all contributions of ^{35}S and ^7Be from the previous year's precipitation have probably decayed to insignificant levels by the time this year's snow starts melting (Cooper *et al.*, 1991).

The second "world-wide" category can be further divided into: (a) lithogenic solutes and gases derived from radioactive decay of elements in minerals in the catchment (e.g. uranium, strontium and argon isotopes); (b) cosmogenic nuclides derived from long-term cosmic ray spallation in the atmosphere or at the land surface (e.g. ^3H, ^{14}C, ^{10}Be and ^{36}Cl) and (c) noble gases derived from external sources and unreactive in the subsurface (e.g. ^3He and ^{85}Kr). Many of these isotopes have both lithogenic, cosmogenic and anthrogenic sources, thus complicating their interpretation. A good review of uses of lithogenic, cosmogenic and noble gas tracers for age-dating and other purposes is Nimz and Hudson (1995).

Groundwater or solute age is usually defined as the time since the material became isolated from the atmosphere. In the case of waters, this is approximately equivalent to the time since the water was recharged. The mean residence time of the "old" water in a catchment can be estimated by comparing the seasonal variations in isotopic compositions of the input (precipitation) and the output (streamflow) in a catchment, and by comparing the long-term input and output functions. Several different modelling approaches are described in Plummer *et al.* (1993).

Oxygen and hydrogen isotopes have frequently been used to estimate residence times in catchments by modelling the dampening of the storm-related or seasonal variations with time. The chapter by Unnikrishna *et al.* in this volume (Chapter 9) provides a thorough review of the conceptual basis of the method, along with several case studies. Pearce *et al.* (1986) used the method of Maloszewski *et al.* (1983) to determine the relationship between the delta values of precipitation and streamflow in a New Zealand catchment. Their analysis yielded an estimated mean residence time for "old" water of 4 months. Turner *et al.* (1987) applied a Kalman filtering technique to their delta values from an Australian watershed to arrive at a mean residence time of 20−50 days for "old" water.

Tritium is also commonly used for determining residence times. Martinec (1976) used tritium data to determine that the mean residence time of "old" water in a Swiss catchment was between 4 and 5 years, using exponential and asymmetrical dispersive age models. Dincer *et al.* (1970) applied a symmetrical dispersive age distribution model for tritium data from their Czechoslovakian watershed to arrive at an "old" water mean residence time of 2.5 years. Rose (1992, 1993) used tritium to show that baseflow in the Georgia Piedmont had a residence time of 15−35 years and should be regarded as a dynamic mixture of waters of different ages rather than as a static reservoir.

One of the most exciting recent advances in groundwater hydrology has been the establishment and/or refinement of methods for dating very young groundwater (less than 50 years old) with an accuracy of 1−3 years. Tracers include $^3H/^3He$, ^{85}Kr and several types of chlorofluorocarbons (CFCs). Publications describing the use of these tracers include Plummer *et al.* (1993), Dunkle *et al.* (1993), Reilly *et al.* (1994) and Ekwurzel *et al.* (1994). ^3He is the daughter product of ^3H decay; thus, in confined systems where dispersion is not significant, the $^3H/^3He$ ratio is a function of the age of the water. ^{85}Kr is a noble gas with a half-life of 10.76 years that is released into the atmosphere by nuclear bombs and reactors; it enters the subsurface by equilibration with air in the unsaturated zone.

CFCs are stable volatile organic compounds produced as refrigerants, aerosol propellants and solvents. Starting in the 1930s, these compounds were released into the atmosphere in ever-increasing concentrations until concern about their contribution to ozone depletion and atmospheric greenhouse gases curtailed their use. These compounds enter the subsurface during recharge and by equilibration with soil gas. Because CFCs are largely inert in the subsurface (Plummer *et al.*, 1993), by knowing the recharge temperature and the atmospheric input function of the particular CFC of interest, the age of the water can be determined. The precision and accuracy of groundwater dating is greatly improved by the measurement of more than one dating tracer in the same water sample because each method has its own advantages and disadvantages. See Plummer *et al.* (1993) for a thorough discussion of the topic.

One powerful potential application of these tracers of young groundwater is to evaluate the impact of changes in agricultural management practices on water quality. For example, by age-dating parcels of groundwater along flowpaths contributing to streamflow, researchers may be able to show that the reason nitrate concentrations in some streams are still high despite changes in management practices designed to eliminate over fertilization of agricultural fields is that the groundwater recharged *since* the changes in management practice has not yet reached the streams. In other words, the old high-nitrate water has not yet been flushed out of the system. By combining nitrogen isotope analyses with groundwater dating, it is possible to trace the sources of non-point contamination, evaluate the extent of natural denitrification and predict when contaminated or remediated groundwaters will reach the streams (Bohlke *et al.*, 1992; Bohlke and Denver, in preparation).

10.5 SOLUTE ISOTOPES

A thorough discussion of the geochemistry of solute isotope tracers is beyond the scope of this book; comprehensive reviews include those by Fritz and Fontes (1980, 1981), Faure (1986) and Rundel *et al.* (1989). This chapter presents a brief synopsis of the fundamentals of five solute isotope systems: carbon, nitrogen, sulphur, strontium and lead. As discussed above, water isotopes often provide relatively unambiguous information about residence times and relative contributions from different water sources; these data can then be used to make hypotheses about flowpaths. Solute isotopes can provide an alternative, independent isotopic method for determining the relative amounts of water flowing along various subsurface flowpaths. However, the least ambiguous use of solute isotopes in catchment research is for tracing the relative contributions of potential solute sources to groundwater and surface water.

10.5.1 Carbon

Carbon has two stable, naturally occurring isotopes, ^{12}C and ^{13}C, and ratios of these isotopes are reported in ‰ relative to the standard V-PDB (Pee Dee belemnite). Carbonate rocks typically have $\delta^{13}C$ values of 0 ± 5‰. There is a biomodal distribution in the $\delta^{13}C$ values of terrestrial plants resulting from differences in the photosynthetic reaction utilized by the plant, with $\delta^{13}C$ values for C3 and C4 plants about -25‰ and -12‰ respectively (Deines, 1980). The $\delta^{13}C$ values of dissolved inorganic carbon (DIC) in catchment waters are generally in the range of -5 to -25‰.

The primary reactions that produce DIC are: (a) weathering of carbonate minerals by acidic rain or other strong acids, (b) weathering of silicate minerals by carbonic acid produced by the dissolution of biogenic soil CO_2 by infiltrating rain water and (c) weathering of carbonate minerals by carbonic acid. The first and second reactions produce DIC identical in $\delta^{13}C$ to the composition of either the reacting carbonate or carbonic acid respectively, and the third reaction produces DIC with a $\delta^{13}C$ value exactly intermediate between the compositions of the carbonate and the carbonic acid. Consequently, without further information, DIC produced solely by the third reaction is identical to DIC produced in equal amounts from the first and second reactions.

If the $\delta^{13}C$ values of the reacting carbon-bearing species are known and the $\delta^{13}C$ of the stream DIC determined, in theory we can calculate the relative contributions of these two sources of carbon to the production of stream DIC and carbonate alkalinity, assuming that: (a) there are no other sources or sinks for carbon and (b) calcite dissolution occurs under closed-system conditions (Kendall *et al.*, 1992). With additional chemical or isotopic information, the $\delta^{13}C$ values can be used to estimate proportions of DIC derived from the three reactions listed above. Other processes that may complicate the interpretation of stream $\delta^{13}C$ values include CO_2 degassing, carbonate precipitation, exchange with atmospheric CO_2, carbon uptake by aquatic organisms, methanogenesis and methane oxidation (Kendall, 1993; Kendall and Mills, submitted). Correlation of variations in $\delta^{13}C$ with chemistry and with other isotope tracers such as $^{87}Sr/^{86}Sr$ and ^{14}C may provide evidence that such processes are insignificant for a particular study.

In recent years, the decline in alkalinity in many streams in Europe and in the northeastern United States as a result of acid deposition has been a subject of much concern (Likens *et al.*, 1979). The concentration of bicarbonate, the major anion buffering the water chemistry of surface waters and the main component of alkalinity in stream water, is a measure of the "reactivity" of the watersheds and reflects the neutralization of carbonic and other acids by reactions with silicate and carbonate minerals encountered by the acidic waters during their residence in the watershed (Garrels and MacKenzie, 1971).

Under favourable conditions, carbon isotopes can be used to understand the biogeochemical reactions controlling alkalinity in watersheds (Mills, 1988; Kendall *et al.*, 1992). An example of this application is Figure 10.9, modified from Kendall (1993). Hunting Creek in the Catoctin Mountains, Maryland, shows a strong seasonal change in alkalinity, with concentrations greater than 600 μeq l^{-1} in the summer and less than 400 in the winter. The $\delta^{13}C$ values for weekly stream samples also show strong seasonality, with lighter values in the summer ($-13\%o$) and heavier values in the winter ($-5\%o$), inversely correlated with alkalinity. Carbonates found in fractures in the altered basalt bedrock have $\delta^{13}C$ values of about $-5\%o$ and soil-derived CO_2 is about $-21\%o$. Therefore, the dominant reaction controlling stream DIC in the winter appears to be strong acid weathering of carbonates, whereas the dominant reaction in the summer appears to be carbonic acid weathering of carbonates (Kendall, 1993). Of course, with just $\delta^{13}C$ values we cannot rule out the possibility that the summer $\delta^{13}C$ values instead reflect approximately equal contributions of DIC produced by strong acid weathering of calcite ($-5\%o$) and carbonic acid weathering of silicates ($-21\%o$). However the presence of carbonate in the bedrock strongly influences weathering reactions, and the stream chemistry (e.g. low Si/Ca) during the summer suggests that silicate weathering has only a minor effect.

Carbon isotopes can also be useful tracers of the seasonal and discharge-related contributions of different hydrologic flowpaths to streamflow (Kendall *et al.*, 1992). In many carbonate-poor catchments, waters along shallow flowpaths in the soil zone have characteristically light $\delta^{13}C$

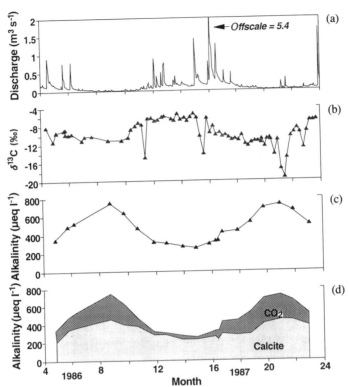

FIGURE 10.9   Determining seasonal changes in the sources of alkalinity by using the $\delta^{13}C$ of stream DIC.
(a) Discharge at Hunting Creek in the Catoctin Mountains, Maryland, 1986−7. (b) $\delta^{13}C$ of stream DIC
collected weekly. (c) Alkalinity. (d) Estimation of the relative contributions of carbon from CO_2 and
calcite to stream alkalinity using $\delta^{13}C$ of calcite $= -5\%o$ and $\delta^{13}C$ of $CO_2 = -21\%o$; shaded areas show
the relative proportions of carbon sources (modified from Kendall, 1993)

values reflecting carbonic acid weathering of silicates, and waters along deeper flowpaths
within less weathered materials have intermediate $\delta^{13}C$ values characteristic of carbonic acid
weathering of carbonates (Bullen and Kendall, 1995).

Figure 10.10 shows that $\delta^{13}C$ can be used to provide more insight into catchment dynamics
than is apparent with $\delta^{18}O$ values alone. The $\delta^{18}O$ values of stream water for a storm at Hunting
Creek show only small, gradual changes in the relative contributions of different water sources
during storms. However, the rapidly oscillating $\delta^{13}C$ values show that the system is much more
dynamic, with the $\delta^{13}C$ values probably reflecting rapid changes in water flowpaths as a
response to subtle changes in rainfall. Analysis of the Sr isotopes in the waters can provide
additional information about solute sources, reactions and flowpaths, as will be discussed later.

The radioactive isotope of carbon, ^{14}C, is continuously being produced by reaction of cosmic
ray neutrons with ^{14}N in the atmosphere and decays with a half-life of 5730 years. ^{14}C concentrations are usually reported as specific activities (disintegrations per minute per gram of carbon
relative to a standard) or as percentages of modern ^{14}C concentrations. Under favourable conditions, ^{14}C can be used to date carbon-bearing materials. However, because the "age" of a mixture

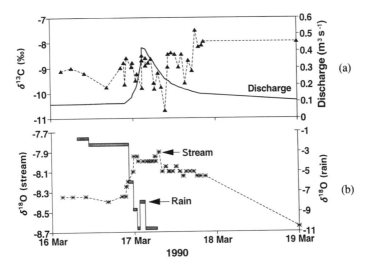

FIGURE 10.10 Rain and stream $\delta^{18}O$ and stream $\delta^{13}C$ for a storm 16–20 March 1990 at Hunting Creek, Maryland. (a) $\delta^{13}C$ values and discharge; (b) $\delta^{18}O$ values of rain and stream water (note difference in the scales). Although the gradual shifts in stream $\delta^{18}O$ suggest equally gradual changes in water sources during the storm, the $\delta^{13}C$ values suggest rapidly fluctuating contributions of old waters of different $\delta^{13}C$ values in response to small changes in rainfall intensity

of waters of different residence times is not very meaningful, the least ambiguous *hydrologic* use of ^{14}C is as a tracer of carbon sources.

^{14}C combined with ^{13}C has been used to study the origin, transport and fate of dissolved organic carbon (DOC) in streams and shallow groundwater in forested catchments (Schiff *et al.*, 1990; Wassenaar *et al.*, 1991; Aravena *et al.*, 1992). They find that DOC in groundwaters is composed of older carbon than surface waters, indicating extensive cycling of DOC. In addition, the ^{14}C of DIC can be a valuable check on conclusions derived using $\delta^{13}C$ values. For example if the $\delta^{13}C$ values suggest that the dominant source of carbon is from carbonic acid weathering of silicates, the ^{14}C activities should be high, reflecting the young age of the soil CO_2 (Schiff *et al.*, 1990).

10.5.2 Nitrogen

There are two stable isotopes of N: ^{14}N and ^{15}N. Since the average abundance of ^{15}N in air is a very constant 0.366% (Junk and Svec, 1958), air is used as the standard for reporting $\delta^{15}N$ values. Most terrestrial materials have $\delta^{15}N$ compositions between -20%₀ and $+30\%$₀. The dominant source of nitrogen in most forested ecosystems is the atmosphere ($\delta^{15}N = 0\%$₀); many plants fix nitrogen and organisms cycle this nitrogen into the soil. Other sources of nitrogen include fertilisers produced from atmospheric nitrogen with compositions of $O \pm 3\%$₀ and animal manure with nitrate $\delta^{15}N$ values generally in the range of $+10 - +25\%$₀; rock sources of N are negligible.

Biologically mediated reactions (e.g. assimilation, nitrification and denitrification) strongly control nitrogen dynamics in the soil, as briefly described below. These reactions almost always result in ^{15}N enrichment of the substrate and depletion of the product. Although precipitation often contains subequal quantities of ammonium and nitrate, because ammonium is preferentially retained by the canopy relative to atmospheric nitrate (Garten and Hanson, 1990), most of the atmospheric nitrogen that reaches the soil surface is in the form of nitrate. Soil nitrate is preferentially assimilated by tree roots relative to soil ammonium (Nadelhoffer *et al.*, 1988).

The $\delta^{15}N$ of total soil N is affected by many factors including soil depth, vegetation, climate, particle size, cultural history, etc.; however, two factors, drainage and influence of litter, have a consistent and major influence on the $\delta^{15}N$ values (Shearer and Kohl, 1988). Soils on lower slopes and near saline seeps have higher $\delta^{15}N$ values than well-drained soils (Karamanos *et al.*, 1981), perhaps because the greater denitrification in more boggy areas results in heavy residual nitrate. The $\delta^{15}N$ values of the soils from valley bottoms at the Walker Branch Watershed in Tennessee (USA) are heavier than for soils from ridges and slopes, consistent with a theoretical model that explains the increase in the $\delta^{15}N$ of inorganic N in soil as a function of the higher relative rates of immobilization and nitrification in these bottom soils (Shearer *et al.*, 1974). Surface soils beneath bushes and trees often have lighter $\delta^{15}N$ values than those in open areas, presumably as the result of litter deposition (Shearer and Koh, 1988). Fractionations during litter decomposition in forests result in surface soils with lighter $\delta^{15}N$ values than deeper soils (Nadelhoffer and Fry, 1988). Gormly and Spalding (1979) attributed the inverse correlation of nitrate $\delta^{15}N$ and nitrate concentration beneath agricultural fields to increasing denitrification with depth.

Although enriched ^{15}N tracer studies have been commonly used in agricultural investigations for decades (Bremmer, 1965), natural abundance studies to trace natural and anthropogenic sources of nitrate started with Kohl *et al.* (1971) but were rare until the last decade. Such studies take advantage of the observation that atmospherically derived nitrogen and fertiliser nitrogen typically have light $\delta^{15}N$ values whereas animal-derived nitrogen (such as manure or septic tank effluents) is typically considerably heavier; hence, under favourable conditions, the end members are distinguishable and the relative contributions of these two sources to groundwater can be estimated. Soil-derived nitrate and fertiliser nitrate usually have overlapping compositions, preventing their separation using $\delta^{15}N$ alone. Applications of ^{15}N to trace relative contributions of fertilizer and animal waste to groundwater (Kreitler, 1975; Kreitler and Jones, 1975; Kreitler *et al.*, 1978; Gormly and Spalding, 1979) are complicated by a number of reactions including ammonia volatilisation, nitrification, denitrification, ion exchange and plant uptake. These processes can modify the $\delta^{15}N$ values of N sources prior to mixing and the resultant mixtures, causing estimations of the relative contributions of the sources of nitrate to be inaccurate.

Denitrification causes the $\delta^{15}N$ of the residual (unreacted) nitrate to increase exponentially as nitrate concentrations decrease. For example, denitrification of fertiliser nitrate that originally had a distinctive $\delta^{15}N$ value of $+1\%_0$ can yield residual nitrate with a $\delta^{15}N$ value of $+15\%_0$; this value is within the range of compositions expected for nitrate from a manure or septic tank source. Measured fractionation factors associated with denitrification range from 10 to $40\%_0$. The N_2 produced by denitrification results in "excess N_2" contents in groundwater. The total N_2 (which consists of air entrained during recharge plus N_2 produced by denitrification) can be collected, analysed for $\delta^{15}N$ and used to estimate the extent of denitrification or initial composition of the nitrate (Vogel *et al.*, 1981; Bohlke and Denver, submitted).

The recent development of methods to analyse nitrate for $\delta^{18}O$ (Amberger and Schmidt, 1987; Silva *et al.*, in preparation) should expand the use of isotope techniques to trace nitrate

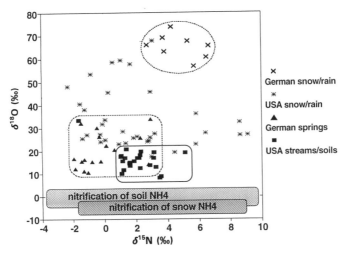

FIGURE 10.11 Nitrogen and oxygen isotopic compositions of nitrate in forested catchments in the USA (Kendall *et al.*, 1995) and Germany (Durka, 1994). Dashed lines enclose the total range of compositions of German precipitation and spring samples, and solid lines enclose the composition of soils and streams in the USA. The shaded areas represent the theoretical range in these catchments for nitrate produced by microbial nitrification of ammonium in soil and snow. Springs and streams have nitrate $\delta^{18}O$ values intermediate between the compositions of atmospheric nitrate and soil-derived nitrate

sources and transformations. Amberger and Schmidt (1987) determined that fertiliser nitrate and nitrate formed from nitrification of organic material have distinctive $\delta^{18}O$ and $\delta^{15}N$ values. All three oxygens in fertiliser nitrate are derived from atmospheric O_2 (+22 to +24‰), and hence the $\delta^{18}O$ of the nitrate is in this range. Two of the oxygens in microbial nitrate derive from the oxygen in water molecules (and presumably reflect the $\delta^{18}O$ of the water) and one oxygen comes from atmospheric O_2. For waters with $\delta^{18}O$ values in the range of -5 to -20‰, the $\delta^{18}O$ of soil nitrate formed from *in situ* nitrification of ammonium, should be in the range of $+5$ to -5‰ respectively. In theory, one might be able to see a seasonal change in soil nitrate $\delta^{18}O$ that mimics the seasonal change in the $\delta^{18}O$ of water; or perhaps nitrate formed from evaporated water might reflect the evaporated signature of the water $\delta^{18}O$. Therefore, relative contributions of nitrate from fertiliser and microbial sources can be estimated by analysing both isotopes of nitrate in samples suspected to be derived from these sources.

Another useful application of the "dual isotope approach" is for determination of the relative contributions of atmospheric and microbial-derived sources of nitrate in shallow groundwater. This problem is intractable using just $\delta^{15}N$ because of overlapping compositions of microbial and atmospherically derived nitrate. Recent data from several German forests (Durka, 1994; Durka *et al.*, 1994) and three catchments in the United States (Kendall *et al.*, 1995; Kendall *et al.* in press) indicate that the $\delta^{18}O$ of nitrate in precipitation is in the range of $+25-75$‰. If such heavy $\delta^{18}O$ values are typical of atmospherically derived nitrate elsewhere, or at least in places where there are significant anthropogenic sources of atmospheric nitrogen, contributions from precipitation should easily be distinguished from soil nitrate. Durka (1994) analysed nitrate in precipitation and spring samples from several German forests for both isotopes (Figure 10.11), and found that the $\delta^{18}O$ of nitrate in springs were correlated with the general health of the forest, with more healthy or limed forests showing lower $\delta^{18}O$ values closer to the composition of microbially produced nitrate and severely damaged forests showing higher $\delta^{18}O$ values indicative of major contributions of atmospheric nitrate to the system. Hence, acid-induced forest decline appears to inhibit nitrate consumption by soil micro-organisms (Durka *et al.*, 1994).

A similar application is for determination of the source of nitrate in early spring runoff. During early spring melt, many small catchments experience episodic acidification because of large pulses of nitrate and hydrogen ions being flushed into the streams when discharge is low. There has been some controversy over the source of this nitrate. The two likely sources are: (a) atmospherically derived nitrate in the snowpack that is eluted during early melt or (b) nitrate developed in the organic horizons of the soil under the insulating snowpack that is flushed into the stream by percolating meltwater.

In three pilot studies conducted in catchments in New York, Colorado and Vermont during the 1994 snowmelt season, Kendall *et al.* (1995) found that almost all the stream samples had nitrate $\delta^{18}O$ and $\delta^{15}N$ values within the range of pre-melt stream and soil waters, and significantly different from the composition of almost all the snow samples (Figure 10.11), indicating that atmospheric nitrate from the 1994 snowpack was not a significant source of nitrate in early runoff in these catchments. Therefore, the nitrate eluted from the snowpack appeared to go into storage and most of the nitrate in streamflow during the period of potential acidification was apparently derived from pre-melt sources. The $\delta^{18}O$ of nitrate in pre- melt soil and stream waters is intermediate between the compositions of atmospheric and soil-derived nitrate (Figure 10.11), indicating a mixture of sources similar to that found in German forests (Durka *et al.*, 1994).

The $\delta^{18}O$ of nitrate in pre-melt waters was 10 to 20‰ heavier than expected (Figure 10.11). Given the range of $\delta^{18}O$ values of catchment waters (-12 to -22‰), the $\delta^{18}O$ of nitrate was expected to be in the range of -5 to 0‰. The more positive values suggest that atmospheric nitrate is actually a dominant source of nitrate to the catchments on a yearly basis and that the nitrate pulse flushed from storage during early melt was actually derived from precipitation from previous years (Kendall *et al.*, 1995).

The main sources of uncertainty in calculating the relative contributions from the two nitrate sources are: (1) the ranges in the end member compositions, especially for atmospheric nitrate, and (2) uncertainty about whether snow ammonium is retained in the soil or is nitrified to nitrate (Kendall *et al.*, 1995). It is unclear why the nitrate in snow and rain in the three USA catchments showed a much larger range of $\delta^{18}O$ and $\delta^{15}N$ values than observed in the German forests. More data are needed to assess processes controlling the spatial and temporal ranges in isotopic composition. Different sources of atmospheric nitrate (e.g., automobile exhaust versus power plant emissions) may have characteristic isotopic signatures that can be used to track different kinds of pollutants. The reason why nitrification of snow ammonium is a problem is that the new nitrate will have a $\delta^{18}O$ value very similar to microbially produced nitrate, causing the atmospheric "signal" to be less distinctive.

Amberger and Schmidt (1987) showed that denitrification results in enrichment in ^{18}O of the residual nitrate, as well as enrichment in ^{15}N. Therefore, analysis of both $\delta^{15}N$ and $\delta^{18}O$ of nitrate should allow denitrification effects to be distinguished from mixing of sources, as shown in Figure 10.12. This dual isotope approach takes advantage of the observation that the ratio of the enrichment in ^{15}N to the enrichment in ^{18}O in residual nitrate during denitrification appears to be about 2:1 (Olleros, 1983; Amberger and Schmidt, 1987; Bottcher *et al.*, 1990). If this

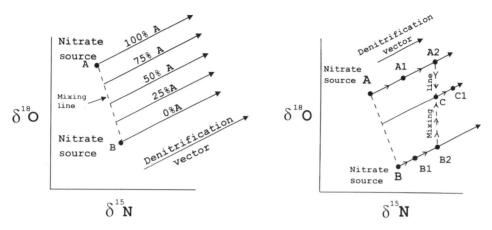

FIGURE 10.12   Determination of the relative contributions of nitrate to groundwater from two sources: fertilizer (A) and animal manure (B) using $\delta^{18}O$ and $\delta^{15}N$ of nitrate. If the sources have distinctive compositions, the effects of mixing can be distinguished from the effects of denitrification. Vectors A→A2 and B→B2 show the compositions of nitrate derived from progressive denitrification of A and B respectively. All mixtures of waters where one-third of the nitrate was derived from B and two-thirds of the nitrate was derived from A plot along vector C→C1, regardless of the timing of denitrification relative to mixing

ratio is constant, and the $\delta^{15}N$ and $\delta^{18}O$ compositions of the two potential sources of nitrate contributing to groundwater are known, are distinguishable and do not show much scatter in composition, the "original" relative contributions of these two sources to the nitrate in any sample of groundwater can be estimated from the $\delta^{15}N$ and $\delta^{18}O$ values of the nitrate. Such estimations are not affected by the extent or timing of denitrification and mixing, or the spatial arrangement of the sources (i.e. point versus non-point sources).

10.5.3   Sulphur

Sulphur has four stable isotopes: ^{32}S (95.02%), ^{33}S (0.75%), ^{34}S (4.21%) and ^{36}S (0.02%) (MacNamara and Thode, 1950). Stable isotope compositions are reported as ratios of $^{34}S/^{32}S$ in ‰ relative to the standard CDT (Canyon Diablo troilite). The general terrestrial range is +50 to −50‰, with rare values much heavier or lighter. Krouse and Grinenko (1991) provide a very comprehensive evaluation of stable sulphur isotopes as tracers of natural and anthropogenic sulphur. Under favourable conditions $\delta^{34}S$ can also be used to determine the reaction mechanisms responsible for sulphide oxidation (Kaplan and Rittenberg, 1964; Fry *et al.*, 1986, 1988) and sulphate reduction (Goldhaber and Kaplan 1974; Krouse, 1980).

In most forest ecosystems, sulphate is derived mostly from the atmosphere (Waring and Schlesinger, 1985); weathering of ore minerals and evaporites also contributes some sulphur. Because sulphur isotope ratios are strongly fractionated by biogeochemical processes, there has been concern over whether ^{34}S could be used to separate sources of sulphur in catchments. Some catchments appear to be affected by isotope fractionation processes (Fuller *et al.*, 1986; Hesslein *et al.*, 1988; Andersson *et al.*, 1992), whereas others seem to show only minor effects of watershed processes on $\delta^{34}S$ in lakes (Caron *et al.*, 1986) or streams (Stam *et al.*, 1992). Andersson *et al.* (1992) explain the decreases in sulphate concentrations and increases in $\delta^{34}S$ of streamflow during the summer by sulphate reduction in stagnant pools. Stam *et al.* (1992) suggest that the extent of fractionation might be a function of water residence time in the catchment, with steep catchments showing less fractionation. They note that increases in $\delta^{34}S$ of stream sulphate during the winter may be a result of micropore flow during the snow-covered period, rather than the more typical macropore flow characteristic of storms.

Intensive investigations of the sulphur dynamics of forest ecosystems in the last decade can be attributed to the dominant role of the sulphur as a component of acidic deposition. Studies in forested catchments include: Fuller *et al.* (1986), Mitchell *et al.* (1989), Stam *et al.* (1992) and Andersson *et al.* (1992). Sulphur with a distinctive isotopic composition has been used to identify pollution sources (Krouse *et al.*, 1984), and enriched sulphur has been added as a tracer (Legge and Krouse, 1992; Mayer *et al.*, 1992, 1993). Differences in the natural abundances can also be used in systems where there is sufficient variation in the $\delta^{34}S$ of ecosystem components. Rocky Mountain lakes thought to be dominated by atmospheric sources of sulphate have been found to have different ^{34}S values from lakes believed to be dominated by watershed sources of sulphate (Turk *et al.*, 1993).

Use of a dual isotope approach to tracing sources of sulphur (i.e. measurement of $\delta^{18}O$ and $\delta^{34}S$ of sulphate) will often provide better separation of potential sources of sulphur and, under favourable conditions, provide information on the processes responsible for sulphur cycling in the ecosystem. The $\delta^{18}O$ of sulphate shows considerable variations in nature, and recent studies indicate that the systematics of cause and effect remain somewhat obscure and controversial (Taylor and Wheeler, 1994; Van Stempvoort and Krouse, 1994). Reviews of applications of $\delta^{18}O$ of sulphate include Holt and Kumar (1991) and Pearson and Rightmire (1980).

The rate of oxygen isotope exchange between sulphate and water is very slow at normal pH levels. Even in acidic rain of pH $\approx$ 4, the "half-life" of exchange is on the order of 1000 years (Lloyd, 1968). Depending on the reaction responsible for sulphate formation, between 12.5 and 100% of the oxygen in sulphate is derived from the oxygen in the environmental water; the remaining oxygen comes from O_2 (Taylor *et al.*, 1984). Atmospheric O_2 is +23.8‰ (Horibe *et al.*, 1973). At isotopic equilibrium at 0°C, aqueous sulphate is about 30‰ enriched in ^{18}O relative to water (Mizutani and Rafter, 1969).

Sulphur-35 is a radioisotope formed from cosmic ray spallation of argon-40 in the atmosphere (Peters, 1959); it has a half-life of 87 days. In the first application of ^{35}S in an aquatic system, Cooper *et al.* (1991) found that sulphur deposited as precipitation in the Arctic is strongly adsorbed within the watershed and that most sulphur released to streamflow is derived from longer-term storage in soils, vegetation or geologic materials. Michel and Naftz (1995) report that the combined use of ^{35}S and tritium shows that meltwater from Wind River Range (Wyoming, USA) glaciers contains water from the current year. Furthermore, although lakes in Colorado fed by snowmelt and precipitation contain recent atmospherically derived sulphate, this atmospherically deposited sulphur takes several months to emerge in springs fed by shallow groundwater (Michel and Turk, 1995).

10.5.4 Strontium

The alkali earth metal Sr has four stable, naturally occurring isotopes: ^{84}Sr, ^{86}Sr, ^{87}Sr and ^{88}Sr. Only ^{87}Sr is radiogenic (i.e. produced by radioactive decay of another isotope) and gradually increases in rocks due to beta decay from the radioactive alkali metal ^{87}Rb, which has a half-life of 48.8 × 10⁹ years. Thus, the ^{87}Sr in any material includes that formed during primordial nucleo-synthesis as well as that formed by radioactive decay of ^{87}Rb. The ratio ^{87}Sr/^{86}Sr is the parameter typically reported in geologic investigations; ^{87}Sr/^{86}Sr in minerals has values ranging from about 0.7 to greater than 4.0.

The utility of the rubidium — strontium isotope system results from the fact that different minerals in a given geologic setting can have distinctly different ^{87}Sr/^{86}Sr as a consequence of different ages, original Rb/Sr values and the initial ^{87}Sr/^{86}Sr. For example, consider the case of a simple igneous rock such as a granite that contains several Sr-bearing minerals including plagioclase feldspar, K-feldspar, hornblende, biotite and muscovite. If these minerals crystallized from the same silicic melt, each mineral originally had the same ^{87}Sr/^{86}Sr as the parent melt. However, because Rb substitutes for K in minerals and these minerals have different K/Ca, the minerals will have different Rb/Sr. Typically, Rb/Sr increases in the order plagioclase, hornblende, K-feldspar, biotite, muscovite. Therefore, given sufficient time for significant production of radiogenic ^{87}Sr, the current ^{87}Sr/^{86}Sr values will be different in the minerals, increasing in the same order.

The important concept for isotope tracing is that Sr derived from any mineral through weathering reactions will have the same ^{87}Sr/^{86}Sr as the mineral. Therefore, differences in ^{87}Sr/^{86}Sr among catchment waters require either (a) differences in mineralogy along contrasting flowpaths or (b) differences in the relative amounts of Sr weathered from the same suite of minerals. This latter situation can arise in several ways. Firstly, differences in initial water chemistry within a homogeneous rock unit will affect the relative weathering rates of the minerals. For example, sections of the soil zone affected by evaporative concentration of recharge waters or by differences in pCO$_2$ can be expected to have different ^{87}Sr/^{86}Sr. Secondly, differences in the relative mobilities of water at scales ranging from inter-grain pores to the catchment scale may also profoundly affect ^{87}Sr/^{86}Sr (Bullen *et al.*, in preparation). For example, the chemical composition and the resultant ^{87}Sr/^{86}Sr in immobile waters at a plagioclase — hornblende grain boundary versus a quartz — mica boundary will be different. Thirdly, a difference in the relative "effective" surface areas of minerals in one part of the rock unit will also cause differences in chemistry and isotopic composition; "poisoning" of reactive surfaces by organic coatings is an example of this kind of process.

For catchments developed on multi-mineralic rocks or soils, ^{87}Sr/^{86}Sr in any water parcel usually represents a mixture of Sr from several sources, and thus the exact contributions from individual minerals are difficult to determine with the Sr isotope data alone. However, when considered in conjunction with water chemistry, the Sr isotopes provide a powerful tool for distinguishing among solute sources. For example, when attempting to develop a reaction model to explain the progressive increase in Mg in groundwater along a flowpath by the dissolution of either hornblende or biotite, the Sr isotopes should provide an effective label because Sr derived from the hornblende will be substantially less radiogenic than that from biotite.

Once in solution, Sr should behave chemically in a manner similar to that of the alkaline earth element Ca. As a cautionary note, however, we point out that Sr can be contributed by both Ca-rich and Ca-poor minerals; thus, ^{87}Sr/^{86}Sr is not in itself a sure indicator of Ca sources in a catchment. For example, although sodic plagioclase and K-feldspar contain little Ca but substantial Sr, because of their high K content and consequent relatively high Rb/Sr, they probably contain more radiogenic Sr than calcic plagioclase. The isotopic composition of Sr derived from the simultaneous weathering of all three types of feldspars would have an intermediate value that could be similar, for example, to that in hornblende. Hence, on the basis

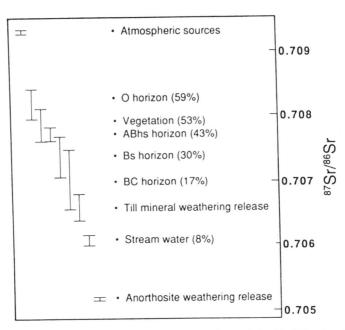

FIGURE 10.13   Weighted average (dots) and 1 sigma error bars of the $^{87}Sr/^{86}Sr$ of cation reservoirs at Whiteface Mountain, New York. The calculated percentage of Sr in each reservoir that is derived from atmospheric sources is indicated in parentheses. (Modified from Miller *et al.*, 1993. Reproduced by permission of *Nature*. © Macmillan Magazines Limited)

of the Sr isotopes alone, one might erroneously conclude that the bulk of Ca in solution was derived from hornblende, even though dissolution of hornblende may have played only a minor role in water chemical evolution. Clearly the Sr isotopes should be closely integrated with water chemistry information in order to identify the important contributing minerals along reaction paths.

Thus far, there have been relatively few studies of applications of Sr isotopes to catchment modelling; a good review is Graustein (1989). At the Tesuque Watersheds in New Mexico (USA), extensive study of Sr systematics have shown that 34% of the Sr in throughfall is botanically cycled and 66% is derived from leaching of airborne dust from the foliage and represents a net input of Sr to the ecosystem. Therefore, 20% of Sr in biomass is derived from the bedrock and 80% from atmospherically transported dust (Graustein, 1989).

Estimates for the release of Ca by weathering can be estimated using strontium isotopes (Aberg *et al.*, 1989). Miller *et al.* (1993) report that $^{87}Sr/^{86}Sr$ can be used to separate the total cations exported from the catchment into the components derived from mineral weathering reactions and cation exchange reactions in the soil. At a forest in the Adirondack Mountains, New York, they found that 70% of the cations in the stream are derived from weathering and 30% from exchange reactions. The $^{87}Sr/^{86}Sr$ of various cation reservoirs was used to determine the per cent of Sr in each that was derived from atmospheric sources; only about 8% of the Sr in stream water was atmospherically derived (Figure 10.13).

Both of the above two studies assumed a constant $^{87}Sr/^{86}Sr$ for the weathering contribution, an assumption that is probably valid only for monomineralic or very "young" bedrocks.

Considerable variability in the weathering contribution has been documented in a northern Wisconsin watershed nested in multi-mineralic sands derived from Precambrian terrain (Bullen *et al.*, in preparation).

10.5.5 Lead

Pb has four stable, naturally occurring isotopes: ^{204}Pb, ^{206}Pb, ^{207}Pb and ^{208}Pb. ^{206}Pb, ^{207}Pb and ^{208}Pb are all radiogenic, and are the end products of complex decay chains that begin at ^{238}U, ^{235}U and ^{232}Th respectively. The corresponding half-lives of these decay schemes vary markedly: 4.47×10^9 years, 7.04×10^8 years and 1.4×10^{10} years respectively. Each radiogenic isotope is reported relative to ^{204}Pb as the typical parameter in geologic investigations. The ranges of isotope ratio values for most natural materials are $14.0-30.0$ for $^{206}Pb/^{204}Pb$, $15.0-17.0$ for $^{207}Pb/^{204}Pb$ and $35.0-50.0$ for $^{208}Pb/^{204}Pb$, although numerous examples outside these ranges are reported in the literature.

The utility of the $U-Th-Pb$ system lies in the fact that in a single rock, individual minerals attain diagnostic Pb isotope signatures due to long-lived differences in U/Pb, Th/Pb and Th/U. Because U, Th and Pb have different geochemical behaviours, the fact that the Pb isotope composition of any material is the composite of the three independent decay chains creates the potential for greater variability of isotope values in minerals of a single rock relative to that for the $Rb-Sr$ system.

Perhaps the most important aspect of Pb isotopes pertinent to catchment hydrology studies is the probability that the isotopic composition of Pb derived from rock weathering will differ significantly from that of Pb in atmospheric deposition. Atmospheric Pb tends to become concentrated in the uppermost, organic-rich soil horizons, presumably as complexes with organic carbon. Dissolved organic carbon mobilized by storm waters will transport some of this Pb through a variety of pathways either to groundwater or to the stream. In contrast, Pb in groundwater should be dominated by that derived from rock weathering. Thus, the relative importance of the two sources in stream water can be determined.

Several studies have used lead isotopes to trace sources. Erel *et al.* (1990) found that the isotopic composition of Pb in stream water was identical to that in the upper 2 cm soil accumulation reservoir, and distinct from that in groundwater; they then used this relation to show that Pb in stream water during snowmelt in a catchment in the Sierra Mountains (California) was delivered along shallow flowpaths. Bullen *et al.* (1994) showed that the isotopic composition of Pb in stream water at a headwater catchment in Vermont varied markedly over the course of a year, with the most radiogenic values (similar to groundwater values) occurring just prior to and during early snowmelt, and the least radiogenic values (similar to atmospheric deposition) occurring during the summer and early fall months (Figure 10.14). These variations suggest correlations with both extent of the forest canopy and its ability to capture atmospheric Pb, as well as with the extent of ground freezing and immobilization of organo-Pb complexes. Determination of the proportion of atmospheric Pb in stream water may help to assess the efficiency with which storm waters are transported to the stream through shallow pathways.

Another useful lead isotope for catchment research is ^{210}Pb, a radiogenic and radioactive isotope with a half-life of 22.1 years. ^{210}Pb can be used to age-date materials formed in the last 100 or so years, with typical precisions of a few years under favourable conditions. ^{210}Pb is produced along the decay chain of ^{238}U; an intermediate member of the decay chain, the noble gas ^{222}Rn degasses from groundwater and escapes to the atmosphere. ^{222}Rn has a half-life of 3.8 days and decays through subsequent decay processes to form ^{210}Pb. ^{210}Pb is rapidly removed from the atmosphere by condensation of water, and becomes incorporated into snow, glaciers, sediments, plants and newly formed minerals (Faure, 1986). The ^{210}Pb "age" reflects the time since the lead became incorporated in the material. The atmospheric source of ^{210}Pb can be augmented by additional ^{210}Pb contributed by decay of uranium in the underlying geologic materials, thus complicating age determinations.

A very powerful application is to use ^{210}Pb to age-date the deposition of layered materials and then to relate the changes in the chemical and isotopic compositions of the layers to changes in ambient environmental conditions. One recent example is the use of ^{210}Pb to age-date speleothems (Baskaran and Iliffe, 1993). Groundwaters can have high concentrations of radon derived from uranium decay. As the ^{222}Rn-containing waters in caves drips from the stalactites, ^{210}Pb accumulates in the growing crystal lattices of the calcium carbonate. Successive bands of carbonate can then be age-dated and the geochemical changes in the bands related to changes in environmental conditions near the land surface.

Bio-indicators such as ostracodes (small freshwater bivalves with calcium carbonate shells) in soils and sediments can also be used as tracers of changes in water chemistry because some ostracode species have narrow anion tolerances (Forester, 1983). For example, as the nitrate or sodium chloride concentration in springs or pools increases, a particular ostracode species may disappear and be replaced by a different species with greater tolerance of the new salt content. Hence, the timing of large-scale climate-related hydrochemical changes or local anthropogenic disturbances such as catchment salinisation because of road salts can perhaps be determined by monitoring changes in ostracode populations and species types with depth in sediments (Taylor and Howard, 1993) and then age-dating ostracodes in the horizons where the changes occurred using ^{210}Pb.

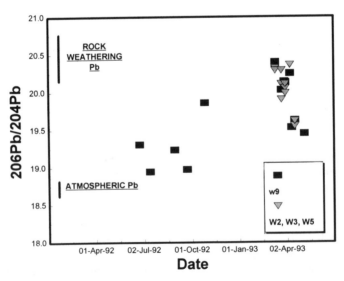

FIGURE 10.14   Temporal variation of ^{206}Pb/^{204}Pb in stream water at Sleepers River Watershed, Vermont. Stream samples are intermediate in composition between the two end members: rock weathering and atmospheric Pb. The streams are the same ones depicted in Figure 10.6

10.6 USE OF A MULTI-ISOTOPE APPROACH FOR THE DETERMINATION OF FLOWPATHS

Flowpaths are the individual runoff *pathways* contributing to surface flow in a catchment. These result from runoff mechanisms that include, but are not limited to, saturation-excess overland flow, Hortonian overland flow, near- stream groundwater ridging, hillslope subsurface flow through the soil matrix or macropores, and shallow organic-layer flow. Knowledge of hydrologic flowpaths in catchments is critical to the preservation of public water supplies and the understanding of the transport of point and non-point source pollutants (Peters, 1994).

Stable isotopes such as ^{18}O and ^{2}H are an improved alternative to traditional non-conservative chemical tracers because waters are often uniquely labelled by their isotopic compositions (Sklash and Farvolden, 1979; McDonnell and Kendall, 1992), often allowing the separation of waters from different sources (e.g. new rain versus old pre-storm water). However, studies have shown that flowpaths cannot be identified to a high degree of certainty using δ^{18}O or δD data and simple hydrograph separation techniques because waters within the same flowpath can be derived from several different sources (Ogunkoya and Jenkins, 1991). Thus, a number of plausible runoff mechanisms can be consistent with the isotope data. The need to incorporate flowpath dynamics is recognised as a key ingredient in producing reliable chemical models (Robson *et al.*, 1992).

Reactive solute isotopes such as ^{13}C, ^{34}S and ^{87}Sr can provide valuable information about flowpaths (not water sources) useful for geochemical and hydrologic modelling precisely because they *can* reflect the reactions characteristic of and taking place along specific flowpaths (Bullen and Kendall, 1991; Kendall *et al.*, 1992; Kendall, 1993). These reactive solute isotopes can serve as additional thermodynamic constraints in geochemical computer models such as NETPATH (Plummer *et al.*, 1991) for eliminating possible geochemical reaction paths (Plummer *et al.*, 1983).

As an example, water isotopes were used to determine the actual flowpaths by which water discharging from Crystal Lake (Wisconsin) mixed with regional groundwater and local rain recharge while flowing across a narrow isthmus separating the lake from Big Muskellunge Lake (Krabbenhoft *et al.*, 1992, 1993; Bullen and Kendall, 1995). Solute isotopes such as ^{87}Sr and ^{13}C, along with chemical data, were then used to identify the geochemical reactions taking place along these flowpaths.

The waters flowing along mineralogically distinctive horizons are sometimes distinctively labelled by their chemical composition and by the isotopic compositions of solute isotopes like ^{13}C, ^{87}Sr, ^{34}S, ^{15}N, etc. For example, waters flowing through the soil zone often have δ^{13}C values that are depleted in ^{13}C relative to deeper groundwaters because of biogenic production of carbonic acid in organic soils (Kendall, 1993).These same shallow waters can also have distinctive Pb and Sr isotopic compositions (Bullen and Kendall, 1995).

A multi-isotope approach can be particularly useful in tracing the reactions specific to a flowpath (Krabbenholt *et al.*, 1992, 1993). The solute isotope tracers can be an especially powerful tool because they usually are affected by a smaller number of processes than chemical constituents, making interpretation of changes in isotopic composition less ambiguous than the simultaneous changes in solute concentrations (Kendall, 1993). In particular the strengths of one isotope may compensate for the weakness of another. For example, one of the concerns with using δ^{13}C to trace carbon sources in catchments is that possible isotopic exchange of subsurface DIC with soil or atmospheric CO_2 could allow the $\delta^{13}C$ of DIC to be "reset" so that the characteristic "geologic" signature of the calcite-derived carbon is lost (Kendall *et al.*, 1992; Kendall, 1993). Since Sr isotopes are not affected by this kind of exchange, the same samples can be analysed for Sr isotopes; if the Sr signature of calcite is distinctive, the correlation of changes in $^{87}Sr/^{86}Sr$ and changes in $\delta^{13}C$ of subsurface waters could provide evidence that calcite dissolution was occurring and that exchange was not a problem.

If the chemical composition of waters along a specific flowpath is related to topography, mineralogy, initial water composition and antecedent moisture conditions, then within a uniform soil layer the degree of geochemical evolution is some function of residence time. In this case, it may be possible to link geochemical evolution of waters along specific flowpaths to hydrologic models such as TOPMODEL (Beven and Kirkby, 1979) by the testable assumption that residence time — and hence degree of rock/water reaction — is primarily a function of topography. Alternatively, because TOPMODEL can predict the different contributions from the different soil horizons along the stream channel as the topographic form of the hillslope changes (Robson *et al.*, 1992), these predictions can then be tested with appropriate isotope and chemical data.

10.7 SUMMARY

The dominant use of isotopes in catchment research in the last few decades has been to trace sources of waters and solutes. Generally such data were evaluated with simple mixing models to determine how much was derived from either of two (sometimes three) constant-composition sources. The world does not seem this simple any more. With the expansion of the field of isotope hydrology in the last decade, made possible by the development and increased availability of automated preparation and analysis systems for mass spectrometers, we have documented considerable heterogeneity in the isotopic compositions of rain, soil water, groundwater and solute sources. We are still grappling with how to deal with this heterogeneity in our hydrologic and geochemical models. A major challenge is to use the variability as signal, not noise, in our models; the isotopes and chemistry are providing very detailed information about sources and reactions in shallow systems, if only we can develop appropriate models to use the data. This integration of chemical and isotopic data with complex hydrologic models constitutes an important frontier of catchment research.

REFERENCES

Aberg, G., Jacks, G. and Hamilton, P. J. (1989). "Weathering rates and $^{87}Sr/^{86}Sr$ ratios: an isotopic approach", *J. Hydrol.*, **109**, 65−78.

Amberger, A. and Schmidt, H.-L. (1987). "Naturliche Isotopengehalte von Nitrat als Indikatoren fur dessen Herkunft", *Geochim. Cosmochim. Acta*, **51**, 2699−2705.

Andersson, P., Torssander, P. and Ingri, J. (1992). "Sulphur isotope ratios in sulphate and oxygen isotopes in water from a small watershed in Central Sweden", *Hydrobiologia*, **235/236**, 205−217.

Aravena, R., Schiff, S.L., Trumbore, S. E., Dillon, P. J. and Elgood, R. (1992). "Evaluating dissolved inorganic carbon cycling in a forested lake watershed using carbon isotopes", *Radiocarbon*, **34(3)**, 636−645.

Baskaran, M. and Illiffe, T. M. (1993). "Age determination of recent cave deposits using ^{210}Pb — a new technique", *Geophys. Res. Lett.*, **2**, 603−606.

Behrens, H., Moser, H., Oerter, H., Rauert, W., Stichler, W., Ambach, W. and Kirchlechner, P. (1979). "Models for the run-off from a glaciated catchment area using measurements of environmental isotope contents", in *Isotope Hydrology 1978*, (Eds. IAEA), pp. 829−846, IAEA, Vienna.

Beven, K. J. and Kirkby, M. J. (1979). "A physically based, variable contributing model of basin hydrology", *Hydrol. Sci. Bull.*, **10**, 43−69.

Bishop, K. H. (1991) 'Episodic increases in stream acidity catchment flow pathways and hydrograph separation", PhD thesis, Department of Geology, Jesus College, University of Cambridge, Cambridge, 246 pp.

Bohlke, J. K. and Denver, J. M. "Combined use of ground-water dating, chemical, and isotopic analyses to resolve the history and fate of nitrate contamination in two agricultural watersheds, Atlantic coastal plain, Maryland", (submitted).

Bohlke, J. K., Denver, J. M., Phillips, P. J., Gwinn, C. J., Plummer, L. N., Busenberg, E. and Dunkle, S. A. (1992). "Combined use of nitrogen isotopes and groundwater dating to document nitrate fluxes and transformations in small agricultural watersheds, Delmarva, Maryland", *EOS Trans. Am. Geophys. Union.* **73**, 140.

Bottcher, J., Strebel, O., Voerkelius, S. and Schmidt, H. L. (1990). "Using isotope fractionation of nitrate-nitrogen and nitrate-oxygen for evaluation of microbial denitrification in a sandy aquifer", *J. Hydrol.*, **114**, 413–424.

Bremmer, J. M. (1965). 'Isotope ratio analysis of nitrogen in nitrogen-15 tracer investigations", in *Methods of Soil Analysis*, (Ed. C. A. Black), pp. 1256–1286, American Society of Agronomy, Madison, Wisconsin.

Brinkmann, R., Eichler, R., Ehhalt, D. and Munnich, K. O. (1963). "Über den Deuterium-gehalt von Niederschlags- und Grundwasser", *Naturwissenschaften*, **19**, 611–612.

Bullen, T. D. and Kendall, C. (1991). "$^{87}Sr/^{86}Sr$ and $\delta^{13}C$ as tracers of interstream and intrastorm variations in water flowpaths, Catoctin Mountain, MD", *Trans. Am. Geophys. Union*, **72**, 218.

Bullen, T. D. and Kendall, C. (1995). "Tracing of weathering reactions and water flowpaths: a multi-isotope approach", in *Isotope Tracers in Catchment Hydrology*, (Eds. J. J. McDonnell and C. Kendall), Elsevier, Amsterdam, 50 pp (in press).

Bullen, T. D., Shanley, J. B. and Clark, S. (1994). "Sr and Pb isotopes as surrogate tracers of water flowpaths in a forested catchment", *Trans. Am. Geophys. Union,*, **75**, 144.

Bullen, T. D., Krabbenhoft, D. P. and Kendall, C. (in preparation). "Kinetic and mineralogic controls on the evolution of ground-water chemistry and $^{87}Sr/^{86}Sr$ in a sandy silicate aquifer, northern Wisconsin".

Buttle, J. M. (1994). "Isotope hydrograph separations and rapid delivery of pre-event water from drainage basins", *Prog. Phys. Geog.*, **18**, 16–41.

Buttle, J. M. (in press). "Fundamentals of small catchment hydrology", in *Isotope Tracers in Catchment Hydrology*, (Eds. J. J. McDonnell and C. Kendall), Elsevier, Amsterdam, 50 pp.

Buttle, J. M. and Sami, K. (1990). "Recharge processes during snowmelt: an isotopic and hydrometric investigation", *Hydrol. Process.*, **4**, 343–360.

Caron, F. A., Tessier, A., Kramer, J. R., Schwarcz, H. P. and Rees, C. E. (1986). "Sulfur and oxygen isotopes in sulfate in precipitation and lakewater, Quebec", *Can. Appl. Geochem.*, **1**, 601–606.

Claassen, H. C. and Downey, J. S. (1995). "A model for deuterium and oxygen-18 isotope changes during evergreen interception of snowfall", *Water Resour. Res.*, **31**, 601–618.

Cooper, L. W., Olsen, C. R., Solomon, D. K., Larsen, I. L., Cook, R. B. and Grebmeier, J. M. (1991). "Stable isotopes of oxygen and natural and fallout radionuclides used for tracing runoff during snowmelt in an arctic watershed", *Water Resour. Res.*, **27**, 2171–2179.

Coplen, T. B. (1993). "Uses of environmental isotopes", in *Regional Ground-water Quality*, (Ed. W. M. Alley), pp. 227–254, Van Nostrand Reinhold, New York.

Craig, H. (1961a). "Isotopic variations in meteoric waters", *Science*, **133**, 1702–1703.

Craig, H. (1961b). "Standard for reporting concentrations of deuterium and oxygen-18 in natural waters", *Science*, **133**, 1833.

Davis, S. N. and Bentley, H. W. Jr (982). "Dating groundwater, a short review", in *Nuclear and Chemical Dating Techniques: Interpreting the Environmental Record*, (Ed. L. A. Cutrie), pp. 187–222, American Chemical Society Symposium Series No. 176, Washington, D.C.

Deines, P. (1980). "The isotopic composition of reduced organic carbon", in *Handbook of Environmental Isotope Geochemistry*, (Eds. P. Fritz and J. Ch. Fontes), Vol. 1, pp. 329–406, Elsevier, New York.

DeWalle, D. R. and Swistock, B. E. (1994). "Differences in oxygen-18 content of throughfall and rainfall in hardwood and coniferous forest", *Hydrol. Process.*, **8**, 75–82.

DeWalle, D. R., Swistock, B.R. and Sharpe, W. E. (1988). "Three-component tracer model for stormflow on a small Appalachian forested catchment", *J. Hydrol.*, **104**, 301–310.

Dincer, T., Payne, B. R., Florkowski, T., Martinec, J. and Tongiorgi, T. (1970). "Snowmelt runoff from measurements of tritium and oxygen-18", *Water Resour. Res.*, **6**, 110–118.

Dunkle, S. A., Plummer, L. N., Busenberg, E., Phillips, P. J., Denver, J. M., Hamilton, P. A. Michel, R. L. and Coplen, T. B. (1993). "Chlorofluorocarbons (CCl_3F and CCl_2F_2) as dating tools and hydrologic tracers in shallow groundwater of the Delmarva Peninsula, Atlantic Coastal Plain, United States", *Water Resour. Res.*, **29**, 3837–3860.

Durka, W. (1994). "Isotopenchemie des Nitrat, Nitrataustrag, Wasserchemie und Vegetation von Waldquellen im Fichtelgebirge (NO-Bayern), PhD thesis, Bayreuth University, Germany, 197 pp.

Durka, W., Schulze, D.-D., Gebauer, G. and Voerkelius, W. (1994). "Effects of forest decline on uptake and leaching of deposited nitrate determined from ^{15}N and ^{18}O measurements", *Nature*, **372**, 765–767.

Ekwurzel, B., Schlosser, P., Smethie, W. M., Jr Plummer, L. N., Busenberg, E., Michel, R. L. Weppernig, R. and Stute, M. (1994). "Dating of shallow groundwater: comparison of transient tracers $^3H/^3He$, chlorofluorocarbons, and ^{85}Kr", *Water Resour. Res.*, **30**, 1693–1708.

Erel, Y., Patterson, C. C., Scott, M. J. and Morgan, J. J. (1990). "Transport of industrial lead in snow through soil to stream water and groundwater", *Chem. Geol.*, **85**, 383–392.

Faure, G. (1986). *Principles of Isotope Geology*, 2nd edition, John Wiley, New York, 589 pp.

Forester, R. M. (1983). "Relationship of two lacustrine ostracode species to solute composition and salinity: Implications for paleohydrochemistry", *Geology*, **14**, 796–798.

Fritz, P. and Fontes, J. Ch. (Eds.) (1980). *Handbook of Environmental Isotope Geochemistry*, Vol. 1, Elsevier, New York, 545 pp.

Fritz, P. and Fontes, J. Ch. (Eds.) (1981) *Handbook of Environmental Isotope Geochemistry*, Vol. 2, Elsevier, New York, 557 pp.

Fritz, P., Cherry, J. A., Weyer, K. V. and Sklash, M. G. (1976). "Runoff analyses using environmental isotopes and major ions", in *Interpretation of Environmental Isotope and Hydrochemical Data in Groundwater Hydrology*, pp. 49–60, IAEA, Vienna.

Fry, B., Cox, J., Gest, H. and Hayes, J. M. (1986). "Discrimination between ^{34}S and ^{32}S during bacterial metabolism of inorganic sulfur compounds", *J. Bacteriol.*, **165**, 328–330.

Fry, B., Ruf, W., Gest, H. and Hayes, J. M. (1988). "Sulfur isotope effects associated with oxidation of sulfide by O_2 in aqueous solution", *Chem. Geol. (Isotope Geoscience Section)*, **73**, 205–210.

Fuller, R. D., Mitchell, M. J., Krouse, H. R., Syskowski, B. J. and Driscoll, C. T. (1986). "Stable sulfur isotope ratios as a tool for interpreting ecosystem sulfur dynamics", *Water, Air and Soil Pollut.*, **28**, 163–171.

Garrels, R. M. and Mackenzie, F. T. (1971). *Evolution of Sedimentary Rocks*, W. W. Norton. 397 pp.

Garten, C. T. Jr and Hanson, P. J. (1990). "Foliar retention of ^{15}N-nitrate and ^{15}N-ammonium by red maple (*Acer rubrum*) and white oak (*Quercus alba*) leaves from simulated rain", *Environmental and Experimental Botany*, **20**, 33–342.

Gat, J. R. (1980). "The isotopes of hydrogen and oxygen in precipitation", in *Handbook of Environmental Isotope Geochemistry*, (Eds. P. Fritz and J. Ch. Fontes), pp. 21–47, Elsevier, Amsterdam.

Gat, J. R. and Gonfiantini, R. (Eds.) (1981). *Stable isotope hydrology — deuterium and oxygen-18 in the water cycle*, Technical Report Series No. 210, IAEA, Vienna, 337 pp.

Gat, J. R. and Tzur, Y. (1967). "Modification of the isotopic composition of rainwater by processes which occur before groundwater recharge", in *Isotope Hydrology, Proceedings of Symposium, Vienna, 1966*, pp. 49–60, IAEA, Vienna.

Genereux, D. P. and Hemond, H. F. (1990). "Naturally occurring radon 222 as a tracer for streamflow generation: steady state methodology and field example", *Water Resour. Res.*, **26**, 3065–3075.

Genereux, D. P., Hemond, H. F. and Mulholland, P. J. (1993). "Use of radon-222 and calcium as tracers in a three-end-member mixing model for streamflow generation on the west fork of Walker Branch watershed", *J. Hydrol.*, **142**, 167–211.

Geyh, M. A. and Schleicher, H. (1990). *Absolute Age Determination*, Springer-Verlag, Berlin.

Goldhaber, M. B. and Kaplan, I. R. (1974). "The sulphur cycle", in *The Sea*, Vol. 5. pp. 569–655, (Ed. E. D. Goldberg), John Wiley, New York.

Gormly, J. R. and Spalding, R. F. (1979). "Sources and concentrations of nitrate-nitrogen in ground water of the Central Platte region, Nebraska", *Ground Water*, **17**, 291–301.

Graustein, W. C. (1989). "$^{87}Sr/^{86}Sr$ ratios measure the sources and flow of strontium in terrestrial ecosystems", in *Stable Isotopes in Ecological Research*. (Eds. P. W. Rundel, J. R. Ehleringer and K. A. Nagy), pp. 491–511, Springer-Verlag, New York.

Harris, D. M., McDonnell, J. J. and Rodhe, A. (1995). "Hydrograph separation using continuous open-system isotope mixing", *Water Resour. Res.*, **31**, 157−172.

Hesslein, R. H., Capel, M. J. and Fox, D. E. (1988). "Stable isotopes in sulfate in the inputs and outputs of a Canadian watershed", *Biogeochemistry*, **5**, 263−273.

Hewlett, J. D. and Hibbert, A. R. (1967). "Factors affecting the response of small watersheds to precipitation in humid areas", Proceedings of 1st International Symposium on *Forest Hydrology*, pp. 275−290.

Hinton, M. J., Schiff, S. L. and English, M. C. (1994). "Examining the contributions of glacial till water to storm runoff using two- and three-component hydrograph separations", *Water Resour. Res.*, **30**, 983−993.

Hoefs, J. (1987) *Stable Isotope Geochemistry*, 3rd edition, Springer-Verlag, Berlin, 241 pp.

Holt, B. D. and Kumar, R. (1991). "Oxygen isotope fractionation for understanding the sulphur cycle", in *Stable Isotopes: Natural and Anthropogenic Sulphur in the Environment*, (Eds. H. R. Krouse and V. A. Grinenko), pp. 27−41, SCOPE 43 (Scientific Committee on Problems of the Environment).

Hooper, R. P. and Shoemaker, C. A. (1986). "A comparison of chemical and isotopic hydrograph separation", *Water Resour. Res.*, **22**, 1444−1454.

Hooper, R. P., Christophersen, N. and Peters, N. E. (1990). "Modelling streamwater chemistry as a mixture of soilwater end- members—an application to the Panola Mountain Catchment, Georgia, USA", *J. Hydrol.*, **116**, 321−343.

Horibe, Y., Shigehara, K. and Takakuwa, Y. (1973). "Isotope separation factors of carbon dioxide−water system and isotopic composition of atmospheric oxygen", *J. Geophys. Res.*, **78**, 2625−2629.

International Atomic Energy Agency (IAEA) (1981). *Statistical treatment of environmental isotope data in precipitation*, IAEA Technical Report Series No. 206, IAEA,l Vienna.

Junk, G. and Svec, H. (1958). "The absolute abundance of the nitrogen isotopes in the atmosphere and compressed gas from various sources", *Geochim. Cosmochim. Acta*, **14**, 234−243.

Kaplan, I. R. and Rittenberg, S. C. (1964). "Microbiological fractionation of sulphur isotopes", *J. Gen. Microbiol.*, **34**, 195−212.

Karamanos, E. E., Voroney, R. P. and Rennie, D. A. (1981). "Variation in natural ^{15}N abundance of central Saskatchewan soils", *Soil, Sci. Soc. of Am. J.*, **45**, 826−828.

Kendall, C. (1993). "Impact of isotopic heterogeneity in shallow systems on stormflow generation", PhD dissertation, University of Maryland, College Park, 310 pp.

Kendall, C. and Cappellato, R. (submitted). "Spatial and species-related variations in the $\delta^{18}O$ and δD of throughfall, 30 pp.

Kendall, C. and Gu, W. (1992). "Development of isotopically heterogeneous infiltration waters in an artificial catchment in Chuzhou, China", in *Isotope Techniques in Water Resources Development 1991*, Proceedings of IAEA Symposium, 11−15 March 1991, pp. 61−73, IAEA, Vienna.

Kendall, C. and McDonnell, J. J. (1993). "Effect of intrastorm heterogeneities of rainfall, soil water and groundwater on runoff modelling", in *Tracers in Hydrology*, (Eds. N. E. Peters *et al.*) International Association of Hydrological Science Publication No. 215, July 11−23 1993, Yokohama, Japan, pp. 41−49.

Kendall, C. and Mills, A. L. (submitted). "Tracing sources of carbonate alkalinity in small streams using stable carbon isotopes", 51 pp.

Kendall, C., Mast, M. A. and Rice, K. C. (1992). "Tracing watershed weathering reactions with $\delta^{13}C$", in *Water−Rock Interaction*, (Eds. Y. K. Kharaka and A. S. Maest) Proceedings of 7th International Symposium, Park City, Utah, 13−18 July 1993, pp. 569−572, Balkema, Rotterdam.

Kendall, C., Campbell, D. H., Burns, D. A., Shanley, J. B., Silva, S. R. and Chang, C. C. Y. (1995). "Tracing sources of nitrate in snowmelt runoff using the oxygen and nitrogen isotopic compositions of nitrate: pilot studies at three catchments", in *Biogeochemistry of Seasonally Snow-covered Catchments*, (Eds. K. Tonnessen, M. Williams, and M. Tranter), International Association of Hydrological Sciences Publication, July 3−14, 1995, Boulder CO, (in press), 15 pp.

Kendall C., Silva, S. R., Chang, C. C. Y., Burns, D. A., Campbell, D. H. and Shanley, J. B. (in press). *Use of the $\delta^{18}O$ and $\delta^{15}N$ of nitrate of determine sources of nitrate in early spring runoff in forested catchments*, International Atomic Energy Agency, Symposium on Isotopes in Water Resources Management, Vienna, Austria, 20-24 March, 1995.

Kendall, C., Silva, S. R., Chang, C. C., Burns, D. A., Campbell, D. H. and Shanley, J. B. (1995). "Tracing sources of nitrate in snowmelt runoff using the oxygen and nitrogen isotopic compositions of nitrate: pilot studies at three catchments", International Association of Hydrological Science Publication, 3−14, Boulder, Colorado, July 1995. (submitted).

Kennedy, V. C., Kendall, C., Zellweger, G. W., Wyermann, T. A. and Avanzino, R. A. (1986). "Determination of the components of stormflow using water chemistry and environmental isotopes, Mattole River Basin, California", *J. Hydrol.*, **84**, 107−140.

Kohl, D. H., Shearer, G. B. and Commoner, B. (1971). "Fertilizer nitrogen: contribution to nitrate in surface water in a corn belt watershed", *Science*, **174**, 1331−1334.

Krabbenhoft, D. P., Bullen, T. D. and Kendall, C. (1992). "Isotopic indicators of groundwater flow paths in a northern Wisconsin aquifer", *Am. Geophys. Union Trans.*, **73**, 191.

Krabbenhoft, D. P., Bullen, T. D. and Kendall, C. (1993). "Use of multiple isotope tracers as monitors of groundwater−lake interaction", *Am. Geophys. Union Trans.*, **74**, 140−141.

Krabbenhoft, D. P., Bowser, C. J., Kendall, C. and Gat, J. R. (1994). "Use of oxygen-18 and deuterium to assess the hydrology of ground-water/lake systems", in *Environmental Chemistry of Lakes and Reservoirs*, (Ed. L. A. Baker), American Chemical Society, Monograph No. 237, pp. 67−90.

Kreitler, C. W. (1975). "Determining the source of nitrate in groundwater by nitrogen isotope studies: Austin, Texas", University of Texas, Austin, Bureau of Economy, Geology Report, of Inves. No. 83, 57 pp.

Kreitler, C. W. and Jones, D. C. (1975). "Natural soil nitrate: the cause of the nitrate contamination of groundwater in Runnels County, Texas", *Groundwater*, **13**, 53−61.

Kreitler, C. W., Ragone, S. E. and Katz, B. G. (1978). "$^{15}N/^{14}N$ ratios of ground-water nitrate, Long Island, NY", *Groundwater*, **16**, 404−409.

Krouse, H. R. (1980). "Sulfur isotopes in our environment", in *Handbook of Environmental Isotope Geochemistry* (Eds. P. Fritz and J. Ch. Fontes), pp. 435−471, Elsevier, Amsterdam.

Krouse, H. R. and Grinenko, V. A. (Eds.) (1991). *Stable Isotopes: Natural and Anthropogenic Sulphur in the Environment*, SCOPE 43 (Scientific Committee on Problems of the Environment), John Wiley, New York, 440 pp.

Krouse, H. R., Legge, A. and Brown, H. M. (1984). "Sulphur gas emissions in the boreal forest: the West Whitecourt Case Study. V: stable sulfur isotopes", *Water, Air and Soil Pollut.*, **22**, 321−347.

Legge, A. H. and Krouse, H. R. (1992). "An assessment of the environmental fate of industrial sulphur in a temperate pine forest ecosystem", in *Critical Issues in the Global Environment*, Vol. 5, Ninth World Clean Air Congress Towards Year 2000, paper 1U22B.01.

Likens, G. E., Wright, R. F., Galloway, J. N. and Butler, T. J. (1979). "Acid rain", *Sci. Am.*, **241**, 43−50.

Lloyd, R. M. (1968). "Oxygen isotope behaviour in the sulfate-water system", *J. Geophys. Res.*, **73**, 6099−6110.

McDonnell, J. J., Bonell, M., Stewart, M. K. and Pearce, A. J. (1990). "Deuterium variations in storm rainfall: implications for stream hydrograph separations", *Water Resour. Res.*, **26**, 455−458.

McDonnell, J. J. and Kendall, C. (1992). "Isotope tracers in hydrology — report to the Hydrology Sections", *EOS Trans. Am. Geophys. Union*, **73**, 260−261.

McDonnell, J. J., Stewart, M. K. and Owens, I. F. (1991). "Effect of catchment-scale subsurface mixing on stream isotopic response', *Water Resour. Res.*, **27**, 3065-3073.

McKnight, D. M. and Bencala, K. E. (1990). "The chemistry of iron, aluminum, and dissolved organic material in three acidic, metal-enriched, mountain streams, as controlled by watershed and in-stream processes", *Water Resour. Res.*, **26**, 3087−3100.

MacNamara, J. and Thode, H. G. (1950). "Comparison of the isotopic composition of terrestrial and meteoritic sulfur", *The Physical Rev.*, **78**, 307−308.

Maloszewzki, P., Rauert, W., Stichler, W. and Hermann, A. (1983). "Application of flow models in an Alpine catchment using tritium and deuterium data", *J. Hydrol.*, **66**, 319−330.

Martinec, J. (1976). "Snow and ice", in *Facets of Hydrology*, pp. 85−118, (Ed. J. C. Rodda), John Wiley, Chichester.

Mayer, B., Fritz, P. and Krouse, H. R. (1992). "Sulphur isotope discrimination during sulphur transformations in aerated forest soils", *Workshop Proceedings on Sulphur Transformations in Soil Ecosystems*, (Eds. M. J. Hendry and H. R. Krouse), National Hydrology Research Symposium No.

11, Saskatoon, Saskatchewan, 5−7 November 1992, pp. 161−172.

Mayer, B., Krouse, H. R., Fritz, P., Prietzel, J. and Rehfuess, K. E. (1993). "Evaluation of biogeochemical sulfur transformations in forest soils by chemical and isotope date", in *Tracers in Hydrology*, IAHS publication No. 215, pp. 65−72.

Michel, R. L. (1989). "Tritium deposition over the continental United States, 1953−1983", in *Atmospheric Deposition*, pp. 109−115, International Association of Hydrological Sciences, Oxford, UK.

Michel, R. L. and Naftz, D. L. (1995). "Use of sulphur-35 and tritium to study runoff from an alpine glacier, Wind River Range, Wyoming", in *Biogeochemistry of Seasonally Snow-covered Catchments*, (Eds. K. Tonnessen, M. Williams, and M. Tranter), International, Association of Hydrological Sciences Publication, July 3−14, 1995, Boulder CO, 8 pp (in press).

Michel, R. L. and Turk, J. T. (1995) "Use of sulphur-35 to study sulphur migration in the Flat Tops Wilderness Area", International Atomic Energy Agency, Symposium on Isotopes in Water Resources Management, Vienna, 20−24 March, 1995, 10 pp, (in press).

Miller, E. K., Blum, J. D. and Friedland, A. J. (1993). "Determination of soil exchangeable-cation loss and weathering rates using Sr isotopes", *Nature*, **362**, 438−441.

Mills, A. L. (1988) "Variations in the delta C-13 of stream bicarbonate: implications for sources of alkalinity", MS thesis, George Washington University, Washington, D.C., 160 pp.

Mitchell, M. J., Driscoll, C. T., Fuller, R. D., David, M. B. and Likens, G. E. (1989). "Effect of whole-tree harvesting on the sulfur dynamics of a forest soil", *Soil Sci. Soc. Am.*, **53**, 933−940.

Mizutani, Y. and Rafter, T. A. (1969). "Oxygen isotopic composition of sulphates: Part 4. Bacterial fractionation of oxygen isotopes in the reduction of sulphate and in the oxidation of sulphur", *NZ J. Sci.*, **12**, 60−67.

Mulholland, P. J., Wilson, G. V. and Jardine, P. M. (1990). "Hydrogeochemical response of a forested watershed to storms: effects of preferential flow along shallow and deep pathways", *Water Resour. Res.*, **26**, 3021−3036.

Nadelhoffer, J. J. and Fry, B. (1988). 'Controls on natural nitrogen-15 and carbon-13 abundances in forest soil organic matter", *Soil Sci. Soc. of Am. J.*, **52**, 1633−1640.

Nadelhoffer, K. J., Aber, J. D. and Melillo, J. M. (1988). "Seasonal patterns of ammonium and nitrate uptake in nine temperature forest ecosystems", *Plant and Soil*, **80**, 321−335.

Nimz, G. J. and Hudson, G. B. (1995). "Lithogenic, cosmogenic, and noble gas nuclides in catchment systems", in *Isotope Tracers in Catchment Hydrology*, (Eds. J. J. McDonnell and C. Kendall) Elsevier, Amsterdam, 50 pp (in press).

Ogunkoya, O. O. and Jenkins, A. (1991). "Analysis of runoff pathways and flow distributions using deuterium and stream chemistry", *Hydrol. Process.*, **5**, 271−282.

Ogunkoya, O. O. and Jenkins, A. (1993). "Analysis of a storm hydrograph and flow pathways using a three-component hydrograph separation model", *J. Hydrol.*, **142**, 71−88.

Olleros, T. (1983). "Kinetische Isotopeneffekte der Arginaseund Nitratedukturase-Reaktion; ein Beitrag zur Aufklarung der entsprechenden Reaktionsmechanismen", Dissertation, Technical University of Munchen-Weihenstephan, 158 pp.

Pearce, A. J., Stewart, M. K. and Sklash, M. G. (1986). "Storm runoff generation in humid headwater catchments. 1. Where does the water come from?", *Water Resour. Res.*, **22**, 1263−1272.

Pearson, F. J. and Rightmire, C. T. (1980). "Sulphur and oxygen isotopes in aqueous sulfur compounds", in *Handbook of Environmental Isotope Geochemistry*, pp. 179−226, (Eds. P. Fritz and J. Ch., Fontes) Elsevier, Amsterdam.

Peters, B. (1959). "Cosmic-ray produced radioactive isotopes as tracers for studying large-scale atmospheric circulation", *J. Atmos. Terr. Phys.*, **13**, 351−370.

Peters, N. E. (1994). "Hydrologic studies", in *Biogeochemistry of Small Catchments: A Tool for Environmental Research*, (Eds. B. Moldan and J. Cerny), SCOPE Report 51, Ch. 9, pp. 207−228.

Pionke, H. B. and DeWalle, D. R. (1992). "Intra- and inter-storm ^{18}O trends for selected rainstorms in Pennsylvania", *J. Hydrol*, **138**, 131−143.

Plummer, L. N., Parkhurst, D. L. and Thorstenson, D. C. (1983). "Development of reaction models for ground-water systems", *Geochim. Cosmochim. Acta.*, **47**, 665−686.

Plummer, L. N., Prestemon, E. C. and Parkhurst, D. L. (1991). *An interactive code (NETPATH) for modelling net geochemical reactions along a flow path*, USGS Water-Resources Inves. Report 91-

4078, 227 pp.

Plummer L. N., Michel, R. L., Thurman, E. M. and Glynn, P.D. (1993). "Environmental tracers for age dating young ground water", in *Regional Ground-Water Quality* (Ed. W. M. Alley), pp. 255–294, V. N. Reinhold, New York.

Reilly, T. E., Plummer, L. N., Phillips, P. J. and Busenberg, E. (1994). "The use of simulation and multiple environmental tracers to quantify groundwater flow in a shallow aquifer", *Water Resour. Res*, **30**, 421–433.

Robson, A., Beven, K. J. and Neal, C. (1992). "Towards identifying sources of subsurface flow: a comparison of components identified by a physically based runoff model and those determined by chemical mixing techniques", *Hydrol. Process.*, **6**, 199–214.

Rodhe, A. (1987). "The origin of stream water traced by oxygen- 18", PhD thesis, Department of Physical Geography, Division of Hydrology, Uppsala University, Sweden, Report Serial A., No. 41, 260 pp.

Rose, S. (1992) "Tritium in groundwater of the Georgia Piedmont: implications for recharge and flow paths", *Hydrol. Process.*, **6**, 67–78

Rose, S. (1993). "Environmental tritium systematics of baseflow in Piedmont Province watersheds, Georgia (USA)", *J. Hydrol.*, **143**, 191–216.

Rundel, P. W., Ehleringer, J. R. and Nagy, K. A. (Eds.) (1989). *Stable Isotopes in Ecological Research*, Springer-Verlag, New York, 526 pp.

Saxena, R. K. (1986). "Estimation of canopy reservoir capacity and oxygen-18 fractionation in throughfall in a pine forest", *Nord. Hydrol.*, **17**, 251–260.

Schiff, S. L., Aravena, R., Trumbore, S. E. and Dillon, P. J. (1990). "Dissolved organic carbon cycling in forested watersheds: a carbon isotope approach", *Water Resour. Res.*, **26**, 2949–2957.

Shanley, J. B., Kendall, C., Smith, T. M. and Wolock, D. M. (submitted). "The effect of catchment scale and land use on the relative contributions of shallow flowpaths to stream discharge during snowmelt", 34 pp.

Shearer, G. and Kohl, D. H. (1988). "$\delta^{15}N$ method of estimating N_2 fixation", in *Stable Isotopes in Ecological Research*. (Eds. P. W. Rundel, J. R. Ehleringer and K. A. Nagy, pp. 342–374, Springer-Verlag, New York.

Shearer, G., Duffy, J., Kohl, D. H. and Commoner, B. (1974). "A steady-state model of isotopic fractionation accompanying nitrogen transformations in soil", *Soil Sci. Soc. of Am. Proc.*, **38**, 315–322.

Silva, S. R., Kendall, C. and Chang, C. C. (in preparation). "Method for preparing nitrate samples for $\delta^{18}O$ analysis".

Sklash, M. G. (1990). "Environmental isotope studies of storm and snowmelt runoff generation", in *Process Studies in Hillslope Hydrology*, (Eds. M. G. Anderson and T. P. Burt) pp. 401–435, John Wiley, Chichester.

Sklash, M. G. and Farvolden, R. N. (1979) "The role of groundwater in storm runoff", *J. Hydrol.*, **43**, 45–65.

Sklash M. G. and Farvolden, R. N. (1982). "The use of environmental isotopes in the study of high-runoff episodes in streams", in *Isotope Studies of Hydrological Processes*, (Eds. E. C. Perry Jr and C. M. Montgomery) pp. 65–73, Northern Illinois University Press, DeKalb, Illinois.

Sklash, M. G., Farvolden, R. N. and Fritz, P. (1976). "A conceptual model of watershed response to rainfall, developed through the use of oxygen-18 as a natural tracer", *Can. J. Earth Sci.*, **13**, 271–283.

Stam, A. C., Mitchell, M. J., Krouse, H. R. and Kahl, J. S. (1992). "Stable sulfur isotopes of sulfate in precipitation and stream solutions in a northern hardwood watershed", *Water Resour. Res.*, **28**, 231–236.

Stichler, W. (1987). "Snowcover and snowmelt process studies by means of environmental isotopes", in *Seasonal Snowcovers: Physics, Chemistry, Hydrology*, (Eds. H. G. Jones and W. J. Orville-Thomas) pp. 673–726, D. Reidel Publishers.

Swistock, B. R., DeWalle, D. R. and Sharpe, W. E. (1989). "Sources of acidic stormflow in an Appalachian headwater stream", *Water Resour. Res.*, **25**, 2139–2147.

Taylor, L. C. and Howard, K. W. F. (1993). "The distribution of *Cypridopsis okeechobel* in the Duffins Creek–Rouge River drainage (Ontario, Canada) and its potential as an indicator of human disturbance", in *Ostracoda in the Earth and Life Sciences*, (Eds. K. G. McKenzie and P. J. Jones) pp. 481−492, Proceedings of 11th International Symposium on *Ostracoda*, Wafrrnambool, Australia, Balkema Press.

Taylor, B. E. and Wheeler, M. C. (1994). "Sulfur- and oxygen- isotope geochemistry of acid mine drainage in the western US: field and experimental studies revisited", in *Environmental Geochemistry of Sulfide Oxidation*, (Eds. C. N. Alpers and D. W. Blowes) pp. 481−514, ACS Symposium Series 550, American Chemical Society, Washington, D. C.

Taylor, B. E., Wheeler, M. C. and Nordström, D. K. (1984). "Oxygen and sulfur compositions of sulphate in acid mine drainage: evidence for oxidation mechanisms", *Nature*, **308**, 538−541.

Thatcher, L. L. (1962). "The distribution of tritium fallout in precipitation over North America". International Association, Hydrological Sciences, Publication No. 7, pp. 48−58, Louvain, Belgium.

Turk, J. T., Campbell, D. H. and Spahr, N. E. (1993). "Use of chemistry and stable sulfur isotopes to determine sources of trends in sulfate of Colorado lakes", *Water, Air and Soil Pollut.*, **67**, 415−431.

Turner, J. V., Macpherson, D. K. and Stokes, R. A. (1987). "The mechanisms of catchment flow processes using natural variations in deuterium and oxygen-18", in *Hydrology and Salinity in the Collie River Basin, Western Australia*, (Eds. A. J. Peck and D. R. Williamson) *J. Hydrol.*, **94**, 143−62.

Unnikrishna, P. V., McDonnell, J. J. and Stewart, M. K. (1995). "Soil water isotopic residence time modelling", in *Solute Modelling in Catchment Systems*, (Ed. S. T. Trudgill), John Wiley, New York. (in press).

Van Stempvoort, D. R. and Krouse, H. R. (1994). "Controls of $\delta^{18}O$ in sulfate: a review of experimental data and application to specific environments", in *Environmental Geochemistry of Sulfide Oxidation*, (Eds. C. N. Alpers and D. W. Blowes) pp. 446−480, ACS Symposium Series 550, American Chemical Society, Washington, D.C.

Vogel, J. C., Talma, A. S. and Heaton, T. H. E. (1981). "Gaseous nitrogen as evidence for denitrification in groundwater", *J. Hydrol.*, **50**, 191−200.

Wallis, P. M. (1979). "Sources, transportation, and utilization of dissolved organic matter in groundwater and streams", Science Series 100, Inland Waters Directorate, Water Quality Branch, Ottawa, Ontario.

Wallis, P. M., Hynes, H. B. N., and Fritz, P. (1979). *Sources, Transportation, and Utilization of Dissolved Organic Matter in Groundwater and Streams*, Science Series 100, Inland Waters Directorate, Water Quality Branch, Ottawa, Ontario.

Waring, R. H. and Schlesinger, W. H. (1985). *Forest Ecosystems Concepts and Management*, Academic Press, London, 340 pp.

Wassenaar, L. I., Aravena, R., Fritz, P. and Barker, J. F. (1991). "Controls on the transport and carbon isotopic composition of dissolved organic carbon in a shallow groundwater system, Central Ontario, Canada", *Chem. Geol.*, **87**, 39−57.

Wels, C., Cornett, R. J. and Lazerte, B. D. (1991). "Hydrograph separation: a comparison of geochemical and isotopic tracers", *J. Hydrol.*, **122**, 253−274.

Wenner, D. B., Ketcham, P. D. and Dowd, J. F. (1991). "Stable isotopic composition of waters in a small Piedmont watershed", in Geochemical Society Special ,Publication No. 3 (Eds. H. P. Taylor Jr, J. R. O'Neil, and I. R. Kaplan, pp. 195−203.

Zimmermann, U., Ehhalt, D. and Munnich, K. O. (1967). "Soil-water movement and evapotranspiration: changes in the isotopic composition of the water", in *Isotopes in Hydrology*, pp. 567−584, IAEA, Vienna.

11

Modelling Upland Stream Water Quality: Process Identification and Prediction Uncertainty

H. S. WHEATER and M. B. BECK

Department of Civil Engineering, Imperial College of Science, Technology and Medicine, London, UK

11.1 INTRODUCTION

Increasing awareness over the last two decades of the regional and transboundary problems of acid deposition and its impact on the terrestrial and freshwater environment in sensitive areas has focused both political and scientific attention on upland water quality. The processes that control stream water quality are complex. They involve the interaction of hydrological, biological and geochemical processes which individually remain poorly understood, despite a relatively large scientific effort in North America and Europe. Indeed, attempts at modelling the process of surface water acidification in upland catchments during the 1980s were to raise profound questions about the role of models in hydrology and aquatic ecology in general (Beck et al. 1990; Beck, 1994). The propagation of contaminants in the subsurface water environment, not least in respect of the containment of hazardous materials, has provoked much self-critical reflection among the community of scientists and engineers working with the development and application of mathematic models (Konikow and Bredehoeft, 1992; McLaughlin et al., 1993).

The use of models can be seen as central to the political and economic decisions that must be made concerning policy on emission controls, as readily apparent in the application of the RAINS model in European negotiations on this subject (Alcamo et al., 1990). In the management context, models provide a predictive capability to simulate response to alternative scenarios, e.g. to define the potential impacts of emission reductions on water quality in sensitive areas.

The scientific role of models is also important, but possibly less obvious. Acidification problems are both complex and multi-disciplinary. This process complexity usually requires the support of models for analysis and interpretation of experimental data, and models are needed as an adjunct to the scientific method, e.g. for hypothesis testing and the reconciliation of concepts with field observations (Beck, 1994). At a basic level, the distillation of hypotheses which is essential to model formulation provides a focus for communication between

Solute Modelling in Catchment Systems. Edited by Stephen T. Trudgill

disciplines; at the very least, the rigours of formulating a model may help to eliminate unsound preconceptions rather more swiftly than might otherwise be the case.

Two key areas of model application have been defined, namely predictive modelling (including the representation of environmental change, as discussed in Beck, 1991, for example) and the use of models in the interpretation of data. However, in practice the modeller is faced with a set of severe difficulties. As noted above, the underlying scientific processes are not well understood, and their model representation will therefore be subject to what is called structural uncertainty as well as to uncertainty due to problems of parameter identification. How can this uncertainty be represented in predictive modelling? In data analysis, limited information is usually available at the scale of observation from which to define a complex set of underlying processes. To what extent, then, can models be used to discriminate between competing hypotheses?

It is in this context that the present chapter seeks to review the modelling of upland stream water quality, focusing in particular on responses to sulphate deposition. However, the issues raised are generic and apply to many problems of the modelling of environmental systems.

11.2 HYDROLOGY AND WATER QUALITY

Stream water quality is determined by the mixing of water that follows different hydrological and chemical pathways (Chen *et al.*, 1983; Mason, 1992). Low flows are commonly associated with a slow stream response and water bearing predominantly inorganic substances that has passed through mineral soil horizons, with water chemistry determined by mineral weathering as well as ion exchange interactions. High flows include more rapid pathways through the upper soil horizons and are generally associated with an increase in organic matter passing to the stream and less opportunity for the neutralisation of acid inputs. The hydrological response of upland streams in acid-sensitive areas to storm (and snowmelt) events is thus associated with rapid changes in water quality, typically a marked reduction of pH and increase in dissolved organic carbon, sulphate and aluminium (Harriman *et al.*, 1990; Seip *et al.*, 1990; Muscutt *et al.*, 1993).

Short-term variations are superimposed on long-term response in which soil chemical changes may occur either over the annual cycle or longer periods. Such changes result from complex biogeochemical processes involving proton transfers which include the role of vegetation in mineral and nutrient cycles and are influenced by climatic and soil water conditions (Reuss and Johnson, 1986). The modelling of long-term change therefore requires appropriate representation of soil and soil water chemistry and, to define stream water quality response, their interaction with runoff processes.

Hence process-based models seek to represent hydrological pathways explicitly, and to associate chemical processes or chemical signatures with these hydrological pathways.

11.3 HYDROLOGICAL MODELS AND A FRAMEWORK FOR MODEL CLASSIFICATION

Given the interdependence of water quality and hydrological processes and the need to represent hydrological processes within models of surface water acidification, the suitability of alternative approaches to hydrological modelling must be considered. In addition, hydrological models have a long history, and this background provides insight into the issues of model identification and uncertainty which were referred to in the introduction (see also Wheater *et al.*, 1993).

11.3.1 Metric models

The history of hydrological modelling can be said to date back to the 1930s and the development of unit hydrograph theory (Sherman, 1932). The stream hydrograph was assumed to comprise two components; a rapid, stormflow response and an underlying baseflow. A simple linear model was sufficient to represent the transfer function linking "hydrologically effective" rainfall and stormflow. Hence the unit hydrograph model, as commonly applied, consists of a rainfall loss model and a linear transfer function. The great strength of this approach is that it allows for analysis of catchment response. Once the effective rainfall and stormflow have been determined for a given event, a unique "unit hydrograph" can be identified. This model was the first of a family of "black box" or metric hydrological models (Wheater *et al.*, 1993) in which the data are allowed to "speak for themselves" (Beck, 1991) with no *a priori* specification of model form. In the intervening period, more sophisticated techniques of time-series analysis have been available with the aim of extracting the maximum information on system response from the input−output signals of catchment rainfall and streamflow, e.g. using the Kalman filter (Beck *et al.*, 1990; Kleissen *et al.*, 1990a) and instrumental variable methods (Whitehead *et al.*, 1979; Jakeman *et al.*, 1990; Young, 1992).

11.3.2 Conceptual models

The advent of digital computers led to the development of more complex hydrological models in which the component hydrological processes were represented by empirical approximations within a pre-specified conceptual framework. The Stanford Watershed Model (Crawford and Linsley, 1966) was a pioneering example, and has continued to have wide-spread use to the present day (it is currently supported by the USEPA as the HSPF model; see Johanson *et al.*, 1980). Conceptual storages within the model are explicitly associated with hydrological features such as upper soil, lower soil and groundwater (and can thus be associated with chemical processes).

Empirical relationships are used to represent the interactions between, and runoff from, the component storages. In general, these are designed to reproduce expected modes of hydrological response, and the relationships are defined by parameters that have physical meaning in a general sense but cannot usually be explicitly linked to properties directly measurable in the field. The resulting model structure is specified by a number of parameters which is likely to be at least 8 or 9, but more typically 10−20.

To operate the models, values must be specified for each of the model parameters. It would be desirable if this could be done on the basis of prior knowledge of the system, but given the indirect link with physically measurable properties, in practice calibration is required to fit the model to data from the system of interest. Model performance must be defined by one or more criteria, selected as appropriate for the model application (e.g. sum of squared errors or the proportion of observed variance explained by the model). Parameters are adjusted to maximise or minimise this criterion (or "objective function").

There is an extensive literature concerning parameter identification in hydrological models (see, for example, Moore and Clarke, 1981; Wheater *et al.*, 1986, 1993). Manual optimisation allows for user experience in guiding the calibration process, but this results in a subjectively determined parameter set. In addition, there is no clear criterion for termination of the optimisation procedure. The alternative is to use automatic methods, which seek to search the parameter space to identify a global optimum model performance, usually defined by a single objective function (although this may be a composite measure).

Traditional approaches to automatic optimisation are based either on a rule-based search algorithm, in which a number of trial parameter sets are evaluated and their relative success is used to direct further search directions (e.g. the simplex method of Nelder and Mead, 1965, or the Shuffled Complex Evolution method of Duan *et al.*, 1992), or on hill-climbing methods. These include gradient-search techniques, which attempt to define the search direction in terms of derivatives of the objective function with respect to the parameters (e.g. the Marquardt algorithm of Marquardt, 1963), and the rotating-coordinate search algorithm of Rosenbrock (1960). Currently, there is considerable interest in the use of controlled random search procedures and their extension in the form of "genetic algorithms" (Wang, 1991).

In practice, severe problems occur. As noted by Wheater *et al.* (1986), the shape of the objective function is determined by the field observations, the model structure and the type of estimator (including assumptions of error structure). With respect to the second of these, it is well known that common structural features such as thresholds, bounding values and parameter interdependence create major problems for optimisation algorithms (Ibbitt, 1970). Thus Pickup (1977), using a 12- parameter version of the Boughton (1965) model, generated a synthetic error-free data series and, using a least-squares objective function, was unable to recover the true parameters using several representative search algorithms.

Wheater *et al.* (1986) and Kleissen (1990) attempted to manipulate the data to improve parameter identification, selecting subsets of the data for which parameter sensitivity is greatest (analogous to the subjective approach of the experienced user). Alternative error criteria were investigated by Chapman (1970), Sorooshian *et al.* (1983) and Sorooshian and Gupta (1983). While some improvement in performance was gained in these various studies and considerable insight into the nature of the identification problem was obtained (Beck *et al.*, 1990; Kleissen *et al.*, 1990a), the essential problem of a complex structure of the objective function response surface remains. In particular, multiple local optima appear to be ubiquitous in models of this level of complexity. It is also commonly the experience that multiple sets of parameters yield essentially similar model performance (e.g. Johnston and Pilgrim, 1976; Beck, 1987; Hooper *et al.*, 1988; Wheater *et al.*, 1989; Beven and Binley, 1992). This obviously presents major problems for any search algorithm but also provides an intractable limitation to model applicability. This is a reflection of the essential problem as discussed by Wheater *et al.* (1986), Beck *et al.* (1990) and Jakeman and Hornberger (1993), for example, that model complexity exceeds the information content of the available data.

If parameters cannot be uniquely identified and alternative sets of parameter values are equally likely, the explicit association between model storages and catchment states is ambiguous. In purely hydrological applications this precludes regional analysis of model parameters, and hence application to ungauged catchments, and renders speculative application to the prediction of effects of catchment change. Where the hydrological model underlies a hydrochemical model, the conceptual association of a given set of chemical processes with hydrological components is similarly undermined.

It will be seen that there is a basic dilemma. The metric approach leads to models that are uniquely identified, but necessarily simple. Usually only two parallel components are identifiable from a rainfall − streamflow input − output data series (Beck *et al.*, 1990; Jakeman and Hornberger, 1993). More complex conceptual models are often intuitively more satisfying, in that they represent features of the catchment considered to be relevant and can usually provide a high quality of fit, but this is achieved at the price of a loss of identifiability. The dilemma has long been recognised (Beck, 1981, 1983); it has led, *inter alia*, to questioning of the usefulness of the concept of model identifiability (Beck, 1994). Studies of upland water quality have been instrumental in these developments, as noted in the introduction and as discussed in more detail subsequently.

11.3.3 Physics-based models

Given sufficient computing power, undoubtedly the case since the late 1970s, more complex representation of component hydrological processes has become possible, and a generation of "physics-based" hydrological models has emerged based on a numerical solution of the relevant equations of motion. Thus a continuum representation is adopted for unsaturated and saturated subsurface flow processes, discretised for numerical solution on a finite difference or finite element grid, with coupling as appropriate to surface flow and vegetation processes (Freeze, 1972; Abbot *et al.*, 1986; Beven *et al.*, 1987; Parissopoulos and Wheater, 1990: Koide and Wheater, 1992; McLaughlin *et al.*, 1993).

Choosing this physical basis of the models intentionally ensures that the model parameters are in principle directly related to observable physical characteristics. For example, in the case of overland and channel flow, surface geometry and roughness are required; for unsaturated subsurface flows, the moisture characteristic curve and unsaturated hydraulic conductivity function are likewise a requirement. However, several important practical and philosophical issues arise, as discussed by Beven (1989) and Wheater *et al.* (1993). Specification of subsurface properties *a priori* necessitates extensive field investigation, and considerable uncertainty in both subsurface geometry and material properties is inevitable, as demonstrated by Stephenson and Freeze (1974).

A scale problem also arises, since the scale of application (the model grid) is usually larger, by orders of magnitude, than the scale on which parameters are measurable. Not only is there insufficient information on upscaling of material properties, but there are also conceptual difficulties in extrapolating a physical representation based on observation of homogeneous materials at small scale to highly heterogeneous large-scale systems. These issues are illustrated by Binley and Beven (1989), who were unable to represent the effects of small-scale variability within an "equivalent" larger-scale grid square. The question arises as to whether the "physics" of the physics- based models is appropriate at the scale of application (Beven, 1989).

The practical implications of these difficulties are that, in application to gauged catchments, calibration of subsurface parameters is essential to reproduce observed streamflow response. Given the distributed representation and complex model formulations, this will potentially result in hundreds if not thousands of parameters that have to be estimated. As with conceptual models, the parameter identification problem becomes indeterminate. Accepted ranges of parameter values are sufficiently large to encompass different modes of response (Stephenson and Freeze, 1974), and inappropriate physical processes may be invoked to represent catchment-scale response (Fawcett, 1992). This obviously presents major problems for hydrochemical representation.

Physics-based models thus have important weaknesses, but also important strengths. They arguably embody our best (albeit imperfect) understanding of component processes, and hence provide an important test-bed with which to explore issues such as catchment change and to quantify, through numerical experiments, issues of aggregation and scale. There is more widely, in physics in general, a move towards the use of highly detailed simulation in the capacity of a third strand of the scientific method, to stand thus alongside theory and experiment (Pool, 1989; Amato, 1991).

11.4 MODELS OF SURFACE WATER ACIDIFICATION

11.4.1 Introduction

The dependence of surface water quality on hydrological flowpaths was noted above, and most of the major hydrochemical models developed to simulate surface water acidification superimpose a chemical model on an underlying hydrological representation. The previous section provided a framework of classification for hydrological models and a perspective on their performance. This framework is extended here to hydrochemical models, and below we discuss their historical development in application to surface water acidification and build on the historical perspective of hydrological models to discuss the strengths and weaknesses of the alternative approaches for hydrochemical application.

11.4.2 First-generation acidification models — the conceptual approach

Historically, the first generation of acidification models was based on a conceptual approach, in which a set of geochemical processes perceived to be the dominant chemical controls were superimposed on a conceptual hydrological framework.

The Birkenes model (Christophersen and Wright, 1981; Christophersen et al., 1982) was developed for the small (0.41 km^2) Birkenes catchment in Norway but has been extensively applied in Scandinavia, the United Kingdom and North America. It comprises two reservoirs which represent upper soil (A) and lower soil (B) respectively. The hydrological response is determined by linear outlets (with thresholds) from the two reservoirs and the hydrological model is specified by eight parameters. The chemical model associates acidic, aluminium-rich stormflow with rapid runoff from the A reservoir, controlled by cation exchange and gibbsite equilibrium. It ignores, therefore, the organic matter content that may be present in this water. Less acidic, more calcium-rich baseflow is supplied by the B reservoir, controlled by mineral weathering reactions and gibbsite equilibrium.

The integrated lake watershed acidification study (ILWAS) model (Gherini et al., 1985) has a more complex structure. The terrestrial component of the hydrological model represents the soil by up to five layers and includes vegetation canopy processes. Vertical exchange between the soil layers and lateral runoff is included, as are overland and streamflow. The model also represent lake processes. The chemistry is based on equilibrium and kinetic expressions. Soil chemistry includes cation exchange and mineral weathering processes, and organic matter decay. Biogeochemical reactions, e.g. nitrification and effects of vegetation growth on ion uptake and release, are also included in the model.

The model of acidification of groundwater in catchments (MAGIC) (Cosby et al., 1985) was designed primarily for long-term applications and runs on an annual time-step (ILWAS and Birkenes typically run on a daily or sub-daily time-step). The focus of the model is thus primarily on long-term chemical changes and the hydrological representation is a simple allocation of fluxes to two soil layers. However, these fluxes can be generated by the use of an ancillary, more complex, hydrological model. The chemical representation includes equilibrium relationships for sulphate sorption, cation exchange, dissolution of gibbsite, aqueous complexation of inorganic aluminium and carbonate speciation, and rate-limited mineral weathering.

The daily time-step enhanced Trickle Down (ETD) model (Nikolaidis *et al.*, 1989) incorporates three soil compartments and represents overland flow, flows between soil compartments and runoff. ANC (acid neutralisation capacity) production is simulated from cation exchange processes in the upper soil compartment, from mineral weathering in the lower soil layers and from sulphate sorption.

11.4.3 Conceptual hydrochemical models — the problem of identifiability

In the discussion of hydrological models, above, the problems of lack of identifiability were raised. For coupled hydrological and chemical models, an important question is whether the additional information provided by chemical signals is able to improve model identifiability or whether, conversely, the additional complexity will exacerbate the underlying problem of lack of identifiability (Beck *et al.*, 1990).

Useful insights can be gained by considering the case of one or more inert chemical tracers, which provide additional information without the burden of an additional set of processes to represent chemical interactions.

11.4.4 *A priori* analysis

Usually, analysis of identifiability is undertaken *a posteriori*, in terms of exploring the features of the objective function used for matching the model's performance with the observed field data (as discussed below). However, Kleissen *et al.* (1990a) present an *a priori* analysis of the identifiability of simple conceptual one- and two-store models using deterministic analysis and stochastic filtering, which illustrates the identifiability problems. In other words, it is possible, before conducting field experiments and implementing a monitoring programme, to determine which observations will enable the model subsequently to be fully identified.

By considering the analytical relationships that constitute the models, the following observations were made.

For a single linear reservoir with no threshold, the model is specified by a single (routing) parameter and an initial condition (the initial storage). These can be identified from precipitation−streamflow data alone. If a threshold storage is incorporated, which must be exceeded for discharge to occur, the model has two parameters and one initial condition. It is evident that the model is not identifiable unless the storage threshold is exceeded and that the threshold cannot be identified from hydrological data alone unless the initial condition is known (which is not usually the case, for it implies measuring an "internal" catchment state variable). However, with the addition of a tracer, the model is fully identifiable.

For two linear reservoirs without thresholds, the model is not identifiable analytically without tracer observations, and then only if the initial conditions are known. A two-store, linear model with thresholds is analogous to the Birkenes model, discussed above. As in the no-threshold case, the model is only identifiable analytically given tracer information and the specification of initial conditions. Furthermore, the identifiability is dependent on the storage states exceeding the threshold values.

In these examples, therefore, chemical tracer data improved model identifiability, but these simple two-store models are even then only identifiable analytically if the initial states are known (which will not in general be the case).

Kleissen *et al.* (1990a) also undertook a stochastic numerical analysis, in which the extended Kalman filter was used to define error variances for model parameters, using synthetic data. The results are thus data-specific, but reinforce the analytical conclusions. In general, the results confirm that the addition of a single tracer improves the identifiability of a two-store model, as demonstrated by a reduction in parameter estimation variance, and show that parameter interdependence is influenced by the characteristics of the precipitation−flow sequence. For example, a single tracer reduces the degradation of parameter identifiability from purely flow data which occurs between storm events. The contribution of a second tracer is limited, but provides some improvement in identifiability results when output concentrations of the first tracer are low.

11.4.5 *A posteriori* identifiability of the Birkenes model

The hydrological identifiability of the Birkenes model using non-linear optimisation methods was investigated by Wheater *et al.* (1986), and the familiar problems of parameter response surface structure, discussed above, were encountered. As a response to this, a novel optimisation method (the segmented optimisation strategy) was developed and did indeed show some improvement in identifiability by selection of subsets of the data series for which parameter sensitivity is greatest. Elsewhere (Beck *et al.*, 1987), work with both conceptual and metric models using the original Birkenes field data showed clearly the lack of identifiability of the hydrochemical hypothesis of the Birkenes model. In particular, the use of a daily sampling frequency to monitor a catchment whose response occurs over just a few hours is problematic.

Analysis of the identifiability of the Birkenes model was extended by de Grosbois *et al.* (1988) who considered the effect of tracer data on model identifiability using synthetic, error-free data. The Marquardt algorithm was used, with multiple seed points, and hence a performance criterion was required that included both hydrological and chemical information. A weighted least-squares objective function was used, and different relative weightings were explored for the hydrological and chemical data. A criterion of success was the percentage of searches for which the final parameters lay within 10% of the "true" values. For hydrological information alone, this gave 33%, and a similar result was obtained for equal weighting of flow and chemistry. However, when the (relatively sparse) tracer data were weighted by a factor of 20, the score improved to 80%, i.e. clearly demonstrating improved identifiability.

This strategy was therefore adopted by Hooper *et al.* (1988), in application to data of streamflow and $\delta^{18}O$ from the Birkenes catchment. Snow-free data from each of 3 years were used. It was found that inclusion of $\delta^{18}O$ reduced the number of aborted searches and, unsurprisingly, the inclusion of $\delta^{18}O$ data in the calibration process improved the model performance with respect of tracer simulation. However, the quality of hydrological simulations was reduced for two years out of the three.

Two important conclusions came from this study. While convergence to a consistent objective function was achieved, a wide range of equally good parameter sets was obtained. Hence the model was non-identifiable. The authors concluded that "only one passive store, not two, is observable from that data", a result consistent with that of Beck *et al.* (1987). Secondly, while the model had been widely used with good results for sulphate, the performance with a conservative tracer was poor. Not only was the model non-identifiable, but it had failed an important performance test.

11.4.6 Second-generation acidification models — mixing model approach

If conceptual models cannot discriminate between alternative parameter sets, it is evident that they cannot be used in a rigorous sense for hypothesis testing. This gives rise to fundamental problems in the application of the scientific method to the interpretation of catchment-scale data.

If thus improved identifiability is the goal, simpler models are required. To use the notation defined for hydrological models, a move towards the metric approach is needed. This can be addressed within the framework of a hybrid metric−conceptual model (Wheater *et al.*, 1993), or more directly in a metric approach based on data analysis.

The analysis of hydrochemical time-series, as for hydrological time-series, requires simplifying assumptions. In the case of the unit hydrograph analysis, the assumption of a two-component representation of streamflow and a linear model allows unique identification of the catchment response. The simplification commonly applied to hydrochemical analysis is the mixing model (e.g. Christophersen *et al.*, 1990; Hooper *et al.*, 1990: Kleissen *et al.*, 1990b; Neal *et al.*, 1990; Robson and Neal, 1990), although, of course, it is perfectly feasible to develop the direct equivalent of a unit hydrograph model, such as the precipitation−stream chemistry models of Beck *et al.* (1987).

The mixing model approach assumes that the water in the stream can be represented as a mixture of waters of fixed chemistry, the so-called "end members", and that this mixing is conservative (or approximately conservative). To identify the flow components q_1 and q_2 associated with two end members of constant concentration c_1 and c_2 respectively, we have

$$q_1 = q_s \left(\frac{c_s - c_2}{c_2 - c_1} \right)$$

and

$$q_2 = q_s \left(\frac{c_1 - c_s}{c_2 - c_1} \right)$$

where q_s is the observed stream discharge and c_s the observed stream concentration (Kleissen *et al.*, 1990b). An analysis based solely on stream data thus required an assumption of the end member concentrations c_1 and c_2. These can be selected from the highest and lowest observed concentrations, but in general q_1/q_2 is not sensitive to the assumed c_1, c_2 concentrations unless c_1 or $c_2 \cong c_s$.

The important element of the analysis is the interpretation of the transient portion of the observed record, when the storm response is occurring. Thus, for example, Kleissen *et al.* (1990b) and Wheater *et al.* (1990b) report an analysis of the 10 km^2 Allt a'Mharcaidh catchment in Scotland where individual chemical species were investigated (Si, Ca, Mg, Na, Cl, total organic carbon (TOC) and alkalinity). The analysis was able to show that event response was primarily due to low-alkalinity, high-TOC water and baseflow from a high-alkalinity source. However, inputs of high-alkalinity water could also be identified during large events (suggesting a displacement process). Results for the other species were more complex, showing variability between species and as a function of event characteristics, thus pointing to the potential significance of spatial heterogeneity and the mode of runoff generation.

Extensions of this concept include multi-component species analysis (Christophersen *et al.*, 1990; Hooper *et al.*, 1990) and investigation of the relationship between end members and observed soil water chemistries. For the Birkenes catchment, Christophersen *et al.* (1990) investigated the mixing of waters from O/H and B/C hillslope soil horizons and from valley bottom locations. In general, stream water inorganic aluminium and hydrogen concentrations could be explained by a mixing of these three components, whereas for calcium and silica an additional, unobserved, source water had to be invoked. Similarly, for Plynlimon (mid Wales), an unobserved high-ANC source was postulated to explain stream water chemistry. However, Hooper *et al.* (1990) were able to explain between 82 and 97% of observed stream water quality variance by mixing of three observed component stream waters (denoted "hillslope', "groundwater" and "organic").

In an application to four small catchments at Llyn Brianne (mid Wales), Robson and Neal (1990) used a similar streamflow analysis to Kleissen *et al.* (1990b), discussed above. Streamflow was clearly a mix of shallow (soil) waters and a deep (unobserved) end member, but a wide range of observed soil waters precluded the simple definition of a single soilwater end member. The analysis also revealed systematic differences between the catchments, associated with land use.

It can be seen that mixing models provide a simple and effective tool for data analysis, and can yield important insights into catchment response. However, their limitations must be borne in mind. It is obvious that chemical interactions cannot be investigated and alternative approaches must be considered. It is also apparent that considerable caution must be exercised in the association of stream water chemistry with implied source waters. There are a number of examples where stream water quality can only be explained in terms of an unobserved water source (e.g. Christophersen *et al.*, 1990; Kleissen *et al.*, 1990b; Neal *et al.*, 1990; and Wheater *et al.*, 1990a). In addition, recent research has demonstrated spatial and temporal complexity which includes important chemical changes along flowpaths. Bishop *et al.* (1990) demonstrate for a Swedish catchment that origins of runoff and sources of acidity are complex and may be highly localised. At the C2 catchment in Plynlimon, central Wales, flows generated from a network of ephemeral natural soil pipes undergo major chemical changes en route to the main stream due to release of base cations and associated consumption of protons within the drift material at the head of the stream (Muscutt *et al.*, 1990, 1993; Chapman *et al.*, 1993a, 1993b). Furthermore, once it has entered the stream, water may in fact pass back into the soils and rock media in the riparian zone adjacent to and below the stream as well as interacting with in-stream and riparian vegetation (Bencala *et al.*, 1993). This points to the need for careful integration of catchment analysis through model application and detailed field process observations.

11.4.7 Third-generation modelling — the "physics-based" approach and synthetic applications

The physics-based approach to hydrological modelling is based on a continuum representation of subsurface processes. Issues of heterogeneity and effective parameters were discussed earlier. In the context of hydrochemical modelling these problems are exacerbated.

The classical continuum-based approach to the modelling of subsurface transport is to combine solution of the flow field with an advection−dispersion representation of transport processes (Bear, 1979) in which solute "particles" are advected with the flow field and disperse due to a molecular diffusion and hydrodynamic mixing processes. However, in the context of groundwater transport, it is widely recognised that there are conceptual difficulties with the conventional representation of dispersivity, which has been shown in practice to be scale-dependent (Anderson, 1979; Gelhar, 1986). In effect, a contaminant plume experiences increasing scales of heterogeneity with distance travelled. These problems can be overcome if effects of small-scale heterogeneity are explicitly recognised (e.g. Tompkins *et al.*, 1994), but the fine detail can only be represented within a stochastic framework that can account for the inevitable uncertainty associated with the definition of fine-scale structure.

Where soil water processes are concerned, the effects of heterogeneity are likely to be more severe, and this is particularly so for upland catchments, which are the primary focus of surface water acidification studies. Upland soils are generally shallow and have complex and heterogeneous physical and chemical structures, for example, associated with the processes of podzolisation. Superimposed on this are other environmental controls, such as biological activity (growth and decay of root systems; burrowing animals and earthworms) and hydrological and climatic effects on soil structure. One of the consequences is the widespread occurrence of macroporosity, reviewed by Beven and Germann (1982). There are a host of field studies that have observed dominant effects of preferred flowpaths, from early studies of Whipkey (1965), Pilgrim *et al.* (1978) and Mosley (1982), to more recent acidification research (e.g. Wheater *et al.*, 1990a; Muscutt *et al.*, 1990, 1993; Chapman *et al.*, 1993a, 1993b).

Fundamental understanding of the flow processes in such systems is limited (although models have been developed to represent macropore flow; (e.g. Jarvis *et al.*, 1991) and the problem of predicting their occurrence and connectivity remains intractable. Potential effects on chemical transport are evident. Macropore flow can provide rapid transmission of flow and solutes, and can bypass soil−solute interactions. However, the interactions between matric soil water and macropore flow and the chemical exchanges that take place between macropore flow and the macropore surface are largely unknown.

The above discussion has considered subsurface flow and transport processes, and has, at least, cast doubt on the adequacy of available physics-based subsurface representations, particularly for upland catchments. However, for catchment-scale applications, a wider set of process interactions must be considered, including surface/subsurface flows. The work of Binley and Beven (1989) demonstrated that the greatest difficulty in defining effective parameters for a physics-based hydrological model occurred where overland flow and subsurface flow interactions took place. The relative occurrence of different flow processes is difficult to predict and is likely to be strongly influenced by small-scale heterogeneity. In a classical study of a small, apparently uniform hillslope plot in California, Pilgrim *et al.* (1978) observed that all of the recognised runoff generation processes occurred, but that their relative importance could not be predicted *a priori* and could only be evaluated by intensive field investigation. However, the relative contribution of different flow mechanisms is fundamental in determining stream chemistry, as clearly shown, for example, by detailed experimental studies at the C2 catchment at Plynlimon (Muscutt *et al.*, 1990, 1993; Chapman *et al.*, 1993a, 1993b). The streamflow consists of transmitted pipe flow, saturation excess overland flow and localised and diffuse throughflow sources in a combination that varies with storm characteristics and antecedent conditions and that determines the streamwater chemistry. In short, description of the many paths taken by water moving through a catchment to the stream outlet may or may not be important for forecasting stream discharge. Such a description, however, is vital to the prediction of stream water quality.

It is evident that physics-based models must be considered with great care. Application to existing catchments must recognise the uncertainty inherent in the representation and parameterisation of component processes and their interactions. They must also accommodate the often dominant influence on stream water quality of "unobserved" source waters. However, as noted earlier, such models have real strengths which can be exploited.

One of the methodological difficulties that underlies model application is the lack of appropriate observational methods at the larger scales of interest (Wheater *et al.*, 1993). One way forward is to use numerical experiments (synthetic modelling) to explore these issues. In such applications, a model is used with assumed error-free parameters to generate an artificial catchment response for subsequent analysis. Relevant hydrological examples include, for example, those of Binley *et al.* (1989), Binley and Beven (1989), Zhu *et al.* (1990) and Wood *et al.* (1990).

In the area of chemical process modelling, the same issues arise. The chemical relationships that are used in the hydrochemical process models introduced above are derived from small-scale observation of homogeneous materials. The validity of such relationships and their parameterisation when applied to heterogeneous field systems at larger scales is open to question. As noted by Neal and Robson (1994), "Presently there is no adequate thermodynamic theory that . . . shows how bulked chemical properties can be linked to equilibrium or kinetic-type equations". However, synthetic modelling has recently been applied and offers some important insights. For example, Neal (1992) takes a set of simple cation exchange relationships for a homogeneous soil water system and considers the ensemble response to a mixture of soil solutions of different strong anion content (including speciation changes). Analysis of the ensemble response shows a systematic deviation from the underlying theoretical relationship. Neal and Robson (1994) show for an acid soil that such lumped response can be described by analogous relationships, at least for long-term changes, if not for short-term response.

The case of real catchments is more complex, since there is a mixing of waters with weathering influence (Taugbol and Neal, 1994), as already noted, but this methodology has obvious considerable further potential.

11.5 FUTURE DIRECTIONS

It has already been noted that models have two distinct types of application, i.e. predictive modelling (in association with questions of environmental management) and data analysis (as part of the scientific method).

11.5.1 Data analysis and the scientific method

It will be evident that the use of conceptual models is of limited benefit for hypothesis testing. Wheater *et al.* (1990b) note that since parameter ambiguity is inevitable in a model which represents a "realistic" conceptualisation, such a model cannot provide a rigorous test. However, a failure of the model can be regarded as a failure of the hypotheses, and success denotes some conditional support for the physical interpretation. Perhaps the major benefit of conceptual models is that they provide a direct and accessible extension to the intellectual process of data analysis (a non-trivial benefit, which in turn may be capable of providing pointers to the reformulation of hypotheses at a finer scale of representation; see Beck, 1987, 1994).

A rigorous test requires an identifiable model, which, as has been seen, can only be achieved by model simplification. Mixing models are only amenable to use with conservative (or approximately conservative) chemical species, but have nevertheless provided powerful insights into catchment hydrochemical response. However, their extension to establish a causal relationship with contributing soil water is subject to possible ambiguity. Detailed observational data of spatial heterogeneity of subcatchment scale is extremely limited, but significant chemical changes along flowpaths have been identified for one of the more detailed catchment studies.

"Physics-based" models have, as yet, only a limited role to play. They are non-identifiable in application to data analysis and can provide incorrect process representation. In particular, probably the greatest doubt over the appropriateness of hydrological model conceptualisation is in application to upland catchments, for which effects such as preferred flowpaths have been observed to predominate (Wheater *et al.*, 1987, 1990a). These effects are important with respect to the modelling of flows (Koide and Wheater, 1992), but are likely to be of greater significance for solute transport. In general, the effect of heterogeneity is known to cast doubt on the applicability of conventional dispersion models for solute transport (Tompkins *et al.*, 1994). In the acidification context, non-observable geochemical processes add further complexity. For example, many study catchments have low flow chemistry which is unexplained by observed source waters (Harriman *et al.*, 1990; Neal *et al.*, 1990; Christophersen *et al.*, 1990). There are also the fundamental questions of an appropriate chemical conceptualisation for heterogeneous systems, discussed above.

The role of "physics-based" models is, at present mainly confined to synthetic simulation. As already noted, this is important for exploring the implications of both hydrological and geochemical heterogeneity, and there is considerable scope for further work in this area. Future applications to gauged catchments is likely to require conditional simulation, in which prior knowledge of hydrological and geochemical response is incorporated, to constrain the range of acceptable parameter values. A framework has been identified by Beven and Binley (1992), but much further research is needed.

11.5.2 Modelling for environmental management

In the introduction, we noted the key role of modelling in the economic and political decisions that must be made concerning environmental management in general and acid rain in particular. The limitations of models have been exposed. Two extreme alternatives are either to reject the use of models as unreliable and potentially misleading or to accept model predictions as a "best guess" of future response (e.g. as formalized within an expert system by Lam *et al.*, 1992). However, a more fruitful approach is to recognise the problems and look for appropriate innovation in modelling methods to minimise prediction uncertainty.

In a classical uncertainty analysis, parameter uncertainty is recognised, and by sampling from parameter distributions, specified *a priori*, this parameter uncertainty can be propagated through the model to define prediction uncertainty. There is, however, a significant risk that the uncertainty will be so great as to render the prediction meaningless in policy terms. In principle, what is required is a method of conditioning the prior estimates of parameter values to reduce prediction uncertainty (Beck, 1983, 1987; Beven and Binley, 1992, for example).

In the context of long-term change of water quality due to acid deposition, appropriate long time-series data to constrain model performance are limited or unavailable. An innovative solution was developed by Hornberger *et al.* (1989), in which extensive spatial data were substituted for the unavailable temporal information in an application of the MAGIC model to Scandinavian data. Simulations were run for the period 1844−1974, using Monte Carlo analysis in which soil parameters were independently sampled from broad ranges of acceptable values, and input (atmospheric deposition) parameters were sampled to retain observed dependence (of chloride and sulphate). Simulations were screened as acceptable or unacceptable, in comparison with the observed distribution of lake water quality from a 1974 sampling of 464 lakes, and 3000 acceptable simulations were generated. Ten re-sampled sets of 1000 simulations were used to define the precision of the estimates, and the re-sampled sets were independently weighted to match the observed joint distribution of alkalinity, calcium and sulphate. These weighted sets represented the conditioned model performance and were then tested against a 1986 spatial data set and used to investigate effects of future scenarios. Minor changes in water quality between 1974 and 1986 were shown by both model and data and, within this limitation, the magnitude of change was quite well reproduced by the model, with the exception of alkalinity. However, an important difference was that the model simulation variance was considerably less than the observed. As noted by the authors, this could reflect non-modelled spatial or temporal variability, data uncertainties or model structural error.

An attempt to include effects of model structural error on prediction uncertainty was made by Rose *et al.* (1991a, 1991b) who undertook an intercomparison of the MAGIC, ILWAS and ETD models. A central problem for this exercise was the production of a consistent set of inputs and parameters across models of widely different complexity, described by Rose *et al.* (1991a). Having achieved this through an "input mapping" procedure, the models were run on two basins in the northeastern United States, with generally similar results.

To represent regional variability, distributions of parameter values were defined, based around the two northeastern United States basins, and hence with some prior conditioning. A Monte Carlo simulation was undertaken (based on Latin hypercube sampling). Comparisons between models under current deposition conditions showed, for example, similar shaped regional cumulative frequency distributions for ANC, but some displacement of the curves for the different models. The exercise was repeated for different atmospheric sulphate loadings. An important result was that all models showed similar shifts in median ANC, but while ILWAS showed a marked increase in variance under high sulphate loading, the MAGIC and ETD models did not. As the predicted responses to sulphate stress were of similar magnitude to the inter-model differences, the absolute differences between models were large. For example, under the high-sulphate deposition scenario MAGIC predicted 97% of lakes with negative ANC, in comparison with 5% using ETD.

The inter-comparison thus powerfully illustrates the strengths and limitations of current modelling practice. A high measure of consistency in the nature of the predicted change is a reassuring commentary on the ability of the different models to represent process response. The absolute differences are an important reminder of the dangers of over-literal reliance on quantitative results.

11.6 CONCLUSIONS

Hydrogeochemical models have been shown to present an essentially similar set of problems to those of purely hydrological models. Only simple models are uniquely identifiable. A conceptual representation of processes that is intuitively "realistic" will be non-identifiable. Chemical tracers provide additional information for hydrological model identification, but the inclusion of interactive chemistry adds substantially to the problems of identifiability, since there is uncertainty with respect to the structure of the chemical components in addition to their parameter values.

Physics-based hydrological models have at present a limited role in catchment analysis. *A priori* model definition is impractical due to uncertainty in subsurface structure and parameter values without some form of *a posteriori* conditioning to constrain parameter values. However, such models provide an important vehicle for the study of effects of heterogeneity and effective parameters, and this approach of synthetic simulation is currently being extended to investigate chemical heterogeneity, with promising results (Neal, 1992; Neal and Robson, 1994).

Despite the problems, models are needed to support decisions in the context of environmental management. They can be regarded as an expert system to guide the decision-making process, or an attempt can be made to quantify uncertainty, which will exist with respect to model inputs, model parameters and model structure. A classical analysis of the effects of parameter uncertainty is likely to be unhelpful, given the anticipated large uncertainty bounds. Some constraints are needed, and in principle these can be included by *a posteriori* conditioning of parameter distributions. In the context of environmental change in general, and long-term effects of acidification in particular, appropriate data for conditioning are limited. An innovative solution to the problem of parameter conditioning has been to substitute spatial data in the analysis of regional distributions, with limited success (Hornberger *et al.*, 1989). A solution to the problems of model structural uncertainty as well as parameter uncertainty has been to undertake model intercomparisons on a regional basis. The results (Rose *et al.*, 1991b) provide important insights into the strengths and limitation of modelling for environmental management. At a general level, three models give a consistent interpretation of the nature of expected regional change; at a specific level of absolute differences in quantitative results, seen in comparison with limit values, large differences between model predictions occurred.

With respect to data analysis and the scientific method, a clear understanding of the limitations of conceptual models has emerged. One solution has been to move towards simple mixing models as a vehicle for data analysis on a catchment scale. While important insights have been gained through the analysis of stream water time-series, the limitations of this approach must also be appreciated. Obviously, mixing models (at least, in their present form) cannot represent interactive chemistry. It is also apparent that care must be taken in the interpretation of catchment processes. Some authors (e.g. Hooper *et al.*, 1990) have been able to explain fully stream water quality in terms of observed source waters within the catchment, but many others have had to invoke an unobserved source (Christophersen *et al.*, 1990; Kleissen *et al.*, 1990b; Neal *et al.*, 1990; *inter alia*). Spatial variability of hydrochemical response is also an important, and largely unexplored, issue on a subcatchment scale. Robson and Neal (1990) note a wide range of observed soil water chemistries at subcatchment scales. Muscutt *et al.* (1993) and Chapman *et al.* (1993a, 1993b) observed important changes in water chemistry along a dominant flowpath and a complex spatial temporal variability of hydrological and chemical response, for a small catchment in mid Wales.

It is evident that there is no neat prescription for scientific progress. Above all, the key is to provide an insight into the dominant hydrochemical processes at the scale of interest. This cannot be achieved without good-quality experimental data, necessarily on a field scale, and at appropriate temporal and spatial resolution. Such data require intensive instrumentation and manpower commitment, and are hard won. Much more work is needed in this area.

The interpretation of data then requires the full range of modelling methods (Beck *et al.*, 1993). The maximum insight must be derived from analysis, which requires the application of simple models. However, model limitations will require more complex formulations to explore issues of process complexibility and spatial heterogeneity. Conceptual models are an important intellectual aid in the formulation and quantification of hypotheses, and, as already noted, can be used to reject hypotheses or provide conditional support (Wheater *et al.*, 1990b). "Physics-based" models must be pursued to develop the theoretical basis for both hydrological and geochemical process representation for heterogeneous systems.

The issues underlying the modelling of upland water quality are common to a wide range of problems of environmental management and environmental science. The political impetus of acid rain has provided an important stimulus, and new insights have been gained into the use of models and the application of the scientific method. However, it will be readily apparent that considerable challenges remain. The hope is that the important recent progress will be continued and that the interplay between models and data (and modellers and experimentalists) can be fruitfully managed to maximise benefits to scientific progress and the quantification of issues of environmental management.

REFERENCES

Abbot, M. B., Bathhurst, J. C., Cunge, J. A., O'Connell, P. E. and Rasmussen, J. L. (1986). "An introduction to the European Hydrology System SHE, 2: structure of a physically-based, distributed modelling system", *J. Hydrol.*, **87**, 61−77.

Alcamo, J., Shaw, R. W. and Hordijk, L. (Eds.) (1990). *The RAINS Model of Acidification: Science and Strategies in Europe*, Kluwer, Dordrecht.

Amato, I. (1991). "Speculating in precious computronium", *Science*, **253**, (23 August), 856−857.

Anderson, M. P. (1979). "Using models to simulate the movement of contaminants through groundwater flow systems", *Crit. Rev. in Environmental Control*, **9**, 97−156.

Bear, J. (1979). *Hydraulics of Groundwater*, 567 pp, McGraw- Hill.

Beck, M. B. (1981). "Hard or soft environmental systems?", *J. Ecological Modelling*, **11**, 237−251.

Beck, M. B. (1983). "Uncertainty, system identification and the prediction of water quality", in *Uncertainty and Forecasting of Water Quality* (Eds. M. B. Beck and G. van Straten), pp. 3−68, Springer, Berlin.

Beck, M. B. (1987). "Water quality modelling: a review of the analysis of uncertainty", *Water Resour. Res.*, **23**(8), 1393−1442.

Beck, M. B. (1991). "Forecasting environmental change", *J. Forecasting*, **10**, 3−19.

Beck, M. B. (1994). "Understanding uncertain environmental systems", in *Predictability and Nonlinear Modelling in Natural Sciences and Economics* (Eds. J. Grasman and G. van Straten), pp. 294−311, Kluwer, Dordrecht.

Beck, M. B., Drummond, D., Kleissen, F. M., Langan, S. J., Wheater, H. S. and Whitehead, P. G. (1987). "Modelling streamwater response to acid deposition: the Birkenes data revisited", in *Systems Analysis in Water Quality Management* (Ed. M. B. Beck), pp. 133−150, Advances in Water Pollution Research Series, Pergamon, Oxford.

Beck, M. B., Kleissen, F. M. and Wheater, H. S. (1990). "Identifying flow paths in models of surface water acidification", *Rev. Geophys.*, **28**(2), 207−230.

Beck, M. B. Jakeman, A. J. and McAleer, M. J. (1993). "Construction and evaluation of models of environmental systems", in *Modelling Change in Environmental Systems* (Eds. A. J. Jakeman, M. B. Beck and M. J. McAleer), pp. 3–35, John Wiley, Chichester.

Bencala, K. E., Duff, J. H., Harvey, J.W., Jackman, A. P. and Triska, F. J. (1993). "Modelling within the stream-catchment continuum", in *Modelling Change in Environmental Systems* (Eds. A. J. Jakeman, M. B. Beck and M. J. McAleer), pp. 163–187, John Wiley, Chichester.

Beven, K. J. (1989). "Changing ideas in hydrology: the case of physically-based models", *J. Hydrol.*, **105**, 157–172.

Beven, K. J. and Binley, A. M. (1992). "The future of distributed models: calibration and predictive uncertainty', *Hydrolog. Process.*, **6**, 279–298.

Beven, K. J. and Germann, P. (1982). "Macropores and water flow in soils", *Water Resour. Res.*, **18**(5), 1311–1325.

Beven, K. J., Calver, A. and Morris, E. M. (1987). *The Institute of Hydrology distributed model*, Report 98, Institute of Hydrology, Wallingford, UK.

Binley, A. M. and Beven, K. J. (1989). "A physically based model of heterogeneous hillslopes. 2. Effective hydraulic conductivities", *Water Resour. Res.*, **25**(6), 1227–1233.

Binley, A. M., Elgy, J. and Beven, K. J. (1989). "A physically based model of heterogeneous hillslopes. 1. Runoff production", *Water Resour. Res.*, **25**(6), 1219–1226.

Bishop, K. H., Grip, H. and O'Neill, A. (1990). "The origins of acid runoff in a hillslope during storm events", *J. Hydrol.*, **116**, 35–61

Boughton, W. C. (1965). *A new simulation technique for estimating catchment yield*, Water Research Laboratory Report 78, University of New South Wales.

Chapman, T. G. (1970). "Optimization of rainfall-runoff model for an arid zone catchment", IAHS Symposium on *Results of Research on Representative and Experimental Basins, Wellington*, Vol. **I**, pp. 126–143.

Chapman, P. J., Reynolds, B. and Wheater, H. S. (1993a). "Hydrochemical changes along stormflow pathways in a small moorland headwater catchment in Mid-Wales, UK", *J. Hydrol.*, **151**, 241–265.

Chapman, P. J., Wheater, H. S. and Reynolds, B. (1993b). "The effect of geochemical reactions along flowpaths on stormwater chemistry in headwater catchments", in *Tracers in Hydrology: Proceedings of the International Symposium, Yokohama, July 1993* (Eds. N.P. Peters *et al.*) pp. 23–30, IAHS, Vienna.

Chen, C. W., Gherini, S. A., Hudson, R. J. M. and Dean, J. O. (1983). "The integrated lake-watershed acidification study. Model principles and application procedures", EPRI EA-3221, Vol. I, Project 1109-5, Electric Power Research Institute, Pala Alto, California.

Christophersen, N. and Wright, R. F. (1981). "Sulphate budget and a model for sulphate concentrations in stream water at Birkenes, a small forested catchment in southernmost Norway", *Water Resour. Res.*, **17**, 377–389.

Christophersen, N., Seip, H. M. and Wright, R. F. (1982). "A model for streamwater chemistry at Birkenes, Norway", *Water Resour. Res.*, **18**(4), 977–996.

Christophersen, N., Neal, C., and Hooper, R. P., Vogt, R. D. and Andersen, S. (1990). "Modelling streamwater chemistry as a mixture of soilwater end-members — a step towards second-generation acidification models", *J. Hydrol.*, **116**, 307–320.

Cosby, B. J., Wright, R. F., Hornberger, G. M. and Galloway, J. N. (1985). "Modelling the effects of acid deposition, estimation of long-term water quality responses in a small forested catchment", *Water Resour. Res.*, **21**, 1591–1601.

Crawford, N. H. and Linsley, R. K. (1966). *Digital simulation in hydrology: Stanford Watershed Model IV*, Technical Report 39, Stanford University, California.

de Grosbois, E., Hooper, R. P. and Christophersen, N. (1988). "A multisignal automatic calibration methodology for hydrochemical models: a case study of the Birkenes model", *Water Resour. Res.*, **24**, 1299–1307.

Duan, Q., Sorooshian, S. and Gupta, V. (1992). "Effective and efficient global optimization for conceptual rainfall-runoff models", *Water Resour. Res.*, **28**(4), 1015–1031.

Fawcett, K. R. (1992). "Hydrological physically based distributed models: parameterisation and process representation", Unpublished PhD thesis, University of Bristol.

Freeze, R. A. (1972). "Role of subsurface flow in generating surface runoff. 2: upstream source areas", *Water Resour. Res.*, **8**(5), 1272–1283.

Gelhar, L. W. (1986). "Stochastic subsurface hydrology from theory to application", *Water Resour. Res.*, **22**(9), 135s–145s.

Gherini, S. A., Mok, L., Hudson, R. J. M., Davis, G. F., Chen, C. W. and Goldstein, R. A. (1985). "The ILWAS model: formulation and application", *Water, Air and Soil Pollut.*, **26**, 425–459.

Harriman, R., Ferrier, R. C., Jenkins, A. and Miller, J. D. (1990). "Long- and short-term hydrochemical budgets in Scottish catchments", in *The Surface Waters Acidification Programme*, (Ed. B. J. Mason) pp. 31–43, Cambridge University Press.

Hooper, R. P., Stone, A., Christophersen, N., de Grosbois, E. and Seip, H. M. (1988). "Assessing the Birkenes model of stream acidification using a multisignal calibration methodology", *Water Resour. Res.*, **24**, 1308–1316.

Hooper, R. P., Christophersen, N. and Peters, N. E. (1990). "Modelling streamwater chemistry as a mixture of soilwater end members — an application to the Panola mountain catchment, Georgia, USA", *J. Hydrol.*, **116**, 321–343.

Hornberger, G. M., Cosby, B. J. and Wright, R. F. (1989). "Historical reconstructions and future forecasts of regional surface water acidification in southernmost Norway", *Water Resour. Res.*, **25**(9), 2009–2018.

Ibbitt, R. P. (1970). "Systematic parameter fitting for conceptual models of catchment hydrology", Unpublished PhD thesis, University of London.

Jakeman, A. J. and Hornberger, G. M. (1993). "How much complexity is warranted in a rainfall-runoff model?", *Water Resour. Res.*, **29**, 2637–2649.

Jakeman, A. J., Littlewood, I. G. and Whitehead, P. G. (1990). "Computation of the instantaneous unit hydrograph and identifiable component flows with application to two small upland catchments", *J. Hydrol.*, **117**, 275–300.

Jarvis, N. J., Jansson, P. E., Dik, P. E. and Messing, I. (1991). "Modelling water and solute transport in a macroporous soil. I. Model description and sensitivity analysis", *J. Soil Sci.*, **42**, 59–70.

Johanson, R. C., Imhoff, J. C. and Davis, H. H. (1980). *Users manual for hydrological simulation program — FORTRAN (HSPF)*, Report EPA - 600/9-80-015, US Environmental Protection Agency.

Johnston, P. R. and Pilgrim, D. H. (1976). "Parameter optimization for watershed models", *Water Resour. Res.*, **12**(3), 477–486.

Kleissen, F. M. (1990). "Uncertainty and identifiability in conceptual models of surface water acidification", Unpublished PhD thesis, Imperial College, University of London.

Kleissen, F. M., Beck, M. B. and Wheater, H. S. (1990a). "Identifiability of conceptual hydrochemical models", *Water Resour. Res.*, **26**(2), 2979–2992.

Kleissen, F. M., Wheater, H. S., Beck, M. B. and Harriman, R. (1990b). "Conservative mixing of water sources: analysis of the behaviour of the Allt a'Mharcaidh catchment", *J. Hydrol.*, **116**, 365–374.

Koide, S. and Wheater, H. S. (1992). "Subsurface flow simulation of a small plot at Loch Chon, Scotland", *J. Hydrolog. Process.*, **6**, 299–326.

Konikow, L. F. and Bredehoeft, J. D. (1992). "Groundwater models cannot be validated", *Adv. Water Resour.*, **15**(1), 75–83.

Lam, D. C. L., Wong, I., Swayne, D. A. and Storey, J. (1992). "A knowledge-based approach to regional acidification modelling", *Environmental Monitoring and Assessment*, **23**, 83–97.

McLaughlin, D. B., Kinzelbach, W. and Ghassemi, F. (1993). "Modelling subsurface flow and transport", in *Modelling Change in Environmental Systems* (Eds. A. J. Jakeman, M. B. Beck and M. J. McAleer), pp. 133–161, John Wiley, Chichester.

Mason, B. J. (1992). *Acid Rain: Its Causes and Effects on Inland Waters*, Clarendon Press, Oxford, 126 pp.

Marquardt, D. W. (1963). "An algorithm for least squares estimation of non-linear parameters", *J. Soc. for Ind. Applic. of Mathematics*, **2**, 431–441.

Moore, R. J. and Clarke R. T. (1981). "A distribution function approach to rainfall-runoff modelling", *Water Resour. Res.*, **17**(5), 1376–1382.

Mosley, M. P. (1982). "Subsurface flow velocities through selected forest soils, South Island, New Zealand", *J. Hydrol.*, **55**, 65–92.

Muscutt, A. D., Wheater, J. S. and Reynolds, B. (1990). "Stormflow hydrochemistry of a small Welsh upland catchment", *J. Hydrol.*, **116**, 239–249.

Muscutt, A. D., Reynolds, B., Wheater, H. S. (1993). "Sources and controls of aluminium in storm runoff from a headwater catchment in mid-Wales", *J. Hydrol.*, **142**, 409−425.

Neal, C. (1992). "Describing anthropogenic impacts on stream water quality: the problem of integrating soil water chemistry variability", *Sci. Total Environ.*, **115**, 207−218.

Neal, C. and Robson, A. J. (1994). "Integrating soil water chemistry variations at the catchment level within a cation exchange model", *Sci. Total Environ.*, **144**, 93−1020.

Neal, C., Smith, C. J., Walls, J., Billingham, P., Hill, S. and Neal, M. (1990). "Hydrogeochemical variations in Hafren Forest stream waters, Mid-Wales", *J. Hydrol.*, **116**, 185−200.

Nelder, J. A. and Mead, R. (1965). "A simplex method for function minimization", *Comput. J.*, **7**(4), 308−313.

Nikolaidis, N. P., Schnoor, J. L. and Georgakakos, K. P. (1989). "Modelling of long-term lake alkalinity responses to acid deposition", *J. Water Pollut. Control Fed.*, **61**(2), 188−199.

Parissopoulos, G. A. and Wheater, H. S. (1990). "Numerical study of the effects of layers on unsaturated−saturated two- dimensional flow", *Water Resour. Managemt*, **4**, 97−122.

Pickup, G. (1977). "Testing the efficiencies of algorithms and strategies for automatic calibration of rainfall-runoff models", *Hydrolog. Sci. Bull.*, **22**, 257−274.

Pilgrim, D. H., Huff, D. D. and Steele, T. D. (1978). "A field evaluation of subsurface and surface runoff. II. Runoff processes", *J. Hydrol.*, **38**, 319−341.

Pool, R. (1989). "Is it real, or is it a Cray?", *Science*, **244**, 1438−1440.

Reuss, J. O. and Johnson, D. W. (1986). *Acid Deposition and the Acidification of Soils and Waters*, Springer-Verlag, 119 pp.

Robson, A. and Neal, C. (1990). "Hydrograph separation using chemical techniques: an application to catchments in mid-Wales", *J. Hydrol.*, **116**, 345−363.

Rose, K. A., Cook, R. B., Brenkert, A. L., Gardner, R. H. and Hettelingh, J. P. (1991a). "Systematic comparison of ILWAS, MAGIC, and ETD watershed acidification models, 1. Mapping among model inputs and deterministic results", *Water Resour. Res.*, **27**(10), 2577−2589.

Rose, K. A., Brenkert, A. L., Cook, R. B., Gardner, R. H. and Hettelingh, J. P. (1991b). "Systematic comparison of ILWAS, MAGIC, and ETD watershed acidification models, 2. Monte Carlo analysis under regional variability", *Water Resour. Res.*, **27**(10), 2591−2603.

Rosenbrock, H. H. (1960). "An automatic method of finding the greatest or least value of a function", *Computer J.*, **3**, 175−184.

Seip, H. M., Blakar, I. A., Christophersen, J., Grip, H. and Vogt, R. D. (1990). "Hydrochemical studies in Scandinavian catchments", in *The Surface Waters Acidification Programme* (Ed. B. J. Mason), pp. 19−29, Cambridge University Press.

Sherman, L.K. (1932). "Streamflow from rainfall by the unit- graph method", *Engng News Record*, **108**, 501−505.

Sorooshian, S. and Gupta, V. J. (1983). "Automatic calibration of conceptual rainfall-runoff models: the question of parameter observability and uniqueness", *Water Resour. Res.*, **19**(1), 260−268.

Sorooshian, S., Gupta, V. J. and Fulton, J. L. (1983). "Evaluation of maximum likelihood parameter estimation techniques for conceptual rainfall-runoff models: influence of calibration data variability and length of model credibility", *Water Resour. Res.*, **19**(1), 252−259.

Stephenson, G. R. and Freeze, R. A. (1974). "Mathematic simulation of subsurface flow contributions to snowmelt runoff, Reynolds Creek watershed, Idaho", *Water Resour. Res.*, **10**(2), 284−294.

Taugbøl, G. and Neal, C. (1994). "Soil and stream water chemistry variations on acidic soils. Application of a cation exchange and mixing model at the catchment level", *Sci. Total Environ.*, **149**, 83−95.

Tompkins, J. A., Gan, K. C. Wheater, H. S. and Hirano, F. (1994). "Prediction of solute dispersion in heterogeneous porous media: effects of ergodicity and hydraulic conductivity discretisation", *J. Hydrol.*, **159**, 105−123.

Wang, Q. J. (1991) "The genetic algorithm and its application to calibrating conceptual rainfall-runoff models", *Water Resour. Res.*, **27**(9), 2467−2471.

Wheater, H. S., Bishop, K. H. and Beck, M. B. (1986). "The identification of conceptual hydrological models for surface water acidification", *Hydrolog. Process.*, **1**, 89−109.

Wheater, H. S., Langan, S. J., Miller, J. D. and Ferrier, R. C. (1987). "The determination of hydrological flow paths and associated hydrochemistry in forested catchments in Central Scotland", *International Symposium on Forest Hydrology and Watershed Management*, Vancouver, August,

IAHS Publication No. 167, pp. 433 – 449.

Wheater, H. S., Bell, N. C. and Johnston, P. M. (1989). "Evaluation of overland flow models using laboratory catchment data III. Intercomparison of conceptual models", *Hydrolog. Sci. J.*, **34**, 319 – 337.

Wheater, H. S., Langan, S. J., Miller, J. D., Ferrier, R. C., Jenkins, A., Tuck, S. and Beck, M. B. (1990a). "Hydrological processes on the plot and hillslope scale", in *The Surface Waters Acidification Programme*, (Ed. B. J. Mason), pp. 121 – 135, Cambridge University Press.

Wheater, H. S., Kleissen, F. M., Beck, M. B., Tuck, S., Jenkins, A. and Harriman, R. (1990b). "Modelling short-term flow and chemical response in the Allt a'Mharcaidh catchment", in *The Surface Waters Acidification Programme* (Ed. B. J. Mason), pp. 455 – 466, Cambridge University Press.

Wheater, H. J., Jakeman, A. J. and Beven, K. J. (1993). "Progress and directions in rainfall – runoff modelling", in *Modelling Change in Environmental Systems* (Eds. A. J. Jakeman, M. B. Beck and M. J. McAleer), 101 – 132, John Wiley.

Whipkey, R. Z. (1965). "Subsurface stormflow from forested slopes", *Hydrolog. Sci. Bull.*, **10**(3), 74 – 85.

Whitehead, P. G., Young P. C. and Hornberger, G. M. (1979). "A systems model of flow and water quality in the Bedford Ouse river system — I. Streamflow modelling", *Water Res.*, **13**, 1155 – 1169.

Wood, E. F., Sivapalan, M. and Beven, K. J. (1990). "Similarity and scale in catchment storm response", *Rev. Geophys.*, **28**(1), 1 – 18.

Young, P. C. (1992). "Parallel processes in hydrology and water quality: a unified time-series approach", *J. Inst. of Water and Environmental Managem.*, **6**(5), 598 – 612.

Zhu, J. L., Mishra, S. and Parker, J. C. (1990). "Effective properties for modelling unsaturated flow in large-scale heterogeneous porous media", in *Field-scale Water and Solute Flux in Soils* (Eds. K. Roth, H. Fluhler, W. A. Jurg and J. C. Parker), Birkhauser, Basel, Switzerland.

SECTION V

SOLUTE MODELS

12

Application of Soil Acidification Models with Different Degrees of Process Description (SMART, RESAM and NUCSAM) on an Intensively Monitored Spruce Site

C. VAN DER SALM, J. KROS, J. E. GROENENBERG, W. DE VRIES
and G. J. REINDS

*DLO Winand Staring Centre for Integrated Land, Soil and Water Research (SC-DLO),
Wageningen, The Netherlands*

12.1 INTRODUCTION

At present various dynamic simulation models exist to predict acidification of soil and surface waters. These models have been designed for use on a continental to national scale, such as MAGIC (Cosby *et al.*, 1985) and SMART (De Vries *et al.*, 1989), or on a national to regional scale, such as RESAM (De Vries *et al.*, 1995a), or for use on a catchment or site scale, such as ILWAS (Chen *et al.*, 1993) and NUCSAM (Groenenberg *et al.*, 1995).

Models designed for regional predictions tend to be more simplified than site-scale models to minimise input requirements. These simplifications can consist of (a) the use of less complex/detailed process formulations, (b) the reduction of temporal resolution, e.g. using an annual time resolution and thereby neglecting variability within a year of both model input and processes and (c) the reduction in vertical resolution, by using a smaller number of soil compartments. All these simplifications may cause errors in the predictions. Georgakos *et al.* (1989), for example, found that natural day-to-day variability in meteorological variables significantly affects long-term predictions of lake and stream acidification.

The objective of this study is to characterise the effect of model simplifications on soil solution response, with emphasis on the influence of temporal and vertical resolution. Therefore we compared the results derived with a one-layer model SMART (simulation model for acidification's regional trends), the multi-layer models RESAM (regional soil acidification model), with a temporal resolution of one year and NUCSAM (nutrient cycling and soil acidification model), with a temporal resolution of one day, with measured concentrations of an intensively monitored spruce site at Solling, Germany. At this site inputs, concentrations and amounts of elements in the soil system have been measured continuously for more than 20 years

---

Solute Modelling in Catchment Systems. Edited by Stephen T. Trudgill
© 1995 John Wiley & Sons Ltd

TABLE 12.1 Overview of the considered processes

Process	SMART	RESAM	NUCSAM
Canopy interaction	−	+	+
Growth uptake	+	+	+
Maintenance uptake	−[a]	+	+
Litterfall	−[a]	+	+
Root decay	−[a]	+	+
Mineralisation	−[a]	+	+
(De)nitrification	+	+	+
Carbonate weathering	+	+	+
Silicate weathering	+	+	+
Al hydroxide weathering	+	+	+
Cation exchange	+	+	+
Sulphate adsorption	+	+	+

[a] Not explicitly included in SMART, instead overall N immobilisation was included.

and were completed by plant physiological, hydrological, micrometeorological and soil biological monitoring programmes during that time.

12.2 MODELS USED

The three considered soil acidification models are dynamic simulation models, SMART is a one-layer model, whereas RESAM and NUCSAM distinguish a litter layer and several soil layers. The temporal resolution of SMART and RESAM is one year, whereas NUCSAM has a temporal resolution of one day. Although RESAM has a temporal resolution of one year, the model uses an internal time step of five days to avoid oscillations. RESAM and NUCSAM simulate the major biogeochemical processes occurring in the canopy, litter layer and mineral horizons. SMART distinguishes only one soil layer and litterfall, root decay and mineralisation are not explicitly taken into account. Only a net uptake of N and base cations and a net N immobilisation rate are included. In RESAM and NUCSAM a description for litterfall, root decay and mineralisation is incorporated in the models. An overview of the considered processes in the three models is given in Table 12.1. Annex 1 gives a brief overview of the model formulations.

12.2.1 SMART

SMART simulates the concentrations of Al, divalent base cations (Ca + Mg), monovalent base cations (Na + K), NH_4, SO_4 and NO_3 in the soil solution. In SMART most of the geochemical processes are incorporated (weathering, cation exchange, sulphate adsorption). However, only a (very) limited number of biological processes are taken into account (Table 12.1).

Nutrient cycling processes are not included because the model is based on the assumption that the amount of organic matter is in an equilibrium state. However, net N immobilisation is taken into account to include the effect of an increase in N content in organic matter due to high N deposition.

Cation exchange, sulphate adsorption, dissolution of carbonates and Al hydroxides are treated as equilibrium reactions (Annex 1). Weathering of base cations and (de)nitrification are described as first-order reactions. An overview of a former version of the model is given in De

TABLE 12.2 Statistical measures for evaluation of model results

Measure	Symbol	Formulation	Optimum
Normalised root mean square error	NMAE	$$\dfrac{\sum\limits_{i=1}^{N}(P_i-O_i)}{N^2\cdot\bar{O}}$$	0
Coefficient of residual mass	CRM	$$\dfrac{\sum\limits_{i=1}^{N}O_i-\sum\limits_{i=1}^{N}P_i}{\sum\limits_{i=1}^{N}O_i}$$	0

P_i is the modelled value, O_i is the observed value, $\bar{O}$ is the average of the observations and N is the number of observations.

Vries *et al.* (1989). Since then the description of N dynamics has been changed (De Vries *et al.*, 1995b).

12.2.2 RESAM

In contrast to SMART, RESAM not only simulates major geochemical processes but biochemical processes occurring in the forest canopy, litter layer and mineral horizons (Table 12.1) as well.

Foliar exudation, litterfall, root decay, nitrification, denitrification, protonation and weathering are described by first-order reactions. Foliar uptake is considered to be a fraction of the dry atmospheric deposition. Root uptake is assumed to be equal to the sum of litterfall, foliar exudation and root decay minus foliar uptake plus a given net growth. Net growth is described by a logistic function. Root uptake per soil layer is assumed to be proportional to the fraction of roots in each soil layer. The dissolution of Ca and Al from carbonates and hydroxides are described as a first-order reaction and an Elovich equation respectively, which are both rate-limited by the degree of undersaturation. If supersaturation occurs, the Ca or Al concentration is set to equilibrium. Cation exchange and sulphate sorption are treated as equilibrium reactions, using Gaines–Thomas exchange equations and a Langmuir isotherm respectively. Speciation/dissolution of inorganic C is computed from equilibrium equations using a constant value for pCO_2. A complete overview of the model structure of RESAM is given in De Vries *et al.* (1994a).

12.2.3 NUCSAM

NUCSAM has been derived from the RESAM model. The main differences between the two models is the temporal resolution used; NUCSAM has a temporal resolution of one day instead of one year. The version of NUCSAM used for this comparison uses practically the same biogeochemical process formulations as RESAM. However, in contrast to RESAM, (a) litterfall, root decay and root uptake are distributed over the year by given monthly coefficients and (b) both upward and downward transport of solutes is taken into account.

12.3 METHODS AND DATA

12.3.1 Methods

General Approach

To compare difference in model predictions, due to differences in process aggregation, temporal or vertical resolution, objectively it is necessary to minimize differences in parameterisation. Data for the models were derived from the Stolling dataset (Bredemeier *et al.*, 1995). In those cases where the models use the same state variables and process parameters with the same vertical or temporal resolution, we simply used the same values for both models. Parameters for SMART were derived by depth-averaging the values that were used for RESAM and NUCSAM (input mapping; Rose *et al.*, 1991). Annual deposition and water fluxes, which are input to the model RESAM and SMART, were derived by accumulating the NUCSAM values to annual values.

Vertical configuration and simulation period

At the Solling site NUCSAM and RESAM considered a litter layer of 7 cm (at the start of the simulations) and seven mineral soil layers up to a depth of 90 cm (see Table 12.4 later). For SMART, in which one mineral soil compartment is distinguished, two simulations were run: (a) with a mineral soil layer of 10 cm thickness and (b) with a layer of 90 cm thickness.

All models were run for the period 1971−90. RESAM and NUCSAM used the period 1961−70 as an initialisation period to estimate solute concentrations in 1970 and to equilibrate solute concentrations with exchangeable cations and adsorbed SO_4. During that period, amounts of exchangeable cations and adsorbed amounts of SO_4 were continuously updated while cation amounts in primary minerals and Al hydroxides were kept constant.

SMART did no use an initialisation period. Initial amounts of base cations were input to the model. Initial adsorbed amounts of SO_4 were calibrated on the amounts in NUCSAM/RESAM in 1970.

Model adaptions

In regional applications, SMART and RESAM use annual average hydrological fluxes which are constant throughout the simulation period. This study focuses on the influence of differences in biogeochemical process descriptions and their vertical and temporal resolution (one day versus one year). Accordingly, for this application SMART and RESAM were slightly adapted to account for variations in hydrological fluxes between the years.

TABLE 12.3 Average drainage fluxes and water contents used in NUCSAM, RESAM and SMART

Soil layer (cm)	Average drainage fluxed (cm a^{-1})			Average moisture content (m^3 m^{-3})	
	NUCSAM	RESAM	SMART	NUCSAM/RESAM	SMART
0−10	73.5	73.6	73.6	0.398	0.398
10−20	70.0	70.1	—	0.394	—
20−30	64.0	64.0	—	0.362	—
30−40	55.6	55.7	—	0.363	—
40−60	47.6	47.7	—	0.367	—
60−80	42.9	43.0	—	0:336	—
80−90^a	40.9	41.0	41.0	0.338	0.362

^{a} SMART soil layer 0−90 cm.

The SMART model is normally applied to calculate concentrations at the bottom on the root zone. To apply the SMART model at a shallow depth (10 cm), the calculation of N immobilisation was slightly adapted. In the standard version of SMART, N immobilisation is supposed to occur in the upper 20 cm of the soil. For the simulation of concentrations at 10 cm depth, the total N immobilisation flux was multiplied by the ratio of the amount of organic C in the considered layer and the amount up to 20 cm depth.

Model comparison

NUCSAM simulates daily concentrations and leaching fluxes, whereas SMART and RESAM simulated flux-weighted annual average concentrations and annual leaching fluxes. A comparison of simulated data of SMART and RESAM with measured data is complicated as flux-weighted annual average concentrations can not be measured. A comparison of results of the three models with measured data, on the same basis, can be made by comparing: (a) measured concentration (once a month) with simulated values (in this case monthly concentrations for RESAM and SMART were derived by linear interpolation between annual values) and (b) estimated flux-weighted annual average measured concentrations (or leaching fluxes) with simulated values. "Measured" leaching fluxes are calculated by multiplying measured concentrations with (monthly) simulated water fluxes. Flux-weighted annual average concentrations are derived by dividing the "measured" leaching flux by the annual water fluxes.

In this study a combined approach was used: simulated concentrations were compared with measured concentrations (according to (a)) and simulated cumulative annual leaching fluxes were compared with (calculated) measured annual leaching fluxes. A comparison of measured concentrations with simulated concentrations gives a good impression of the performance of the models and the ability of the models to simulate trends and extreme values. A comparison of cumulative fluxes shows whether the models tend to underestimate or overestimate total leaching fluxes for the simulation period.

To give more objective information concerning the performance of the models two statistical measures were calculated: the normalized mean absolute error (NMAE) and the coefficient of residual mass (CRM) (Table 12.3). NMAE quantifies the average deviation between model

TABLE 12.4 Soil properties used for NUCSAM, RESAM and SMART

Soil layer (cm)	ρ (kg m^{-3})	CEC (mmol$_c$ kg^{-1})	$ctAl_{ox}$ (mmol$_c$ kg^{-1})	SSC
NUCSAM and RESAM				
0–10	930	132.14	96.5	0.99
10–20	1140	78.95	96.5	4.46
20–30	1190	57.98	185.3	4.46
30–40	1390	45.32	185.3	4.46
40–60	1390	56.12	185.3	4.46
60–80	1690	56.12	175.3	6.70
80–90	1690	75.90	93.7	6.70
SMART				
0–10	930	132.14	96.5	0.99
0–90	1389	78.95	140.0	5.05

prediction and measurements. CRM gives an indication of the tendency of the model to under- or overestimate (negative value) the measured data. NMAE and CRM for the three models were calculated using monthly concentrations for model results and measurements.

12.3.2 Hydrological data

For all models used, hydrological fluxes and water contents were calculated by the model SWATRE (Belmans *et al.*, 1983). This model provides a finite difference solution to the Richards equation. The version that is used here (Groenenberg *et al.*, 1995) differs from the original with respect to the formulation of interception evaporation and root uptake. Furthermore, a snow module was added. Root uptake was divided over the different soil layers according to a given fixed root distribution.

Drainage fluxes, root uptake fluxes and water contents calculated by SWATRE were directly used by NUCSAM. RESAM and SMART use annual values and discard year to year changes in storage of soil water. Annual root uptake fluxes were derived by accumulating the daily root uptake fluxes to annual values. To keep water contents constant throughout the simulation period, annual drainage fluxes were calculated by subtracting the root uptake fluxes from the input flux for each layer. For use in RESAM, water contents for each layer were averaged over the simulation period. The data for SMART were derived by depth-averaging the water contents that were used for RESAM. An overview of the main hydrological fluxes and water contents is given in Table 12.3.

12.3.3 Geochemical data

Data used in NUCSAM and RESAM

Geochemical data for NUCSAM and RESAM (Groenenberg *et al.*, submitted; Kros *et al.*, submitted) were directly derived from the Solling dataset (Bredemeier *et al.*, 1994). An overview of the main parameters is given in Tables 12.4 to 12.6. Gaines–Thomas exchange

TABLE 12.5 Elovich constants for Al dissolution, base cation weathering rate constants and Gaines–Thomas exchange contants, SO_4 sorption parameters used in the simulation by NUCSAM and RESAM

Soil layer (cm)	$kr\text{Elo}1$[a] (10^{-7} m^3 kg^{-1} a^{-1})	$kr\text{Elo}2$[b] (10^2 kg mol$_c^{-1}$)	Weathering rate[c] constants (10^{-3} a^{-1})				Exchange constants[d] (mol l^{-1})$^z X^{-2}$						$Ke\text{SO}_{4ad}$[e] (1 mol^{-1})
			Ca	Mg	K	Na	H	Al	Mg	K	Na	NH$_4$	
0–10	0.58	7.5	6.5	93.6	0.011	0.021	5180	0.97	1.60	647	8.40	1.05	0.5×10^{-3}
10–20	2.00	7.5	6.0	73.2	0.008	0.015	57.5	26.2	2.56	3660	29.1	6.53	4.2×10^3
20–30	5.10	7.5	5.6	66.9	0.007	0.013	15.3	8.75	65.3	7470	21.2	30.7	8.2×10^3
30–40	5.10	7.5	5.4	63.7	0.006	0.010	15.3	7.37	0.42	18700	32.0	30.7	8.2×10^3
40–60	5.10	7.5	5.3	61.8	0.005	0.011	15.3	7.37	1.25	16900	36.2	30.7	8.2×10^3
60–80	5.10	7.5	6.2	51.7	0.005	0.011	15.3	26.2	1.25	16900	36.2	30.7	2.2×10^3
80–90	5.10	7.5	10.9	25.8	0.003	0.011	15.3	26.2	1.25	16900	36.2	30.7	2.2×10^3

[a] Derived from average soil solution concentrations of H$^+$ and Al^{3+} in 1983, assuming KAl$_{ox}$ = 3.5×10^8 and $kr\text{Elo}2$ = 7.5×10^{-2}.

[b] The average of values given in De Vries et al. (1995a)

[c] Based on total analysis and weathering fluxes of base cations from Wesselink et al. (1994) and average H$^+$ concentration in 1983.

[d] Based on average soil solution concentration measurements in 1983 and solid-phase analyses in the same year except for NH$_4$ which is taken from De Vries et al. (1994a)

[e] Derived from Meiwes (1979)

TABLE 12.6　Values for soil layer independent model parameters used in the simulation based on the Solling dataset (Bredemeier *et al.*, submitted)

Process	Parameter	Unit	Value	Model
Foliar uptake[a]	$frNH_{4fu}$	—	0.11	NUCSAM, RESAM
	frH_{fu}	—	0.33	NUCSAM, RESAM
Foliar exudation[a]	$frCa_{FE}$	—	0.49	NUCSAM, RESAM
	$frMg_{fe}$	—	0.09	NUCSAM, RESAM
	frK_{fe}	—	0.42	NUCSAM, RESAM
Tree growth[b]	$krgrl$	a^{-1}	0.09	NUCSAM, RESAM, SMART input
	$A_{st,max}$	kg ha^{-1}	2.5×10^5	NUCSAM, RESAM, SMART input
	t_{05}	a	66.0	NUCSAM, RESAM, SMART input
Litterfall[c]	k_{lf}	a^{-1}	0.19	NUCSAM, RESAM
Nitrification[d]	$k_{ni,max}$	a^{-1}	40.0	NUCSAM, RESAM
	$frni$	—	0.88	SMART 10 cm
		—	0.98	SMART 90 cm
Denitrification[e]	$frde$	—	0.10	SMART
N immobilisation[f]	C/N	—	19.5	SMART
Al dissolution[g]	KAl_{ox}	$l^2\,mol^{-2}$	3.5×10^8	NUCSAM, RESAM

[a] Based on average throughfall and deposition data over the period 1974–90.
[b] Derived by curve-fitting of the biomass measurements.
[c] Average needlefall rate over the period 1967–73, taking into account that 92.5% of the litterfall is needlefall.
[d] $k_{ni,max}$ is derived from average throughfall and mineralization fluxes over the period 1970–85, assuming that all mineralised N is released as NH_4^+. *frni* is derived from average throughfall and average drainage fluxes and calculated average root uptake fluxes for the period 1973–90.
[e] Derived from De Vries *et al.* (1994b).
[f] Based on 1973 data for C_{org} and N_{org}.
[g] Average IAP for Al(OH)$_3$ at 90 cm over the period 1973–91. The value given is the value at 25°C, which is derived from the value at field temperature (10°C).

constants (for all three models) were based on average soil solution concentration measurements in 1983 and solid-phase analyses in the same year. Sulphate adsorption constants for NUCSAM and RESAM were directly derived from Meiwes (1979).

　　Weathering fluxes of primary minerals in NUCSAM and RESAM are described by a first-order equation (Annex 1). Rate constants for this equation (Table 12.5) were derived from a budget study (Wesselink *et al.*, 1994). Dissolution parameters of Al hydroxides in RESAM and NUCSAM, described by an Elovich equation (Annex 1), are given in Table 12.5 together with their derivation.

SMART

Most data for SMART were derived by depth-averaging the data that were used for NUCSAM and RESAM (Tables 12.4 and 12.5). Some parameters that are only used in SMART were directly obtained from the Solling dataset. Soil properties that were used in SMART, i.e. bulk density, CEC, sulphate sorption capacity (SSC) and Al (hydr)oxide content (Table 12.4) were derived by depth-averaging the data used in NUCSAM and RESAM (Table 12.4 and 12.5). To

TABLE 12.7 Geochemical parameters for SMART

Parameter	Unit	Values	
		10 cm	90 cm
$KAl_{ox}{}^a$	$l^2 mol^{-2}$	4.0×10^7	2.0×10^9
$FBC_{we}{}^b$	$mol_c m^{-3} a^{-1}$	0.039	0.043
$FBC1_{we}{}^b$	$mol_c m^{-3} a^{-1}$	0.011	0.012
KAl_{ex}	$l mol^{-1}$	0.7	3.5
KH_{ex}	$mol l^{-1}$	4786	1862
KSO_{4ad}	$l mol^{-1}$	4.2×10^3	3.9×10^3

[a] Average IAP for $Al(OH)_3$ at 10 and 90 cm, based on measured Al and H concentrations in the period 1973−90.
[b] For 10 cm based on NUCSAM weathering rates and average H concentration at 10 cm depth for the period 1973−90; for 90 cm depth directly based on weathering fluxes from Wesselink *et al.* (1994).

calculate Gaines−Thomas exchange constants for SMART (Table 12.7), concentrations and solid-phase analyses were depth averaged for the 10 and 90 cm soil compartments. A depth-weighted sulphate adsorption constant for SMART was derived as follows: first adsorbed amounts of sulphur were calculated for all layers, considered in NUCSAM/RESAM, using a Langmuir equation (Annex 1) and the sulphate adsorption constants from Meiwes (1979). In calculating these amounts we assumed the same range in SO_4 concentrations in the soil solution with depth. Next the calculated adsorbed amounts were depth-weighted. Finally the depth-weighted sulphate adsorption constant was derived by fitting the depth-weighted adsorbed SO_4 amounts against the SO_4 concentration range considered.

In SMART, weathering fluxes are input to the model and were directly derived from the above-mentioned budget study. In SMART, dissolution of Al hydroxide is described by equilibrium with an Al hydroxide. Solubility products for the Al hydroxide at 10 and 90 cm depth were derived from average soil solution concentrations of H and Al in 1983 at these depths. The solubility product for Al hydroxide at 90 cm depth was also used in RESAM and NUCSAM to calculate the Al concentration at equilibrium.

12.3.4 Biological data

An overview of the biological data and their derivation is given in Table 12.6. The parameters for N cycling in NUCSAM/RESAM and SMART were derived independently from the Solling dataset as the process description in the models is different. An important difference does exist in the parameterisation of the nitrification process between RESAM/NUCSAM and SMART, although parameters for both models were based on an input−output budget (Table 12.6). RESAM/NUCSAM use an overall nitrification rate which is reduced by moisture content, pH and organic matter content. For the simulations with SMART, separate nitrification fractions, based on input−output budgets, were used for the topsoil and the subsoil. The relationship between moisture content, pH, organic matter content and nitrification rate that was used in NUCSAM/RESAM was not calibrated on the site data.

Growth uptake in NUCSAM and RESAM was calculated by multiplying a given (logistic) growth rate (Annex 1) by the element content in 1968 in stems and branches respectively. Element contents were assumed constant with the exception of the N content. N content is calculated with a linear relationship between N content and N deposition. N content is minimal at a N deposition of 1500 $kmol_c$ ha^{-1} a^{-1} and maximal at a N deposition 7000 $kmol_c$ ha^{-1} a^{-1}. The growth uptake fluxes for SMART are input to the model. Growth uptake fluxes at 90 cm depth were derived by multiplying the growth rates of stems and branches with the element content in stems and branches, using the same values and the formulations as used by NUCSAM/RESAM. Growth uptake fluxes at 10 cm were derived by multiplying the total growth uptake fluxes by the fraction of roots in the upper 10 cm.

12.4   RESULTS AND DISCUSSION

To characterise the effects of differences in vertical and temporal resolution and process aggregation in the models, the simulated concentrations and leaching fluxes were compared with measured concentrations and leaching fluxes in the topsoil (10 cm) and subsoil (90 cm). Results were limited to major anions and cations, i.e. SO_4, Cl, NO_3, NH_4, Al and BC (divalent base cations). Simulated and measured concentrations are shown in Figure 12.1 (SO_4 and Cl), Figure 12.2 (NO_3 and NH_4) and Figure 12.3 (Al and BC). An overview of the calculated values of NMAE and CRM for the various concentrations in topsoil and subsoil is given in Table 12.8. The figures and the statistical measures show that all models were able to reasonably simulate the measured concentrations during the examined period. Differences between RESAM and NUCSAM, the multi-layer models, were rather small. Somewhat larger differences did occur between the concentrations simulated by SMART and those simulated by the multi-layer models. A more detailed discussion on the performance of the models to simulate the individual ions is held in the following sections, where the influence of the model differences is presented.

12.4.1   Influence of vertical resolution

The influence of vertical resolution is most clearly shown by the SO_4 concentrations and leaching fluxes, as SO_4 concentrations are mainly governed by deposition and adsorption, which are described in all models in practically the same way. Measured and simulated concentrations and leaching fluxes of SO_4 are shown in Figure 12.1. The trends in SO_4 concentrations, as simulated by NUCSAM and RESAM, were generally in good agreement with measured data. SMART, however, overestimated SO_4 concentrations at 90 cm depth during the period 1972−8 in which a strong rise in SO_4 concentrations took place at this depth. This overestimation is caused by a larger dispersion of the SO_4 front in a one-layer system compared to a multi-layer system. In a multi-layer system the rise in SO_4 input initially leads to a rise in the adsorbed amounts in the upper soil layers, whereas in the subsoil adsorbed amounts remain unchanged. In one-layer system, a rise in the input immediately leads to a (small) rise in the adsorbed amounts and concentrations at greater depth. For all models, the performance for SO_4 in the topsoil was comparable. NMAE values were somewhat higher for NUCSAM compared to the other models (Table 12.8), as the simulated variation within the year was larger than the measured variation. Cumulative leaching fluxes at 10 cm depth were somewhat underestimated by all models in the period 1985−9, due to an underestimation of SO_4 concentrations in this period with high water fluxes. Although concentrations were overestimated by SMART in the subsoil, during the period 1973−5, the overall performance was comparable with the multi-layer models. Total leaching fluxes in 1989, simulated by SMART, were comparable with the measurements.

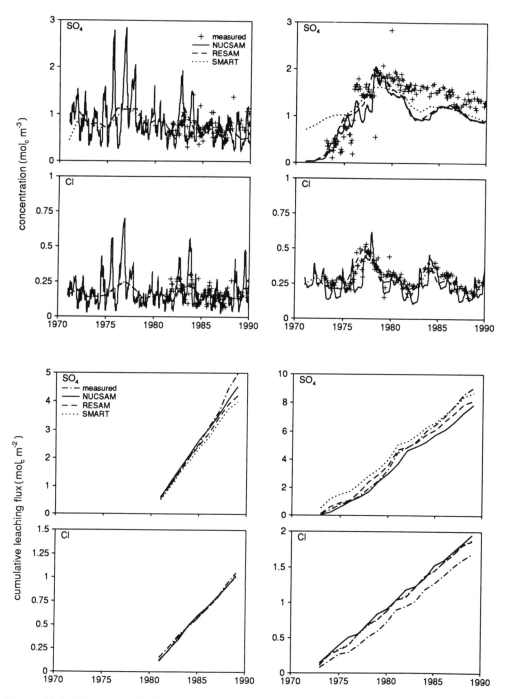

FIGURE 12.1 Measured and simulated SO₄ and Cl concentrations and leaching fluxes at 10 (left) and 90 cm depth (right

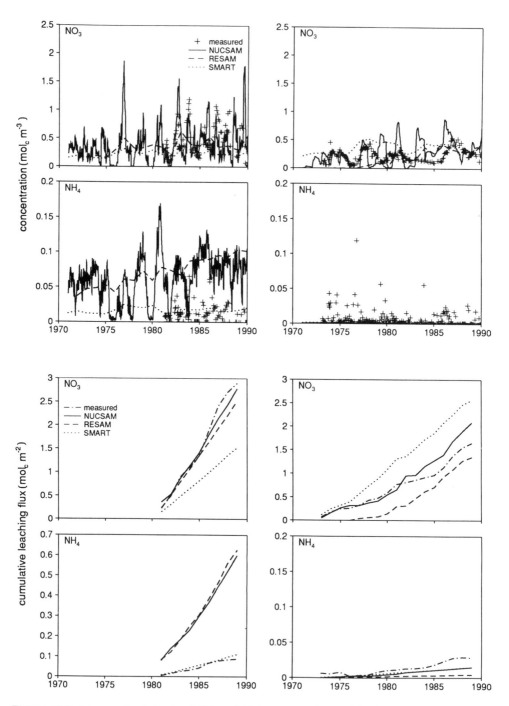

FIGURE 12.2   Measured and simulated NO_3 and NH_4 concentrations and leaching fluxes at 10 (left) and 90 cm depth (right)

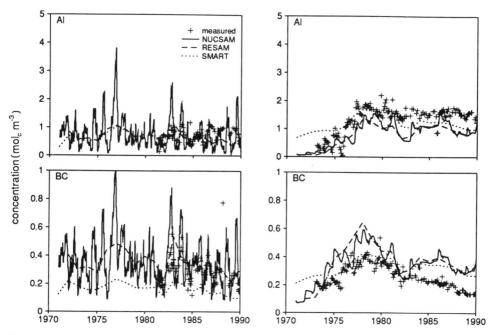

FIGURE 12.3 Measured and simulated Al and BC concentrations at 10 (left) and 90 cm depth (right)

TABLE 12.8 Normalized mean absolute error (NMAE) and coefficient of residual mass (CRM) for simulated concentrations

Component	Depth	NMAE			CRM		
		SMART	RESAM	NUCSAM	SMART	RESAM	NUCSAM
SO_4	10	0.23	0.25	0.37	0.10	0.05	−0.01
	90	0.26	0.24	0.25	0.06	0.12	0.18
NO_3	10	0.56	0.50	0.62	0.48	0.10	−0.04
	90	0.84	0.63	0.76	−0.79	0.36	−0.25
NH_4	10	1.00	6.17	5.17	−0.34	−5.96	−4.92
	90	0.89	0.89	0.89	0.73	0.93	0.80
BC	10	0.47	0.25	0.41	0.46	0.21	−0.21
	90	0.29	0.16	0.45	0.27	0.02	−0.43
Al	10	0.32	0.33	0.52	0.20	0.13	0.02
	90	0.24	0.37	0.33	0.01	0.34	0.30
H	10	0.45	0.47	0.53	0.41	0.45	0.51
	90	0.38	0.49	0.48	0.30	0.48	0.47
Cl	10	0.27	0.28	0.40	0.05	0.03	−0.04
	90	0.11	0.16	0.23	0.04	0.06	0.16

As a result of the smoothed SO_4 front, the rise in Al due to weathering in the period 1972—8 is less pronounced in SMART. This causes a lower exchange of adsorbed base cations against Al compared to the other models. This lower BC desorption in turn leads to a lower in the rise of the BC concentrations in the subsoil, as simulated by SMART.

12.4.2 Influence of process description

The main differences in process description between the models occur in the description of processes involving nutrient cycling. In SMART a net N immobilisation flux is calculated, whereas RESAM and NUCSAM account for storage of N in the litter layer and for mineralisation. Furthermore, nitrification is described in a different way in SMART.

Figure 12.2 shows that SMART overestimated NO_3 concentrations in the subsoil (negative CRM) during the entire period, whereas concentrations in the topsoil tended to be slightly underestimated (positive CRM). The deviation between measured and simulated concentrations in the topsoil is due to the neglect of mineralisation in the topsoil. To obtain a closer fit between simulated and measured NO_3 concentrations, a slight change in the parameters describing N immobilisation in the topsoil would have been useful.

NH_4 concentrations (Figure 12.2) were generally overestimated in the topsoil. However, the overestimation of the concentrations in the topsoil by SMART was small compared to NUCSAM and RESAM (see CRM values, Table 12.8). In the subsoil all models simulated comparable NH_4 concentrations, which were underestimated with respect to the measurements (see CRM values, Table 12.8). The relatively good agreement between measured concentrations and concentrations simulated with SMART, in the topsoil, is partly due to the fact that in SMART different nitrification constants at 10 and 90 cm depth were used which were directly derived from the Solling dataset. RESAM/NUCSAM, however, use one overall nitrification parameter which is adapted for each horizon depending on pH, moisture content and organic matter content. Furthermore, in SMART, NH_4 concentrations in the topsoil tend to be lower due to the neglect of mineralisation. The influence of mineralisation is also shown by the simulation of divalent base cations (BC) by the models (Figure 12.3). In the topsoil SMART simulated lower BC concentrations compared to the other models.

Another difference in process description between SMART and RESAM/NUCSAM is the way in which Al concentrations are calculated. In SMART, Al concentrations are calculated from equilibrium with Al hydroxide, whereas RESAM and NUCSAM use a kinetic description (Annex 1). Figure 12.3 shows that results for the simulation of Al (main cation) were comparable with those for the main anion SO_4. The way in which Al concentrations were calculated appears to have hardly any influence on the results for the chosen period. When applying the models for long-term predictions deviations between the concentrations predicted by NUCSAM/RESAM and SMART may occur, particularly in the topsoil where the dissolution of Al compounds is far from equilibrium with gibbsite. A decrease in the Al dissolution rate, due to exhaustion of certain Al compounds, will lead to a lower simulated concentration by NUCSAM/RESAM whereas concentrations simulated by SMART will remain constant.

12.4.3 Influence of temporal resolution

The influence of neglecting seasonal fluctuations in the considered process fluxes can best be derived by comparing RESAM and NUCSAM, models with a comparable process description and a difference in temporal resolution. The most direct influence of the chosen temporal resolution is found in the simulation of Cl concentrations and fluxes. For example, NUCSAM used daily upward and downward water fluxes to calculate solute transport. Accordingly, stronger fluctuations in concentrations (Figure 12.1) were simulated by NUCSAM compared to other models. NMAE values for the Cl concentrations, however, showed that the simulation of the Cl concentrations by NUCSAM was not better than for the other models. In the topsoil the simulated fluctuation of the Cl concentration was sometimes out of phase with the measured fluctuation. In the subsoil NUCSAM underestimated Cl concentrations in wet periods (Table 12.8).

The influence of the chosen temporal resolution on model performance can particularly be expected for the simulated concentrations of NO_3, NH_4 and base cations which are strongly influenced by seasonal processes such as nutrient cycling and mineralisation. NO_3 concentrations (Figure 12.2 and Table 12.8) simulated with NUCSAM and RESAM were in close agreement with the measurements in the topsoil. Although NUCSAM simulated the seasonal peaks in NO_3 concentrations, NMAE values in the topsoil were somewhat higher for NUCSAM compared to RESAM. NO_3 concentrations in the subsoil were poorly simulated by RESAM up to 1980. From 1980 onwards concentrations simulated by NUCSAM and RESAM were in the same range as measured values (relatively low NMAE and CRM). However, fluctuations in simulated concentrations by NUCSAM occurred more frequently than the measured multi-year fluctuations in concentrations. The differences in simulated NO_3 concentrations in the subsoil between NUCSAM and RESAM are caused by the fact that in NUCSAM total N uptake is lower. N uptake in NUCSAM is lower due to a restriction of the N uptake to the growing season, which leads in certain years to a higher N demand than available in the soil solution, causing a lower total N uptake in that year.

Cumulative leaching fluxes for NO_3 in the topsoil (Figure 12.2) were in close agreement with measured leaching fluxes both for NUCSAM and RESAM. In the subsoil, cumulative leaching fluxes were underestimated (-0.3 mol_c m^{-2}) by RESAM, due to the underestimation of the concentrations (positive CRM) in the period up to 1980 and overestimated ($+0.4$ mol_c m^{-2}) by NUCSAM due to the overestimation of seasonal peak concentrations (negative CRM).

The correspondence between simulated and measured NH_4 concentrations (Figure 12.2) was meagre for RESAM and NUCSAM. The periodical fluctuations in concentrations in the subsoil were not simulated by NUCSAM and generally concentrations were overestimated in the topsoil. Although both measured and simulated NH_4 concentrations were relatively low, the deviation between measured and simulated values leads to a serious overestimation (circa 0.5 mol_c m^{-2}) in the period $1983-9$.

Base cation concentrations (Figure 12.3) are both influenced by processes with a strong seasonal variation, such as mineralisation, solute transport and ion-exchange and by more constant processes such as weathering and ion exchange. The general trend in divalent base cation concentrations in the topsoil was reasonably simulated both by NUCSAM and RESAM. NUCSAM and RESAM overestimated the rise in BC concentrations in the subsoil up to 1978. From 1982 onwards concentrations were overestimated by all models, which is probably due to an underestimation of tree growth during this period. RESAM simulated a somewhat stronger rise and fall in BC concentrations in the subsoil than NUCSAM. This is caused by a stronger desorption of BC in RESAM. The same phenomenon, however somewhat weaker, can be observed by SO_4 (Figure 12.1). RESAM simulated slightly higher SO_4 concentrations than NUCSAM during the period $1975-80$. The deviation between RESAM and NUCSAM in induced by slight differences in hydrology, as reflected by the differences in simulated Cl concentrations in the subsoil.

12.5 CONCLUSIONS

Although clear differences in process description do exist between SMART, RESAM and NUCSAM, all models were able to reasonably simulate most of the concentrations during the study period. The description of the dissolution of Al hydroxides either by a rate-limited reaction or by an equilibrium equation did not lead to differences in modelled Al concentrations during the study period. Large differences in complexity of the description of N cycling do exist between the models. In SMART, mineralisation is not included in the model, which led to lower concentrations of NH_4 and divalent base cation in the subsoil compared to the other models. NH_4 concentrations simulated by SMART were closer to the measurement than in the other models; RESAM and NUCSAM underestimated NH_4 concentrations in the topsoil and overestimated these concentration in the subsoil. The better results for SMART are a consequence of the use of separate nitrification parameters in SMART for the topsoil and the subsoil, which were directly derived from the measurements, whereas in RESAM/NUCSAM the nitrification rate was dependant on pH, water content and organic matter content. The relationship between these environmental parameters and nitrification parameters was not calibrated on the site data.

The influence of vertical resolution of the models was clearly shown by the simulation of SO_4 and base cations in the subsoil. All models were able to simulate a rise in SO_4 concentration, between 1975 and 1980, due to a decrease in sulphate adsorption. However, the one-layer model, SMART, tended to overestimate the initial rise in SO_4 concentration, due to a larger dispersion of the sulphur front in a one-layer system compared to a multi-layer system.

A strong influence of temporal resolution was expected in the simulation of NO_3 by NUCSAM compared to RESAM. In the topsoil, NO_3 concentrations simulated by the models were in the same range as the measurements. In the subsoil, NO_3 concentrations were underestimated by RESAM, as RESAM simulated a higher N uptake compared to NUCSAM. In the subsoil, NO_3 concentrations simulated by NUCSAM were in the range of the measurements; however, fluctuations were poorly simulated. The NMAE values for the NO_3 concentrations in the top- and the subsoil were higher for NUCSAM than for RESAM. In the topsoil the higher NMAE values resulted from the fact that simulated fluctuations were sometimes out of phase with the measured fluctuations. In the subsoil fluctuations occurred more frequently than measured fluctuations.

REFERENCES

Belmans, C., Wesseling, J. G., and Feddes, R. A., (1983). "Simulation model of the water balance of a cropped soil providing different types of boundary conditions (SWATRE)", *J. Hydrol.*, **63**, 61–62.
Bredemeier, M. A., Tiktak, A. and van Heerden, C. (1995) "The Solling spruce stand. Back-ground information on the dataset", *Ecological Modelling* (in press).
Chen, C. W., Gherini, S. A., Mok, L., Hudson, R. J. M. and Goldstein, R. A. (1983). *The integrated Lake-Watershed Acidification Study.* Volume 1. *Model principles and application procedures.* EPRI EA-3221, Vol. 1, Project 1109-5, Final Report.
Cosby, B. J., Wright, R. F., Hornberger, G. M. and Galloway, J. N. (1985). "Modelling the effects of

acid deposition: estimation of long-term water quality responses in a small forested catchment", *Water Resour. Res.*, **21**, 1591–1601.

De Vries, W., Posch, M. and Kamari, J. (1989). "Simulation of the long-term soil response to acid deposition in various buffer ranges", *Water, Air and Soil Pollut.*, **48**, 349–390.

De Vries, W., Kros, J. and Van der Salm, C. (1995a). "Modelling the impact of acid deposition and nutrient cycling in forest soils", *Ecological Modelling* (in press).

De Vries, W., Posch, M., Reinds, G. J. and Kamari, J. (1995b). "Simulation of soil response to acidic deposition scenarios in Europe", *Water, Air and Soil Pollut.*, **78**, 215–246.

Georgakos, K. P. Valle-Filho, G. M., Nikolaidis, N. P. and Schnoor, J. L. (1989). "Lake-acidification studies: the role of input uncertainty in long-term predictions", *Water Resour. Res.*, **25**, 1511–1518

Groenenberg, J. E., Kros, J. E., Van der Salm, C. and De Vries, W. (1995). "Application of the model NUCSAM to the Solling spruce site", *Ecological Modelling* (in press).

Kros, J., Groenenberg, J. E., De Vries, W., and Van der Salm, C. (submitted). "Uncertainties in long-term predictions of forest soil acidification due to neglecting interannual variability", *Water, Air and Soil Pollut*, **79**, 353–375.

Meiwes, K. J. (1979). "Der Schwefelhaushalt eines Buchenwald- und eines Fichtenwaldokosystem im Solling", *Gottinger Bodenkundliche Berichte*, **60**.

Rose, K. A., Cook, R. B., Brenkert, A. L., Gardner, R. H. and Hetterlingh, J. P., (1991). "Systematic comparison of ILWAS, MAGIC and ETD watershed acidification models. 1. Mapping among model inputs and deterministic results", *Water Resour. Res.*, **27**(10), 2577–2589.

Wesselink, L. G., Van Grinsven, J. J. M. and Grosskurth, G. (1994). "Measuring and modelling mineral weathering in an acid forest soil. Solling, Germany", in *Quantitative Modeling of Soil Forming Processes* (Eds. R.B. Bryant and R.W. Arnold) *Soil Sci. Soc. of Am.*, Special Publ. No. 39.

ANNEX 1

Description of the most important processes included in RESAM and NUCSAM

1. *Foliar uptake and foliar exudation*
   RESAM and NUCSAM:
   $$FNH_{3/fu} = frNH_{3fu} \cdot FNH_{3dd}$$
   $$FX_{fe} = krX_{fe} \cdot A_{lv} \cdot ctX_{jv}; \qquad X = Ca, Mg, K$$
   SMART: not included

2. *Litterfall and root decay*
   RESAM and NUCSAM:
   $$FX_{lf} = krX_{lf} \cdot A_{lv} \cdot ctX_{lv}; \qquad X = N, S, Ca, Mg, K$$
   $$FX_{rd} = kr_{rd} \cdot A_{rt} \cdot ctX_{rt}; \qquad X = N, S, Ca, Mg, K$$
   SMART: not included

3. *Mineralization*
   RESAM and NUCSAM:
   $$FX_{mi} = FX_{mi\,lf} + FX_{mi\,lt} + FX_{rd}; \qquad X = N, S, Ca, Mg, K$$
   $$FX_{mi\,lf} = (fr_{le} + fr_{mi} \cdot (1 - fr_{le})) \cdot X_{lf}; \qquad X = N, S, Ca, Mg, K$$
   $$FX_{mi\,lt} = kr_{mi\,lt} \cdot A_{lt} \cdot ctX_{lt}; \qquad X = N, S, Ca, Mg, K$$
   $$FX_{mi\,rn} = kr_{mi\,rn} \cdot A_{rn} \cdot ctX_{rn}; \qquad X = N, S, Ca, Mg, K$$
   Rate constants and fractions describing mineralization are given as maximum values, which are reduced for a high ground water level. Mineralization rate constants for N are reduced at low N contents.
   SMART: not explicitly included; instead net N immobilisation is calculated from the increase in N content in organic matter. Between a critical (C/N$_{cr}$) and a minimal C/N ratio (C/N$_{min}$) the immobilization rate N$_{im}$ is linearly related to the prevailing C/N ratio (C/N):

$$FN_{im} = \begin{cases} 0, & \text{if } C/N \leq C/N_{min} \\[2ex] (FN_{td} - FN_{gu} - FN_{le,min}) \cdot \dfrac{C/N - C/N_{min}}{C/N_{cr} - C/N_{min}}, & \text{if } C/N_{min} \leq C/N \leq C/N_{cr} \\[2ex] FN_{td} - FN_{gu} - FN_{le,min} & \text{if } C/N \geq C/N_{cr} \end{cases}$$

4. *Net growth*
   RESAM and NUCSAM:

$$dAst = krgrl \cdot Ast \cdot \left(1.0 - \frac{Ast}{Astmx}\right)$$

$dAbr = frbrst \cdot dAst$

in which $dAst$ and $dAbr$ are the growth of stems and branches respectively.
$FX_{gu} = dAst \cdot ctXst + dAbr \cdot ctXbr$
SMART: input to the model

5. *Root uptake*
   RESAM and NUCSAM:
   $FX_{ru} = FX_{gu} + FX_{lf} + FX_{fe} - FX_{fu} + FX_{rd};$ $X = N, S, Ca, Mg, K$
   Distribution of N over NO_3^- and NH_4^+

$$FNH_{4ru} = fr_{pr}NH_{4ru} \cdot \frac{cNH_4}{cNH_4 + cNO_3} \cdot FN_{ru}$$

$FNO_{3ru} = FN_{ru} - FNH_{4ru}$
SMART:
$FX_{ru} = FX_{gu};$ $X = N, BC (Ca+Mg), K+Na$
Distribution of N over NO_3^- and NH_4^+

$$FNH_{4gu} = FN_{gu} \cdot \frac{FNH_{3td}}{FN_{td}}$$

$$FNO_{3gu} = FN_{gu} \cdot \frac{FNO_{x\,td}}{FN_{td}}$$

6. *Nitrification and denitrification*
   RESAM and NUCSAM:
   $FNH_{4ni} = \theta \cdot D \cdot kr_{ni} \cdot cNH_4$
   $FNO_{3de} = \theta \cdot D \cdot kr_{de} \cdot cNO_3$
   SMART:
   $FNH_{4ni} = fr_{ni}(FNH_{3td} - FNH_{4gu} - FNH_{4im})$
   $FNH_{3de} = fr_{de}(FNO_{x\,td} + FNH_{4im} - FNO_{3gu} - FNO_{3im})$

7. *Protonation*
   RESAM and NUCSAM:
   $FRCOO_{pr} = \theta \cdot D \cdot kr_{pr} \cdot cRCOO$
   SMART:

$$cROO^- = cRCOO \cdot \frac{K_{pr}}{K_{pr} + cH}$$

in which $cRCOO$ is the sum of dissociated and non-dissociated organic acids.

8. *Carbonate dissolution/precipitation*

   RESAM and NUCSAM:

   $$FCa_{we\,cb} = \rho \cdot D \cdot krCa_{we\,cb} \cdot ctCa_{cb} \cdot (cCa_e - cCa)$$

   $$cCa_e = KCa_{cb} \cdot \frac{pCO_2}{cHCO_3}$$

   with cCa_e = equilibrium concentration

   pCO_2 = partial CO_2 pressure

   SMART:

   $$FBC_{we\,cb} = \theta \cdot D \cdot (cAl_t - cAl_{t-1})$$
   $$cAl = KAl_{ox} \cdot cH^3$$

9. *Weathering of primary minerals*

   RESAM and NUCSAM:

   $$FX_{we\,pm} = \rho \cdot D \cdot krx_{we\,pm} \cdot ctx_{pm} \cdot cH^{\alpha(x)}; \qquad X = Ca, Mg, K, Na$$
   $$FAl_{we\,pm} = 3FCa_{me\,pm} + 0.6FMg_{we\,pm} + 3FK_{we\,pm} + 3FNa_{we\,pm}$$

   (congruent weathering of equal amounts of anorthite (Ca), chlorite (Mg), Microcline (K) and albite (Na))

   SMART:

   $$FBC_{we} = \text{input value}$$
   $$FAl_{we\,pm} = 2FBC_{we}$$

10. *Aluminium hydroxide dissolution/precipitation*

    RESAM and NUCSAM:

    $$FAl_{we\,ox} = \rho \cdot D \cdot krElo1 \cdot \exp\,(krElo2 \cdot ctAl_{ox}) \cdot (cAl_e - cAl)$$
    $$cAl_e = KAl_{ox} \cdot cH^3$$

    with cAl_e = equilibrium concentration

    and $krElo1$ and $krElo2$ are Elovich constants

    SMART:

    $$FAl_{we\,ox} = \theta \cdot D \cdot (cAl_t - cAl_{t-1})$$
    $$cAl = kAl_{ox} \cdot cH^3$$

11. *Cation exchange*

    RESAM, NUCSAM and SMART:

    $$\frac{frX_{ac}}{frBC_{ac}^{z_x}} = KX_{ex} \cdot \frac{cX^2}{cBC^{z_x}}$$

    $X = H, Al, Mg, K, Na, NH_4$ for RESAM and NUCSAM

    $X = H, Al$ for SMART

    with

    $$frX_{ac} = \frac{ctX_{ac}}{\text{CEC}}$$

    $BC = Ca$ for RESAM and NUCSAM

    $BC = Ca+Mg$ for SMART

    z_x valence of cation X

12. *Sulphate adsorption*

    RESAM, NUCSAM and SMART:

$$ctSO_{4ad} = \frac{SSC \cdot KSO_{4ad} \cdot cSO_4}{1 + KSO_{4ad} \cdot cSO_4}$$

13. *Dissolution/speciation of inorganic C*
    RESAM and NUCSAM:

$$cHCO_3 = KCO_2 \cdot \frac{pCO_2}{cH}$$

	Entity	Constituent	Process		Compartment	
A	amount (kg ha^{-1})	N	dd	dry deposition	ac	adsroption complex
c	concentration in the	NO_2	de	denitrification	ad	sorption site
	soil solution (mol$_c$ m^{-3})	NO_3	dw	wet deposition	cb	carbonates
ct	content (mmol$_c$ kg^{-1})	NH_3	ex	exchange	lv	leaves/needles
CEC	cation exchange	NH_4	fe	foliar exudation	ox	oxides
	capacity (mmol$_c$ kg^{-1})	S	fu	foliar uptake	pm	primary minerals
D	layer thickness (m)	SO_2	gu	net (growth) uptake	rt	roots
fr	fraction ($-$)	SO_4	le	leaching	st	stems
fp	preference factor ($-$)	Ca	lf	litterfall		
F	flux (mol$_c$ ha^{-1} yr^{-1})	Mg	mi	mineralization		
kr	rate constant (yr^{-1})	K	ni	nitrification		
K	equilibrium constant (mol^x l^y)	Na	pr	protonation		
ρ	bulk density (kg m^{-3})	Cl	rd	root decay		
SSC	sulphate sorption	H	ru	root uptake		
	capacity (mmol$_c$ kg^{-1})	Al	we	weathering		
θ	volumetric moisture	HCO_3	td	total deposition		
	content (m^3 m^{-3})	RCOO				
		CO_2				

13

Linking Mixing Techniques to a Hydrological Framework — An Upland Application

A. J. ROBSON, C. NEAL

Institute of Hydrology, Wallingford, Oxfordshire, UK

and

K. J. BEVEN

Centre for Research on Environmental Systems and Statistics, University of Lancaster, UK

13.1 INTRODUCTION

A remarkable characteristic of upland streams is the rapid alteration in chemistry caused by a rise or fall in flow. Stream hydrology and stream chemistry are clearly inextricably linked and ideally they need to be studied alongside one another. Modelling provides a means of trying to extend understanding of catchment processes, yet few models make full use of both chemical and hydrological information. Many applications of hydrological models only make use of hydrological data and little is known about how well these models relate to chemical processes in upland catchments. Hydrochemical models frequently function at a time-scale which is too crude to pick up the full dynamics of the episodic changes seen in small upland streams over minutes and hours. Furthermore, many of the current hydrological and hydrochemical models are overparameterized or ill-defined relying on too few data signals (Beck, 1987; Hooper *et al.*, 1988). Inaccuracies in the way the models represent physical processes must affect reliability and will inevitably have important consequences for management issues such as critical load estimation and deposition control strategies (Klemes, 1986). There is thus a need to develop approaches that bring chemical and hydrological information together into a coherent whole, and which allow catchment hydrology and hydrochemistry to be modelled within a better defined structure.

Ideally, a wide spectrum of data should be used in modelling work, including different types of hydrological and hydrochemical responses, with spatially distributed measurements at appropriate time-scales. Technological advances bring us closer and closer to this ideal and

Solute Modelling in Catchment Systems. Edited by Stephen T. Trudgill
© 1995 John Wiley & Sons Ltd

modelling techniques must move forward to make use of the increasingly detailed data that new equipment can provide. Nowadays it is usual for streamflow, pH and conductivity to be recorded every 10 or 15 minutes over periods of months or years (e.g. UK Acid Waters Monitoring Network, 1991; Neal *et al.*, 1992). Continuous data records allow significantly longer time spans to be studied in a detail that is simply not practical for spot sampled data. For small upland catchments this degree of resolution is crucial; the longer term changes that occur over months and years need to be studied against the background of the highly dynamic and flow-dominated system.

In this chapter hydrological models are developed that are based on a chemical mixing framework but incorporate a quasi-physically based hydrological structure. The primary objective underlying the work has been to obtain well-defined models linking mixing components to a hydrological description of the system. For this it has been necessary to keep the representation of processes as simple as is practically possible and to minimise the number of parameters used in the models. An important component of this work has been the ability to work at an hourly time-scale by incorporating continuous chemical information. The work is illustrated by application to chemical data for the acidic upland Plynlimon catchments.

13.2   THE PLYNLIMON CATCHMENTS

The Plynlimon catchments are located in mid Wales, 24 miles (40 km) inland of the Irish Sea. They experience a cool wet maritime-mediated climate. The area is typical of much of the British uplands with thin acidic soil overlying relatively impermeable rock. Stagnopodzols are common on the steeper slopes whereas the tops of the hills are coated in eroding blanket peat. Hilltop peats support moorland vegetation while parts of the lower slopes are forested. Streams respond rapidly to rainfall and experience episodic acidification. Chemical data on rainfall stream waters and soil waters have been collected over a number of years. In addition continuous monitoring of pH and conductivity for stream and rainwater has been undertaken. Full details on hydrology and chemistry of the Plynlimon catchments are given in Kirby *et al* (1991) and Neal *et al* (1992). The modelling applications presented here centre on the Hafren stream waters. This acidic stream drains 347 ha, of which the upper half is moorland and the remainder is conifer plantation. Two monitoring sites the upper and lower Hafren sites are used. The upper site is located just below the forest/moorland divide.

13.3   A HYDROLOGICAL FRAMEWORK

For hydrochemical studies it is necessary to find hydrological models that summarize the *essential* hydrological characteristics of catchments. At the same time, a balance must be struck between the number of parameters in the model and its hydrological meaningfulness (Hornberger *et al.*, 1985). Two levels of modelling complexity are undertaken here, both based on the quasi-physically based hydrological model TOPMODEL (Beven and Kirkby, 1979; Beven *et al.*, 1984; Quinn and Beven, 1993). TOPMODEL is selected because it represents a compromise between the lumped models and the extravagant detail of the fully distributed models. The model requires the measurement/calibration of relatively few characteristics, yet retains spatial information about the catchment. A lumped model, TOP1D, is used in addition to, and for comparison with, TOPMODEL. In TOP1D the catchment is crudely represented as

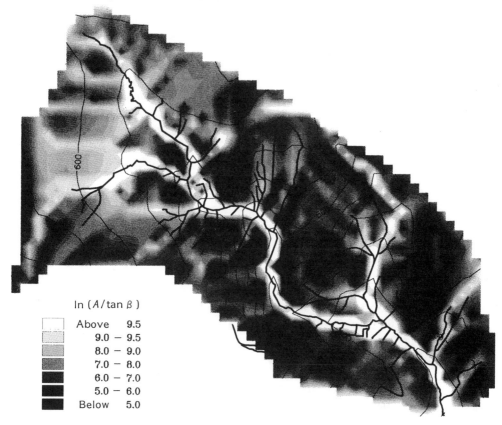

FIGURE 13.1   Map of the topographic index for the Hafren catchment, Plynlimon

one lumped reservoir but the main physical basis of TOPMODEL is retained. TOP1D has the advantage of extreme simplicity it requires very few parameters to be calibrated.

13.3.1   TOPMODEL

TOPMODEL's most distinctive feature is its use of topographical information within the model. The incorporation of topography is important because of its influence on local vegetation, soil type, soil wetness and stream chemistry (Beven *et al.*, 1988; O'Loughlin, 1986). An index function is used to give this topographical summary, and all points with the same index value are considered to be hydrologically similar. The index is calculated from a digital terrain map (DTM) across a grid covering the catchment (Quinn *et al.*, 1991) using the index proposed by Kirkby (1975), $\ln(a/\tan \beta)$. For this, a is the area draining through a grid square per unit length of contour, and $\tan \beta$ is the average outflow slope from the square. The index aims to give good distinction between wet and dry parts of the catchment (Figure 13.1).

In its basic form, TOPMODEL distinguishes between two sources of stream water: water draining from subsurface saturated zones and water displaced quickly from saturated and near-saturated parts of the catchment. The second of these sources, termed here "quickflow", is generated by rainfall landing on saturated contributing areas and causing a rapid response in the stream. It probably includes more than one type of flow process: overland flow from saturated areas near the stream, return flow, macropore flow or linkage via an extended stream network for saturated areas at a distance from the stream. The local predicted water table depth determines the local rate of vertical recharge through the unsaturated zone. Vertical movement is modelled by grouping together hydrologically similar areas, as determined from the $\ln(a/\tan \beta)$ index. Thus, the modelled hydrological state of the catchment depends directly on this index and so it is only necessary to make calculations for the index distribution function. It is also possible to map the predictions back into space, so that each of the modelled components allows some interpretation in terms of physical processes.

At each hourly time step, the movement of water through each part of the catchment is calculated. Rainfall enters an interception store from which evaporation takes place (Figure 13.2). When the interception store is full, the water progresses into the unsaturated zone and is allowed to flow vertically down to the saturated zone. The changes in the water table levels are updated at each time step and the flow to the stream from subsurface zones is estimated. Areas where saturation to the soil surface has occurred are found, and rainfall landing on these areas gives rise to the quick component of flow. A full mathematical description of TOPMODEL may be found elsewhere (Beven, 1986; Robson, 1993; Quinn and Beven, 1993). Here only a brief outline of the most important aspects of the model structure are outlined.

An exponentially decreasing profile of soil hydraulic conductivity is assumed in the model. Thus,

$$K(z) = K_0\, e^{-zf} \tag{1}$$

where $K(z)$ is the hydraulic conductivity at depth z (m h^{-1}), K_0 is the hydraulic conductivity at the surface (m h^{-1}) and f is a parameter describing the exponential decrease in transmissivity with depth (m^{-1}), and where all flow-related concepts, such as transmissivity, flow and

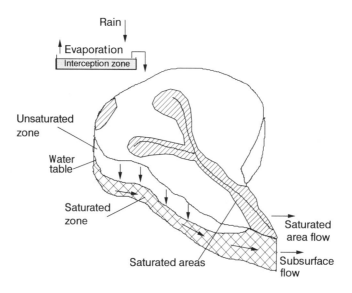

FIGURE 13.2 Schematic diagram of TOPMODEL

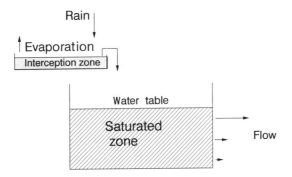

FIGURE 13.3   Schematic diagram of TOPID

conductivity, are given per unit length of contour. This provides a profile of increased conductivity near the surface — as is commonly observed in the field (Beven, 1984).

Using Darcy's law and assuming that the local hydraulic gradient is approximately parallel to the local surface slope, the total subsurface flow per unit area of catchment, qs (m h^{-1}), is given by

$$qs = T_0 \, e^{-fz-\lambda} \qquad (2)$$

where T_0 is the transmissivity to the surface (m^2 h^{-1}), λ is the area-weighted average of the ln(a/tan β) index values and z is the average depth to the water table. Furthermore, assuming that the saturated zone can be represented as always being in a state of quasi-equilibrium, then local water depth, z_i, is related to the mean depth to the water table by

$$z_i = z + (\lambda - \lambda_i)/f \qquad (3)$$

where λ_i is the local value of the index ln(a/tan β).

This information is used to determine both the spatial distribution of wetness across the whole catchment and those areas where the water table is near the surface and which contribute to the quick component of total flow.

13.3.2 TOP1D: a simplification of TOPMODEL

TOP1D is designed to be a simplified version of TOPMODEL: it retains the main hydrological features of TOPMODEL but it is not semi-distributed. The essence of the hydrological structure in TOPMODEL lies in the exponential transmissivity curve which determines the saturated flow characteristics (eqn. (1)). TOP1D uses this transmissivity curve to derive the flow out of the catchment. The idea is very simple; the catchment is thought of as one unit (Figure 13.3), with rain as input and flow as output. Rainfall is fed into an interception store and evapotranspiration is allowed to occur from this store as in TOPMODEL. Any water exceeding the interception store is added into the saturated store. The flow, Qs (m h^{-1}), out of the catchment is related to a relative mean storage deficit, s, by

$$Qs = e^{-Fs-\lambda} \qquad (4)$$

where s (m) is the storage deficit relative to the storage at which flow equals $e^{-\lambda}$ and F (m^{-1}) is the rate of decay of hydraulic conductivity with storage deficit.

This equation is analogous to the subsurface flow, qs, in TOPMODEL. The use of λ is not essential but allows easier comparison with TOPMODEL. Note that F is related to the parameter f used in TOPMODEL by $F = f/\Delta\Theta$, where $\Delta\Theta$ is the effective storage capacity of the soil. The expression of flow in relation to storage deficit, rather than water depth, is not new, but is convenient for this application; it has been used in many TOPMODEL applications (e.g. Beven *et al.*, 1984; Hornberger *et al.*, 1985). Changes in mean storage deficit are related to changes in the water table using $\Delta\Theta$.

Where water tables rise to the surface, the flow grows exponentially, effectively extending the exponential hydraulic conductivity profile above the surface. This avoids using the surface transmissivity parameter, T_0. The extension of the hydraulic conductivity profile above the soil surface still allows physical interpretations to be made. It can be thought of as describing the flow characteristics of routes that have faster flow velocities than the soil matrix, e.g. saturated overland flow, flow along flushes and macropore flow. As the catchment becomes increasingly wet, faster flow routes will increasingly dominate. For example, when saturation occurs, water may begin to flow by overland flow routes. The velocity of this water is higher than for water moving through the upper soils (Kirkby *et al.*, 1991). If the volume of water increases still further, then flow along ephemeral steams and flushes will occur with yet higher flow velocities.

13.3.3 Model application

Model calibration has been undertaken using the Simplex optimisation algorithm (Nash, 1979) with a least-squares criterion. This choice of criterion has been shown to work well for TOPMODEL (Hornberger *et al.*, 1985). Most of the parameters and constants required in TOPMODEL and TOP1D have been estimated from suitable field measurements or from separate data analysis (see Robson, 1993, for full details). Direct calibration of the exponential decay parameter (both models) and the surface transmissivity (TOPMODEL only) has been used in fitting the models. Both models are found to be sensitive to f, but the parameter T_0 used in TOPMODEL, is poorly defined. T_0 influences the modelled proportion of the catchment that becomes saturated: low transmissivity values tend to increase the chance of saturation and of quickflow. For this application, it appears that it is difficult to distinguish clearly between subsurface and quickflow components.

Three data sets were used to test the models; these were for the periods December 1987 to February 1988, mid October to mid December 1989 and February to May 1992. An example of the modelled fit is shown in Figure 13.4. R^2 values (corrected for the mean flow) ranged between 0.91 and 0.93 for calibration and 0.78 and 0.88 for cross-validation.

In general, the modelled hydrological responses given by the two models are very similar and compare well with the true hydrological response. The models give good fits in terms of the R^2 values, although a visual examination of the modelled fits indicates that there are some flaws in the predictions. These are especially notable at the beginning of a series of rainfall events, i.e. when the catchment is wetting up. There is a dilemma between choosing a model which represents processes known to be occurring or choosing a model with well-defined parameters (Hornberger *et al.*, 1985). However, rather than introduce further parameters, it seems preferable to keep to a minimal hydrological model and, if the linking of chemical information allows better identification of the model structure, to refine the hydrological structure at a later point. The development of a more elaborate hydrological model, albeit with a likely improvement in fit, is consciously avoided at this stage.

Since the difference in fit between TOPMODEL and TOP1D is small, TOP1D is in some ways the better of the two models for making hydrological predictions. Other studies have also found that simple hydrological models can explain flow just as well as complex models (Loague and Freeze, 1985; Beven, 1989). Here, the quick component of flow does not need to be distinguished from the subsurface flow in order to predict flow; it is preferable to model the flow from saturated areas simply by extrapolating the hydraulic conductivity profile above the surface. The added complexity of TOPMODEL can only be justified if spatial information (e.g. on soil chemistry) is to be incorporated for describing chemical characteristics.

13.3.4 Modelling soil water contributions

The major changes in stream chemistry which occur in response to changes in flow are linked to changes in the sources of water, and specifically to fluctuations in the water table level. The representation of subsurface flows used in TOPMODEL and TOP1D allows contributing volumes from different soil depths to be identified within the models. For TOPMODEL, this is

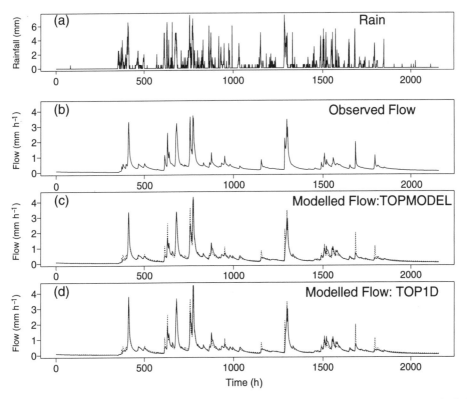

FIGURE 13.4 Hydrological results for the period December 1987 to February 1988. In (c) and (d) the modelled results are shown by a solid line and observed flow is denoted by a dotted line

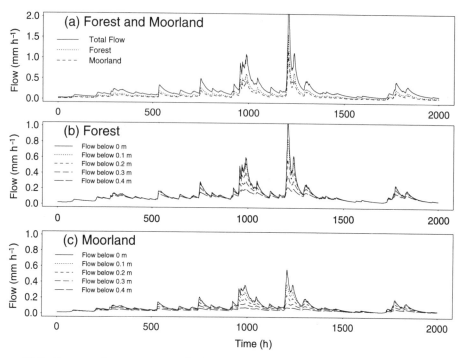

FIGURE 13.5   Sources of stream water at the lower Hafren modelled using TOPMODEL and identified (a) by vegetation thype, (b) by depth for contributions from forested areas and (c) by depth for contributions from moorland areas

achieved by estimating the depth to the water table in the vicinity of the stream banks and by integrating the bank contributions along the length of the stream. In other words, it is assumed that the soils nearest the stream are likely to be the most relevant to the chemical behaviour of stream waters. This assumption is based on the view that soil structure and chemical characteristics evolve as a result of soil water pathways (Feibig *et al.*, 1989; Mulder *et al.*, 1991) and that the soils near the stream may be thought of as having integrated out the upslope variations in soil water chemistry. Figure 13.5 shows the results of the subsurface flow calculations for the lower Hafren stream for part of 1989 using 0.1 m depth bands, chosen to cover the main contributions to stream flow. These contributions are also labelled according to the vegetation type (which influences soil chemistry; Reynolds *et al.*, 1988), i.e. whether they originate from the forest or moorland. The link between water table level and the surface of the soils is not well defined, so these bands are effectively relative to an arbitrary zero. Note that the quick component of flow has been incorporated into the estimated flow components by assuming that it is a mixed water with the same proportions as the modelled subsurface flow. This assumption is selected because (a) the quick component is small for this application and (b) the chemical signals in the stream are highly damped relative to rainfall inputs (Reynolds and Pommeroy, 1988; Neal and Rosier, 1990). These suggest that the quick component does not equate to either rainwater or soil water (see Robson *et al.*, 1992). The upper layers only make significant contributions during episodes, but they account for the major increase in flow at such times.

13.4 pH, ANC AND END MEMBER MIXING

Having set up a hydrological framework, the next stage in the process is to incorporate the chemical mixing components. The essential concept behind chemical mixing is one of describing stream water as a mixture of chemically distinctive water types called end members (Christophersen *et al.*, 1990; Hooper *et al.*, 1990; Neal and Rosier, 1990; Robson *et al.*, 1990). These end members relate to particular regions of the catchment which generate waters of recognisable chemical characteristics, perhaps as the result of different geologies, residence times and vegetation types.

In this section, the mixing application is centred around the information recorded in the continuous stream pH signal. In practice, soil and stream water pH are affected by processes such as degassing and solution/precipitation reactions, so it is preferable to work with the closely associated acid neutralisation capacity (ANC), defined as the charge difference between the strong cations and the strong anions in solution (Neal *et al.*, 1991). The detailed weekly chemical data have been used to calculate ANC as a function of total aluminium, Gran alkalinity, dissolved organic carbon and pH. Using these data, a non-linear relationship was fitted between ANC and pH ($R^2 = 0.8$). This relationship was used to transform the continuous pH to give continuous ANC. The use of ANC in the modelling work means that, as a first approximation, (a) the end members may be assumed to be of fixed composition (the pH/ANC response to increased flow is relatively consistent and is independent of rainfall) and (b) the end members can be taken to mix conservatively because ANC is a conservative chemical characteristic with respect to degassing and aluminium solubility speciation effects.

13.4.1 End member selection

At least two end members are needed to characterise the stream ANC variations seen for the Plynlimon catchments. A rainwater end member is not considered to be necessary because stream pH is largely unaffected by rainfall chemistry, even in large events (Robson, 1993). A soil water end member provides a high acidic and aluminium rich water, similar to the high-flow stream water chemistry. A further end member provides waters of a type similar to the calcium-rich and bicarbonate-bearing baseflow stream water. Soil water chemistry data are available, but for the baseflow end member it has been necessary to use observed stream baseflow waters to characterise the chemistry. Work to locate and collect the baseflow water type from the catchment has begun with the drilling of boreholes, but detailed chemical information is not yet available.

The soil end member ANC is estimated using average soil water chemical data (Reynolds *et al.*, 1988); soil ANC values for the peats and the forest stagnopodzols are -62 and -111 μeq^{-1} respectively. The baseflow estimate is found from the weekly chemical samples by fitting a regression line through the low-flow ANC data and then estimating the ANC for a selected exceptionally low flow (0.02 mm h^{-1}). For both sites, the estimated baseflow ANC value is 36 μeq l^{-1}.

13.4.2 Inserting end members into TOPMODEL

A formal link between the above end members and the catchment hydrological processes is attempted by using end member mixing principles together with the subsurface components predicted by TOPMODEL (Section 13.3.4). As a starting point, three end members are used (the moorland peat, the forest stagnopodzols and a baseflow end member). A single parameter is calibrated and this is used to describe the "threshold" depth; water flowing above this threshold depth is assumed to have a soil water chemistry (i.e. either peat or stagnopodzol chemistry), whilst water flowing beneath this depth is assumed to have a baseflow type chemistry. For simplicity, this threshold depth is initially assumed to be constant across the whole catchment. Results show a good fit to the lower Hafren data when run in calibration mode (Figure 13.6(a)) but a poor cross-validation performance. It is not clear whether this difference indicates that the assumption of fixed end member composition has been violated or whether it relates to fluctuations in the accuracy of the pH electrodes.

Predictions are made for the upper Hafren using the calibrated threshold depth from the lower Hafren application (Figure 13.6(b). The results are poor; the problem appears to originate in the method of baseflow estimation. Baseflow was estimated on the basis of all available data — but the longer data series for the lower Hafren contains some unusually dry years and it seems that the relative values of the lower and upper Hafren baseflow estimates are affected by this. Direct comparison of weekly spot sampled values for 1990−2 for the two sites shows that the upper Hafren baseflow ANC is consistently higher than the lower Hafren. To allow for this, separate baseflow chemistries were calibrated in the model and this resulted in much improved fits (Figure 13.6(c) and (d); $R^2_{lower} = 0.87$; $R^2_{upper} = 0.83$). The TOPMODEL application indicates that the baseflow ANC difference between the two sites is related to the end member composition; it does not fit in with differences caused by the modelled flow routing, i.e. differences in the end member proportions for the two catchments.

An alternative approach to linking stream chemistry and hydrology is to try and identify the ANC of each subsurface component by calibration. For this multiple end member approach, six end members are used, one for each of the six TOPMODEL subsurface contributions, and their ANC values are used as calibration parameters. If the model is good, the predicted soil end member profile should resemble the field soil data. The method gave good stream chemistry predictions $R^2_{lower} = 0.90$; $R^2_{upper} = 0.92$), but the ANC estimates were found to be very poorly defined, indicating overparameterisation. The soil ANC estimates showed some similarity to the soil chemistry profile — but not enough to be conclusive.

13.4.3 Discussion

The end member mixing and TOPMODEL approaches are compatible, to the extent that it is possible to assign an ANC to the hydrological components identified in TOPMODEL, and thereby predict stream ANC. As the approach works reasonably well for the Hafren catchment, it seems that some of the assumptions made in the end member mixing approach are justified. For example, the assumption that virtually all rainfall is chemically modified before it reaches the stream appears sensible. However, the poor results from cross-validation runs may indicate that an assumption of constant end member chemistry cannot be applied across wide time periods. It will be necessary to achieve a better standard of continuous pH measurement before this point can be resolved.

The reasons for the problems in simultaneously fitting the model to both the upper and lower Hafren are open to various interpretations. The fit can easily be improved by making a few modifications to the end members. This may indicate that the model is a reasonable representation, but that the end members are not yet well enough identified, especially in

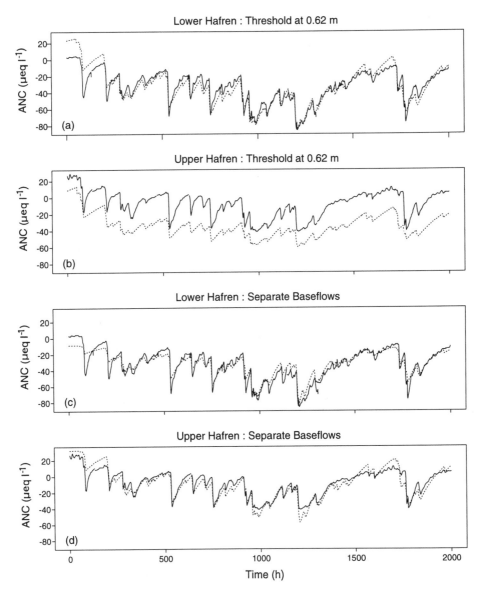

FIGURE 13.6   Observed and modelled ANC for the upper and lower Hafren. Observed values are shown by a solid line and modelled values by a dotted line. (a) Calibration to lower Hafren; (b) validation on upper Hafren using parameters from (a); (c) and (d) joint calibration on upper and lower Hafren with separate baseflow end members

relation to spatial distribution. However, it may instead mean that the basic approach is correct, but that a structure with only two or three components is too much of an oversimplification. Care must be taken in interpreting modelling results since, within reason, almost any response can be predicted if enough parameters are included. There is a danger that the introduction of extra parameters results in an overparameterisation that explains the results but is no longer a representation of the true processes.

In a sense it is reasonable, at least in retrospect, that the hydrological response of TOPMODEL can be linked fairly easily with the ANC signal. As ANC is closely related to flow, any model that explains flow must also stand a very good chance of explaining ANC. If ANC had not been well explained by the model, then there would have been clear evidence that something was incorrect somewhere. As it is, it can only be said that there is no clear evidence against the mixing hypothesis. ANC is distinct from flow, but not sufficiently independent of it; ANC responds in a very predictable way to flow. This dependency, which is seen in many determinands other than ANC, is a major reason for it being so difficult to establish the true underlying processes that operate in the catchment.

13.5 STREAM CONDUCTIVITY

Stream conductivity response differs from pH response in an important aspect; unlike pH, its dependency on flow is very variable and is significantly affected by the composition of rainfall inputs (Robson *et al.*, 1993; see Figure 13.7). Qualitative study of rainfall conductivity inputs and of stream conductivity outputs suggests the following:

1. During large events, rainfall can have a direct impact on stream conductivity. However, much of the rainwater remains in the catchment for some time to come and may affect stream response in the coming weeks and months.
2. For small rainfall events, little of the generating rainfall reaches the stream; instead relatively recent rainfall contributes.
3. Soil structure/flow pathways affect response. There appear to be marked gradients between different within-catchment sources of water. These gradients cause strong dependencies between flow and conductivity which may be maintained over long periods of time and can be enhanced by evaporation/dry deposition. Large inputs of rainwater can substantially alter these gradients.

These interpretations can be investigated by modelling the stream conductivity response. The aims are to see whether a simple hydrological model can explain stream conductivity variations and to consider the properties of this type of model and the implications for residence times. Hence, this is attempted using the simple TOP1D model. The flow of water through different soil layers is calculated and is used to predict stream conductivity from the rainfall inputs.

13.5.1 Conservative conductivity-based measures

Conductivity cannot be used as a direct tracer of water movement unless processes such as ion exchange, weathering and degassing are allowed for. To represent such processes within TOP1D is not possible at present — there are too many unknowns with regards to the chemical reactions. In order to proceed, it is necessary to transform the conductivity so as to generate a near-conservative chemical species. This involves an adjustment for the main chemical processes which affect conductivity, namely ion exchange (where H^+ is involved), weathering and dry deposition/evaporative effects. The end result can be thought of as a measure of the total number of dissolved ions in solution.

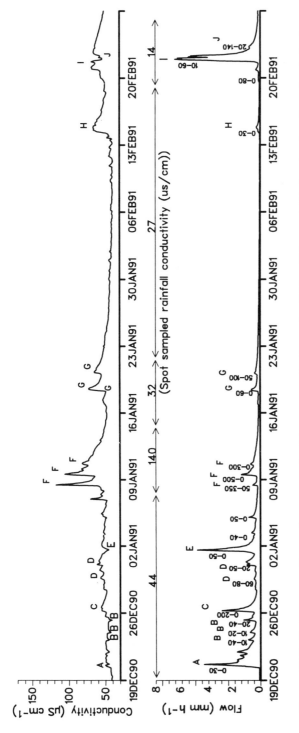

FIGURE 13.7 Hafren stream flow and stream conductivity response for December 1990 to February 1991 showing the effect of variation in rainfall conductivity. The numbers written alongside the flow hydrographs give summaries of the rainfall conductivity detected by the continuous monitors. A: large low-conductivity event on relatively dry catchment; B: low-conductivity rainfall events — stream conductivity response is inverted relative to usual storm response; C: high-conductivity rainfall event; D: water from C is flushed out; E: large low-conductivity event on wet catchment — inversion of conductivity response; F: medium flow events but high-rainfall conductivity landing on a wet catchment causes very high stream conductivity response; G, H: low-conductivity events, but still flushing out water from F; I: initial response as for H, but low conductivity of the rainfall for this large event takes over (J)

Ion exchange reactions primarily affect conductivity by causing changes in the concentration of the highly conductive hydrogen ion. The specific conductivities of all other ions in solution are substantially lower than hydrogen, and any exchange between non-hydrogen ions will only have a relatively small effect on conductivity. The effects of ion exchange involving hydrogen ions can be compensated for, by modifying the hydrogen contribution to an adjusted conductivity, so that it has a contribution similar to the other ions in solution. Weathering reactions usually lead to the formation of bicarbonate. For each molecule of bicarbonate produced from weathering, a matching base cation is generated, adding to the conductivity. The net effect of weathering on conductivity may therefore be compensated for by subtracting the conductivity attributable to the bicarbonate ion and the accompanying base cation. This gives the correction formula

$$\text{Cond}_{pa} = \text{Cond} - (0.35 - 0.07)H^+ - (0.045 + 0.07)HCO_3^- \tag{5}$$

where Cond is the measured conductivity and Cond_{pa} is the process-adjusted conductivity (Robson, 1993). The constants 0.35 and 0.045 are the respective specific conductance values for hydrogen and bicarbonate ions, and 0.07 is the average specific conductance for non-hydrogen ions in solution. For rainfall, the bicarbonate contribution is very low, whilst for the stream waters it can be estimated as a function of pH using average pCO_2 levels. Unfortunately, the stream pH records were unreliable and incomplete for the periods of interest. Because of this, it was necessary to estimate the hydrogen concentration from flow using the more accurate weekly spot sampled data; a non-linear relationship was fitted ($R^2 = 0.83$).

The effects of dry deposition/evapotranspiration need not be directly corrected for; dry deposition is modelled by incorporating an additional input, whilst evaporation is handled directly from the hydrological information in the model. Other processes affecting conductivity are expected to be unimportant relative to the above processes (see Robson, 1993).

An alternative approach to estimation of a conservative quantity is to try and relate stream chloride and stream conductivity. This is appealing because chloride is generally considered to be conserved within the catchment and because high-conductivity events are most commonly a result of sea-salt events which are high in chloride. The relationship between chloride and conductivity for the stream waters is estimated from spot sampled data (although a high level of scatter is present). The resulting estimate is found to compare very closely with the process-adjusted conductivity (Robson, 1993). Only results for process-adjusted conductivity are presented here.

13.5.2 Incorporating conductivity within TOP1D

The conductivity is modelled here by viewing the catchment as a series of layers. TOP1D provides the hydrological component of the model and is used as the basis for routing the water through the layers (Figure 13.8). At each stage, a tally is kept of the volumes and conductivities of the waters in different parts of the soil profile.

The layers are defined in terms of a range of soil moisture storage deficits. Each layer represents 10 mm of active storage per unit area of catchment, i.e. the difference in storage capacity between a layer being saturated and a layer being drained is equivalent to 10 mm of evaporation compensated rainfall. In each layer, a reservoir store is also depicted (Figure 13.8). This represents chemical/water storage which is not removed from the soils when the soil reverts from a saturated state to an unsaturated state. It is not possible to distinguish what proportion of this storage is volumetric, e.g. water held tightly within the soils, and what proportion is chemical, e.g. salt deposits building up on the soil surfaces as the soils dry out. However, a large reservoir will effectively mean that an input of rainwater has only a small immediate effect on the chemical composition of the water in the layer and that the rainwater signal is well damped. A small reservoir will mean that rainwater inputs have a very rapid effect on the soil water conductivity.

Within TOP1D there is no explicit reference to the position of the upper soil surface. The uppermost modelled soil layer may therefore represent a combination of fast flow routes, including near surface flows, macropore flows and overland flow. For the upper soil layer, the water storage is allowed to increase above 10 mm, as far as is required to generate the maximum stream flow. Within the model optimisation, a parameter is used to allow the model to select the active depth of the upper soil layer. In this application of the model, no chemical distinction is made between the interception store and the upper soil layer (although for the purposes of calculating evaporation, the interception store is still required within the model). Chemical effects of dry deposition and evaporation can then be taken account of within the upper layer. Separation of the two zones into chemically distinct entities did not improve the model structure and led to ambiguities in the allocation of dry deposition and evaporation effects between the two zones. After allowing for evaporation, approximately 73% of the process-adjusted conductivity output (and 85% of the chloride output) is accounted for by the bulk rainfall. The remainder is assumed to result from dry deposition, and these quantities are in line with estimates of dry deposition for sulphate for the area.

At each time step, water is considered to mix fully between the active water and the reservoir water in each layer (Figure 13.8). All water flowing laterally from a layer into the stream is assigned the concentration of that layer. Within each layer, infiltrating water is assumed to mix before any water is displaced. All water flowing vertically down from a layer has the

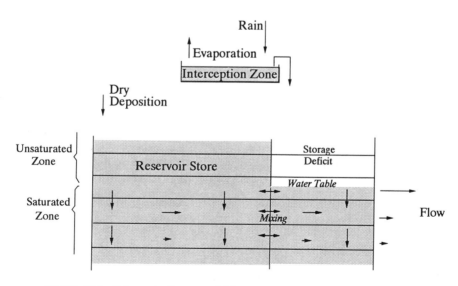

FIGURE 13.8 Schematic diagram of TOPID incorporating chemical mixing

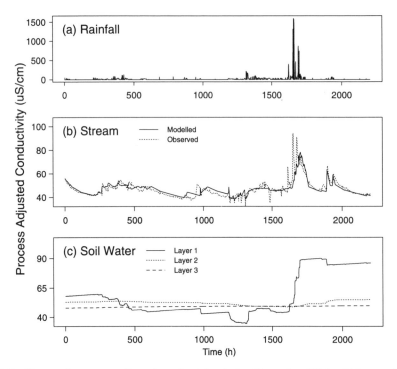

FIGURE 13.9   Changes in process-adjusted conductivity during December 1989 to February 1990 for (a) rain, (b) stream and (c) modelled soil waters. Graph (c) shows how the process-adjusted conductivity is increasingly damped with depth

concentration of the layer it is leaving. The vertical flow replenishes water that flows laterally from lower layers.

A number of non-hydrological parameters are required in the model, despite the simplistic approach used here. Parameters are required to describe the initial conditions within the soils, because the chemical compositions of the soil layers change only slowly. All soil layers are taken to have the same reservoir storage: no significant improvement is achieved when they are allowed to vary from layer to layer. Parameters are also used to locate the position of the upper soil layer and to set the initial baseflow conductivity level more finely. The last of these is allowed to vary during cross-validation runs, so as to compensate for the effects of the different initial conditions in the two data sets.

Two data sets were used to test the model, December 1989 to February 1990 and September 1991 to March 1992 (Figures 13.9 and 13.10). For the first of these periods, some exceptionally high conductivity inputs were recorded following high sea-salt events (though there were some difficulties with the continuous rainfall conductivity measurements for this period). For the second, the rainfall conductivity inputs were relatively low — so in turn the stream response is somewhat less variable. A first run of the model is used to optimise the hydrological parameter F. This is followed by a second run to locate the optimum values of parameters affecting the stream process-adjusted conductivity. Optimised runs for 1990−1 and 1992−3 gave R^2 values of 0.71 and 0.40 respectively (Figures 13.9 and 13.10). The lower R^2

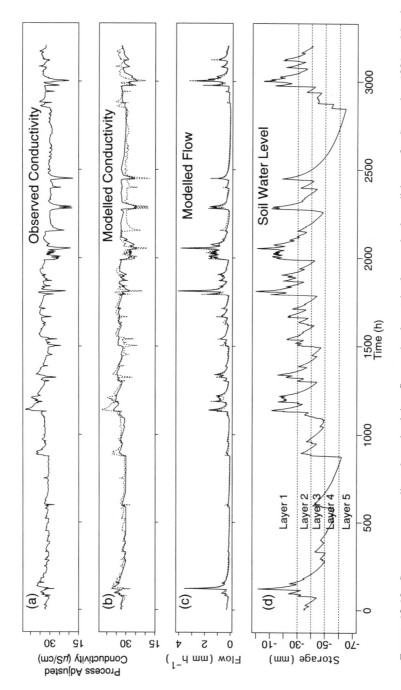

FIGURE 13.10 Stream process-adjusted conductivity, flow and volumetric changes in the soil layers for September 1991 to March 1992. In (b) and (c) solid lines show modelled results and dotted lines show observed data. In (d) the volume of water in the soils is shown

for 1992–3 is, in part, expected because there is less variation in the observed stream conductivity for this period than for 1990–1. The results mimic the general conductivity trends in the stream waters well, and the shorter-term dynamics less well. The rainfall chemistry becomes progressively more damped in the lower soil layers (Figure 13.9); a large variation occurs in the top layer, but below the third layer very little change is seen. The modelled stream chemistry variations are the result of the alterations in the chemical compositions of these stores, and also the flow-linked variations in the relative contributions from the stores (Figure 13.10).

In all the runs, the calibration results in very large fitted reservoir stores which greatly damp the rainwater signal on its route to the stream. Taking, for example, the 1990–1 run, the storage reservoir is 280 mm relative to the active storage of 10 mm. In comparison, the maximum possible water storage per layer is estimated to be about 80 mm. Thus, if the result is to be believed, it means that the storage reservoir must include chemical storage as well as volumetric water storage. Alternatively, it may point to inaccuracy in the model structure. For example, the high reservoir storage parameter may, in part, be a symptom of the use of a vertically structured model for what is in reality a horizontally dominated catchment (Neal *et al.*, 1988). A lateral transfer of water though the catchment will involve much more contact with the soils than a purely vertical movement and it would therefore be expected to increase the damping within each layer.

Cross-validation runs show that there is a significant deterioration in performance when calibrated parameters are transferred to a new data set. The observed differences in the calibrated parameter values are probably related to the differences in the rainfall conductivity inputs observed during the two periods. Such differences alter the relative importance of the various model component processes and thereby influence calibrated values. For instance, for the 1990–1 period, the high conductivity rainfall event dominates the fitting of the model. It may also be that part of the variation in parameter values is related to the inaccuracies in both input and output signals. For example, the exact timing of the very high rainfall conductivity inputs in 1990–1 is likely to have a significant effect on model calibration.

13.5.3 Investigating residence times

The time taken for rainfall inputs to reach the stream outlet will follow a probability distribution that is dependent on the physical characteristics of the catchment and on the climatic conditions before, during and following a rainfall event. Any chemical inputs occurring shortly before a prolonged dry spell will, on average, remain in the catchment for a long time, whereas chemical inputs occurring during large, multiple events will probably leave the catchment within a short period of time. Note that if chemical storage occurs within the catchment, then the distribution of residence times for process-adjusted conductivity need not be the same as that of a water molecule. It is only possible here to assess the probable storage times of the modelled signal, i.e. process-adjusted conductivity.

The time taken for a chemical input to move through the catchment is modelled by using a synthetic rainfall chemistry signal corresponding to a tracer being added to the rainfall input during a storm (Figure 13.11). As much as one-third of the tracer appears to remain in the catchment after nearly 3 months (Figure 13.11(b)). The highest tracer concentrations in the stream occur during the initial storm (labelled A) and during the next storm; concentrations at peak flows gradually tail off in subsequent storms. The amount of tracer seen at baseflow

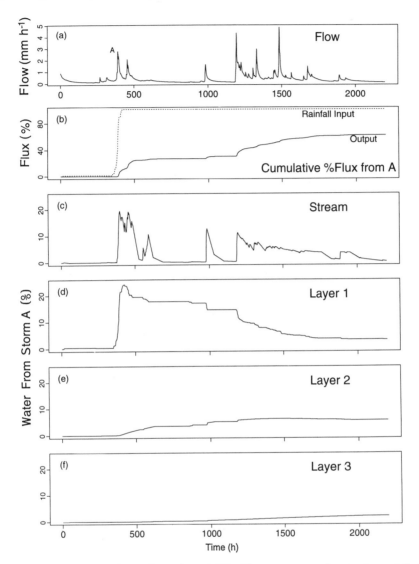

FIGURE 13.11 Modelled residence times for rainfall. The movement of water is modelled using a synthetic rainfall conductivity signal which mimics a tracer having been added to the storm labelled A (rainfall volume 80 mm; time 350−425 hours). Graph (b) shows the cumulative flux of tracer input and output from the system. The rainfall tracer input can be seen to arrive during a short period (storm A). The output of tracer via the stream takes much longer to emerge; after two months only 60 % of the tracer has left the catchment. The bottom four graphs show the percentage of water which is from storm A (this is equivalent to tracer concentration) for the stream and the three uppermost soil layers

remains low throughout, but increases very slowly over the period. There is a noticeable difference between the contribution of the tracer at baseflow and at high flows, even 2−3 months after the event. This indicates that the labelled rainfall water is poorly mixed with the water originally present in the catchment. The layered structure of the soils explains why this is occurring. The upper soil layer responds quickly, whilst lower layers respond increasingly slowly.

13.5.4 Discussion

The results presented here indicate that the stream conductivity response is a result of the chemical inputs in rainwater being stored for differing lengths of time. The layered modelling approach can largely explain the damped stream behaviour, providing there exist sizeable residual stores in the soil. As these stores are larger than can simply be accounted for by water storage, it seems that ions are effectively being stored within the catchment. This means that there may be important chemical concentration gradients within the soils, perhaps related to uptake of water by trees. If this is the case then it may, in the future, be preferable to use a different formulation for the control of mixing between ions in the reservoir stores and ions in the active storage area. Given the similarity between process-adjusted conductivity and the chloride estimate, it may also imply that chloride is involved in storage processes within the catchment. The alternative is that the process-adjusted conductivity is not behaving in as conservative a manner as is hoped (and that the chloride estimate is in error). This problem could be resolved by future continuous measurement of chloride inputs and outputs and of ^{18}O or deuterium. The isotope data would provide a test of whether chloride is fully conservative (where "conservative" is taken to mean that it has no chemical interaction with the catchment — all changes are purely related to hydrological effects). Such data would also be invaluable in trying to address the finer dynamics of the stream water variations which were not predicted particularly well by TOP1D. It is likely that some of this unexplained variation is due to chemical reactions (not involving chloride or isotopes) which affect the adjusted conductivity.

The results show how flow-related changes can occur, apparently in the absence of dynamic chemical reactions, because of the chemical gradients that result from the layered structure. In other words, a very simple explanation for stream chemical response seems possible. However, although the model is able to explain the basics of the stream response, the model structure is not particularly well defined. For example, a modification of the model structure to allow for bypass flow gave a very similar fit — but with altered parameter values. The long residence times and slow responses of the lower soil layers to rainfall inputs means that residence times for rainwaters are very long tailed and that very long data records would be needed to determine what is happening in the lowest soil layers. If model structure is to be better identified, then future efforts will have to work towards expanding the data resource and towards making some internal measures of the system — a non-trivial task. However, the first step in improving the model should be the direct measurement of conservative characteristics.

The results highlight the need for the continuing collection of both continuous and spot-sampled data. Long records of data will be required before catchment processes can be fully understood. The continuous data provide the detail and the spot-sampled data the breadth. To expect to have unbroken, accurate continuous data taken over years is unrealistic and costly. In the future, it may be best to develop methods of combining short- and long-term data sets efficiently for modelling work. For example, it might be possible to run a "starter" on bulk rainfall data to generate the initial conditions and then to use this to provide more robust initial conditions for the detailed modelling work presented here.

The results have an important bearing on chemical hydrograph separation techniques. A standard approach in techniques such as isotope separation is to use baseflow chemistry as the sole measure of within-catchment stores. The validity of this assumption must be questioned. The results found here suggest that stream water response is largely the result of within-catchment gradients. If the observed changes are attributed solely to the generating rainfall, then large inaccuracies will result. Observations in this vein have recently been made by D. R. DeWalle (personal communication) who noted a seasonality in storm ^{18}O isotope response which was a lagged function of the variations in the rainfall inputs. In other words, the isotopic response to an event was determined by the gradients within the catchment, and not just by rainfall and baseflow chemistry. Isotope separations which assume that baseflow is representative of all water in the catchment could seriously over- or underestimate the rainwater contribution, depending on the state of the catchment before the storm. It is therefore concluded that long series of isotope data are required before hydrograph separation can be sensibly undertaken.

13.6 SUMMARY AND CONCLUSIONS

Mixing techniques have been applied within a hydrological framework and used to model pH/ANC and conductivity for the Hafren stream. For ANC, good predictions are possible, but they do not provide conclusive evidence about mixing hypotheses — the ANC is too highly correlated with flow to provide sufficiently independent evidence. Even so, the study of ANC has been beneficial in that it has prompted the drilling of boreholes within the catchment. This investigation may in itself provide greater insight into the hydrology of the uplands.

For conductivity, the data have been modelled using a layered structure which allows the rainfall conductivity signal to be traced as it moves through the catchment. The model is able to reproduce the broad features of the stream conductivity response — although it fails to match all of the short-term dynamics. The model implies that there are significant chemical stores with which rainwater mixes in the soil zones. These stores account for the high damping of the rainfall inputs. The conductivity results suggest that the ranges of residence times for rainwater are very wide. The oldest stores of water, seen mainly at baseflow, change in composition only very slowly. The upper soil waters change more rapidly, but are still highly damped relative to rainfall. The results indicate that water within the catchment cannot be treated as having a uniform chemistry. This means that isotope separations using only rainwater and baseflow components could be in error.

Taken together, it appears that the dynamic chemical response of the stream may be explained in terms of alterations in flow routing, without the need for including dynamical chemical reactions. However, even with such a straightforward strategy, the model structure is still not well defined and overparameterisation is still a problem. For the future it will only be possible to achieve better defined models if (a) further chemical measures with independent characteristics are monitored on a short-term basis, (b) conservative tracers such as isotopes are monitored and (c) internal hydrological and chemical measures of the system are taken.

REFERENCES

Beck, M. B. (1987). "Water quality modelling: a review of the analysis of uncertainty", *Water Resour. Res.*, **23**(8), 1393−1442.

Beven, K. J. (1984). "Infiltration into a class of vertically non-uniform soils", *Hydrol. Sci. J.*, **29**(4), 425−434.

Beven, K. J. (1986). "Runoff production and flood frequency in catchments of order *n*: an alternative approach", in *Scale Problems in Hydrology: Runoff Generation and Basin Response* (Eds. V. K. Gupta, and I. Rodriguez-Iturbe), pp. 107−131, Reidel, Dordrecht.

Beven, K. J. (1989). "Interflow", in *Unsaturated Flow in Hydrological Modelling Theory and Practice* (Ed. H. J. Morel-Seytoux), pp. 191−219, Kluwer Academic Publishers.

Beven, K. J. and Kirkby, M. J. (1979). "A physically based variable contributing area model of basin hydrology", *Hydrolog. Sci. Bull.*, **24**(1), 43−69.

Beven, K. J., Kirkby, M. J., Schoffield, N. and Tagg, A. (1984). "Testing a physically based flood forecasting model TOPMODEL for three UK catchments", *J. Hydrol.*, **69**, 119−143.

Beven, K. J., Wood, E. F. and Sivapalan, M. (1988). "On hydrological heterogeneity — catchment morphology and catchment response", *J. Hydrol.*, **100**, 353−375.

Christophersen, N., Neal, C. and Hooper, R. P. (1990). "Modelling streamwater chemistry as a mixture of soil water endmembers, a step towards second generation acidification models", *J. Hydrol.*, **116**, 307−321.

Feibig, D. M., Lock, M. A. and Neal, C. (1990). "Soil water in the riparian zone as a source of carbon for a headwater stream", *J. Hydrol.*, **116**, 217−238.

Hooper, R. P., Stone, A., Christophersen, N., De Grosbois, E. and Seip, H. M. (1988). "Assessing the Birkenes model of stream acidification using a multisignal calibration methodology", *Water Resour. Res.*, **24**, 1308−1316.

Hooper, R. P., Christophersen, N. and Peters, J. (1990). "Endmember mixing analysis (EMMA): an analytical framework for the interpretation of streamwater chemistry", *J. Hydrol.*, **116**, 321−345.

Hornberger, G. M., Beven, K. J., Cosby, B. J. and Sappington, D. E. (1985). "Shenandoah watershed study: calibration of a topography based, variable contributing area hydrological model to a small forested catchment", *Water Resour. Res.*, **21**(12), 1841−1850.

Kirkby, M. S. (1975). "Hydrograph modelling strategies", in *Process in Physical and Human Geography* (Eds. R. F. Peel, M. D. Chisholm and P. Haggett), pp. 69−90, Heinmann.

Kirkby, C., Newson, M. D. and Gilman, K. (1991). *Plynlimon research: the first two decades*, Institute of Hydrology Report, 109, pp. 1−187.

Klemes, V. (1986). "Dilettantism in hydrology: transition or destiny", *Water Resour. Res.*, **22**(9), 1776−1886.

Loague, K. M. and Freeze, R. A. (1985). "A comparison of rainfall runoff modelling techniques on small upland catchments", *Water Resour. Res.*, **21**(2), 229−248.

Mulder, J., Pijpers, M. and Christophersen, N. (1991). "Water flow paths and the spatial distribution of soils and exchangeable cations in an acid rain impacted and a pristine catchment in Norway", *Water Resour. Res.*, **27**, 2919−2928.

Nash, J. C. (1979). *Compact Numerical Methods for Computers*, Hilger, Bristol.

Neal, C. and Rosier, P. T. W. (1990). "Chemical studies of chloride and stable oxygen isotopes in two conifer afforested and moorland sites in the British uplands", *J. Hydrol.*, **115**, 269−283.

Neal, C., Christophersen, N., Neale, R., Smith, C. J., Whitehead, P. G. and Reynolds, B. (1988). "Chloride in precipitation and streamwater for the upland catchment of the river Severn, mid- Wales", *Hydrolog. Process.*, **2**, 156−165.

Neal, C., Robson, A. J. and Smith, C. J. (1991). "Acid Neutralisation Capacity Variations for Hafren forest streams: inferences for hydrological processes", *J. Hydrol.*, **121**, 85−101.

Neal, C., Smith, C. J. and Hill, S. (1992). *Forestry impact on upland water quality*, Institute of Hydrology Report 119, pp. 1−50.

O'Loughlin, E. M. (1986). "Prediction of surface saturation zones in natural catchments by topographic analysis", *Water Resour. Res.*, **22**(5), 794−804.

Quinn, P. and Beven, K. J. (1993). "Spatial and temporal predictions of soil moisture dynamics, runoff, variable source areas and evapotranspiration for Plynlimon, Mid-Wales", *Hydrolog. Process.*, **7**, 425−448.

Quinn, P., Beven, K., Chevallier, P. and Planchon, O. (1991). "The prediction of hillslope flow paths for distributed hydrological models using digital terrain models", *Hydrolog. Process.*, **5**(1), 59−80.

Reynolds, B. and Pommeroy, A. B. (1988). "Hydrogeochemistry of chloride in an upland catchment in mid-Wales", *J. Hydrol.*, **99**, 19−32.

Reynolds, B., Neal, C., Hornung, M., Hughes, S. and Stevens, P. A. (1988). "Impact of afforestation on the soil solution chemistry of stagnopodzols in mid-Wales", *Water, Air and Soil Pollut.*, **38**, 55 – 70.

Robson, A. J. (1993). "The use of continuous measurement in understanding and modelling the hydrochemistry of the uplands", PhD thesis, Lancaster University, pp. 1 – 278.

Robson, A. J., Neal, C. and Smith, C. J. (1990). "Hydrograph separation using chemical techniques: an application to catchments in Mid-Wales", *J. Hydrol.*, **116**, 345 – 365.

Robson, A. J., Beven, K. and Neal, C. (1992). "Towards identifying sources of subsurface flow: a comparison of components identified by a physically based runoff model and those determined by chemical mixing techniques", *Hydrolog. Process.*, **6**, 199 – 214.

Robson, A. J., Neal, C., Hill, S. and Smith, C. S. (1993). "Linking variations in short- and medium-term stream chemistry to rainfall inputs — some observations at Plynlimon, mid-Wales", *J. Hydrol.*, **114**, 219 – 310.

UK Acid Waters Monitoring Network (1991). *Site Descriptions and Methodology*, Ensis, London.

14

The MAGIC Model Approach to Assessing Environmental Impacts of Land-Use Change and Atmospheric Pollution

P.G. WHITEHEAD

Department of Geography, The University of Reading, UK

14.1 INTRODUCTION

Acidification may be regarded as essentially a problem over two very different time-scales. Short-term fluctuations in acidification caused by the flushing of near-surface waters, or snowmelt, are generally driven by the hydrological processes operating in the catchment. The time-scales of these events are in the order of hours, or at most days, and the level of acidity will be largely controlled by the ability of the catchment to neutralise incoming acidity within the catchment hydrological response time. On the other hand, there are long-term changes in catchment acidity that occur over decades and even longer periods of time. These trends arise from the chemical processes operating on the catchment soils which slowly reduce the catchment buffering capacity.

From a modelling point of view it is important to recognise the two differing time-scales so that appropriate models can be developed and applied.

14.1.1 Short-term models

Most of the models have been designed to investigate the short-term responses of catchments on an hourly or daily time-scale. The simplest of the models is the time-series approach described by Whitehead *et al.* (1986a) in which H^+ or other variables of interest are forecast from flow, as illustrated in Figure 14.1. However, these models are derived empirically and are therefore catchment-specific and cannot be used for long-term prediction.

At a second level of sophistication is the Birkenes model (Christophersen *et al.*, 1982, 1984) and the PULSE model (Bergstrom *et al.*, 1985). The Birkenes model is illustrated in Figure 14.2 and shows a two-box hydrological representation coupled with chemical equations that describe the principal processes thought to be operating. All the major cations and anions are simulated. The Birkenes model has been evaluated in a number of studies (Seip and Rustad, 1983; Grip *et al.*, 1986; Seip *et al.*, 1986) and has been used to investigate the effect of changing hydrology on stream chemistry (Whitehead *et al.*, 1986b). For example, Figure 14.3

Solute Modelling in Catchment Systems. Edited by Stephen T. Trudgill
© 1995 John Wiley & Sons Ltd

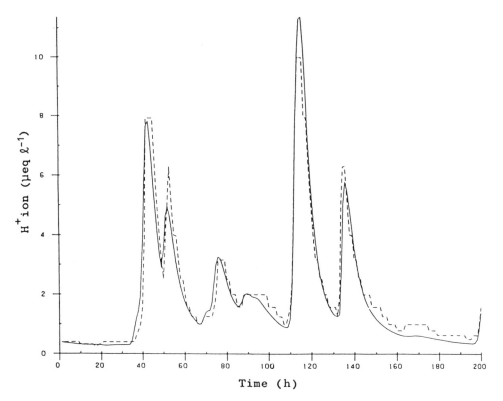

FIGURE 14.1   Simulated (——) and observed (-----) H⁺ ion in the White Laggan, Loch Dee, Scotland, based on the flow model

shows simulated H⁺ and Al³⁺ stream chemistry as a function of changing baseflow in a catchment. Thus land-use change involving a change in drainage patterns does significantly affect catchment hydrology and hence stream chemistry.

14.1.2   Long-term trend models

There are three models that have been designed specifically to simulate long-term trends in acidification. These include the Trickle Down model of Schnoor *et al.* (1984), the RAINS (regional acidification system) model of Kamari *et al.* (1984) and the MAGIC (model of acidification of groundwaters in catchments) model of Cosby *et al.* (1985a, 1985b, 1985c).

The Trickle Down model simulates long-term weathering processes but again has not been calibrated extensively against field data. The RAINS model makes use of the Birkenes model coupled to an air pollution model to simulate how changes in emission levels will affect soil and water chemistry. The model simulates monthly deposition, water and soil chemistry on a 50 × 50 km grid and can be used to map the effect of control strategies throughout Europe. Again it has not been fully tested against historical data.

The MAGIC model is perhaps the most sophisticated of the "trend" models in terms of hydrochemistry. A detailed description is presented in the next section and this model has been extensively applied to a wide range of catchments (see Cosby *et al.*, 1985a. 1985b or 1985c, 1986; Whitehead *et al.*, 1988).

14.2 MAGIC (MODEL OF ACIDIFICATION OF GROUNDWATERS IN CATCHMENTS)

In order to estimate surface water chemistry in response to historical and future deposition patterns, mathematical models must (a) be based on physical, chemical and biological processes that control catchment response, (b) treat interacting processes simultaneously and (c) be capable of representing long-term responses. MAGIC (model of acidification of groundwater in catchments) provides a tool by which soil processes can be simultaneously and quantitatively linked to examine the impact of acid deposition on surface water chemistry over time-scales of several years to several decades. The model was originally developed and tested for catchments in Shenandoah National Park, Virginia (USA), and has recently been adapted for catchments in Scotland (Loch Dee, Loch Grannoch), in Wales (Plynlimon, Llyn Brianne) and in Norway

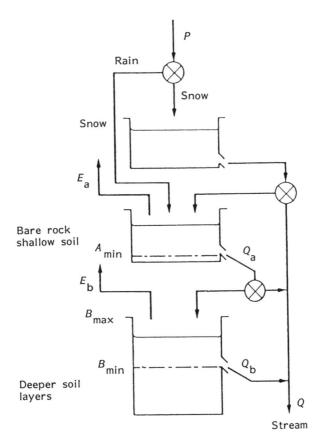

FIGURE 14.2 Hydrological model used for Harp Lake catchment and main processes operationg: *P*, precipitation; *E*, evapotranspiration; *Q*, water flux

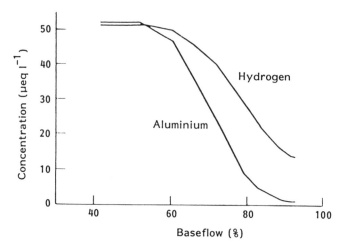

FIGURE 14.3   Maximum H^+ and Al^{3+} concentrations in the stream, showing variation over a range of baseflow conditions (three-year simulation 1977–80)

(Lake Hovvatn, and the RAIN project catchments at Sogndal and Risdalsheia). The model has also been applied in regional studies of Norway, Wales and Scotland. Figure 14.4 shows a typical simulation for MAGIC for Loch Grannoch and compares the MAGIC response with the reconstruction of pH from palaeo-ecological data (Battarbee *et al.*, 1985).

The processes on which the model is based are:

(a)   anion retention by catchment soils (e.g. sulphate adsorption);
(b)   adsorption and exchange of base cations and aluminium by soils;

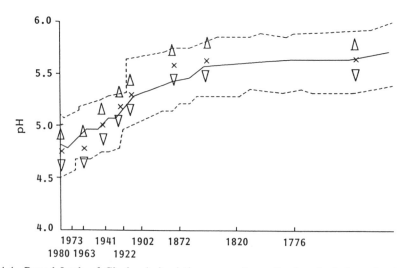

FIGURE 14.4   Round Loch of Glenhead simulation comparison of palaecological results (x MAGIC simulation, △ upper 95% confidence bound, ▽ lower 95% confidence bound, —— palaeoecological pH reconstruction, ----- mean and 95% confidence bounds

(c) alkalinity generation by dissociation of carbonic acid (at high CO_2 partial pressures in the soil) with subsequent exchange of hydrogen ions for base cations;
(d) weathering of minerals in the soil to provide a source of base cations;
(e) control of Al^{3+} concentrations by an assumed equilibrium with a solid phase of $Al(OH)$;

MAGIC simulates these processes using:

(a) a set of equilibrium equations which quantitatively describe the equilibrium soil processes and the chemical changes that occur as soil water enters the stream channel;
(b) a set of mass balance equations which quantitatively describe the catchment input−output relationships for base cations and strong acid anions in precipitation and stream water;
(c) a set of definitions which relate the variables in the equilibrium equations to the variables in the mass balance equations (a detailed description of the model is given by Cosby *et al.*, 1985a, 1985b, or 1985c), and a summary of its conceptual basis is presented in the Appendix).

The principal data requirements for MAGIC are a knowledge of rainfall and runoff quantity and quality, the sequence of deposition over time (taken from emission records or from atmospheric transport modelling studies), information on soil characteristics (such as bulk density, depth, cation exchange capacity, sulphate adsorption rates and base saturation), temperature and CO_2 levels. Parameters in the models such as selectivity coefficients, weathering rates, nitrate and ammonia uptake rates are determined by calibrating the model against the catchment data. Land use is an important factor in acidification because of the scavenging effect of dry particles and acid mist by vegetation. For example, forested catchments enhance the acid loading by up to 80% compared to moorland catchments. Such effects are incorporated into MAGIC and their importance is illustrated in the next section.

14.3 USING MAGIC AS A DIAGNOSTIC AID

As previously discussed, acidification processes are complex, interacting and not easily identified from field observation. A model such as MAGIC can assist in understanding the various processes operating. The model allows separation of the differing factors and the establishment of their relative importance. In this section we consider a few of theses factors, including the relative effects of sea salt and atmospheric pollution and the effect of afforestation.

A case study of a small Scottish catchment, Dargall Lane (a subcatchment of Loch Dee in Galloway, South West Scotland), is used to illustrate the effects.

14.3.1 Long-term acidification trends for Dargall Lane

Figure 14.5 shows a simulation of long-term acidity for the Dargall Lane catchment together with the sulphate deposition history which "drives" the MAGIC model. The historical simulation of pH shown in Figure 14.5 is similar to the values obtained from the diatom records of lochs in the region, in that a significant decrease in pH from 1900 onwards is inferred (Battarbee *et al.*, 1985). The steeper decline from 1950 to 1970 follows from the increased emission levels during this period. The model can also be used to predict future stream water acidity given different future deposition levels. For Dargall Lane, stream acidity trends are investigated assuming two scenarios for future deposition. Firstly, assuming deposition rates

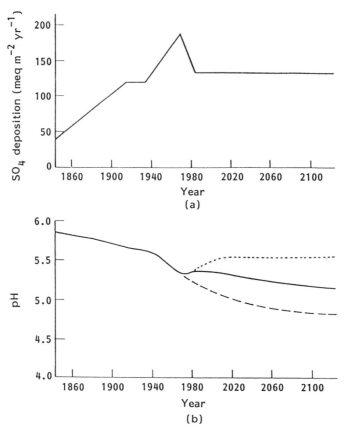

FIGURE 14.5 Sulphate deposition history used as input for the MAGIC reconstruction of pH in the Dargall Lane moorland catchment, (b) Simulation of the pH of stream water in the Dargall Lane moorland catchment assuming three sulphate deposition scenarios; — historical levels to 1984 and constant 1984 levels thereafter (see (a)); --- historical levels to 1984 and 1984 levels reduced by 50% by the year 2000, and constant thereafer; ⎯⎯⎯ historical levels to 1970 and constant 1970 levels thereafter

are maintained in the future at 1984 levels, the model indicates that, following initial recovery, the annual average stream pH is likely to decline below presently measured values. Secondly, assuming deposition rates are reduced by 50% from 1985 levels (between 1985 and 2000), the results indicate that stream water acidity will improve. Further details of the application of this model are given elsewhere (Cosby *et al.*, 1986). Note the increase in stream water pH about 1980; this follows a significant drop in sulphur emissions during the 1970s. Note also that an earlier increase in stream water acidity is predicted if there had been no reductions in emissions since 1970.

14.3.2 Afforestation

Afforested systems are more complex to model than grassland systems because the introduction of the forest perturbs a grassland ecosystem, which in itself is difficult to model. The effects of the forest root system, leaf litter layer and drainage ditches will change the hydrological pathways and this will control the nature and extent of the chemical reactions in the soil and bedrock. Further, the additional filtering effect of a tree on the atmosphere will enhance occult/ particle deposition, and evapotranspiration will increase the concentration of dissolved components entering the stream. The magnitude of these different effects varies considerably; for example, evapotranspiration from forests in the British uplands is typically of the order of 30% of the precipitation, which is almost twice the figure for grassland at 16%. This will have the consequence that the total anion concentrations within the stream and soil waters with increase following afforestation. The forest will also increase anion and cation loading due to the enhanced filtering effect of the trees on air and occult sources. The filtering effects will apply both to marine and pollutant aerosol components. To illustrate the effects of afforestation simply in terms of increased concentrations from both enhanced dry deposition and evapotranspiration, the MAGIC model has been applied to the Dargall Lane catchment assuming that a forest is developed over the next forty years. It should be noted that no allowance has been made for the effects of cation and anion uptake by the trees during their development; the incorporation of base cations into the biomass would result in an enhanced acidification effect during this period.

14.3.3 Sea salt and pollutant effects

Of critical importance is the relative and absolute contribution of marine and pollutant inputs from dry and occult deposition. Figure 14.6 shows the effects of increasing evapotranspiration from 16 to 30% over the forest growth period with varying levels of marine, pollutant and marine plus pollutant inputs. Increasing either marine of pollutant components leads to enhanced stream acidity, the greatest effects being observed when both components are present; the effect of simply increasing evapotranspiration from 16 to 30% is similar, but the changes are much smaller. The important features of these results are that the enhanced acidic oxide inputs from increased scavenging by the trees result in a marked reduction in pH levels and that there is an additive effect when both processes are combined. These reductions are much greater than the effect of evapotranspiration.

14.3.4 Varying pollutant loads

An important factor in determining stream acidity in the uplands is the level of acidic oxide deposition; rates of deposition (non-marine wet deposition and dry deposition) can vary from 0.5 to over 6 g S m^{-2} yr^{-1} and from 0.1 to over 0.5 g N m^{-2} yr^{-1}. Figure 14.7 shows the effects of such variations for both moorland and forested catchments; the highest level corresponds to areas with high atmospheric acidic oxide rates (three times the 1984 deposition levels observed in the Southern Uplands of Scotland). With increasing atmospheric acidic oxide pollution, the decline in stream pH is accelerated, the changes occur much earlier and the final pH of the stream water is lower.

14.3.5 Implications

The modelling enables assessment of the relative effects of atmospheric oxide pollution and conifer afforestation, as well as highlighting some of the topics that need further consideration.

For example, the long-term trends in stream water acidification for the grassland catchment suggest that for at least part of the upland United Kingdom, acidic oxide pollutant inputs are the dominant source of increased stream water acidity. The model predictions are similar to observations of stream acidity found in southern Scandinavia and add weight to the conclusion that such pollutant inputs are a major source of stream acidification in those countries as well.

14.4 APPLYING MAGIC IN A REGIONAL SURVEY OF WALES

MAGIC has been applied to several specific sites in Wales (e.g. Llyn Brianne; see Whitehead *et al.*, 1988). These applications demonstrate why significant acidification has occurred in moorland and forest streams in Wales. In order to evaluate acidification across Wales a regional modelling study has been undertaken.

Between October 1983 and September 1984 the Welsh Water Authority undertook a regional survey of streams and rivers within an area of Wales believed to be sensitive to acid deposition. One hundred and twenty streams were sampled on a weekly basis. During the same period

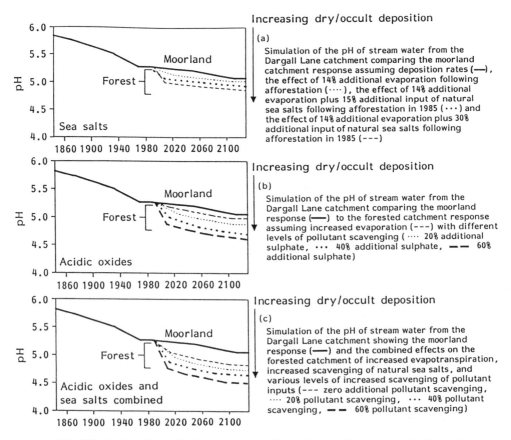

FIGURE 14.6 Effects of marine and pollutant inputs on dry and occult distribution: (a) effect of sea salts, (b) effect of acidic oxides, (c) effect of sea salts and acidic oxides combined

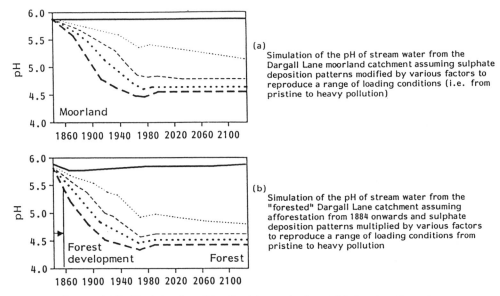

FIGURE 14.7 Variation in acidity for (a) moorland and (b) forest catchments

rainfall quality was monitored at fifty sites throughout Wales. A breakdown of the results by region shows the most acidic rainfalls falling in the uplands of mid and north Wales (Donald *et al.*, 1986).

The study area is underlain by rocks of the Cambrian, Ordovician and Silurian Periods. In the upland areas the soils are thin and base-poor. The combination of slowly weathering rocks and thin base-poor soils gives the region its vulnerability to acid deposition. Prolonged acid deposition may induce a leaching of the base cations from the soil, reducing the buffering ability and allowing the surface waters to become acidic. The study area lies mainly in the region of high rainfall acidity and is consequently of prime interest in the acidification in Wales.

14.4.1 Regional modelling procedure and simulation results

In the regional approach, the MAGIC model is repeatedly run using different sets of parameter values chosen randomly from given distributions. The ensemble of model runs is then evaluated and compared with the observed distributions of water quality across the region obtained from the analysis of the Welsh Water Survey data. this technique is termed Monte Carlo analysis and the technique allows the region as a whole to be simulated. This regional approach has been developed by Cosby *et al.* (1990) and Hornberger *et al.* (1986, 1987) in an analysis of 700 Norwegian lakes. A similar approach has been by Musgrove *et al.* (1990) in an analysis of the Galloway region of Scotland. In order to calibrate the regional model, the input parameter distributions are adjusted until the observed stream quality distributions are matched. The model can then be used to evaluate changes in distribution over time. Figure 14.8 shows the simulated distribution of calcium, magnesium, alkalinity and sulphate against the observed distribution. In all cases the simulated distribution is close to the observed distribution, suggesting that the model has captured the principal features of water quality across the region.

14.4.2 Regional changes in chemistry over time

Of particular interest is the question of how the water quality distributions for the region have changed over time. Since the model simulates the pre-industrial conditions the data from the ensemble runs in 1840 can be investigated. As shown in Figure 14.9, the 1840 distributions for sulphate and alkalinity are very different from present-day distributions. Sulphate levels are much lower and alkalinity is significantly higher. The temporal changes in the entire regional distributions have also been investigated under an assumed driving deposition sequence of 30% reduction in excess sulphate deposition linearly between present day and 2000. Figure 14.9 shows the effects of changing excess sulphur concentrations in rainfall on the stream sulphate and alkalinity distributions. As shown, whilst sulphate chemistry in the streams is reduced there is not a major shift in alkalinity. This is probably because the base saturation levels on the soils are low and even a 30% reduction in sulphate is insufficient to achieve a significant recovery on the base-poor Welsh soils.

14.5 USING MAGIC TO INVESTIGATE BIOLOGICAL CHANGES

Biological models have been developed by Ormerod *et al.* (1987, 1988) for predicting the effects of acidity on fisheries and invertebrates in Welsh streams. In the case of fisheries, trout density can be predicted from a knowledge of aluminium, hardness and streamflow.

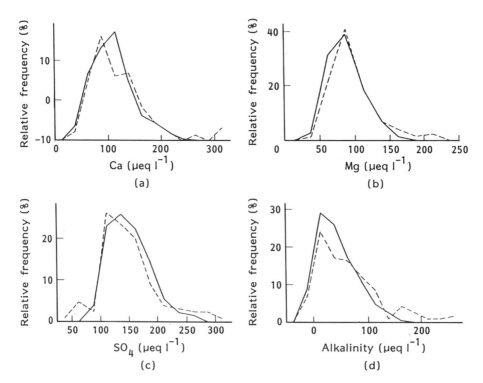

FIGURE 14.8 Simulated distribution (——) against observed distributions (-----) of (a) calcium, (b) magnesium, (c) sulphate and (d) alkalinity

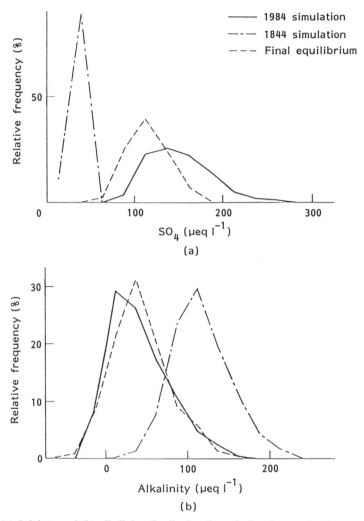

FIGURE 14.9 (a) Sulphate and (b) alkalinity distribution through time in a regional survey of Wales

The effects of acidity on trout density and stream invertebrate groups has been explored under four scenarios investigated using MAGIC applied to three streams at Llyn Brianne (Whitehead *et al.*, 1988). The scenarios involved in each catchment were:

1. Moorland with sulphate deposition constant at 1984 levels
2. Moorland with sulphate deposition progressively reduced to 50% of 1984 levels
3. Conifer forest from 1958 onwards and sulphate deposition at 1984 levels
4. Conifer forest from 1958 onwards, with sulphate deposition at 50% of 1985 levels

The MAGIC simulation for each scenario provides predicted stream chemistry which is used in the biological models to investigate trout density and invertebrate assemblages over time.

The patterns generated by the models indicate a considerable reduction in salmonid abundance in soft-water streams in the upper Tywi between the nineteenth century and the present day, a feature supported by changes in the likely survival of brown trout over this period. The acidifying influences of conifer forest severely exacerbate aluminium concentrations in the soft- water streams, almost totally eliminating fish populations and causing change amongst invertebrates by 1984. Even though the density model indicated that some fish would be present under the forest scenario, their survival would be severely limited. Moreover, the influences of forests on fish and invertebrates would not be reversed, according to the model, in spite of deposition reduction by 50%. Finally, the models also suggest that, despite reduced deposition, pronounced increases in fish abundance will not occur in moorland streams in the upper Tywi. However, the absence of forest with reduced deposition could at least support some salmonids in all the streams studied. If extrapolated to other areas, this feature could be important because of the nursery functions fulfilled by Welsh headwaters.

14.6 PROBLEMS AND PREDICTIONS WITH MAGIC

Any model is a simplification of reality and must therefore be subject to parametric uncertainty. Also, many inputs cannot be accurately measured. For example, although acid deposition is of prime importance in MAGIC, real data on historic patterns of deposition are generally lacking, and absent for the Welsh and Scottish uplands. The temporal sequence of changing deposition is, therefore, an estimate. Additionally, soil chemistry is lumped on a catchment basis, which might not truly mimic chemical behaviour in the different soil types and soil horizons that occur at upland sites. The chemical processes themselves are subject to considerable debate. In MAGIC the release of aluminium is controlled largely by reactions that assume the solubility of gibbsite ($Al(OH)_3$), which is now known to be a scarce mineral in many catchments to which MAGIC has been applied. The ecological models would be particularly sensitive to errors in aluminium chemistry, in view of its prominent place in the empirical relationships. Recently Neal (1988) has shown that aluminium chemistry can be adequately predicted using a two-component mixing model and such ideas can be tested by a new two-layer model of MAGIC. However, despite the many cautions governing the use of MAGIC, it has been able to reproduce trends in pH (see Figure 14.4) and regional water quality with reasonable accuracy and it is able to predict the changes in anions and cation chemistry following a land-use change such as afforestation.

The use of MAGIC to model acidification due to forestry does, however, carry the major assumption that the key acidifying process in these cases is due to increased evapotranspiration and dry/occult deposition. However, other acidifying processes are possible in afforested catchments. The reduced residence time of water because of plough furrows at afforested sites may be particularly important for MAGIC, altering the time available for weathering reactions. Additionally, MAGIC appears to be particularly sensitive to changes in soil pCO_2, for which field data are available for the uplands (Neal and Whitehead, 1988). the Net effect of these factors might indicate that the scenarios of change are probably underestimates of the real change.

As with any model, confidence in the prediction will only be increased following a wide range of applications. MAGIC has proved to give a good representation of chemistry and trends in many catchments in the United Kingdom, Scandinavia and the United States and has been successfully applied in three regional studies (Wales, Scotland and Norway). It is without doubt the most sophisticated model developed to date but must be used with caution. As a tool to evaluate acidification processes and interactions it is particularly valuable.

REFERENCES

Battarbee, R. W., Flower, R. J., Stevenson, B. C. and Rippey, B. (1985). "Lake acidification in Galloway: a palaeoecological test of competing hypotheses", *Nature*, **134**, (6009), 350−352.

Bergstrom, S., Carlsson, B., Sandberg, G. and Maxe, L. (1985). "Integrated modelling of runoff, alkalinity and pH on a daily basis", *Nord. Hydrol.*, **16**, 89−104.

Christophersen, N., Seip, H. M. and Wright, R. G. (1982). "A model for stream-water chemistry at Birkenes, a small forested catchment in southernmost Norway", *Water Resour. Res.*, **18**, 977−966.

Christophersen, N., Rustad, S. and Seip, H. M. (1984). "Modelling stream-water chemistry with snowmelt", *Phil. Trans. R. Soc. B.*, **305**, 427−439.

Cosby, B. J., Wright, R. F., Hornberger, G. M. and Galloway, J. N. (1985a). "Modelling the effects of acid deposition: assessment of a lumped parameter model of soil water and streamwater chemistry", *Water Resour. Res.*, **21**, 51−63.

Cosby, B. J., Wright, R. F., Hornberger, G. M. and Galloway, J. N. (1985b). "Modelling the effects of acid deposition: assessment of a lumped parameter model of soil water and streamwater chemistry", *Water Resour. Res.*, **21**, 1591−1601.

Cosby, B. J., Hornberger, G. M., Galloway, J. N. and Wright, R. F. (1985c). "Fresh water acidification from atmospheric deposition of sulphuric acid: a quantitative model", *Environ. Sci. Technol.*, **19**, 1144.

Cosby, B. J., Whitehead, P. G. and Neale, R. (1986). "A preliminary model of long-term changes in stream acidity in south western Scotland", *J. Hydrol. (Amst.)*, **1984**, 381−401.

Cosby, B. J., Hornberger, G. M. and Wright, R. G. (1990). "A regional model of surface water acidification in southern Norway: calibration and validation using survey data", in *Impact Models to Assess Regional Acidification* (Ed. J. Kamari), IIASA, Kluwer Academic Publishers.

Donald, A., Underwood, J. and Stone, J. (1986). *Results of a survey into deposition quality in Wales during 1984*, Welsh Water Authority Scientific Services Report.

Gaines, G. L. and Thomas, H. C. (1953). "(1) Adsorption studies on clay minerals. (2) A formulation of the thermodynamics of exchange adsorption", *J. Chem. Phys.*, **21**, 714−718.

Grip, H., Jannson, P. E., Jonssonm, H. and Nilsson, S.I. (1986). "Application of the Birkenes model to two forested catchments on the Swedish west coast", *Ecol. Bull.*, **37**, 176−192.

Hornberger, G. M., Cosby, B. J. and Rastetter, E. B. (1986). "Regionalization of predictions of effects of atmospheric deposition on surface waters", in *Proceedings of the International Conference on Water Quality Modelling in the Inland Natural Environment*, Bournemouth, England, 10−13 June 1986, BHRA, The Fluid Engineering Centre, Bedford.

Hornberger, G. M., Cosby, B. J. and Wright, R. F. (1987). "Analysis of historical surface water acidification in southern Norway using a regionalized conceptual model (MAGIC)", in *Systems Analysis in Water Quality Management* (Ed. M. B. Beck), Pergamon Press, New York.

Kamari, J., Posach, M. and Kauppi, L. (1984). "Development of a model analysing surface water acidification on a regional scale: application to individual basins in southern Finland", in *Proceedings of the Nordic h.p. Workshop*, nhp Report No. 10, Uppsala.

Musgrove, T. J., Whitehead, P. G. and Cosby, B. J. (1990). "Regional modelling of acidity in the Galloway region in south east Scotland", in *Impact Models to Assess Regional Acidification* (Ed. J. Kamari) IIASA, Kluwer Academic Publishers.

Neal, C. (1988). "Aluminium−hydrogen ion relationships for streams draining an acidic and acid-sensitive spruce forest", *J. Hydrol.*, **104**, 141−159

Neal, C. and Whitehead, P. G. (1988). "The role of CO_2 in long term stream acidification processes: a modelling viewpoint", *Hydrolog. Sci. J.*, **33**, 103−107.

Ormerod, S. J., Boole, P., McCahon, C. P., Weatherley, N. S., Pascoe, D. and Edwards, R. W. (1987). "Short-term experimental acidification of a Welsh stream: comparing the biological effects of hydrogen ions and aluminium", *Freshwater Biol.*, **17**, 341−356.

Ormerod, S. J., Weatherly, N. and Gee, A. S. (1988). "Modelling the ecological impact of changing acidity in Welsh streams", in *Acid Waters in Wales* (Ed. R. W. Edwards), Kluwer Academic Publishers.

Reuss, J. O. (1980). "Simulations of soil nutrient losses resulting from rainfall acidity", *Ecol. Modelling*, **11**, 15−38

Reuss, J. O. (1983). "Implications of the Ca−Al exchange system for the effect of acid precipitation on soils", *J. Environ. Qual.*, **12**, 591−595.

Reuss, J. O. and Johnson, D. W. (1985). "Effect of soil processes on the acidification of water by acid deposition", *J. Environ. Qual.*, **14**, 26−31.

Schnoor, J. L., Palmer, W. D. Jr and Glass, G. E. (1984). "Modelling impacts of acid precipitation for northeastern Minnesota", in *Modelling of Total Acid Precipitation Impacts* (Ed. J.L. Schnoor) Vol. 9, pp. 155−173, Butterworth, Boston, Mass.

Seip, H. M. and Rustad, S. (1983). "Variations in surface water pH with changes in sulphur deposition", *Water, Air and Soil Pollut.*, **21**, 217−223.

Seip, H. M., Seip, R., Dillon, P. J. and de Grosbols, E. (1986). "Model of sulphate concentration in a small stream in the Harp Lake catchment, Ontario", *Can. J. Fish. Aquat. Sci.*, **42**, 927.

Singh, B. R. (1984). "Sulphate sorption by acid forest soils, 1. Sulphate adsorption isotherms and comparison of different adsorption equations in describing sulphate adsorption", *Soil Sci.*, **138**, 189−197.

Whitehead, P. G., Neal, C., Seden-Perriton, S., Christophersen, N. and Langan, S. (1986a). "A time series approach to modelling stream acidity", *J. Hydrol. (Amst.)*, **85**, 281−304.

Whitehead, P. G., Neal, C. and Neale, R. (1986b). "Modelling the effects of hydrological changes on stream acidity", *J. Hydrol. (Amst.)*, **84**, 353−364.

Whitehead, P. G., Bird, S., Hornung, M., Cosby, J., Neal, C. and Paricos, P. (1988). "Stream acidification trends in the Welsh Uplands: a modelling study for the Llyn Brianne catchments", *J. Hydrol. (Amst.)*, **101**, 191−212.

APPENDIX: CONCEPTUAL BASIS OF MAGIC

The most serious effects of acidic deposition on catchment surface water quality are thought to be decreased pH and alkalinity and increased cation and aluminium concentrations. In keeping with an aggregated approach to modelling whole catchments, a relatively small number of important soil processes — processes that could be treated by reference to average soil properties — could produce these responses. In two papers, Reuss (1980, 1983) proposed a simple system of reactions describing the equilibrium between dissolved and adsorbed ions in the soil and soil water system. Reuss and Johnson (1985) expanded this system of equations to include the effects of carbonic acid resulting from elevated CO_2 partial pressure in soils and demonstrated that large changes in surface water chemistry would be expected as either CO_2 or sulphate concentrations varied in the soil water. MAGIC has its roots in the Reuss−Johnson conceptual system, but has been expanded from their simple two-component (Ca−Al) system to include other important cations and anions in catchment soil and surface waters.

In MAGIC it is assumed that atmospheric deposition, mineral weathering and exchange processes in the soil and soil water are previously responsible for the observed surface water chemistry in a catchment. Alkalinity is generated in the soil water by the formation of bicarbonate from dissolved CO_2 and water:

$$CO_2 + H_2O = H^+ + HCO_3^-$$ (1)

Bicarbonate ion concentrations in soil water are calculated using the familiar relationships between the partial pressure CO_2 (P_{CO_2} atm) aznd hydrogen ion activity in the soil water:

$$[HCO_3^-] = K_C \frac{P_{CO_2}}{[H^+]} \qquad (2)$$

where the combined constant K_C is known for a given temperature.

The free hydrogen ion produced (eqn (1)) reacts with an aluminium mineral in the soil:

$$3\ H^+ + Al(OH)_3\ (s) = Al^{3+} + 3\ H_2O \qquad (3)$$

The MAGIC model assumes a cubic equilibrium relationship between Al^{3+} and H^+. The equilibrium expression for this reaciton is

$$K_{Al} = \frac{[Al^{3+}]}{[H^+]^3} \qquad (4)$$

where the brackets indicate aqueous activities.

Classically this relationship describes $Al(OH)_3$ solubility controls. However, as in most previous modelling studies where a cubic relationship is used, it represents potentially a variety of chemical reactions. As such, the equilibrium constant does not need to have the value for the solubility product for gibbsite; e.g. several aqueous complexation reactions of Al^3 are included in the model (Cosby *et al.*, 1985;, 1985b). These reactions are temperature-dependent and appropriate corrections for temperature and ionic strength are made in the model.

Generally, the cation exchange sites on the soil matrix have a higher affinity for the trivalent aluminium cation than for di or monovalent base cations. An exchange of cations between the dissolved and adsorbed phase results:

$$Al^{3+} + 3\ BCX(s) = AIX_3(s) + 3\ BC^+ \qquad (5)$$

where X is used to denote an adsorbed phase and BC^+ represents a base cation. The net result of these reactions is the production of alkalinity (e.g. $Ca(HCO_3)_2$). As CO_2 partial pressure or the availability of base cations on the soil exchange sites increases, the equilibrium reactions proceed further to the right-hand side of eqn (5) in each case, resulting in higher alkalinity.

When the solution is removed from contact with the soil matrix and is exposed to the atmosphere (i.e. when soil water enters the stream channel), the CO_2 is lost to the atmosphere. Because the solution is no longer in contact with the soil matrix, cation exchange reactions no longer occur. The alkalinity and base cation concentrations are thus unchanged.

If the exchangeable base cations on the soils become depleted, less aluminium is exchanged from the soil water (eqn (3)) and the Al^{3+} concentration in the water entering the stream is higher. As the stream water loses CO_2 and the pH begins to rise, the solubility of aluminium species in the stream is exceeded and a solid phase of aluminium precipitates. These aluminium precipitation reactions retard the increase of stream water pH for the case where exchangeable cations are less available.

Less adsorption of aluminium by the soils also decreases the soil and surface water alkalinity. Consider an abbreviated definition of the alkalinity if soil and surface waters:

$$Alk = (HCO_3^-) - (H^+) - 3(Al^{3+}) \qquad (6)$$

where the parentheses indicate molar concentrations. It is apparent that as the ability of the catchment soils to exchange Al^{3+} declines and aluminium and hydrogen ion concentrations increase, the alkalinity of the solution must decline, even though the source of HCO_3^- is not affected.

The process of acidification is controlled in part by the rate at which the exchangeable base cations on the soil are depleted. This in turn is affected by the rate of re-supply through weathering of base cations from primary minerals and the rate of loss through leaching of base cations from the soil. Leaching of base cations is affected mainly by the concentration of strong acid anions (i.e. SO_4^{2-}, meq m^{-3}) in soil water is assumed to follow a Langmuir isotherm (Singh, 1984):

$$E_s = E_{mx} \frac{(SO_4^{2-})}{C + (SO_4^{2-})} \tag{7}$$

where E_{mx} = maximum adsorption capacity of the soils (meq kg^{-1}) and C = half-saturation concentration (meq m^{-3}).

If anions derived from atmospheric deposition are accompanied by H$^+$, as is the case for acid deposition, the excess H$^+$ will initially displace base cations from the soil exchange sites. As the base saturation declines, aluminium and hydrogen ions become increasingly important in maintaining the ionic charge balance in solution. The water delivered to the stream becomes more acidic as the acidic deposition persists.

The model assumes that only Al^{3+} and four base cations are involve din cation exchange between soil and soil solution. The exchange reactions are modelled assuming an equilibrium-like expression (Gaines and Thomas, 1953):

$$S_{AlBC} = \frac{[BC^{2+}]^3 \, E_{Al}^2}{[Al^{3+}]^2 \, E_{BC}^3} \quad \text{or} \quad S_{AlBC} = \frac{[BC^{2+}]^3 \, E_{Al}^2}{[Al^{3+}]^2 \, E_{BC}^3} \tag{8}$$

For divalent or monovalent base cations respectively, where the brackets indicate aqueous activities, S_{AlBC} is a selectivity coefficient (Reuss, 1983) and the E_{xx}'s indicate exchangeable fractions of the appropriate ions on the soil complex. If the amount of Ca^{2+} on the soil of a catchment is given by X meq kg^{-1}, then

$$E_{Ca} = \frac{X}{CEC} \tag{9}$$

where CEC is the cation exchange capacity of the soil (meq kg^{-1}.

The base saturation (BS) of the soil is then the sum of the exchangeable fractions of base cations:

$$BS = E_{Ca} + E_{Mg} + E_{NA} + E_K = 1 - E_{Al} \tag{10}$$

If the aluminium – base cation exchange equations in the model (eqn (8)) are combined with the aluminium solubility equation (eqn (4)), the resultant equations are the Gaines – Thomas expressions for hydrogen ion – base cation exchanges. The parameters describing the cation exchange process in the model are the selectivity coefficients, S_{AlBC} (one coefficient for each base cation, Ca^{2+}, Mg^{2+}, Na$^+$, K$^+$), and the soil cation exchange capacity, CEC.

Further details of the equations and the model structure have been given by Cosby *et al.* (1985a).

15

Integrating Field Work and Modelling — The Birkenes Case

HANS M. SEIP

Department of Chemistry, University of Oslo, Norway

NILS CHRISTOPHERSEN

Department of Informatics, University of Oslo, Norway

JAN MULDER

Department of Soil Science and Geology, Agricultural University, Wageningen, The Netherlands

and

GEIR TAUGBØL

Department of Chemistry, University of Oslo, Norway

15.1 INTRODUCTION

In the 1960s measurements indicated that the precipitation in Europe was becoming more acidic. At that time serious loss of fish in Norwegian rivers and lakes had been observed. It seemed natural to conclude that acid precipitation was the cause of fish death, although fish loss, presumably caused by acid water, had been reported in Norway as early as around the turn of the century. Concern about acidification led to intense research activities. In 1972 the OECD launched the "Cooperative Technical Programme to Measure the Long-Range Transport of Air Pollutants" and in Norway the Ministry of the Environment and two research councils launched the SNSF Project ("Acid Precipitation — Effects on Forest and Fish"). It was confirmed that the loss of fish populations was due to acid water. (Later it was shown that increased concentrations of inorganic aluminium species were more important than acidity *per se*). However, during the SNSF Project conflicting opinions about the major processes controlling stream water chemistry became evident. In the report after the first phase of the SNSF Project (Braekke, 1976), it was not questioned that acid precipitation was the dominant cause of surface water acidification and loss of fish. Rosenqvist (1977, 1978), however, argued that altered land

Solute Modelling in Catchment Systems. Edited by Stephen T. Trudgill
© 1995 John Wiley & Sons Ltd

use causing vegetation changes (e.g. forest regrowth) had led to more acid soils, which in turn acidified the water. He stated that stream water in the Birkenes catchment in southernmost Norway (cf. Section 15.2) would be acid at high discharge irrespective of the acidity of the rain storm causing the flow. This was confirmed by Nordø (1977).

Rosenqvist's views resulted in a scientific debate that persists to this day (see, for example, Krug, 1991; Renberg *et al.*, 1993, Seip, 1993) and forced a more rigorous scientific assessment of catchment behaviour. One result of this controversy was stronger emphasis on process-oriented studies and modelling of stream water chemistry in acidified catchments in the second phase of the SNSF Project. This was reflected in the final SNSF report (Overrein *et al.*, 1980). Although it was stated that the change in the composition of the precipitation has played an important role in the acidification of freshwaters, possible contributions due to land-use changes were not ruled out.

Much of the modelling work in Norway was concentrated on the Birkenes catchment where precipitation and stream water had been analysed since 1971−2 (cf. Section 15.2). The model that was developed was therefore commonly referred to as "the Birkenes model", although it has, in various versions, been applied to several catchments, as will be described later. We will here refer to the model as BIM.

The BIM was initially developed primarily as a research tool to test hypotheses about processes affecting stream water acidity. Although modelling results cannot prove any hypothesis to be correct, they are very useful in distinguishing between reasonable assumptions and those to be rejected. Fair agreement for a period of some years between observed and simulated values for discharge and concentrations of important chemical species in stream water greatly strengthens the confidence in the main assumptions. The BIM has also to some extent been used for predictions of future trends in stream water chemistry under various deposition scenarios.

15.2 DESCRIPTION OF THE BIRKENES CATCHMENT AND INITIAL FIELD WORK

The Birkenes catchment is situated 200−300 m above sea level about 15 km inland north of Kristiansand on the south coast of Norway. The total area is 0.41 km² and the catchment comprises two small connected valleys surrounded by steep hills rising 50−100 m (Figure 15.1). The vegetation is about 90% coniferous forest of pine (*Pinus sylvestris*) and spruce (*Picea abies*) with some birch (*Betula pubescens*). Ground vegetation is 58% blueberry-bracken community (*Eu- Picetum myrtillotosum/dryopteridetosum*), 28% pine community (*Barbilophozia-Pinetum*) and 7% fen and bog forest. The soils consist of 79% podzols, 17% peaty soils and 4% bare rock (Dale *et al.*, 1974; Braekke 1980). The category "podzols" is loosely defined here and, as will be discussed in more detail in Section 15.8, also includes organic topsoil directly on unweathered gravel or bedrock. The podzols have developed on glacial till with a humus layer seldom exceeding 10 cm. About half of the catchment has a soil depth less than 20 cm. Bedrock is biotite-granite. The brooks are surrounded by peaty soils which are generally more than 50 cm in depth.

Birkenes is in the part of Norway with the most acid precipitation and the highest sulphur deposition. Average concentrations of major species in precipitation and stream water are given in Table 15.1. The concentrations of sulphate in precipitation have decreased in recent years, presumably due to reduced sulphur emissions in Europe. The annual average concentration of

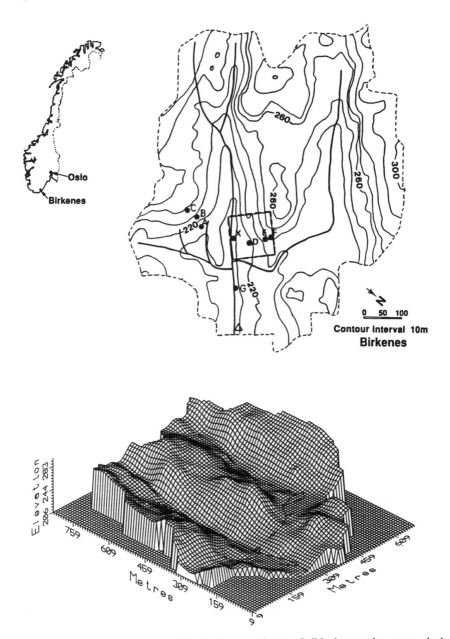

FIGURE 15.1 Map and topographic view of the Birkenes catchment. Soil lysimeter plots are marked with a filled circle and the weir with a triangle. The square represents the grid where detailed soil studies were conducted (Section 15.8)

excess (i.e. sea-salt corrected) sulphate was about 1.2 mg S l^{-1} around 1980 and about 0.75 mg S l^{-1} in 1991 (SFT, 1992). This decrease is reflected in lower streamwater concentrations of sulphate, but no clear trend in pH is seen so far (Christophersen *et al.*, 1990a). At present the S deposition is about 1.2 g m^{-2} a^{-1} (SFT, 1992).

TABLE 15.1 Amounts of and concentrations in precipitation and discharge at Birkenes (SFT, 1992). The values are averages weighted according to volume for five years (1 June 1986–31 May 1991). Amounts are in mm, "reactive aluminium" in μmol l^{-1}, total organic carbon (TOC) in mg C l^{-1}; otherwise the unit is μeq l^{-1}

	Amount of TOC	H$^+$	Ca^{2+}	Mg^{2+}	Na$^+$	K$^+$	NH$_4^+$	Al	So$_4^{2-}$	Cl$^-$	NO$_3^-$	
Precipitation	1688	51	10	13	52	6	47	—	63	60	44	—
Discharge	1263	24	55	31	120	7	—	16	115	137	12	4.9

Monitoring of amounts and composition of precipitation and discharge was initiated by the Norwegian Institute for Air Research (NILU) and the Norwegian Institute for Water Research (NIVA) in 1971–2. During events precipitation sampling occurred every day. Stream water samples were routinely collected weekly, but during some events much more frequent sampling was carried out. The samples were analysed for major cations (Ca^{2+}, Mg^{2+}, H$^+$, Na$^+$, K$^+$, NH$_4^+$) and anions (Cl$^-$, SO$_4^{2-}$, NO$_3^-$). In the first years only total Al was analysed; later, fractionation has been carried out. In some samples dissolved organic carbon (DOC) was analysed. Soil mapping was carried out by Frank (1980.

15.3 DEVELOPMENT OF THE BIRKENES MODEL (BIM) — AN OUTLINE

During the initial phase of model development only data on precipitation and discharge were available. However, as the need for more information became evident, largely through the modelling work, the sampling programme was extended (see Section 15.8).

As mentioned above, it had been shown that stream water, at least at Birkenes, tends to be acid at high flow irrespective of the acidity of the rain causing the discharge. This indicated strongly the importance of the water pathways within the catchment (Rosenqvist, 1978). The routing of the water and the residence time in the different soil layers for various hydrological conditions determine the interactions between the percolating water and the soil. Reasonable simulation of the hydrology is therefore a prerequisite for modelling the effects of various processes in the soil on soil water chemistry.

As a starting point for BIM, a hydrological model developed by Lundquist (1977) was chosen. It simply consists of two soil reservoirs meant to represent chemically different compartments of the soil column (see Figure 15.2), i.e. the upper organic soil horizon and deeper mineral (E, B, C) horizons. The approximation of using just two soil reservoirs is discussed in Sections 15.7 and 15.9. When necessary there is also a snow reservoir. The chemical features of the model reservoirs were intended to correspond to the most dominating chemical processes taking place in the soil. These are described in Section 15.4.

Initially much emphasis was put on the fate of sulphate in the soils (Christophersen and Wright, 1981) as this major component of acid rain was considered a vehicle for transport of cations through soils. This is in accordance with the mobile anion concept (Seip, 1980; Reuss and Johnson, 1986). An increase in the deposition of sulphate, which is a relatively mobile ion, at least in the soils of main concern in Norway, raises the level of anions through the total soil column and in the stream. This increase has to be balanced by a corresponding increase in the level of cations. In areas with low content of base cations in the soils, this increase will to a considerable degree comprise H^+ and inorganic Al species (e.g. Al^{3+}) which may cause biological damage.

The modelling of sulphate in stream water at Birkenes, including wet and dry deposition and production of sulphate by mineralization or other oxidation processes, was quite successful (Christophersen and Wright, 1981) and the model was soon extended to simulate other major species (Christophersen *et al.*, 1982). Main processes included were cation exchange in the upper reservoir and weathering in the lower one. The concentration of Al^{3+} was obtained assuming equilibrium with some form of gibbsite. (Further description of processes is given in Section 15.4). Results obtained by Christophersen *et al.* (1982) are shown in Figure 15.3. It is seen that major trends in stream water chemistry are captured quite well by the model. Both observations and simulations show the most acid periods during stormflow; at the same time calcium is low. During the period shown no fractionation of aluminium had been carried out. Hence, only total Al concentrations were known, while total Al may be assumed to consist of

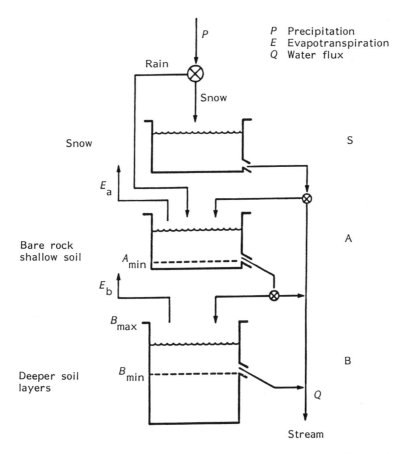

FIGURE 15.2 The Birkenes model (modified after Lundquist, 1977). The model comprises a snow reservoir and two soil reservoirs (A and B). Each soil reservoir has a threshold value (A_{min}, B_{min}), while the lower reservoir also has a maximum storage, B_{max}

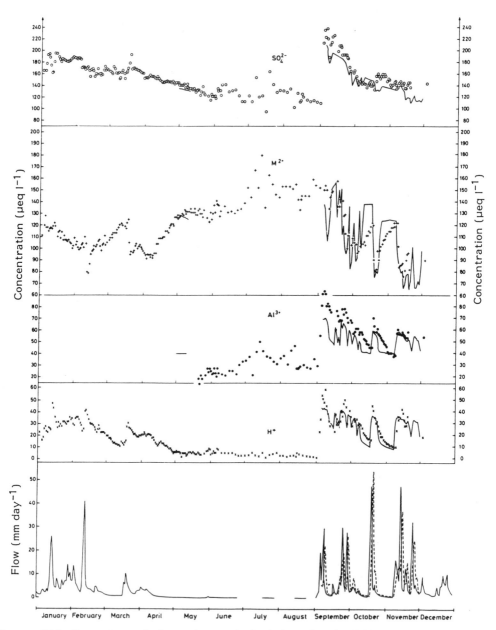

FIGURE 15.3 Comparison of observations and simulations for the Birkenes stream for 1974 (Christrophersen *et al.*, 1982). The upper panels show observed (points) and simulated (curves) concentrations of SO_4^{2-}, Ca^{2+} + Mg^{2+} (M^{2+}), reactive aluminium expressed as Al^{3+} and H^+. The lowest panel shows observed (solid curve) and simulated discharge (dashed curve). Due to difficulties in modelling snowmelt, the version of the model used at that time did not include periods with snow. Periods with no or negligible flow were also discarded. Therefore simulated values are given only for a short period in May and for most of the autumn. For additional results, including snowmelt, see Stone and Seip (1989, 1990)

colloidal Al, organically bound complexes (Al_o) and Al^{3+} + inorganic complexes (Al_i). Rough assumptions were made to estimate the concentrations of Al^{3+}, which is the main inorganic species at the prevailing pH. Although the simulations underestimated the Al^{3+} concentrations to some degree, the trends seem fairly well reproduced. However, further simulations showed that the aluminium concentrations are in general difficult to simulate.

The study described by Christophersen *et al.* (1982) did not include simulation of Cl^- and Na^+. Due to the short distance to the sea, the deposition of Cl^- and Na^+ is high at Birkenes, as are the concentrations in stream water (Table 15.1). However, since the concentrations of these species in precipitation are not much different and since Na^+ is more mobile in soils than other important cations, it was assumed that these ions roughly balance. This approximation may be fairly satisfactory for most periods, but it fails during episodes with particular high sea-salt concentrations. Under such conditions Na^+ is slightly retained in the catchment relative to Cl^-, presumably due to ion exchange. In the acidic Birkenes soils this exchange process results in release of significant amounts of acid cations, i.e. H^+ and Al_i, into soil solutions and stream water. An extreme example of the effect of elevated inputs of NaCl on the acidity of soil solutions and streams at Birkenes was discussed by Mulder *et al.* (1990). The omission of Cl^- was unfortunate also for another reason. In Figure 15.4 simulations of Cl^- using the original BIM are compared to observations (Stone and Seip, 1989). Chloride is considered as a conservative tracer which closely follows the water due to little interaction with soil or vegetation. The observations are clearly much more damped than the simulations, indicating that the hydrological model is not correct, even if the discharge was well reproduced.

The above-mentioned deficiency in the hydrological model was actually first discovered in an attempt to model the $^{18}O/^{16}O$ ratio in stream water (Christophersen *et al.*, 1985). Like Cl^-, the stable isotope ^{18}O in water may be considered as a conservative tracer. Information about

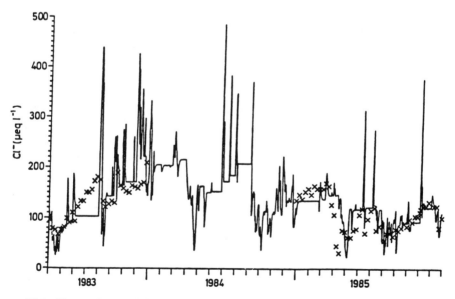

FIGURE 15.4   Observed (x) and simulated (curve) concentrations of Cl^- in stream water at Birkenes for 1983–5 using the original BIM. Cl^- was not measured in stream water in 1984 (from Stone and Seip, 1989)

flowpaths may be obtained using concentrations of such tracers in precipitation, stream water and soil water. During the early development of BIM, studies using ^{18}O data reported that most of the discharge during an event (typically 70%) consisted of pre-event water residing in the catchment before the rain storm or snowmelt (Sklash and Farvolden, 1979; Rodhe, 1981). To accommodate the damped behaviour of the conservative species in the stream while still reproducing a "spiky" hydrograph, Seip et al. (1985) introduced piston flow where a part of the runoff from the upper reservoir pushes out an equivalent amount of water from the lower reservoir. This feature proved important in several later applications (Stone and Seip, 1989; Taugbøl et al., 1994a). The approach is intuitively sensible and has been supported by recent work (Lindström and Rodhe, 1992).

The first version of BIM was only applicable to water with low concentrations of dissolved organic carbon (DOC). The omission of DOC is a fairly good approximation at Birkenes, but during some episodes the DOC concentration may increase to about 15 mg C l^{-1}. Rustad et al. (1986) had introduced a simple submodel for DOC, but no real study of the importance of these species for the modelling results was carried out until Taugbøl et al. (1994a) improved the submodel and applied it to a stream with very high DOC.

The first version of BIM was developed based on data for precipitation and stream water (amounts and concentrations of chemical species) as well as a general knowledge of presumably important soil processes. However, as modelling work proceeded it became clear that additional data were needed. The modelling work highlighted the need in particular for more information on water flowpaths and on aluminium chemistry. A new field program initiated in 1983–4 is described in Section 15.8.

15.4 MAIN PROCESSES IN THE BIM

The structure and the main features of the BIM have been thoroughly described in the literature (Christophersen et al., 1982, 1984; Rustad et al., 1986; Stone and Seip, 1989, 1990; Stone et al., 1990b). Taugbøl et al. (1994a) described the processes involving dissolved organic carbon (DOC). The model has three conceptual compartments (Figure 15.2) representing the snow cover, organic soil and mineral soil.

To run the model there is need for daily input values for mean air temperature, the amount of precipitation and its concentrations of SO_4^{2-} and Cl^- as well as dry deposition of sulphur and Cl^-. (In some applications Na^+ has been treated as conservative in the upper soil reservoir. In these cases input of Na^+ is needed as well.) To calibrate and verify the model, daily streamflow and at least weekly stream water composition are needed for a period of not less than 2–3 years.

Calibration of all model parameters is conducted by use of runoff data only. If appropriate soil and soil water data are available, they may be used to estimate initial parameter values and to assess if the calibrated values are reasonable. An automatic calibration routine is built into the model (deGrosbois et al., 1988; Hooper et al., 1988) enabling optimization of model parameters by a least-squares method.

The three-reservoir structure relies on the assumption that the catchment may be divided into three chemically and hydrologically distinct compartments. The stream water is a mixture of these contributions. By using the best estimate of model parameters, obtained through calibration, or preferably through reliable measurements, appropriate mixing of water affected by different chemical reactions can be simulated. The modelled runoff water quality will

TABLE 15.2   The main chemical processes governing the species modelled by the Birkenes model (BIM)

A reservoir

Cl^- and ^{18}O conservative
HCO_3^- is in equilibrium with a fixed CO_2 pressure in soil air, usually 10 times that in the atmosphere
SO_4^{2-} determined through reduction and mineralization, adsorption according to a linear isotherm
Ca^{2+}, Mg^{2+}, Na^+ exchanged with other cations on exchange surfaces with constant composition according to a Gapon-type equilibrium
Al^{3+} exchanged on surfaces and complexed by organic anions
H^+ in equilibrium with HCO_3^- and organic acids, and exchanged with Ca^{2+}, Mg^{2+}, Na^+ and Al^{3+}
DOC (dissolved organic carbon) produced from humic material

B reservoir

Cl^- and ^{18}O conservative
HCO_3^- is in equilibrium with CO_2 in soil air, enriched compared to the atmosphere according to calculated soil temperature
SO_4^{2-} determined by adsorption or release towards an equilibrium level, accounting for reduction/mineralization and adsorption/desorption processes
Ca^{2+}, Mg^{2+} and Na^+ produced by weathering of primary soil minerals
Al^{3+} in equilibrium with gibbsite or determined by ion exchange; equilibrium with organically complexed aluminium (Al_o)
H^+ in equilibrium with HCO_3^- and organic acids; consumed by weathering of minerals and exchange with Ca^{2+}, Mg^{2+} and Na^+. Equilibrium with Al^{3+} through gibbsite. Adsorption along with SO_4^{2-}
DOC (dissolved organic carbon) adsorbed towards an equilibrium level

indicate whether the inherent structure and processes represent a fair approximation of the catchment.

Routing of water between the compartments is crucial to the simulated hydrochemistry. When water storage in the catchment is low, a large portion of water is directed from the A reservoir to the B reservoir. In addition to the routing factor, a piston flow factor directs a portion of water leaving the A reservoir into the B reservoir, pushing out an equal amount of water from the B reservoir into the stream. This increase in stream water of the relative amount of water with B reservoir quality dampens the chemical response without affecting the timing and magnitude of the hydrological response (Seip *et al.*, 1985).

The chemical features included in the soil reservoirs are summarized in Table 15.2. The calculations of ionic concentrations are based on charge balance for the included ions.

15.4.1   The snow cover

This reservoir has only physical processes. Snow accumulation is difficult to simulate if mean daily temperatures are around 0°C. BIM uses a "sleet interval" where the precipitation changes linearly from snow to rain as the mean daily temperature increases from 0°C to a set value, usually 1.6°C. The daily amount of meltwater is computed by the commonly used degree-day method (Christophersen *et al.*, 1984). The model accounts for the preferential release of chemical components with the first meltwater (Johannessen and Henriksen, 1978). The rather simple functions used in the snow submodel have been found satisfactory for the small catchments for which they were developed.

15.4.2 The organic soil

The A-reservoir in the model represents the upper soil horizons, dominated by organic-rich ion exchange complexes. The sulphate submodel is crucial for determining the level of anions in solution. Sulphur is assumed to exist in a solid phase or as sulphate in the soil solution. The sulphate concentration (C_A) in the discharge from the A reservoir (Q_A) is calculated as $C_A = F_A K_A$, where F_A is the amount of water-soluble sulphate in the solid phase in reservoir A and K_A is a constant. Mineralization of organic S compounds adds sulphate to F_A during dry periods and when there is a thick snowpack. During wet periods sulphate may be removed from F_A (e.g. by reduction). The rate of reduction/mineralization depends on water saturation and temperature in the soil compartment (Christophersen and Wright, 1981; Seip *et al.*, 1985; Stone and Seip, 1989).

Chloride is assumed to be conservative (Stone and Seip, 1989) and is modelled using a simple mass balance equation. This is probably a fair approximation at Birkenes where the concentrations are quite high.

The interrelationships between the remaining ions, i.e. protons, Al species, base cations, carbonate species and organic anions, should ideally be determined by true thermodynamic constants for ion exchange, ionization and complexation. However, such constants are not readily transformed to "functional average" constants for a total catchment (cf. Section 15.7), and reported constants from the literature are generally modified through calibration. If no reported data exist, constants have to be entirely determined by calibration.

The ion exchange processes, which determine the cation composition in this layer, are assumed to proceed fast enough to establish equilibrium within the model's time resolution of one day. Gapon-type ion exchange equilibria are assumed, i.e.

$$\frac{(C^{n+}/n)^{1/n}}{(C^{m+}/m)^{1/m}} K_G = \frac{Y_{n+}}{Y_{m+}} \tag{1}$$

The C values should strictly be activities, but they are approximated by concentrations (μeq l^{-1}) in the model, n and m are valencies, K_G is the Gapon constant and the Y values are the amounts of ions on the exchange surface. In BIM it is assumed that the exchange surfaces do not change during the period covered by model simulations, i.e. the Ys are constants, implying that the base saturation and cation exchange capacity (CEC) are constant during the simulation period. The ratio between the concentrations of ions in solution is therefore fixed. This is an approximation which is reasonably valid for short-term modelling (a few years) as the Ys in the upper organic layer presumably are large compared to the leaching of ions (except possibly for Na$^+$). However, if long-term forecasts (or hindcasts) are to be modelled, changes in the amounts of exchangeable cations on the soil exchange surfaces must be included. Seasonal variations in CEC and base saturation due to changes in the organic layer are also ignored. As long as the adsorbed phase is assumed not to change, the use of Gapon exchange formulas gives the same results as the Gaines – Thomas equations.

Bicarbonate, which generally is the dominant ionic carbonate species at the pH levels of interest, results from dissolution of CO_2 from the soil atmosphere. In the A reservoir a constant CO_2 pressure ten times the atmospheric pressure is applied (see eqn (4) below).

Organic compounds are considered to be negatively charged sites on macromolecules, and the concentration of sites, represented by A$^-$, is the entity modelled. H$^+$ and Al^{3+} compete for these sites, governed by ionisation constants and stability constants respectively:

$$HA \leftrightarrow H^+ + A^- \qquad K_a = [H^+][A^-]/[HA] \tag{2}$$

$$Al^{3+} + A^- \leftrightarrow AlA^{2+} \qquad K_k = [AlA^{2+}]/([A^-][Al^{3+}]) \tag{3}$$

where $[HA] + [A^-] + [AlA^{2+}]$ is the total concentration of organic sites present, calculated as $N[DOC]$, where N is a conversion factor (μeq mg C^{-1}). K_a is pH- dependent and Taugbøl *et al.*, (1994a) used the empirical relationship of Oliver *et al.* (1983), i.e. $pK_a = c + dpH - epH^2$. This approach introduces additional parameters which have to be calibrated. Taugbøl *et al.* (1994b) showed later that fixed values both for K_a and for K_k are satisfactory approximations for normal ranges of H^+, Al and DOC and should be used in BIM.

The input of organic substances to the system is modelled as degradation of humic material by a zero-order reaction. An Arrhenius-type dependence between the reaction rate constant and temperature is assumed, i.e. $r = Pe^{-dE/RT}$. The constant P here includes the total amount of humic matter, while dE denotes the activation energy for the reaction, R the molar gas constant and T soil temperature in K, calculated as an average of air temperature over the last 20 days (cf. Smith *et al.*, 1964).

In the A reservoir the concentration of Al^{3+} is assumed to be determined by ion exchange. This gives the same equation for the relationship between $[Al^{3+}]$ and $[H^+]$ as a gibbsite equilibrium (see eqn (5)), but with a different, presumably lower, constant. In the model Al is speciated into only two forms, inorganic Al^{3+} ions and organically complexed AlA^{2+}. The model could be extended to incorporate inorganic hydroxide and fluoride species, as well as several organic complexes, to achieve a more complete Al representation. However, apart from the failure to simulate the specific Al compounds, the approximation of BIM is probably satisfactory in most cases (see Taugbøl *et al.*, 1994a, 1994b).

15.4.3 The mineral soil

The deeper soil layers contain more weatherable minerals and have lower soil solution DOC concentrations than the top layer. In the model, weathering of mineral surfaces is the main source for release of base cations in the lower reservoir. Weathering is a slow process, and the rate of base cation release depends on both minerals present and pH (Sverdrup and Warfvinge, 1988). The pH, in turn, is modified through consumption of H^+ by the weathering reactions. All equilibria involving H^+ will consequently be influenced by the weathering rate, which means that components depending on equilibria with H^+ also are affected by this rate. In the model, weathering is expressed as consumption of H^+ and an equivalent release of base cations (Ca^{2+}, Mg^{2+}, Na^+). The rate is proportional to the difference between the present level of H^+ and an equilibrium level which is determined by the observed H^+ level at base flow, as well as to the amount of water in the compartment. The proportionality factor is the weathering rate constant. This is a calibrated parameter, as are also the parameters determining the relative amounts of different base cations released (see Christophersen *et al.*, 1982). Cation exchange reactions are not included in the B reservoir.

As in the A reservoir, sulphate in the B reservoir is adsorbed or desorbed principally as an anion exchange reaction. Electroneutrality is maintained through equivalent uptake of H^+, which corresponds to release of OH^- by anion exchange. (This was different in the earlier versions). The solution phase sulphate is modelled as a fraction of a large sulphate storage. The rate of transfer between fixed and soluble sulphate is made dependent on deviation in the solution phase from a set equilibrium level. In both soil reservoirs the initial storage of sulphate has to be estimated. The initial sizes of this storage must be carefully calibrated due to its direct influence on the solution phase concentration of sulphate and therefore on the ionic strength of the solution.

Chloride is assumed to be conservative.

Soil P_{CO_2} is regulated through soil temperature, which thereby affects the bicarbonate concentration, i.e.

$$[HCO_3^-] = P_{CO_2} K_C/[H^+] \tag{4}$$

In the B reservoir $[Al^{3+}]$ is assumed to be in equilibrium with gibbsite:

$$Al(OH)_3(s) + 3 H^+ \leftrightarrow Al^{3+} + 3 H_2O \Rightarrow K_g = [Al^{3+}]/[H^+]^3 \tag{5}$$

The constant K_g is calibrated. It should be checked that the value obtained is in the range expected for gibbsite. Al^{3+} is simultaneously in equilibrium with AlA^{2+} through a complexation reaction as in the upper reservoir.

The effective acidity and aluminium complexation constants, K_a and K_k, for organic acids in the B reservoir are modelled by fixed values. This approximation is probably even better than for the A reservoir considering the more stable chemistry in the B reservoir. Thus the variation in these constants with, for example, pH, is probably quite small and assuming a constant value introduces only minor errors, while computational complexity is substantially reduced.

DOC is assumed not to be produced in the B reservoir, but to be supplied from the A reservoir and adsorbed/desorbed on surfaces. Adsorption is assumed to be directed towards an equilibrium concentration which is determined from the observed baseflow concentrations for DOC.

15.4.4 Stream

Water is mixed from the reservoirs to give the discharge and the stream water quality. Normally there are just two water types although the model allows a third type, i.e. meltwater, assumed to reach the stream unchanged. The CO_2 pressure in the stream is assumed to be lower than in the soil reservoirs (e.g. constantly equal to four times that in the atmosphere). The degassing of CO_2 when the water leaves the soil results in a pH increase. In the model SO_4^{2-}, Cl^- and base cations behave conservatively during mixing, while equilibria govern the ions dependent on H^+, namely Al^{3+}, HCO_3^-, A^-, AlA^{2+} and H^+ itself. Normally Al is assumed not to interact with any solid phase in the stream; therefore oversaturation relative to Al minerals, e.g. gibbsite, may occur. However, the precipitation of $Al(OH)_3$ may be simulated by adjusting the parameter governing the $[H^+]/[Al^{3+}]$ relation in stream water (cf. Section 15.10).

15.5 USE OF THE BIM

Stone and Seip (1990) tested BIM by using model parameters for Birkenes determined in previous work (Stone and Seip, 1989) to simulate data for two years that were meteorologically quite different and different from the years modelled earlier. The results contributed to increased confidence in the description of the most important processes. In addition to the simulations of the Birkenes stream, the BIM has been applied to several other sites, e.g. a tributary to Harp Lake, Ontario (Seip et al., 1985; Rustad et al., 1986), the Allt a'Mharcaidh catchment in Scotland (Stone et al., 1990b) and Svartberget in Sweden (Taugbøl et al., 1994a). The model developed by Lam and co-workers for sites in Canada used the BIM as a starting point (see, for example, Lam et al., 1988).

In the applications mentioned above the agreement between observations and simulations has in general been quite good. However, the simulation of aluminium concentrations has often proved difficult. This is not surprising since the mechanisms controlling the Al concentrations in stream and soil water are not well understood. The number of dissolved Al species also complicates the modelling. Taugbøl (1993) found that inclusion of DOC and organically bound $Al(Al_o)$ in the model improved the fit for Al and H^+ concentrations in the Birkenes stream for a two month period that previously was not modelled successfully (Stone and Seip, 1989). The DOC concentration during this period varied between 4 and 13 mg C l^{-1}. It should be noted that DOC may be important for mobilisation and complexation of aluminium in soil solutions even if the DOC concentration in stream water is low.

Although BIM has been primarily designed as a research tool, it has also been used for predictions. Seip *et al.* (1986) predicted large changes in pH (up to 0.9 units) in discharge during snowmelt events for a doubling (or halving) of sulphate concentrations in stream water, even if the amounts of exchangeable cations on the soil were assumed constant. Stone *et al.* (1990a) used MAGIC to predict soil changes and BIM to look at changes in stream water chemistry for various discharge patterns at Birkenes for several deposition scenarios. The results indicated that a 90% reduction in the deposition of excess sulphate may be necessary to achieve conditions in the stream suitable for trout (cf. Christophersen *et al.*, 1990b).

Applications of BIM have illustrated the usefulness of the mobile anion concept. In many catchments the sulphate concentration is decisive for acidity and Al concentrations in discharge. However, as shown by Taugbøl *et al.* (1994a), organic acids must be included in streams rich in DOC.

Seip (1980) found it illustrative to consider three mechanisms of water acidification which in general work together:

1. A direct effect of acid deposition reaching the rivers or lakes essentially unchanged.
2. Increased ionic levels due to increased deposition of sulphate ions which are mobile in many soils (e.g. at Birkenes).
3. Soil acidification which in turn causes water acidification. The soils may be acidified by acid deposition and/or by other changes (e.g. in vegetation).

The modelling work with BIM strengthens this view, in particular the importance of mechanism 2. If the S deposition in a catchment changes, this mechanism will usually manifest itself after a short period (perhaps a few years or less except when the soils adsorb sulphate strongly). Mechanism 3 will cause a delayed response. The immediate effect on runoff acidity will be similar for a salt of a strong acid and for the acid itself, but the effect on the soil, and thus the long-term effect on discharge chemistry, will be different.

Suggestions for improving the model are given in Section 15.11.

15.6 COMPARISON WITH OTHER MODELS

Reuss *et al.* (1986) compared a number of hydrogeochemical models. A more recent discussion of methods for projecting future changes has been given by Thornton *et al.* (1990). Both these overviews discuss, BIM, MAGIC (Cosby *et al.*, 1985) and the ILWAS model (Gherini *et al.*, 1985) in addition to other models. Several key processes are modelled in a similar way in BIM as in MAGIC and ILWAS. However, there are also important differences, partly related to the scope. Thus MAGIC was designed for modelling changes over long periods (decades or even centuries). Changes in the stores of exchangeable cations in the soil must then be modelled. On the other hand, less emphasis is put on the description of hydrology since the concentration changes during single events are not simulated. The ILWAS model is more comprehensive than BIM and includes, for example, nitrogen processes (cf. Section 15.10). However, ILWAS requires more input information than BIM, at least to fully use all features. The Trickle Down model, using alkalinity as the key variable (Schnoor and Stumm, 1985), may be mentioned as an example of a different modelling approach.

15.7 SOME GENERAL PROBLEMS IN HYDROGEOCHEMICAL MODELLING

15.7.1 Lumping of processes

The parameters in BIM represent average or lumped values of processes, spatially and temporally, for the whole catchment. The same is true to various degrees for all hydrogeochemical models. The lumping makes it difficult to transfer field observations from small plots to model parameter values. The errors introduced by the lumping have not been much discussed. Some support for the applicability of lumping was obtained by Taugbøl and Neal (1994). They looked at cation exchange processes. These reactions take place in micropores in the soils. The differences in the cation activity on the adsorption surfaces, as well as in the exchange selectivity constants, result in solution phase equilibria that differ between micropores. The macropore water, comprising the soil solution that drains into the stream, is a mixture of different micropore waters. They found that for waters that mix into macropores, with cationic compositions obeying exchange relationships, the mixture will follow an "average" exchange relationship if the mixing waters have the same ionic strength and mix in fixed proportions. This may be a good approximation for a certain soil horizon on a short time-scale, e.g. in each soil reservoir of BIM within the hydrological response time for small catchments. Furthermore, the composition of a mixture is not very sensitive to reasonable deviations from this ideal mixing pattern.

Also, Kirchner *et al.* (1993) concluded that the assumption of spatial uniformity, while clearly unrealistic, does not by itself compromise the predictions of chemical response to changes in acid anion concentrations under a given hydrological regime. However, as discussed in Section 15.3, the hydrological regime varies strongly with flow at Birkenes and in many other catchments. Seip and Rustad (1984) showed by using a simple model with two water types that the expected shift in pH for a doubling (or halving) of the sulphate concentration in stream water varied dramatically with the mixing ratio. Furthermore, basic assumptions used by Kirchner *et al.* (1993) are only good approximations for conservative mixing. Thus, important deviations may occur, for example, for H^+ in waters with pH around 5.

Clearly more work is needed to clarify the validity of lumping.

15.7.2 Parameter identifiability and model testing

Modelling hydrochemical processes in catchments is a question of scale. On the one hand, one has the model "splitters" wanting to depict the complex interactions taking place at a detailed level. The "lumpers", on the other hand, aim more at describing average processes at the catchment scale. In both cases, the challenge is to strike a balance between the model complexity and the available data. One desirable goal in modelling is parameter identifiability, meaning that, for the given data, there should be one single best set of optimized parameter values, according to a set criterion. In this case, the model is better posed for testing, and comparisons across applications to different catchments are facilitated. However, strict parameter identifiability is a stringent requirement for simulation models. Hooper *et al.* (1988) showed that even for the hydrologic submodel of the Birkenes model, the six parameters controlling the hydrology could not be uniquely determined given runoff amount and O^{18}/O^{16} values. Some improvement was obtained by including Cl^- values and using a longer series of observations (cf. Stone and Seip, 1989). Beck *et al.* (1990) also argued that the information content in the runoff is only sufficient to identify parameters in very simple models. The consequences of these findings are a matter of debate within the modelling community and a consensus has yet to be reached. Clearly, a model may be useful for understanding the complex interactions in a catchment even if some parameters are not identifiable. It may happen that two or more sets of parameter values give nearly the same fit (Stone, 1990). It is then permissible to choose between the sets on the basis of other information or even intuition. See Christophersen *et al.* (1993) for a more detailed discussion of identifiability and related modelling issues.

The Birkenes model was developed from the point of view of observed stream water variations and must be categorised as a lumped model. The basic idea is that there are a few soil environments (generally only two) that control the runoff quality. This general picture is in agreement with results using end member mixing analysis (EMMA) (see Section 15.9), but the outcome of that analysis suggests that three compartments are more realistic than two.

In testing and evaluating a hydrogeochemical model it is important to have sufficiently long periods for calibration and verification. Preferably each period should be several years and comprise years of variable meteorological conditions (i.e. both dry and wet years) to ensure that all important processes are "activated" during the periods.

Assessment of model performance is not trivial. There are many signals to be simulated, some of them highly variable. The fit must be assessed at a certain point of time. Simple criteria, as used in least-squares optimisation, may be misleading. A small difference between observed and simulated timing of a peak, for example in discharge, may subjectively be considered as a minor error, particularly during snowmelt. However, this may lead to a large difference also in concentrations with a corresponding large contribution to the sum of least squares. Better objective criteria for assessing performance of complex models would be very useful.

15.8 RECENT INTENSIVE FIELD INVESTIGATIONS

The early modelling work identified areas where more detailed information was needed. One important addition in the measurement programme was analysis of the $^{18}O/^{16}O$ ratio as mentioned earlier. However, a number of other extensions and improvements were also made in the field work.

Although episode studies were carried out in the 1970s, it was felt that the modelling work by the middle of the 1980s had given a better basis for designing such studies. Analytical methods for determining concentrations of various fractions of aluminium had also become available.

15.8.1 Soil water studies

To get information on what happens "inside" the catchment, soil water samples (lysimeters) were installed. Initially porous tension lysimeters with a 0.2 μm Teflon membrane were used. The locations are given in Figure 15.1. A suction of 0.5 bar was applied and samples were

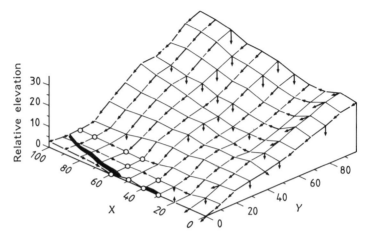

FIGURE 15.5   Three-dimensional map of the grid used in the soil study (cf. Figure 15.1). The *X* axis is parallel and the *Y* axis perpendicular to the stream (indicated by the bold solid line). Units along the axes are in metres. The elevation (*Z* axis) is relative to grid point (0,0). Arrows indicate the direction of steepest descent from each grid point to one of its nearest neighbours; their length indicates the angle of the slope (1 mm equals 10°). Solid circles indicate slopes less than 5°

collected 6−12 hours later. These lysimeters sample water only under saturated conditions. In 1987 the sampling was extended by installing ceramic cup lysimeters (Mulder *et al.*, 1990), which also allowed for water sampling in unsaturated soils.

Results from the years 1984−6 were presented by Seip *et al.* (1989). Despite sampling under quite variable hydrological conditions, soil solution concentrations of H^+ and Al_i generally showed small variations both between and within seasons, as compared with variations found in stream water. Neither in stream water nor in soil water were the concentrations of aluminium found to be in equilibrium with $Al(OH)_3$ (gibbsite), $Al_2Si_2(OH)_4$ (kaolinite, halloysite) or $Al(OH)SO_4$ (jurbanite). At least in the organic horizon, ion exchange seems the most likely process controlling Al concentrations. It was concluded that most probably different controlling mechanisms predominate in different soil layers, producing different chemical signatures which are transferred to the stream to varying degrees depending on hydrological conditions. These findings form the basis for the end member mixing analysis (EMMA) described in Section 15.9.

15.8.2   Topography and soils

The dominant soil types and their distribution in a representative, 1 ha area of the Birkenes catchment (Figure 15.1) were studied by Mulder *et al.* (1991). The same area was the focal point for intensive hydrochemical work in the terrestrial environment. The soil study was designed such that description and sampling occurred at grid points, which were at 10 m intervals (Figure 15.5).

Of the 121 grid points in Figure 15.5, five were located on paths or the root system of wind-thrown trees. Soils of the remaining 116 grid points were classified on the basis of characteristics observed in the field. For our specific purpose (i.e. the role of soils in the

TABLE 15.3   Soil types distinguished in the Birkenes catchment, including their diagnostic criteria. The relative proportions of the soil types are also given

Type	Description	% of area	Hydrology
0	Rock outcrop	4	
1	Peat; >25 cm deep organic deposit to gravel or bedrock	22	Return flow
2	Shallow organic; <25 cm deep organic deposit to gravel or bedrock	31	Return flow/ infiltration
3	Podzol; mineral soil with a >5 cm thick bleached layer (E horizon)	16	Infiltration
4	Acid brown soil; mineral soil with a <5 cm thick bleached layer (E horizon)	18	Infiltration
5	Rest group; all soils that do not fit in any of the above soil types	6	Return flow/ infiltration
	Paths, windthrown tress, etc.	3	

hydrochemical response of catchments) a new classification scheme was designed (Table 15.3), which gave a good subdivision of the soils and emphasised the local hydrology (potential infiltration versus return flow areas). Note that this scheme is not in accordance with existing soil classification systems (e.g. Soil Survey Staff, 1975).

The Birkenes catchment has a large relatively flat area in the valley bottom which increases in width from about 0 to 30 m (Figures 15.1 and 15.5). The stream, also shown in Figure 15.5, typically has 10−15 m wide stretches of flat terrain on both sides. In addition, a few minor, relatively flat spots occur on the slope (around grid points (20,50), (50,60) and (90,70), indicated by the short arrows in Figure 15.5). The grid has an average slope of about 20°, with about 40° at the steepest parts.

The flat valley bottom is a homogeneous peat area (Figure 15.6). Relatively flat spots on the slope also have peat soils (around grid points (20,50), (50,60) and (90,70); Figures 15.5 and 15.6). However, shallow organic deposits and mineral soils (podzols and acid brown earths) dominate on the slope. The proportional distribution of soil types indicates that organic soils (i.e. shallow organic deposits and deep peats) cover 53% of the area, which underlines the importance of these soils in the catchment. By contrast, mineral soils (podzols and brown earths) are only found at 34% of the grid points. Minor fractions of the grid points belonged to the rest group (6%), or were located on bedrock (4%), or paths, windthrown trees, etc. (3%) (Table 15.3).

15.8.3 Soil chemistry

Here we will focus on pH and cation exchange characteristics of the dominant soils, as these properties are believed to be most relevant with respect to the hydrochemical behaviour of the soils at Birkenes. Soil pH, cation exchange capacity (CEC) and exchangeable cations were determined using 0.1 M $BaCl_2$ (Hendershot and Duquette, 1986; Mulder *et al.*, 1991).

Soils at Birkenes are predominantly acidic (Table 15.4). Podzols and acid brown earths have extremely low pH($BaCl_2$) values in the organic surface (O) horizon; the values increase somewhat with depth in the mineral horizons. Also peat soils are strongly acidic throughout.

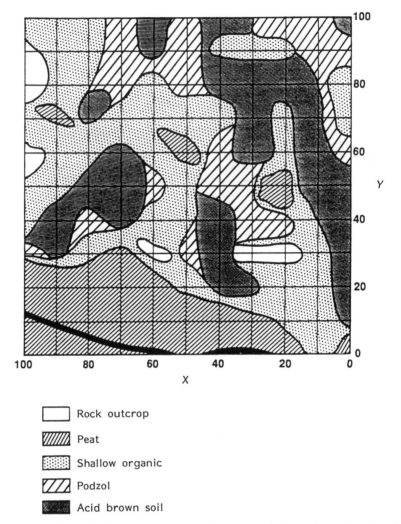

Rock outcrop

Peat

Shallow organic

Podzol

Acid brown soil

FIGURE 15.6   Soil map of the grid at Birkenes, according to the classification criteria given in Table 15.1

However, pH values are relatively high in the mineral sediment below the peat in the valley bottom. It is particularly this environment below the peat deposits that may have a dominant influence on stream water chemistry at base flow (cf. Section 15.9).

Because the mineral soils (podzols and acid brown soils) are sandy and thus relatively poor in clay minerals, most of the negative charge in these soils is associated with organic matter. Consequently, the mineral soil horizons have significantly lower CEC values than the organic O horizons. Mulder *et al.* (1995) reported CEC values of 20 meq kg^{-1} (E horizon), 43 meq kg^{-1} (Bhs horizon) and 22 meq kg^{-1} (Bs horizon) respectively. For similar reasons the CEC of the mineral sediments below the peat is low (15 meq kg^{-1}).

Base saturation (BS) values are below 20% in all but the O horizons on the hillslopes (podzols, acid brown earths and shallow organic deposits) and the mineral sediment below the

TABLE 15.4 Chemical properties of the major soils (layers) at Birkenes. Numbers are median values (Mulder *et al.*, 1991, 1994)

Soil horizon	pH(H$_2$O)^a	pH(BaCl$_2$)	CEC (meq kg^{-1})	Base saturated (%)	Aluminium saturated (%)
Podzol, brown earth					
O		2.8	220	64	11
E	4.2		20	15	62
Bhs	4.7	3.9	43	9	84
Bs	4.8		22	8	87
Peat soil					
0−10 cm		3.1	211	19	69
50 cm	3.7	2.9	183	9	89
100 cm			106	7	88
Sediment^a	5.1		15	42	49

^{a} Based on a few observations taken from Mulder *et al.* (1990).

peat in the valley bottom. The high value of BS in the O horizon (forest floor) is probably related to cycling of nutrient base cations. By contrast, the relatively high BS levels in the mineral sediments below the peat are more likely due to increased mineral weathering, possibly as a result of the omnipresence of weathering-enhancing organic acids and long residence times of soil water.

The CECs of the surface organic (O) horizons have median values around 220 meq kg^{-1} and are not significantly different for the five soil types (Mulder *et al.*, 1991). Even though the cation exchange sites in the O horizons of all soils are dominated by Al and base cations, the composition of the exchangeable cations differs widely between soil types (Table 15.5). For example, (median) base saturation values of the O horizons increase significantly in the order deep peat (16%) < shallow organic deposit (43%) < podzol (62%) ≈ acid brown soil (65). Conversely, the Al saturation is highest in the surface layers of the peats (median 73%) and is significantly lower in the O horizons of all other soil types (37% in the shallow organic soils and 11% in podzols and brown earths). Our data indicate that the elevated aluminium saturation of the cation exchange sites in the podzol B horizon (Table 15.4) is of similar magnitude as that in the surface layer of the peat. The reason for this phenomenon is discussed below.

15.8.4 Water flowpaths

Interpreting the arrows in Figure 15.5 (representing the direction of steepest descent) as the major flow direction, water entering the upper half of the grid (i.e. more than 50−60 m from the X axis) is largely routed through the flat area around grid point (20,50). Subsequently, this water may flow via a threshold between the grid points (10,40) and (20,40) to the lower part of the slope and out into the stream. Generally the transport of water takes place as subsurface flow at Birkenes. Only in the case of extremely intense storms has surface runoff been observed in the form of ephemeral flow channels, notably one starting below the threshold (i.e. between grid points (20,10) and (20,20)), and several others in the flat area close to the stream. Considering the relatively thin soils at Birkenes (with the exception of the deep peats, which can

TABLE 15.5 The fractional composition of the O horizon cation exchange complex with Al (assumed as Al^{3+}, H^+ and base cations (Na^+, K^+, Ca^{2+} and Mg^{2+}) for the various soil types (see Table 15.3) in the Birkenes grid. Numbers are medians and 25 and 75 percentiles. Significant difference in a soil parameter for the various soil types is indicated by different letters (95% confidence interval; Mann–Whitney two-tailed test). Values for the Bhs horizon are means taken from Mulder et al. (1990)

Soil type	% Al				% H^+				%Base cations			
	25%	Median	75%	Mann–Whitney	25%	Median	75%	Mann–Whitney	25%	Median	75%	Mann–Whitney
1	45	73	82	a	4	8	21	a	10	16	33	a
2	12	37	57	b	7	18	22	a	25	43	64	b
3	7	11	14	c	21	26	28	b	56	62	68	c
4	6	11	35	c	13	19	22	a,b	50	65	71	c
5	10	11	11	b,c	22	24	24	a,b	54	65	66	b,c

TABLE 15.6   The average solubility of aluminium, expressed as $pK_g = pAl - 3pH$ (cf. eqn (5)) between 8 October and 16 October 1989. Values are averages for four to five replicate lysimeters per soil horizon (Mulder *et al.*, 1991)

Horizon	pAl − 3pH
O podzol	− 6.31
O peat	− 8.25
B podzol	− 8.80

be up to 3 m in depth), we believe that steepest descents, as measured at the soil surface, are a valid approximation for the surface of the bedrock.

From Figures 15.5 and 15.6 it is evident that peat soils are associated with flat areas. All other soil types are more common on the slopes. In this respect is should be noted that the formation of peat, which is characteristic for (local) depressions due to convergence of transported water (Figure 15.5), has further enhanced the flatness of these areas. Our data suggest that the peats may be considered as potential return flow areas, where water transported from the upslope subsoils emerges.

The importance of return flow in the peat soils is further supported by the soil chemistry data, which indicate a significant enrichment in exchangeable Al. Because mosses, which are the major source of peat accumulation at Birkenes, grow without contact with the mineral soil, most of the exchangeable Al must originate from outside the peat area, most probably from B horizons upslope. By contrast, the relatively low levels of exchangeable Al in O horizons of the mineral soils (brown earths and podzols) suggest a predominantly vertical transport of aluminium-poor water.

The similarity in exchangeable Al between the surface layer of the peat and the podzol B horizon suggests a lateral extension of the zone of elevated aluminium solubility from the mineral B horizon into the peat. This hypothesis is supported by the aluminium solubilities as calculated from soil solution composition data collected by lysimetry (Mulder *et al.*, 1991; Table 15.6). The data in Table 15.6 indicate that the solubility of Al is only slightly lower in the surface peat soil than in the mineral B horizons upslope. By contrast, the O horizons of the mineral soils have a significantly lower solubility of aluminium. Our data suggest that at present water originating in the mineral soil B horizons may reach the brook virtually unaltered. We hypothesise that the elevated level of exchangeable aluminium in the peats has resulted from a continuous influx of aluminium-rich B horizon water. Subsurface translatory flow through the O horizons on the slope may bypass the peat and reach the stream via ephemeral stream channels during high-intensity storms. The latter mechanism was report by Mulder *et al.* (1990), who monitored the transport of Cl⁻ in soils and stream water during a period with extremely high inputs of sea salt.

15.9   END MEMBER MIXING ANALYSIS (EMMA)

The great importance of knowing the water pathways for different hydrological conditions had become evident both through applications of BIM and through the intensive field work, which in addition had increased the knowledge of soil properties and soil water concentrations in the Birkenes catchment. The modelling work indicated that mixing of waters from two reservoirs described the main features in stream water chemistry reasonably well under most, but not all, conditions. These findings led to a new modelling approach — end member mixing analysis (EMMA).

In EMMA, the stream is viewed as a mixture of soil waters from different horizons (the end members) where the contributions from a certain horizon change over events depending on the hydrological conditions. Preliminary work using the mixing concept was carried out by Seip and Rustad (1984). However, the research tool now known as EMMA was initiated by Neal and Christophersen (1989) and developed for Birkenes and catchments in Plynlimon, Wales, and at Panola Mountain, Georgia, USA (Christophersen *et al.*, 1990c; Hooper *et al.*, 1990; Neal *et al.*, 1990).

A key aspect is that the chemical compositions of the soil waters are considered approximately constant over the time of interest, compared to the variations in stream water. This is found at Birkenes for certain species. The basic EMMA approach thus rests on two main assumptions:

1. There are chemical species in the soil solutions that are controlled by fast (equilibrium) reactions with the solid phase and therefore constitute a fingerprint of that horizon.
2. Leaving the soils with the water on its way to the stream and in the stream itself, these same species mix conservatively and do not take part in chemical reactions.

In practice, one may ask how realistic such assumptions are, in particular the second one. An experiment in which NaCl was injected into parts of the peat soils along the Birkenes stream that are hydrologically saturated during rain storms indicated that deviations occur at least under some conditions (Müller *et al.*, 1993). Elaborations and refinements of the basic EMMA approach are clearly possible. However, it seems clear that EMMA in several cases, including Birkenes, has been useful in the qualitative and quantitative study of catchment hydrochemistry, and thus may provide a framework that could be of more general value in such studies.

The theoretical background of EMMA, utilising principal component analysis and elements of linear algebra, is given by Christophersen and Hooper (1992). Here we will only treat the specific application at Birkenes.

The EMMA analysis usually starts by making so-called mixing diagrams as exemplified in Figure 15.7. Here, inorganic aluminium (Al_i) is plotted against H^+ for both single stream samples and soil waters. For the latter, only the median values as well as the inter-quartile ranges derived from individual lysimeter samples are indicated. The three horizons are the organic O/H horizon, the mineral podzolic B/C horizon and the deep groundwater in the valley bottom (VB). The median is meant to represent typical concentrations whereas the quartiles reflect spatial as well as temporal variability. The number of lysimeters in each horizon is in the interval 3 (VB) to 20 (B/C) and the number of corresponding samples per horizon ranges from 15 to 200. Using the variability in stream water as the yardstick, one may conclude that the soil waters display a reasonably stable chemistry with respect to the two species and that the soil waters are distinctly different between horizons. An important feature is further that the soil waters approximately circumscribe the stream water samples, which is a necessary condition for stream water to be considered as a mixture. The flowpath picture emerging from Figure 15.7 shows that the highflow stream water samples, located in the upper right part of the figure, are either fairly close to the organic O/H end member or are a mixture of O/H and B/C, indicating that the dominant flowpaths are in the upper soil horizons. The low-flow samples (lower left corner) are close to the deep groundwater in the valley bottom.

This interpretation of the mixing diagram relies on the assumption that Al_i and H^+ mix conservatively, which is generally not the case. At Birkenes, however, the pH in stream water and in the end members, except the VB, is typically in the range 4.1—5.3, where such an assumption is reasonable. However, at low-flow the VB contributes significantly and the deviation from conservative mixing increases.

When applying EMMA, an important part is the selection of the chemical species to enter the analysis. The more species one can consider simultaneously, the more reliable are the results. At Birkenes, Ca^{2+} was also found to be horizon-specific and conservative mixing is probably a better assumption for this element than for Al_i and H^+ (Christophersen *et al.*, 1990c). Calcium fits the general mixing picture in Figure 15.7 in that the Ca^{2+} concentration increases in stream water from high flow to low flow, which is consistent with increased soil water Ca^{2+} concentrations from O/H to B/C and VB. However, the soil water Ca^{2+} levels were too low to span the stream water samples satisfactorily. Therefore a terrestrial source of Ca^{2+} seems to exist that is not covered by the current lysimeter installations.

Subsequently, an EMMA analysis was done based on three fractions of dissolved organic carbon (DOC), which were again horizon-specific (Easthouse *et al.*, 1992). The DOC fractions gave a mixing picture in agreement with Figure 15.7. Easthouse *et al.* (1992) could therefore carry out the final step in the complete EMMA analysis, i.e. computation of the contributions from each end member to the stream for each stream water sample, using data both for organic and inorganic species (Figure 15.8).

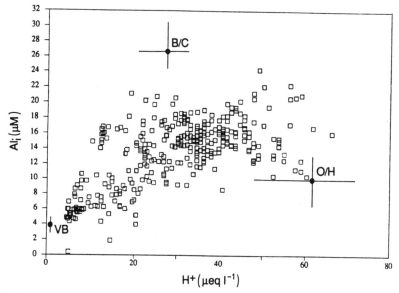

FIGURE 15.7 Mixing diagram of Al_i versus H^+ for the Birkenes catchment. Stream water samples are shown by open squares whereas the soil water end members are represented by medians and upper and lower quartiles respectively. VB represents the valley bottom

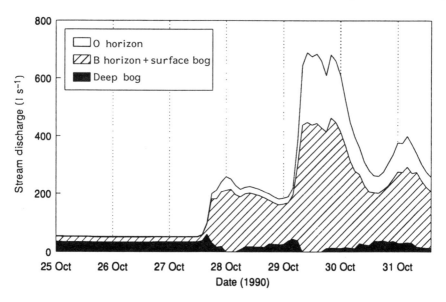

FIGURE 15.8   Estimated end member contributions to the Birkenes stream during an autumn study based on both organic and inorganic data (from Easthouse *et al.*, 1992). The three end members are: podzol O horizon; podzol B horizon and/or surface peat; deep bog (valley bottom)

An advantage of EMMA in model development is that the model compartments have been identified in the field, allowing direct measurements from the soil horizons — and not only stream water — to be used in modelling. Main problems are related to the identification of end members and to variations in end member composition.

15.10   SUGGESTED IMPROVEMENTS OF BIM

BIM has been applied quite successfully to several streams, indicating that processes of major importance have been reasonably well described. However, a number of areas where improvements are recommended may be identified:

1. *Hydrological representation*. The importance of water pathways has been confirmed both by comparing simulated and observed data and by direct observations. We have, for example, found that subsurface translatory flow through the O horizons on the slopes of the catchment may bypass the peat and reach the stream via ephemeral stream channels during high-discharge periods. There may be different opinions about how detailed one should make a hydrological model. For Birkenes it seems clear that an improved model should include a third soil reservoir representing deep soil layers, as indicated by EMMA.

2. *Mobilisation of aluminium into the solution phase*. The mechanisms controlling the levels of aluminium species in stream and soil water are not satisfactorily known and therefore Al concentrations cannot be predicted well by any model at present. This gap in our knowledge is quite critical due to the harmful biological effects of aluminium.

3. *The role of nitrogen compounds in the system*. The neglect of nitrogen processes is another limitation of BIM, NH_4^+/NH_3 and NO_3^- are often deposited in significant amounts.

However, biological uptake of these compounds reduces the concentrations in many streams to very low values (cf. Table 15.1). In such cases the contribution to the ion balance in the discharge may be disregarded. Soil processes may still be affected by pH changes caused by reactions involving the nitrogen compounds, but since the uptake of NH_4^+ and NO_3^- is often similar, effects on pH tends to cancel (van Breemen *et al.*, 1984). If the aim of the model is the effects on vegetation, these compounds should definitely be considered. However, the present model simulates runoff water composition where the ionic contribution from nitrogen compounds often amounts only to a few per cent.

4. *Sulphur dynamics in soils.* The main problems in the sulphur submodel are related to transformations between species, in particular between sulphate and organic sulphur compounds. As for nitrogen, microbiological processes are essential and difficult to model. During periods of increasing S deposition, both reduction and adsorption may delay effects on stream water, while desorption and mineralization will delay improvements if depositions are reduced. In the Birkenes catchment mineralization may be a decisive factor during periods of recovery.

5. *Weathering rates.* In BIM weathering rates play a role only in the deeper soils; preferably weathering should also be included in the upper soils. In particular, for models used to predict changes in soil properties, weathering rates are critical. Lumped values applicable in models like BIM are difficult to obtain, although the Sr method described by Jacks *et al.* (1989) is promising. In the model the weathering rate must be adjusted to compensate for biological uptake/release of cations. It would be more realistic to treat these processes separately.

6. *Stream processes.* The chemical parameters governing in-stream reactions appear to be crucial to the model simulations; hence the assumptions made for the stream are very important. The model performance will be very sensitive to any possible deviation from the equilibrium situation which is postulated for protolysis and complexation reactions. Although not required by the model, no solid phase interactions were assumed for Al compounds in the stream in most applications of BIM. Henriksen *et al.* (1988) showed that precipitation/dissolution reactions may strongly affect the water quality in some cases. The conservative behaviour of base cations may also be questioned. Stream substrates are possible sources of additional Ca^{2+} as well as Al, as pointed out by Henriksen *et al.* (1988). Although description of non-equilibrium situations is difficult, an effort to improve the modelling of stream reactions seems justified.

15.11 CONCLUSIONS

Integration of modelling and field work is essential for understanding hydrogeochemical processes. In our work modelling has played an important role in designing new field studies which in turn have stimulated further model development. At the same time the complexity of the system has become evident. It has been suggested that there may be fundamental limitations to the degree hydrogeochemical properties of catchments and other complicated ecosystems can be quantified by process-based models (Christophersen *et al.*, 1993). It is worth to have in mind Kalman's (1985) version of the "uncertainty principle", i.e. *inexact data imply a non-unique model.*

However, as with all catchment models, the details of how the hydrochemical processes are quantified and incorporated into the model are to some extend based on subjective decisions.

This is perhaps the main reason why many research groups in this field develop their own models. Detailed data from inside catchments are important in further model development, but the interpretation is not straightforward. This is a problem of scale, since soil and soil water data generally represent point measurements in a heterogenous environment, while the models are based on catchment averages. Further work on averaging procedures and the uncertainties they introduce is recommended.

Obviously good fit between observed and simulated values does not prove a model to be correct. One might, for example, envisage similar results for the studied period for two models in which differences in cation exchange parameters are compensated by differences in weathering rates. These models might deviate considerably regarding future predictions of stream water chemistry. The confidence in a model increases if the agreement is satisfactory for many signals (discharge, conservative tracers, chemically active species). Since BIM can be made to reproduce both discharge and runoff chemistry over many years for several catchments, the process descriptions are feasible. We are confident that the model captures main features of the hydrological pathways, with stormflow chemistry being determined by the upper organic soil horizons while the low-flow quality is controlled by deeper mineral layers. Main chemical processes such as cation exchange, sulphate adsorption/desorption and CO_2 degassing are also accounted for. To improve model reliability further, submodels preferably should be tested separately on data collected for that purpose.

Process-based models are necessary tools to obtain an integrated picture of catchment response based on detailed knowledge and a broad range of field data. Furthermore, such models are very useful for testing hypotheses. One can learn as much from the failure of a model simulation as from its success, since disagreements require an explanation that may lead to new ideas. Development of better objective criteria for assessing the performance of complex models would enhance their usefulness and make model comparison simpler.

In spite of the uncertainties, it seems unlikely that further research will change our present conceptual picture of catchments dramatically. Although present models only give tentative predictions, there is no better way to predict changes corresponding to future deposition scenarios. Process-based models are therefore useful management tools provided that due considerations are given to the uncertainties.

REFERENCES

Beck, M. B., Kleissen, F. M. and Wheater, H. S. (1990). "Identifying flow paths in models of surface water acidification", *Rev. Geophys.*, **28**, 207–230.

Braekke, F. H. (Ed.) (1976). "Impact of acid precipitation on forest and freshwater ecosystem in Norway", The SNSF-project, FR6/76. Available from Norwegian Institute for Water Research, PO Box 69, Korsvoll, 0808 Oslo.

Braekke, F. H. (1980). "Hydrochemistry in low-pH-soils of South Norway. 1. Peat and soil water quality", *Rep. of the Norwegian Forest Res. Inst.*, **36**(11), 32 pp.

Christophersen, N. and Hooper, R. P. (1992). "Multivariate analysis of stream water chemical data: the use of principle components analysis for the end-member mixing problem", *Water Resour. Res.*, **28**, 99–107.

Christophersen, N. and Wright, R. F. (1981). "Sulphate budget and a model for sulphate concentrations in streamwater at Birkenes, a small forested catchment in southernmost Norway", *Water Resour. Res.*, **17**, 377–389.

Christophersen, N., Seip, H. M. and Wright, R. F. (1982). "A model for streamwater chemistry at Birkenes, Norway", *Water Resour. Res.*, **18**, 977–996.

Christophersen, N., Rustad, S. and Seip, H. M. (1984). "Modelling streamwater chemistry with snowmelt", *Phil. Trans. R. Soc. London.*, **B305**, 427–439.

Christophersen, N., Kjaernsrod, S. and Rodhe, A. (1985). "Preliminary evaluation of flow patterns in the Birkenes catchment using natural ^{18}O as tracer", in *Hydrological and Hydrogeochemical Mechanisms and Model Approaches to the Acidification of Ecological Systems*, (Ed. I. Johansen), pp. 29–40, Nordic IHP-Report No. 10, Oslo.

Christophersen, N., Robson, A., Neal, C., Whitehead, P. G., Vigerust, B. and Henriksen, A. (1990a). "Evidence for long-term deterioration of streamwater chemistry and soil water acidification at the Birkenes catchment, southern Norway", *J. Hydrol.*, **116**, 63–76.

Christophersen, N., Neal, C., Seip, H.M. and Stone, A. (1990b). "Hydrochemical models for simulation of present and future short-term changes in stream chemistry: development and status", in *The Surface Water Acidification Programme* (Ed. B. J. Mason), pp. 445–452, Cambridge University Press.

Christophersen, N., Neal, C., Hooper R. P., Vogt, R. D. and Andersen, S. (1990c). "Modelling streamwater chemistry as a mixture of soilwater end-member — a step towards second-generation acidification models", *J. Hydrol.*, **116**, 307–320.

Christophersen, N., Neal, C. and Hooper, R. P., (1993). "Modelling the hydrochemistry of catchments: a challenge for the scientific method", *J. Hydrol.*, **152**, 1–12.

Cosby, B. J., Wright, R. F., Hornberger, G. M. and Galloway, J. N. (1985). "Modelling the effects of acidic deposition: assessment of a lumped-parameter model of soil water and stream water chemistry", *Water Resour. Res.*, **21**, 51–63.

Dale, T., Henriksen, A., Joranger, E. and Krog, S. (1974). "Vann- og nedbørkjemiske studier i Birkenesfeltet for perioden 20 juli 1972 til 30 april 1973 (Chemical investigations of precipitation and runoff in the Birkenes basin 20 July 1972–30 April 1973)", The SNF-project, TN74, 45 pp. Available from Norwegian Institute for Water Research, PO Box 69, Korsvoll, 0808 Oslo.

DeGrosbois, E., Hooper, R. P. and Christophersen, N. (1988). "A multisignal automatic calibration methodology for hydrochemical models: a case study of the Birkenes model", *Water Resour. Res.*, **24**, 1299–1307.

Easthouse, K. B. Mulder, J., Christophersen, N. and Seip, H. M. (1992). "Dissolved organic carbon fractions in soil and stream water during variable hydrological conditions at Birkenes, southern Norway", *Water Resour. Res.*, **28**, 1585–1596.

Frank, J. (1980). "Undersøkelse av jordsmonnet i Birkenesfeltet i Aust-Agder fylke (Soil survey at Birkenes, a small catchment in Aust-Agder county. Southern Norway)", The SNSF-project, TN60/80, 41 pp. Available from Norwegian Institute for Water Research, PO Box 69, Korsvoll, 0808 Oslo.

Gherini, S. A., Mok, L., Hudson, R. J. M., Davis, G. F., Chen, C. W. and Goldstein R. A. (1985). "The ILWAS model: formulation and application", *Water, Soil Pollut.*, **26**, 425–459.

Hendershot, W. H. and Duquette, A. (1986). "A simple barium chloride method for determining cation exchange capacity and exchangeable cations", *Soil Sci. Soc. of Am. J.*, **50**, 605–608.

Henriksen, A., Wathne, B. M., Røgeberg, E. J. S., Norton, S. A. and Brakke, D. F. (1988). "The role of stream substrates in aluminium mobility and acid neutralization", *Water Res.*, **22**, 1069–1073.

Hooper, R. P., Stone, A., Christophersen, N., DeGrosbois, E. and Seip, H. M. (1988). "Assessing the Birkenes model of stream acidification using a multisignal calibration methodology", *Water Resour. Res.*, **24**, 1308–1316.

Hooper, R. P., Peters, N. E. and Christophersen, N. (1990). "Modelling streamwater chemistry as a mixture of soilwater end-members—an application to the Panola Mountain catchment, Georgia, USA", *J. Hydrol.*, **116**, 321–343.

Jacks, G., Åberg, G. and Hamilton, P. J. (1989). "Calcium budgets for catchments as interpreted by strontium isotopes", *Nord. Hydrol.*, **20**, 85–96.

Johannessen, M. and Henriksen, A. (1978). "Chemistry of snowmelt water: changes in concentration during melting", *Water Res.*, **14**, 615–619.

Kalman, R. E. (1985). "Identification of noisy systems", *Russian Math. Surveys*, **40**, 25–42.

Kirchner, J. W., Dillon, P. J. and LaZerte, B. D. (1993). "Predictability of geochemical buffering and runoff acidification in spatially heterogeneous catchments", *Water Resour. Res.*, **29**, 3891–3901.

Krug, E. (1991). "Review of acid-deposition–catchment interaction and comments on future research needs", *J. Hydrol.*, **128**, 1–27.

Lam, D. C.L ., Bobba, A. G., Jeffries, D. S. and Craig, D. (1988). "Modelling stream chemistry for the

Turkey Lakes Watershed: comparison with 1981−84 data", *Can. J. Fish. Aquat. Sci.*, **45**, (Suppl. No. 1), 72−80.

Lindström, G. and Rodhe, A. (1992). "Transit times of water in soil lysimeters from modelling of oxygen-18", *Water, Air and Soil Pollut.*, **65**, 83−100.

Lindquist, D. (1977). "Modellering av hydrokjemi i nedbørfelter (Hydrochemical modelling of drainage basins)", The SNSF-Project, IR31/77, Norwegian Institute for Water Research, Oslo, 32 pp.

Mulder, J., Christophersen, N., Hauhs, M., Vogt, R. D., Andersen, S. and Andersen, D. O. (1990). "Water flow paths and hydrochemical controls in the Birkenes catchment as inferred from a rainstorm high in sea salts", *Water Resour. Res.*, **26**, 611−622.

Mulder, J., Pijpers, M. and Christophersen, N. (1991). "Water flow paths and the spatial distribution of soils and exchangeable cations in an acid rain-impacted and a pristine catchment in Norway", *Water Resour. Res.*, **27**, 2919−2928.

Mulder, J., Christophersen, N., Kopperud, K. and Fjeldal, P.H. (1995). "Water flow paths and the spatial distribution of soils as a key to understanding differences in streamwater chemistry between three catchments (Norway)", *Water, Air and Soil Pollut.*, **81**, 67.

Müller, D. -I., Wohlfeil, I. C., Christophersen, N., Hauhs, M. and Seip, H. M. (1993). "Chemical reactiveness of soil water pathways investigated by point source injections of chloride in a peat bog at Birkenes", *J. Hydrol.*, **144**, 101−125.

Neal, C. and Christophersen, N. (1989). "Inorganic aluminium−hydrogen ion relationships for acidified streams; the role of water mixing processes", *Sci. Total Environ.*, **80**, 195−203.

Neal, C., Smith, C. .J., Walls, J., Billingham, P., Hill, S. and Neal, M. (1990). "Hydrogeochemical variations in Hafron forest stream waters, Mid Wales", *J. Hydrol.*, **116**, 185−200.

Nordø, J. (1977). "En statistisk undersøkelse av surheten i en bekk nær Birkenes i Aust-Agder (A statistical investigation of the acidity of a stream near Birkenes in Aust-Agder.)", in *Sur jord−surt vann (Acid soil−acid water)* (Ed. I. T. Rosenqvist), pp. 106−110, Ingeniørforlaget, Oslo.

Oliver, B. G., Thurman, E. M., and Malcolm, R. (1983). "The contribution of humic substances to the acidity of colored natural waters", *Geochim. Cosmochim. Acta*, **47**, 2031−2035.

Overrein, L. N., Seip. H. M. and Tollan, A. (1980). *Acid precipitation — effects of forest and fish*, Final Report of the SNSF-project 1972−1980, The SNSF-project FR19/80, 175 pp. Available from Norwegian Institute for Water Research, PO Box 69, Korsvoll, 0808 Oslo.

Renberg, I., Korsman, T. and Birks, J. H. B. (1993). "Prehistoric increases in the pH of acid-sensitive Swedish lakes caused by land-use changes", *Nature*, **362**, 824−826.

Reuss, J. O. and Johnson, D. W. (1986). *Acid Deposition and the Acidification of Soils and Waters*, Ecological Studies 59, Springer-Verlag, New York, 119 pp.

Reuss, J. O., Christophersen, N. and Seip, H. M. (1986). "A critique of models for freshwater and soil acidification", *Water, Air and Soil Pollut.*, **30**, 909−930.

Rodhe, A. (1981). "Spring flood. Meltwater or groundwater?", *Nord. Hydrol.*, **12**, 21−30.

Rosenqvist, I. T. (1977). *Sur jord — surt vann. (Acid soil−acid water)*, Ingeniørforlaget, Oslo, 123 pp.

Rosenqvist, I. T. (1978). "Alternative sources for acidification of river water in Norway", *Sci. Total Environ.*, **10**, 39−49.

Rustad, S., Christophersen, N., Seip, H. M. and Dillon, P. J. (1986). "Model for stramwater chemistry of a tributary to Harp Lake, Ontario", *Can. J. Fish. Aquat. Sci.*, **43**, 625−633.

Schnoor, J. L. and Stumm, W. (1985). "Acidification of aquatic and terrestrial systems", in *Chemical Processes in Lakes*, (Ed. W. Stumm), pp. 311−338, John Wiley, New York.

Seip, H. M. (1980). "Acidification of freshwater — sources and mechanisms", in *Ecological Impacts of Acid Precipitation*, (Eds. D. Drabløs and A. Tollan) pp. 358−366, SNSF-project, Proceedings of International Conference, Sandefjord, Norway.

Seip, H. M. (1993). "Review of acid deposition−catchment interactions and comments on future research needs — comment", *J. Hydrol.*, **142**, 483−492.

Seip, H. M. and Rustad, S. (1984). "Variations in surface water pH with changes in sulphur deposition", *Water, Air and Soil Pollut.*, **21**, 217−223.

Seip, H. H., Seip, R., Dillon, P. J. and deGrosbois, E. (1985). "Model of sulphate concentration in a small stream in the Harp Lake catchment, Ontario", *Can. J. Fish. Aquat. Sci.*, **42**, 927−937.

Seip, H. M., Christophersen, N. and Rustad, S. (1986). "Predicted changes in chemistry following reduced-/increased sulphur deposition using the 'Birkenes model'", *Water, Air and Soil Pollut.*, **31**,

239–246.

Seip. H. M., Andersen, D. O., Christophersen, N., Sullivan, T. J. and Vogt, R. D. (1989). "Variations in concentrations of aqueous aluminium and other chemical species during hydrological episodes at Birkenes, southernmost Norway", *J. Hydrol.*, **108**, 387–405.

SFT (1992). "Overvåkning av langtransportert forurenset luft og nedbør. Årsrapport 1991 (Monitoring of long-range transported air and precipitation. Annual report 1991)", Statens forurensningstilsyn, Oslo.

Sklash, M. G. and Farvolden, R. N. (1979). "The role of groundwater in storm runoff", *J. Hydrol.*, **43**, 45–65.

Smith, G. D., Newhall, F. and Robbinson, L. H. (1964)., "Soil-temperature regimes, their characteristics and predictability", SCS-TP-144, Soil Conservation Survey, US Department of Agriculture, Washington, DC.

Soil Survey Staff (1975). *Soil Taxonomy: A Basic System of Soil Classification for Making and Interpreting Soil Surveys*, USDA-SCS Agricultural Handbook 436, US Government Printing Office.

Stone, A. (1990). "Use of the Birkenes model to study the effects of stream and soil acidification", Thesis, University of Oslo.

Stone, A. and Seip, H. M. (1989). "Mathematical models and their role in understanding water acidification: an evaluation using the Birkenes model as an example", *Ambio*, **18**, 192–199.

Stone, A. and Seip, H. M. (1990). "Are mathematical models useful for understanding water acidification?", *Sci. Total Environ.*, **96**, 159–174.

Stone, A., Christophersen, N., Seip, H. M., and Wright, R. F. (1990a). "Predictions of stream acidication by using the Birkenes and MAGIC models", in Mason, B. J. (ed.) *The Surface Water Acidification Programme*, (Ed. B. J. Mason) pp. 495–499, Cambridge University Press.

Stone, A., Seip., H. M., Tuck, S., Jenkins, A., Ferrier, R. C. and Harriman, R. (1990b). "Simulations of Hydrogeochemistry in a Highland Scottish catchment using the Birkenes model", *Water, Air and Soil Pollut.*, **51**, 239–259.

Sverdrup, H. and Warfvinge, P. (1988). "Weathering of primary silicate minerals in the natural soil environment in relation to a chemical weathering model", *Water, Air and Soil Pollut.*, **38**, 387–408.

Taugbøl, G. (1993). "Hydrogeochemical modelling of acid waters", Thesis, University of Oslo, Norway.

Taugbøl, G. and Neal, C. (1994). "Soil and stream water chemistry variations on acidic soils. Application of a cation exchange and mixing model at the catchment level", *Sci. Total Environ.*, **149**, 83–95.

Taugbøl, G., Seip, H. M., Bishop, K. and Grip, H. (1994a). "Hydrochemical modelling of a stream dominated by organic acids and organically bound aluminium", *Water, Air and Soil Pollut.*, **78**, 103–139.

Taugbøl, G., Seip, H. M. and Christophersen, N. (1994b). "Interactions of organic substances with Al and H^+ in soil- and surface waters: two equilibrium models and a concept for statistical testing of models", in G. Taugbøl, "Hydrogeochemical modelling of acid waters", Thesis, University of Oslo, Norway.

Thornton, K. W., Mamorek, D., Ryan, P. F., *et al.* (1990). "Methods for projecting future changes in surface water acid-base chemistry", NAPAP-report 14, The US National Acid Precipitation Program, 722 Jackson Place, NW, Washington, DC, 20503.

Van Breemen, N., Driscoll, C. T. and Mulder, J. (1984). "Acidic deposition and internal proton sources in acidification of soils and waters", *Nature*, **307**, 599–604.

16

Contaminant Transport Component of the Catchment Modelling System SHETRAN

JOHN EWEN

Water Resource Systems Research Unit, Department of Civil Engineering, University of Newcastle upon Tyne, UK

16.1 INTRODUCTION

The SHETRAN modelling system for water flow and sediment and contaminant transport in river catchments has been developed by the Water Resource Systems Research Unit at the University of Newcastle upon Tyne. The contaminant transport component of the system is described here.

Currently there is world-wide interest in environmental quality issues, including surface and groundwater pollution and the effects on water resources of changes in climate (e.g. global warming) and land use (e.g. deforestation). The aim in developing SHETRAN was to create a flexible tool which could be used in research and planning decision support to simulate the basic processes and pathways for flow and transport in river catchments. It is designed to have the capability to predict the consequences of given changes in climate and land use and to have a modular basis that will allow it to be upgraded in the future as our understanding of the processes improves.

SHETRAN has three main components: one each for water flow, sediment transport and contaminant transport. The water flow component is an updated version of the Système Hydrologique Européen (SHE) (Abbott *et al.*, 1986a, 1986b); the sediment component is an updated version of SHESED (Wicks, 1988), and the scientific basis for the contaminant transport component is given in Ewen (1990) and Purnama and Bathurst (1991).

The philosophy behind the SHE is that there are good physically based hydraulic models available for the main processes of water movement in catchments, and these can be integrated into a flexible spatially distributed (rather than lumped) catchment modelling system which can be successfully applied to a diverse range of catchment types and sizes (Abbott *et al.*, 1986a). Since the process models are physically based rather than conceptual or black box types, the parameters for the system can be based directly on field measurements, or literature values, for physical properties (e.g., the saturated hydraulic conductivity of soil). It is argued, then, that the effects of changes in climate and land use on the water flow in a catchment can be predicted

Solute Modelling in Catchment Systems. Edited by Stephen T. Trudgill
© 1995 John Wiley & Sons Ltd

TABLE 16.1 Parameters for the contaminant transport component of SHETRAN

Variables	Description
D	Dispersion coefficient for porous media (varies with contaminant and media type and flow velocity)
D_s	Dispersion coefficient for river channels (varies with river type, size, and flow rate; nb dispersion with surface and horizontal subsurface flows is not accurately represented in SHETRAN)
f	Fraction of the adsorption sites in porous media associated with mobile water (varies with contaminant and media type)
K_d'	Reference adsorption distribution coefficient for porous media (varies with contaminant and media type)
$K_d^{\#'}$	Reference adsorption distribution coefficient for sediments (varies with contaminant type and sediment type and size)
n	Exponent for Freundlich adsorption equation for porous media and sediments (varies with contaminant type, porous media type and sediment type and size)
α	Coefficient for contaminant exchange between mobile and immobile water in porous media (varies with contaminant and media type)
α_{bd} and α_{bs}	Coefficients for contaminant exchange between the bed surface layer and the bed deep layer, and between the river water and the bed surface layer (varies with contaminant and river type)
λ	Radioactive decay constant (varies with contaminant type)
ϕ	Fraction of pore water which is mobile in porous media (varies with contaminant and media type)
ζ	Efficiency for plant uptake of mobile water (varies with plant type)

provided the changes in the physical properties are known or can be estimated. SHESED and SHETRAN were developed according to the same philosophy.

SHETRAN is basically a three-dimensional transient finite difference model written in a very general modular form which allows the user, via input data files, to use it as a flexible modelling system to build a model of any catchment to the appropriate level of detail. The input data files give values for the physical properties. Many of these properties will vary with location in the catchment and the depth below the surface, and this is allowed for using simple map and array data structures. (The full parameter list for the contaminant transport component of SHETRAN is given in Table 16.1.)

A catchment is modelled as an ensemble of columns and stretches of channels. As shown in Figure 16.1, each column comprises a rectangular area of the ground surface and vegetation canopy, and the parts of the unsaturated and saturated zones lying directly below that area. The main drainage channels are represented as networks of stretches of channels (called river links), with each river link running along the edge of a column.

The soil and rock in each column is divided, by horizontal slicing, into several (sometimes up to one hundred) parallelepipedal finite difference cells; each river link has three finite difference cells: one for river water and two for the river bed. Thin columns, typically only 10 metres wide, are positioned on either side of each link to ensure that some of the complexity of

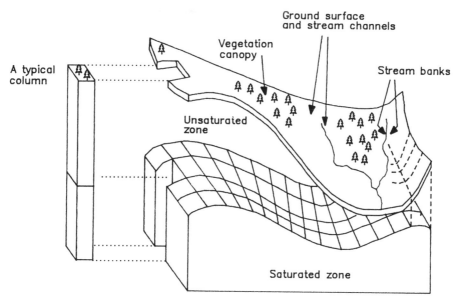

FIGURE 16.1 A typical SHETRAN column

the flow and transport paths close to channels is represented. (Ideally, some form of full grid refinement would be used near channels; however, at present, this is not practical.) When SHETRAN is running, a fully coupled three-dimensional description of the distribution and movement of water, sediment and contaminants is predicted in the usual finite difference fashion (i.e. the simulation steps through time, taking one time-step after another, giving a full three-dimensional picture of the catchment at the end of each time-step).

The water flow processes that are represented include: canopy interception of rainfall; evaporation and transpiration; infiltration; surface runoff (overland, overbank and in channels); snowpack development and snowmelt; soil moisture storage and flow in the unsaturated zone; and groundwater storage and flow in the saturated zone. Surface water flow, both overland and in channels, is modelled using the diffusion form of the Saint Venant equation; flow in the unsaturated zone is modelled using Richards' equation; and flow in the saturated zone is modelled using the two-dimensional Boussinesq equation for phreatic aquifers.

For each time-step, Richards' equation is applied at each column in turn and is solved to give the moisture content within each unsaturated cell, and the vertical flow rates into and out of each unsaturated cell. (Lateral flow is not allowed for in the Richards solution, so flow and transport in the unsaturated zone of the catchment are restricted to the vertical direction only.) The Boussinesq equation is applied to the entire saturated zone, and the Richards and Boussinesq solutions are coupled in a way that ensures water mass is conserved within each column. To allow the simulation of three-dimensional contaminant transport, the results from the Boussinesq solution are used to generate a three-dimensional velocity field for the saturated zone (following the approach of Connorton, 1985). (A fully three-dimensional saturated/unsaturated flow model is being developed for SHETRAN at present. This new model will allow three-dimensional water flow and contaminant transport to be simulated for the entire subsurface, including combined systems of unsaturated regions, and phreatic, perched and confined aquifers.)

The simultaneous transport of several classes of sediment (usually based on particle size) can be simulated. The sediment transport processes that are represented include: erosion by raindrop and leaf drip impact and overland flow; deposition and storage of loose sediments on the ground surface; total-load convection with overland flow; overbank transport; erosion of river beds and banks; deposition on the river bed; down-channel convection; and infiltration of fine sediments into the river bed.

16.1.1 Contaminant transport component

Since the SHE philosophy is to be followed, the contaminant transport component must be physically based and spatially distributed; the transport equations for the contaminant transport component must therefore be partial differential equations with parameters that are physical properties. The convection−dispersion equation has been adopted, as it is the only transport equation that is well established and satisfies these requirements (see, for example, the review for contaminant transport in soil of Addiscott and Wagenet, 1985). The transport equations are developed and disucssed in Sections 16.2 to 16.5 below. The preferred form of the equation for soil and rock is the version developed by Kasten *et al.* (1952) and extended by Coats and Smith (1964) and van Genuchten and Wierenga (1976); for transport in river channels the preferred approach follows Shih and Gloyna (1969) and Onishi (1981).

SHETRAN can be used to simulate the simultaneous transport of several contaminants. At present, however, radioactive decay and generation (as required for a parent−daughter radioactive decay chain) are the only contaminant transformations allowed for; models for more complex transformations, including those for the main species of nitrogen involved in the soil phase of the nitrogen cycle, are being developed for incorporation in SHETRAN in the future.

In the current version of SHETRAN (Version 3.4), it is assumed that water flow and sediment transport are not influenced by the presence of the contaminants; so, when SHETRAN is running, no information on contaminant distribution or transport need be passed back from the contaminant transport component to the water flow or sediment transport components. A cold-region heat-transport component, based on the contaminant transport component, is being developed for a future version of SHETRAN. Since temperature has a strong effect on water flow (especially during frozen periods), the three-dimensional temperature distribution calculated in the heat-transport component will be fed back, in the future version, to the water flow component at the end of every time-step, where it will be used in determining the current status and hydraulic properties of the surface and subsurface waters.

16.1.2 General

The computer memory and processing times required by SHETRAN are quite large. For example, to simulate five years of combined water flow and sediment and contaminant transport for a catchment modelled using 200 columns (each with 50 cells) and 60 links requires an overnight run on a standard computer workstation (64 Mb memory; 20 Mflops processing).

The computer code for the contaminant-transport component of SHETRAN was written in FORTRAN 77. It was developed and tested within a quality assurance system for software development, based on BS 5750, and has been in place since 1991.

16.2 PHYSICS OF TRANSPORT PROCESSES

The transport of a contaminant in a catchment is effected by three main processes — convection, molecular diffusion and mechanical dispersion — and is affected by two other main processes — adsorption and absorption. Two other contaminant processes are modelled by the current version of SHETRAN: radioactive decay and plant uptake.

16.2.1 Convection

The convection equation below describes the distribution of contaminant in a one-dimensional steady flow of water down a hillslope, channel or column of porous media, provided there is no dispersion. It is simply an expression of the law of conservation of mass:

$$\frac{\partial c}{\partial t} = - \frac{\partial (uc)}{\partial x} \tag{1}$$

where c $[ML^{-3}]$ is the concentration in solution, t $[T]$ time, u $[LT^{-1}]$ the magnitude of the average water flow velocity and x $[L]$ distance. (A list of symbols used is given in the Appendix.)

16.2.2 Molecular diffusion and mechanical dispersion

Molecular diffusion is a microscopic process involving molecule/molecule collisions, and therefore takes place even if there is no flow. Mechanical dispersion arises through mechanical mixing, and takes place only if there is flow. For soil and rock, continuous mechanical mixing occurs within the pores in the medium, as previously adjacent packets of solution move relative to one another under the velocity gradients near the solid walls of the pores; and separation and mixing of packets occurs as a variety of tortuous flowpaths are taken by the water as it moves from pore to pore. In surface water flows there is mechanical mixing associated with turbulence and the velocity gradients within planes perpendicular to the flow direction. In general it is not possible to distinguish between the contributions of molecular diffusion and mechanical dispersion, so both are usually considered together under the general term dispersion.

Taylor (1953) found that the dispersion of a contaminant injected into a solvent flowing through a tube can be accurately modelled using a diffusion equation. Strictly, dispersion does not act like diffusion and a diffusion equation should only be applied to dispersion after the transport paths are well developed, i.e. when all the molecules have had a good chance of sampling the full range of velocities seen in the flow.

Adding a term for dispersion to the convection equation gives the convection−dispersion equation for one-dimensional steady flow:

$$\frac{\partial c}{\partial t} = \frac{\partial}{\partial x}\left(D\,\frac{\partial c}{\partial x}\right) - \frac{\partial (uc)}{\partial x} \tag{2}$$

where D $[L^2 T^{-1}]$ is the effective longitudinal dispersion coefficient.

Dispersion is a complex process which is not limited to the pore and turbulent eddy scale as discussed above. For example, there is dispersion when a contaminant migrates with water flow through an interconnected network of fractures in rock; for such a case the dispersion process should be viewed on a large physical scale, covering a volume of rock at least sufficient to contain a fully representative sample of the network. (Dispersion in rock is discussed in Jefferies *et al.*, 1993.)

In the equations developed below for contaminant transport in SHETRAN columns and river links, a diffusion term, as in eqn (2) above, is used to model dispersion. The magnitude of the dispersion coefficients used with SHETRAN should therefore be consistent with the physical scale of the columns and river links.

16.2.3 Adsorption

Contaminants in solution can be adsorbed to the microscopic surfaces of soil, rock and sediments. Adsorption can be positive or negative, as the contaminant can accumulate preferentially either at surfaces or in the bulk solution. In general the rate at which adsorption takes place will depend on the concentration in solution and the amount adsorbed (Kasten *et al.*, 1952). However, under nearly all circumstances, an appropriate description for the kinetics is not known. A practical approach is to assume that the adsorbed contaminant is in thermodynamic equilibrium with that in solution. At the low concentrations expected when contaminants migrate through the near-surface and surface zones, the Freundlich equation is probably the most appropriate (Bohn *et al.*, 1985):

$$s = K_d c \tag{3}$$

where s $[ML^{-3}]$ is the mass adsorbed per unit volume of the medium and K_d [1] the (dimensionless) distribution coefficient, which varies with the concentration in solution:

$$K_d = K_d' (c/c')^{n-1} \tag{4}$$

where K_d' is the distribution coefficient at some arbitrarily chosen reference solution concentration c' and n [1] is an empirical constant.

If the empirical constant n in the Freundlich equation is set to one, the Freundlich equation is reduced to a linear equilibrium adsorption equation. For most simulations, the linear equation will be the better choice: full data on non-linear adsorption is often not available for all the materials in a catchment, or is available only with limited accuracy, and the results of simulations can often be shown to be insensitive to the exact form of the adsorption equation used.

A convenient and compact form of the convection–dispersion equation for porous media including the effects of adsorption can be written by following the approach of Hashimoto *et al.* (1964). When there is adsorption, if the volumetric moisture content is θ[1] the total effective concentration at a point is $c+s\theta$, yet the concentration for convection and dispersion with water flow is simply c. Therefore, for mass balance, in one-dimensional steady flow:

$$\frac{\partial(c+s\theta)}{\partial t} = \frac{\partial}{\partial x}\left(D\frac{\partial c}{\partial x}\right) - \frac{\partial(uc)}{\partial x} \tag{5}$$

giving

$$\frac{\partial(Rc)}{\partial t} = \frac{\partial}{\partial x}\left(D\frac{\partial c}{\partial x}\right) - \frac{\partial(uc)}{\partial x} \tag{6}$$

where

$$R = 1 + K_d\theta \tag{7}$$

For transport in porous media, adsorption tends to retard the rate of transport of contaminant relative to the rate of water flow. The factor R [1] is therefore called the retardation factor when applied to contaminant transport in soil and rock. An equation with the same form as eqn (6) also applies to contaminant transport in surface water if there is adsorption to stationary surface soils and sediments.

The factor R can also be used in the convection−dispersion equation for contaminant transport in flowing water that is carrying adsorbing sediments. The name retardation factor is best avoided in such a case, however, as the contaminant adsorbed on the sediments will move with the sediments, and, provided the sediments move with the flow velocity, the rate of contaminant transport will not be retarded relative to the rate of flow. For transport in one-dimensional steady flow, with adsorption to suspended sediments that are moving with the flow velocity:

$$\frac{\partial(Rc)}{\partial t} = \frac{\partial}{\partial x}\left(D\,\frac{\partial c}{\partial x}\right) - \frac{\partial(Ruc)}{\partial x} \tag{8}$$

where the coefficient D accounts for the combined dispersion of the dissolved and adsorbed phases.

16.2.4 Absorption in porous media

Soil and rock have a strong tendency to become wetted, and a dry sample placed in a reservoir of solution will absorb the solution into its pores by capillary action. Also, if there is a change in the concentration in the solution flowing past a point in saturated or unsaturated soil or rock, that change will not be transmitted instantaneously to all the solution: there are dead spaces in which there is little or no flow, and into which (at least under steady flow conditions) a contaminant can be transported only by diffusion (Nielsen and Biggar, 1961). In general terms, the transfer of contaminant into dead-space is a form of absorption.

One of the most successful approaches to modelling the absorption of contaminant into dead-space involves assuming that the pore water can be divided into two parts: mobile water and immobile water. The mobile water is in the "dynamic region" of the medium and the immobile water is in "dead-space"; convection through the medium and mechanical dispersion take place only in the dynamic region. This division into the dynamic region and dead-space is in some ways a modelling simplification. However, the value for the fraction of pore water which is mobile has been measured in field soils (e.g. Gvirtzman and Magaritz, 1986), and the division is clearly physical if there is significant preferential flow through macropores in soil (Beven and Germann, 1982) or fractures in rock (Jefferies *et al.*, 1993).

Following the approach of Deans (1963) and Coats and Smith (1964), the convection−dispersion equations for one-dimensional steady flow in saturated porous media, without adsorption, extended to include the effect of absorption into dead-space, are

$$\frac{\partial(\phi c)}{\partial t} + \frac{\partial[(1-\phi)c^*]}{\partial t} = \frac{\partial}{\partial x}\left(\phi D\,\frac{\partial c}{\partial x}\right) - \frac{\partial(\phi u c)}{\partial x} \tag{9}$$

and

$$\frac{\partial[(1-\phi)c^*]}{\partial t} = \alpha(c - c^*) \tag{10}$$

where ϕ [1] is the fraction of the pore water which is in the dynamic region (i.e. the fraction that is mobile), c^* [ML^{-3}] is the concentration of contaminant in solution in dead-space and α [T^{-1}] the coefficient for mass transfer between the dynamic region and dead-space.

16.2.5 Sources

Any direct gain or loss of contaminant can be modelled using a source term added to the right-hand side of the convection–dispersion equation. The convection–dispersion equation without adsorption or absorption into dead-space but with a spatially distributed time-varying source of strength B [$ML^{-3}T^{-1}$] is

$$\frac{\partial c}{\partial t} = \frac{\partial}{\partial x}\left(D\frac{\partial c}{\partial x}\right) - \frac{\partial(uc)}{\partial x} + B(x,t) \tag{11}$$

The important sources are for radioactive decay, plant uptake and lateral convection. For radioactive decay, the source term for the parent contaminant is

$$B = -\lambda c \tag{12}$$

where the decay constant λ [T^{-1}] is related to the half-life of the contaminant. The decay of a contaminant results in the generation of its daughter products; for conservation of mass, the combined strength of generation of daughter products at any point must equal the strength of decay of the parent contaminant at that point.

The second important source is plant uptake. If plants take up contaminant from the rooting zone and return some contaminant with recycled dead plant material, the net strength of source involved can be calculated using a plant uptake and recycling model, and this can be coupled to the convection–dispersion equation via a spatially distributed time-varying source term.

The final important source is lateral convection. If contaminant transport in a river or a column is modelled using a one-dimensional convection–dispersion equation, transport into or out of the river or column with lateral flow can be modelled using a spatially distributed time-varying source term. The strength of this source will be proportional to the product of the rate of lateral flow and the concentration in solution in the lateral flow.

16.3 EQUATIONS FOR CONTAMINANT TRANSPORT IN SHETRAN COLUMNS

The convection–dispersion equation for one-dimensional flow with adsorption and dead-space was adapted for non-steady flow conditions in unsaturated porous media by van Genuchten and Wierenga (1976):

$$\frac{\partial(\phi\theta c)}{\partial t} + \frac{\partial[(1-\phi)\theta c^*]}{\partial t} + \frac{\partial(fs)}{\partial t} + \frac{\partial[(1-f)s^*]}{\partial t}$$

$$= \frac{\partial}{\partial x}\left(\phi\theta D \frac{\partial c}{\partial x}\right) - \frac{\partial(\phi\theta uc)}{\partial x} \tag{13}$$

In this equation the superscript * indicates dead-space, and the adsorption sites are assumed to be divided between the dynamic region and dead-space, with fraction f [1] of the total being associated with the dynamic region.

The general equation for absorption into dead-space which complements eqn (13) was not developed by van Genuchten and Wierenga (1976). To account for both diffusional and convective exchanges between the dynamic region and dead-space, this equation must be of the form:

$$\frac{\partial[(1-\phi)\theta c^*]}{\partial t} + \frac{\partial[(1-f)s^*]}{\partial t} = \alpha(c-c^*) + 0.5 \frac{\partial[(1-\phi)\theta]}{\partial t}(c+c^*)$$

$$+ 0.5 \left|\frac{\partial[(1-\phi)\theta]}{\partial t}\right|(c-c^*) \tag{14}$$

where the vertical lines in the final term indicate that the absolute value is taken. The final two terms in this equation account for the transfer of contaminant with water moving between the dynamic region and dead-space. It is assumed that the concentration in water entering dead-space will be c whereas the concentration in water leaving dead-space will be c^*; the rate of transfer of water simply depends on the rate of change of the moisture content in dead-space.

To create the equations for transport within the soil and rock part of a SHETRAN column, the van Genuchten and Wierenga equations were rewritten to include the retardation coefficients R and R^*, and were extended by the addition of terms for radioactive decay, plant uptake and convection with lateral water flow to and from the four neighbouring columns. The full equations for the transport of a single contaminant in the soil and rock in a SHETRAN column are

$$\frac{\partial}{\partial t}\{\theta[\phi Rc + (1-\phi)R^*c^*]\} = \frac{\partial}{\partial z}\left(\phi\theta D \frac{\partial c}{\partial z}\right) - \frac{\partial(\phi\theta uc)}{\partial z}$$

$$- \lambda\theta[\phi Rc + (1-\phi)R^*c^*]$$

$$+ e_d + e_{ds} + \sum_{j=1}^{4} (v\bar{c})_j/A \tag{15}$$

and

$$\frac{\partial}{\partial t}[(1-\phi)\theta R^*c^*] = \alpha(c-c^*) - \lambda(1-\phi)\theta R^*c^* + e_{ds}$$

$$+ (\gamma+|\gamma|)c/2 + (\gamma-|\gamma|)c^*/2 \tag{16}$$

where

$$R = 1 + fK_d/(\phi\theta) \tag{17}$$

$$K_d = K_d'(c/c')^{n-1} \tag{18}$$

$$R^* = 1 + (1-f)K_d^*/[(1-\phi)\theta] \tag{19}$$

$$K_d^* = K_d^{*'}(c^*/c')^{n-1} \tag{20}$$

and

$$\gamma = \frac{\partial[(1-\phi)\theta]}{\partial t} + (1-\zeta\phi)\eta \tag{21}$$

The boundary conditions, based on Danckwerts (1953), are

$$c = c_b(t) \quad \text{or} \quad Q_b(t) = \phi\theta\left(uc - D\frac{\partial c}{\partial z}\right) \text{ at } z = 0 \tag{22}$$

and

$$c = c_{gs}(t) \qquad\qquad \text{at } z = z_{gs} \text{ if } (d_{sw} + d_{ls}) > 0 \tag{23}$$

$$Q_{gs}(t) = \phi\theta\left(uc - D\frac{\partial c}{\partial z}\right) \text{ at } z = z_{gs} \text{ if } (d_{sw} + d_{ls}) = 0$$

The new subscripts introduced in these equations are: b for bottom boundary, d for the dynamic region (in which mobile water can flow), ds for dead-space, gs for the ground surface, ls for loose sediments lying on the surface of the column and sw for free water lying on the surface of the column. A bar over a symbol for a variable indicates that the variable is evaluated at one of the vertical faces of the column. The new variables are: A $[L^2]$, the plan area of the column; d $[L]$, depth; j $[1]$, the index for the four vertical faces of the column; e $[ML^{-3}T^{-1}]$, the combined rate of input from plant uptake and recycling, and generation by decay of parent contaminants; Q $[ML^{-2}T^{-1}]$, the rate of transfer (positive upwards) of contaminant at the ground surface or the bottom of the column, with rainfall, irrigation, direct application, dry deposition or regional flow of groundwater; v $[L^2T^{-1}]$, the volumetric lateral flow rate into a face of the column, per unit depth; z $[L]$, elevation within the column; ζ $[1]$, the efficiency of water uptake, by plants, from the dynamic region (a fraction between 0 and 1); and η $[T^{-1}]$, the volumetric rate, per unit volume of soil, of uptake of water by plants.

16.3.1 Overland transport

At the ground surface, there are exchanges of contaminant between the subsurface and surface waters. Equations (15) to (23) must therefore be coupled (at $z = z_{gs}$) to the overland transport equations.

The basic equation for overland transport of contaminant is a two-dimensional version of the convection–dispersion equation. It is assumed that the surface water, sediments and associated contaminants are in direct contact with the underlying soil and rock. It is convenient here, therefore, to write the overland convection–dispersion transport equation, for a single contaminant, in the form in which it applies to the top of a soil and rock column. In effect, this is the equation for the ground surface concentration, c_{gs} in eqn (23):

$$\frac{d}{dt} [(d_{sw} R_{sw} + d_{ls} \theta_{ls} R_{ls})c_{gs}] = \left[\phi\theta\left(uc - D\frac{\partial c}{\partial z}\right) \right]_{z=z_{gs}} - \lambda(d_{sw} R_{sw} + d_{ls} \theta_{ls} R_{ls})c_{gs}$$

$$+ v[\theta(\phi Rc + (1-\phi)R^*c^*)]_{z=z_{gs}}$$

$$- Q_{gs} + e_{ss} + \sum_{j=1}^{4} (\bar{q}_{sw} \bar{R}_{sw} \bar{c}_{gs} + \bar{b})_j/A \tag{24}$$

where

$$R_{sw} = 1 + \sum_{l=1}^{ns} (\delta K_d^{\#})_l \tag{25}$$

$$R_{ls} = 1 + \sum_{l=1}^{ns} (\beta K_d^{\#})_l/\theta_{ls} \tag{26}$$

and

$$K_{d_l}^{\#} = K_{d_l}^{\#'} (c_{gs}/c')^{n-1} \tag{27}$$

In order, the terms in eqn (24) are for: change in amount of contaminant stored; exchange of contaminant with the underlying soil and rock; radioactive decay; erosion of surface soils; loss of contaminant to the atmosphere; net input of contaminant as the result of plant uptake and generation by the decay of parent contaminants; and net input with lateral convection and dispersion with surface water flow and sediment transport to and from the four neighbouring columns and river links.

One new subscript has been introduced here: ss is for ground surface water and suspended sediments combined. The new variables introduced are: b $[MT^{-1}]$, the rate of dispersion into the column with surface water flow; ns [1], the total number of sediment classes (e.g. ns will be 3 if the clay, silt and sand fractions are treated separately); l [1], the index for the sediment classes; q $[L^3T^{-1}]$, the volumetric rate of flow; β_l [1], the fractions of the local loose (i.e. lying on the surface) sediments that are in class l; δ_l [1], the relative density of suspended sediments in class l; and v $[LT^{-1}]$, the rate of vertical erosion of the ground surface. Only two extra parameters have been introduced in these equations: the reference distribution coefficients for adsorption to the sediments in the different classes, $K_{d_l}^{\#'}$ [1], and the Freundlich equation exponent, n, for the sediments. (In the current version of SHETRAN, the amount of the soil and rock and in a column does not change when soil is eroded or loose sediments consolidate; eqn (15) therefore does not have a term corresponding to the erosion term in eqn (24).)

16.3.2 Plant model

A plant model for contaminant uptake, storage and recycling has been incorporated in SHETRAN. Uptake is modelled using a linearised version of the Michaelis−Menten equation

(Barber, 1984), and each plant type on each column is modelled using two computational storage compartments, one of which grows and decays in size with the canopy leaf area index prescribed for the plant type. The plant model is not described here.

16.3.3 Simultaneous transport of several contaminants

In simulations where several contaminants are transported simultaneously, the full set of column and river link transport equations are solved for each contaminant.

16.4 EQUATIONS FOR CONTAMINANT TRANSPORT IN RIVER CHANNELS

A three-layer description of contaminant transport in river channels has been developed for SHETRAN. At every point along a channel it is assumed that there are three distinct layers: a river water layer (subscript s is used to denote this layer) which includes the suspended and bed-load sediments; the bed surface layer (subscript bs, this layer is assumed to have a small fixed thickness) which contains the materials of the bed which are in direct contact with the river water; and the bed deep layer (subscript bd) which contains the bulk of the river bed and includes deposited materials and, if appropriate, some parent bed materials. This three-layer approach makes it possible to represent both quick and slow transfers of contaminant between the river water and the bed, and both long- and short-term storage of contaminant in the bed. (For a single river link, the two bed layers are, in effect, dynamic finite difference cells for a one-dimensional convection – dispersion equation for contaminant transport down through the river bed.)

The equations below are for the transport of a single contaminant down a river, and include a term for inputs from tributaries and losses to distributaries. The simultneous transport of several contaminants in an entire river network can be modelled using simultaneous sets of these equations.

At any point along a channel, there are three concentrations, one for each layer. Three equations are therefore required. The first equation (eqn (28)) is for overall mass balance within the three layers; the second (eqn (29)) is for mass balance within the bed surface layer on its own; and the third (eqn (30)) for mass balance within the bed deep layer on its own.

$$\frac{\partial}{\partial t}(a_s R_s c_s + a_{bs}\theta_{bs} R_{bs} c_{bs} + a_{bd}\theta_{bd} R_{bd} c_{bd}) = \frac{\partial}{\partial x}\left(a_s D_s \frac{\partial c_s}{\partial x}\right)$$

$$-\frac{\partial}{\partial x}[(q_s + q_D)R_s - q_D]c_s$$

$$-\lambda(a_s R_s c_s + a_{bs}\theta_{bs} R_{bs} c_{bs} + a_{bd}\theta_{bd} R_{bd} c_{bd})$$

$$+\sum_{nk=1}^{2}\left[\left[\int_0^{z_{gs}} v_{bk}(Fc + Gc^*)_{bk}\,dz\right]_{nk}\right.$$

$$+\hat{Q}_s + \hat{e}_s + \hat{e}_{bs} + \hat{e}_{bd} + \psi_s$$

$$+ 0.5 \sum_{nk=1}^{2} \left\{ \int_{0}^{z_{gs}} [(\hat{q}_{bk} + |\hat{q}_{bk}|)c + (\hat{q}_{bk} - |\hat{q}_{bk}|)c_s]dz \right.$$

$$+ (\hat{q}_{sw} + |\hat{q}_{sw}|)R_{sw}\,c_{gs} + (\hat{q}_{sw} - |\hat{q}_{sw}|)R_s\,c_s$$

$$\left. + (\hat{q}_{sb} + |\hat{q}_{sb}|)c + (\hat{q}_{sb} - |\hat{q}_{sb}|)c_{bd} \right\}_{nk} \tag{28}$$

In order, the terms in eqn (28) are for: change in amount of contaminant stored; dispersion; down-channel convection with water and sediments; radioactive decay; erosion of the river banks; inputs with rainfall, irrigation and dry deposition; net input of contaminant with plant uptake and generation by decay of parent contaminants (the three e terms); net contaminant input with tributary and distributary water flow and sediment transport; and net input with convection with water flow through the river bank, over the river bank and through the bed.

$$\frac{d}{dt}[a_{bs}\,\theta_{bs}\,R_{bs}\,c_{bs}] = \alpha_{bs}\,gU\,(c_s - c_{bs}) - \alpha_{bd}\,g\,(c_{bs} - c_{bd}) - \lambda(a_{bs}\,\theta_{bs}\,R_{bs}\,c_{bs})$$

$$+ 0.5[(a_{bd})_t + |(a_{bd})_t|](\theta_{sd}\,R_{sd}\,c_s - \theta_{bs}\,R_{bs}\,c_{bs})$$

$$+ 0.5[(a_{bd})_t - |(a_{bd})_t|](\theta_{bs}\,R_{bs}\,c_{bs} - \theta_{bd}\,R_{bd}\,c_{bd})$$

$$+ (1 - U)\hat{Q}_s + i_{bs} - i_{bd} + \hat{e}_{bs}$$

$$+ 0.5 \sum_{nk=1}^{2} [(\hat{q}_{sb} + |\hat{q}_{sb}|)\,(c_{bd} - c_{bs}) + (\hat{q}_{sb} - |\hat{q}_{sb}|)\,(c_{bs} - c_s)]_{nk} \tag{29}$$

$$\frac{d}{dt}[a_{bd}\,\theta_{bd}\,R_{bd}\,c_{bd}] = \alpha_{bd}\,g\,(c_{bs} - c_{bd}) - \lambda(a_{bd}\,\theta_{bd}\,R_{bd}\,c_{bd})$$

$$+ 0.5[(a_{bd})_t + |(a_{bd})_t|]\theta_{bs}\,R_{bs}\,c_{bs}$$

$$+ 0.5[(a_{bd})_t - |(a_{bd})_t|]\theta_{bd}\,R_{bd}\,c_{bd} + i_{bd} + \hat{e}_{bd}$$

$$+ 0.5 \sum_{nk=1}^{2} [(\hat{q}_{sb} + |\hat{q}_{sb}|)\,(c - c_{bd}) + (\hat{q}_{sb} - |\hat{q}_{sb}|)\,(c_{bd} - c_{bs})]_{nk} \tag{30}$$

In order, the terms in eqns (29) and (30) are for: change in amount of contaminant stored; inter-layer transfers by diffusion (the α terms); radioactive decay; two terms for transfers with bulk movement of bed materials (associated with deposition and erosion of the bed, and consistent with the fact that the thickness of the bed surface layer is fixed); input from the atmosphere (in eqn (29) only); infiltration of fine sediments into the bed (the i terms); net input of contaminant with plant uptake and generation by decay of parent contaminants; and convection with water flow through the bed.

The R factors in eqns (28) to (30) are evaluated in the same way as R_{sw} and R_{ls}. The new subscripts introduced above are: bk for river bank; D for sediment dispersion; s for river water; sb for river bed; sd for newly deposited sediments; and t for time derivative. The new variables are: a $[L^2]$, the cross-sectional area of a layer (measured on a plane perpendicular to the river direction); $\hat{e}$ $[ML^{-1}T^{-1}]$, combined rate (per unit length of channel) of plant uptake and generation by decay of parent contaminants; F and G [1], factors accounting for contaminant adsorbed to eroded bank materials; g $[L]$, the effective width of the bed surface layer (measured following its contour across the width of the channel); i $[ML^{-1}T^{-1}]$, rate (per unit length of channel) of contaminant infiltration with sediments infiltrating from the river water; nk [1], river bank number (1 or 2); $\hat{q}$ $[L^2T^{-1}]$, lateral volumetric flow rate per unit length of channel; $\hat{Q}$ $[ML^{-1}T^{-1}]$, rate (per unit length of channel) of surface input of contaminant with rainfall, irrigation or dry deposition; U [1], unitary function (zero when there is no river water, one when there is); ψ $[ML^{-1}T^{-1}]$, rate (per unit length of river) of contaminant input with tributary and distributary flow.

16.5 SOLUTION METHODS

Within SHETRAN there are two main computational algorithms (solvers) for contaminant transport: COLM, for the equations for the transport of a single contaminant in a column (including the water and sediments lying at the ground surface), and LINK, for the equations for the transport of a single contaminant in one river link. COLM and LINK are both single subroutines written in FORTRAN 77. To create these solvers, the differential equations describing contaminant transport were scaled, reduced to finite difference equations and an algorithm chosen which solves the finite difference equations. (The fact that the transport equations were scaled — i.e. reduced to non-dimensioned form — is invisible to users of SHETRAN, but was of benefit in developing and testing the solvers.)

Fully implicit finite difference approximations were used to reduce the scaled process equations to finite difference equations. Such approximations give great numerical stability to the solvers, even under extreme storm conditions, but can lead to (overly damped) unphysical solutions if very large time-steps are taken. In practice, however, time-steps much larger than two hours cannot be taken, as they result in inaccuracy in the water flow component of SHETRAN.

16.5.1 COLM

It is assumed that every column is areally homogeneous, and at every depth in every column the contaminant is uniformly distributed over the plan area. As described in Section 16.1, each column is divided, by horizontal slicing, into parallelepipedal finite difference cells; this allows the vertical distribution of contaminant to be simulated.

For compatibility with the water flow solvers used in the SHE, a mesh-centred approach is used (i.e. cell boundaries lie halfway between nodes). A 'staggered mesh' approach is not required, so the cells for contaminant transport are the same as those for water flow. The mesh need not be uniform; this allows the possibility of using a finer mesh (giving higher resolution) within the plant rooting zone and thin high-permeability or strongly adsorbing layers of the soil and rock.

To ensure strong coupling (which reflects strong physical coupling) between the equations for contaminant transport in the soil and rock and in surface waters and sediments, the top cell of the column holds the top few centimetres of the soil and rock and all the surface waters and sediments. The finite difference equations for the top cell satisfy both the column porous media transport equations, (15 to 23), and the overland transport equations, (24 to 27). With this approach, if there is surface water on a column, the concentration of contaminant in solution in the top few centimetres of soil and rock is always equal to that in solution in the surface water.

A fully implicit hybrid finite difference approximation (upwind/central, with harmonic mean values for the dispersion coefficients) has been used for the terms describing vertical convection and dispersion within a column. This ensures numerical stability and, provided the cell Peclet number is less than two for all cells, gives very accurate results for vertical convection and dispersion in both the saturated and unsaturated zones.

To ensure numerical stability, fully implicit upwind finite difference approximations have been used for the terms describing horizontal convection (including downstream convection in river channels). Since the upwind approximations introduce numerical dispersion which is usually stronger than the expected physical dispersion, finite difference representations of the physical dispersion terms for horizontal transport should not be, and are not, included in the contaminant transport component.

If there is linear adsorption, the finite difference equations for the contaminant contained in the dynamic region in each column form a tridiagonal set of simultaneous linear algebraic equations. This set is solved at each time level using the Thomas algorithm (see, for example, Ames, 1977). The finite difference equations for dead-space are a set of independent linear algebraic equations. These equations are solved after those for the dynamic region, using direct calculation. If there is non-linear adsorption, all the algebraic equations are non-linear, and the full set is solved using a simple iteration procedure which involves the repeated use of the Thomas algorithm and direct calculation.

16.5.2 LINK

There are three finite difference cells for each river link: one each for the river water, the bed surface layer and the bed deep layer. The thickness of the cell for the bed surface layer does not vary with time; to ensure that the concentration in the cell can respond quickly to changes in the concentration in river water, this thickness should be a few centimetres at most. The thickness of the river water cell will vary with time as the level of water in the river rises and falls. Similarly, the thickness of the cell for the bed deep layer will vary when there is erosion or deposition.

The finite difference equations for each river link form a set of three simultaneous non-linear algebraic equations, which is solved, at each time level, using an iteration method based on Demidovich and Maron (1987).

16.5.3 Coupling at catchment scale

When SHETRAN is running, the overall water, sediment and contaminant simulation progresses step by step from time level to time level. At each time level, the contaminant calculations are carried out only after the water and sediment calculations have been completed and their results are available for use. In finite difference terminology, water and sediment data at time level $m+1$ are passed to the contaminant component, and are then used in the calculation of the contaminant concentrations at time level $m+1$.

It is important to have very robust, mass-conserving methods of coupling the solutions for the transport in any column or link to the transport in neighbouring columns and links, and for coupling the transport of one contaminant to the transport of other contaminants. Within the contaminant component, when a solver is called it returns the final values for the concentrations at time level $m+1$. There is, therefore, no iteration involving repeated calls to the solvers COLM and LINK. At each time level, the whole catchment is swept once. Columns and river links are processed one by one; for each column or river link, one call is made to COLM or LINK, as appropriate, for each contaminant. The order of processing is an important feature of the overall approach to obtaining a stable solution for the entire catchment. The basic rules for ordering are:

1. Columns and river links are processed in decreasing order of the elevation of the free surface of the surface waters associated with them (there is a secondary ordering system for the columns on which there is no surface water).
2. For each solver, calls for the first contaminant are always made before calls for the second, etc.

As a result of the first rule, when a column or link is being processed, the upstream concentration of contaminants is known at time level $m+1$ for all the surface water flowing into the column or river link from neighbouring columns and river links. This makes it possible to use a fully implicit upwind approach for modelling contaminant convection with surface water flows and sediment transport. This method of coupling columns and river links ensures the stability of the solutions, even during extreme storms. It does, however, as noted earlier, introduce numerical dispersion.

The second rule in the list above makes it possible to use a mass-conserving approach to coupling the transport of daughter contaminants to the transport of parent contaminants.

16.6 VERIFICATION

The solvers COLM and LINK have been tested independently of the rest of the SHETRAN computer code. They were embedded in specially written FORTRAN harnesses which produce synthetic water flow and sediment data, and were verified against analytic solutions and self-consistent physical descriptions of contaminant distribution and transport. A programme of testing requiring one man-year of time was designed and successfully carried out. This involved testing the description of each process for contaminant transport on its own and in association with the other processes. Many of the tests on COLM used the analytic solution given by van Genuchten and Wierenga (1976) for the convection−dispersion equation with adsorption and absorption into dead-space. The tests on LINK used specially derived solutions.

Both COLM and LINK were found to be numerically stable under all the conditions considered, and both were found to be accurate provided the time-step was kept to a reasonable level. The recommended time-step (for a typical catchment) is two hours during quiet periods, dropping to 15 minutes during storms. It is not possible to give an overall figure for accuracy, which will apply under all conditions. However, typical mass balance errors are extremely small and are usually best measured in hundredths of a per cent; typical results are well within 1% of analytic solutions. A paper on the verification of SHETRAN is in preparation. This includes full details on the testing of COLM and LINK, and on the catchment simulations described briefly in the next section.

16.6.1 Catchment simulations

Analytic solutions are not available for the full set of coupled equations for three-dimensional integrated surface and subsurface contaminant transport. The only criteria for testing the full system are therefore self-consistency and physical reasonableness. As part of a verification programme, SHETRAN has been run for several simple hypothetical catchments. The main test catchment is a V-shaped catchment, 14 km N−S by 3 km E−W, drained by a single central channel running N−S. This catchment is shown, schematically, in Figure 16.2; each rectangle on the figure represents the top of a column. The figure is not drawn to scale: the main columns are 1 km N−S by 0.25 km E−W, so are not square in shape, and the 1 km N−S by 10 m E−W stream bank columns located on either side of the river are relatively much thinner than they appear in the figure. At all points on the ground surface there is a N−S slope and a slope towards the river. The subsurface has a simple layered structure, with 20 m of Quaternary deposits, including sandy and clayey layers, overlying sandstone. A variety of heterogeneous surface soil and vegetation covers have been considered. The main meteorological data used are from a several-year-long set for a catchment in Finland.

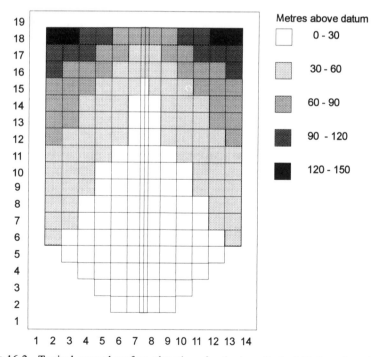

FIGURE 16.2 Typical ground surface elevations for the hypothetical V-shaped catchment

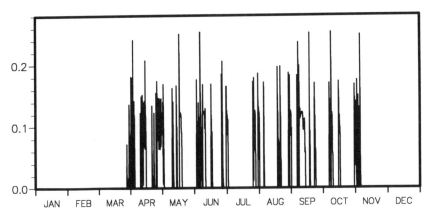

FIGURE 16.3   Typical relative concentration in river water as the result of erosion of highly adsorbing contaminated soils

To date, in the hypothetical catchment simulations, every process of contaminant transport that can be simulated using SHETRAN has been simulated (except for infiltration of fine sediments into the river bed) and the simulation of each process and each combination of processes has been self-consistent and physically reasonable. There is not space here for anything other than a few representative results which give a flavour of the types of results that can be obtained using SHETRAN. These are shown in Figures 16.3 to 16.6.

Figure 16.3 shows pulses of contaminant being discharged into the river during storms, with highly adsorbing sediments. The sediments were eroded from column 5,17 (see Figure 16.2) and washed down the hillslope into the channel. The only significant process for contaminant transport under these high adsorption conditions is convection with sediment. Between storms, therefore, when there is little or no input of sediment, the relative concentration in solution in the river is negligible. (Relative concentration is the ratio of the local concentration to the reference concentration c'; for the simulation resulting in Figure 16.3, the concentration in solution in the soil water in the surface soils at column 5,17 was maintained at c'.)

Figure 16.4 shows results for a special type of simulation designed to test whether contaminant mass is conserved within a SHETRAN simulation. The initial concentrations at all points within the catchment and the concentration in all incoming water (including rain and snowmelt water) were set to the reference value c'. Under these circumstances, the only physical process that should cause the concentration at any point in the catchment to change from the reference value is evaporation, which should cause an increase. The figure shows the relative concentration of contaminant in solution at a point on the ground surface; for the meteorological data used there is no evaporation until late April, and, as it should, the relative concentration stays at one until this time. This test is run every time a dataset is assembled for a new catchment and, to date, every test has been successful, showing that SHETRAN conserves contaminant mass.

Figure 16.5 shows typical relative concentrations on a river bank and in river water. The source of contaminant is groundwater (at the reference concentration c') flowing steadily upwards, through the base of the modelled aquifer, into the catchment at some distance from the river. From the gradual rise in concentration over the nine years, it is clear that it will take

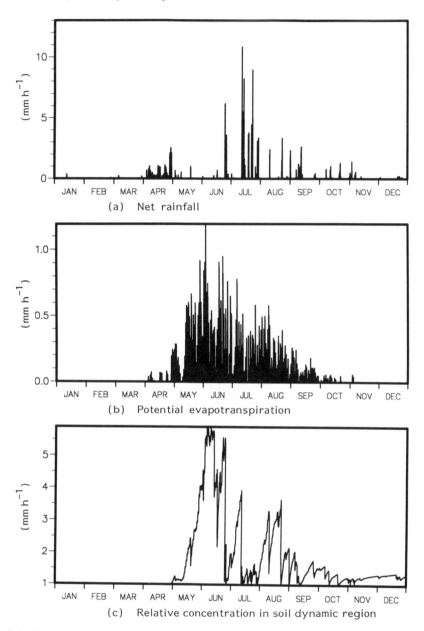

FIGURE 16.4   Typical mass balance verification result: (a) combined effective rainfall and snowmelt rate at the ground surface; (b) potential evapotranspiration rate; (c) relative concentration in solution at a point on the ground surface

many years, or decades, for near-steady transport conditions to develop. The figure shows that there are two distinct dilution regimes: the curves are smooth when there is a snowpack and little or no melt, and rough when there is dilution by rainfall and meltwater and reconcentration by evaporation.

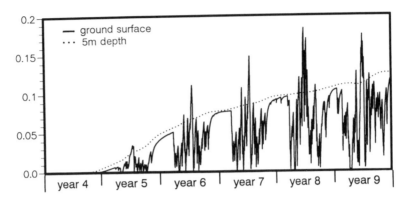

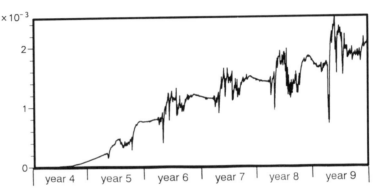

FIGURE 16.5   Typical relative concentration in solution for a non-adsorbing contaminant rising from the subsurface: (a) on the river bank; (b) in the river water

Figure 16.6 shows typical relative concentrations resulting in the subsurface when contaminated groundwater enters the catchment from below and flows to the river through a layer of permeable sand.

16.7 VALIDATION METHOD

The strength of SHETRAN and the SHE lies in their physically based spatially distributed nature; the cost of this strength, however, is complexity and many parameters. As it is for any complex model which has many parameters, the process of calibrating SHETRAN or the SHE is fraught with practical and philosophical difficulties. Yet calibration has been used in all the SHE application studies published to date (e.g. Bathurst, 1986; Refsgaard *et al.*, 1992; and Jain *et al.*, 1992), so these studies are clearly open to criticism. A new approach to the validation of catchment models against field data has therefore been developed (Ewen and Parkin, in press). The new method involves testing on research catchments. It is based on the idea that a model should be tested under conditions similar to those under which it will be used, and is designed to show whether a model is fit for purpose. When applying the new method, appropriate features like weekly runoff and the frequency with which a particular agricultural field will flood are chosen, and bounding values for the magnitude of these features are set using the results from "blind" simulations. By "blind" it is meant that the measured data for the catchment (e.g. the data describing water storage and flow) are not available to the modeller in any form until all the predictions have been made (so calibration is not possible). The quality of the predictions must also be specified by the modeller. For example, the modeller may say that, for success, the measured data must lie within the predicted bounds 90% of the time.

Since the new method involves "blind" validation, it is especially appropriate for validating models which are to be used on ungauged catchments or for the prediction of the effects of future changes in climate and land use.

In practice it is impossible to validate a complex modelling system like SHETRAN for use at all catchments under all conditions. Also, the performance in a particular test will depend on the quality and experience of the modeller as well as the quality of the data and model. The new method therefore involves testing the model, whenever possible, throughout its life, and keeping a full record of its performance. Potential users can then look at this record and, bearing in mind the amount and quality of the catchment data they have available, judge for themselves whether or not the model and modeller can give them useful results for their particular catchments and problems.

The new validation method has been used to test the capability of SHETRAN to predict the flow discharge for Rimbaud, a subcatchment of the Real Collobrier research catchment in Mediterranean France. A paper giving the results of this test is in preparation.

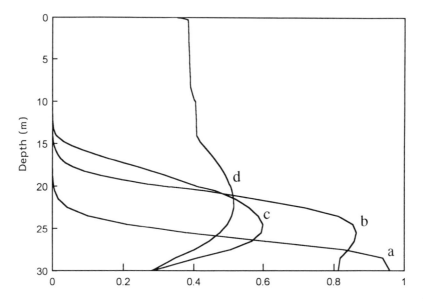

FIGURE 16.6 Typical relative concentrations in solution in subsurface waters, for a non-adsorbing contaminant, as a result of contaminated upflow through a region of the base of the modelled aquifer: a is for the column above the entry point; b and c are for the columns lying between a and the river bank at d. There is a very permeable sandy layer at 20–25 m depth

16.8 DISCUSSION

The main advantage of following a physically based spatially distributed approach in the development of SHETRAN is that the results from SHETRAN simulations are traceable to basic physical descriptions and physical property data that are independent of SHETRAN, and the prediction of the effects of changes in climate and land use on water resources is therefore possible. Since catchments are naturally complex, however, following the physically based spatially distributed approach has led to a contaminant transport component which is quite large and complex; it has twelve parameters, some of which vary spatially and with contaminant type. Despite this complexity, only one significant compromise has been made in the approximation of the governing partial differential equations by finite difference equations: a falsely dispersive upwind approximation has been used to represent the combined terms for horizontal convection and dispersion.

SHETRAN is a fully detailed fully integrated three-dimensional modelling system for combined water flow and sediment and contaminant transport in catchments. The contaminant transport component of the system has been fully verified and gives self-consistent and physically reasonable simulations of the simultaneous transport of several contaminants in, and between, the surface and subsurface zones. Catchments modelled using SHETRAN can be represented in quite fine physical detail, accounting for the heterogeneity of both surface and subsurface conditions; a typical ten-year simulation can show both storm responses and the slow development of contaminant transport pathways.

The new "blind" validation method is a practical scientific approach to the validation of catchment models against field data. We look forward to putting the method to use in validating the contaminant transport component of SHETRAN.

ACKNOWLEDGEMENTS

Work on the development of the contaminant transport component of SHETRAN was initiated in 1988 by Enda O'Connell, James Bathurst and Rae MacKay. A considerable amount of preparatory work (including the upgrading of the water flow component) was carried out before the contaminant transport component was installed. The major part of this preparatory work was undertaken by Geoff Parkin; a contribution was made by Anton Purnama. (Papers giving a general introduction to SHETRAN and describing the preparatory work are in preparation.) Significant contributions to the verification studies were made by Geoff Parkin, Stephen Anderton and Carl Ishemo. The work was funded by UK Nirex Ltd.

REFERENCES

Abbott, M. B., Bathurst, J. C., Cunge, J. A., O'Connell, P. E. and Rasmussen, J. (1986a). "An introduction to the European Hydrological System — Système Hydrologique Européen 'SHE', 1: history and philosophy of a physically-based, distributed modelling system", *J. Hydrol.*, **87**, 45−59.

Abbott, M. B., Bathurst, J. C., Cunge, J. A., O'Connel, P. E. and Rasmussen, J. (1986b). "An introduction to the European Hydrological System — Système Hydrologique Européen, 'SHE', 2: structure of a physically-based, distributed modelling system", *J. Hydrol.*, **87**, 61−77.

Addiscott, T. M. and Wagenet, R. J. (1985). "Concepts of solute leaching in soils: a review of modelling approaches", *J. Soil Sci.*, **36**, 411−424.

Ames, W. F. (1977). *Numerical Methods for Partial Differential Equations*, 2nd edition, p. 52, Academic Press, New York.

Barber, S. A. (1984). *Soil Nutrient Bioavailability*, Wiley-Interscience, New York.

Bathurst, J. C. (1986). "Physically-based distributed modelling of an upland catchment using the Système Hydrologique Européen", *J. Hydrol.*, **87**, 79−102.

Beven, K. and Germann, P. F. (1982). "Macropores and water flow in soils", *Water Resour. Res.*, **18**, 1311−1325.

Bohn, H. L., McNeal, B. L. and O'Connor, G. A. (1985). *Soil Chemistry*, 2nd edition, John Wiley, New York.

Coats, K. H. and Smith, B. D. (1964). "Dead-end pore volume and dispersion in porous media", *Soc. Pet. Engng J.*, **4**, 73−84.

Connorton, B. J. (1985). "Does the regional groundwater-flow equation model vertical flow?", *J. Hydrol.*, **79**, 279−299.

Danckwerts, P. V. (1953). "Continuous flow systems: distribution of residence times", *Chem. Engng Sci.*, **2**, 1−13.

Deans, H. A. (1963). "A mathematical model for dispersion in the direction of flow in porous media", *Trans. AIME*, **228**, 49−52.

Demidovich, B. P. and Maron, I. A. (1987). *Computational Mathematics*, p. 484ff, Mir Publishers, Moscow, USSR.

Ewen, J. (1990). *Basis for the subsurface contaminant transport components of the catchment water flow, sediment transport, and contaminant transport modelling system SHETRAN-UK*, NSS/R229, UK Nirex Ltd, Harwell, UK, 85 pp.

Ewen, J. and Parkin, G. (in press). "Validation of catchment models for predicting land-use and climate change impacts", *J. Hydrol.*

Gvirtzman, H. and Magaritz, M. (1986). "Investigation of water movement in the unsaturated zone under an irrigated area using environmental tritium", *Water Resour. Res.*, **22**, 635−642.

Hashimoto, I., Deshpande, K. B. and Thomas, H. C. (1964). "Peclet numbers and retardation factors for ion exchange columns", *Ind. Engng Chem. Fundamentals*, **3**, 213−218.

Jain, S. K., Storm, B., Bathurst, J. C., Refsgaard, J. C. and Singh, R. D. (1992). "Application of the SHE to catchments in India. Part 2. Field experiments and simulation studies with the SHE on the Kolar subcatchments of the Narmada River", *J. Hydrol.*, **140**, 25−47.

Jefferies, N. L., Lever, D. A. and Woodwark, D. R. (1993). *NSARP Reference Document: radionuclide transport through the geosphere: January 1992*, NSS/G118, UK Nirex Ltd, Harwell, UK, 108 pp.

Kasten, P. R., Lapidus, L. and Amundson, N. R. (1952). "Mathematics of adsorption in beds. V. Effect of intraparticle diffusion in flow systems in fixed beds", *J. Phys. Chem.*, **56**, 683−688.

Nielsen, D. R. and Biggar, J. W. (1961). "Miscible displacement in soils: 1. Experimental information", *Soil Sci. Soc. of Am. Proc.*, **25**, 1−5.

Onishi, Y. (1981). "Sediment-contaminant transport model", *J. Hydraul. Div., ASCE*, **107** (HY-9), 1089−1107.

Purnama, A. and Bathurst, J. C. (1991). *Basis for the ground surface and stream channel radionuclide transport components of the catchment modelling system SHETRAN-UK, including a review of sorption (DRAFT)*, NSS/R233, UK Nirex Ltd, Harwell, UK, 136 pp.

Refsgaard, J. C., Seth, S. M., Bathurst, J. C., Erlich, M., Storm, B., Jorgensen, G. H. and Chandra, S. (1992). "Application of the SHE to catchments in India. Part 1. General results". *J. Hydrol.*, **140**, 1−23.

Shih, C. S. and Gloyna, E. F. (1969). "Influence of sediments on transport of solutes", *J. Hydraul. Div., ASCE*, **95** (HY-4), 1347−1367.

Taylor, G. I. (1953). "Dispersion of soluble matter in solvent flowing slowly through a tube", *Proc. R. Soc., A*, **219**, 186−203.

van Genuchten, M. Th. and Wierenga, P. J. (1976). "Mass transfer studies in sorbing porous media. 1. Analytical solutions", *Soil Sci. Soc. of Am. J.*, **40**, 473−480.

Wicks, J. M. (1988). "Physically-based mathematical modelling of catchment sediment yield", Unpublished PhD thesis, University of Newcastle upon Tyne, 238 pp.

APPENDIX: SYMBOL LIST

a $\quad L^2$ $\qquad$ cross-sectional area of a river channel layer, measured on a plane

		perpendicular to the flow direction
A	L^2	plan area of a column
b	MT^{-1}	rate of dispersion into a column
B	$ML^{-3}T^{-1}$	strength of source
c	ML^{-3}	concentration of contaminant in solution (for the column and link equations, c without a subscript is for the solution in a column)
d	L	depth
D	L^2T^{-1}	dispersion coefficient
e	$ML^{-3}T^{-1}$	combined rate of input from plant uptake and recycling, and from generation by decay of parent contaminants
$\hat{e}$	$ML^{-1}T^{-1}$	as e, but for channels
f	—	fraction of the adsorption sites in porous media which are associated with the mobile water
F	—	factor accounting for contaminant adsorbed to eroded river bank materials
g	L	effective width of the bed surface layer
G	—	factor accounting for contaminant adsorbed to eroded river bank materials
i	$ML^{-1}T^{-1}$	rate (per unit length) of contaminant infiltration, with fine sediments, into the river bed
j	—	index for the vertical faces of a column ($j = 1, 2, 3$ or 4)
K_d	—	adsorption distribution coefficient (ratio of the mass adsorbed per unit volume of the porous material or surface water to the concentration in solution)
l	—	index for sediment classes
n	—	exponent for Freundlich adsorption equation
nk	—	river bank number ($nk = 1$ or 2)
ns	—	total number of sediment classes
q	L^3T^{-1}	volumetric flow rate
$\hat{q}$	L^2T^{-1}	volumetric lateral flow rate per unit length of channel
Q	$ML^{-2}T^{-1}$	vertical (positive upwards) rate of transport of contaminant at the ground surface or the bottom of a column
$\hat{Q}$	$ML^{-1}T^{-1}$	rate of input of contaminant to the surface of a channel
R	—	factor accounting for the contaminant adsorbed to porous media or sediments (for porous media only, called the retardation factor)
s	ML^{-3}	concentration of contaminant adsorbed
t	T	time
u	LT^{-1}	velocity of flow (for the column equations, u is the pore water velocity)
U	—	unitary function (0 when there is no water in a channel, 1 when there is)
v	L^2T^{-1}	volumetric lateral flow rate, per unit depth, into a face of a column
x	L	distance
z	L	elevation within a column
α	T^{-1}	coefficient for exchange between the dynamic region and dead-space, and between the layers of a river channel
β	—	fraction of the loose sediments which are in a given class
γ	T^{-1}	volumetric rate, per unit volume of porous media, of water flow into

		dead-space
δ	—	relative density of the suspended sediments in a given class
ζ	—	efficiency of water uptake by plants from the dynamic region (a number between 0 and 1)
η	T^{-1}	volumetric rate of uptake of water by plants, per unit volume of porous media
θ	—	volumetric moisture content
λ	T^{-1}	radioactive decay constant
ϕ	—	fraction of the pore water which is in the dynamic region (i.e. the fraction which is mobile)
v	LT^{-1}	rate of erosion
ψ	$ML^{-1}T^{-1}$	rate (per unit length) of contaminant input to a stretch of a river with tributary and distributary flow

Special symbols and subscripts

*	dead-space
′	reference value
—	evaluated at the face of a column
$\lvert h \rvert$	absolute value of variable h
b	bottom boundary of column
bd	bed deep layer
bk	river bank
bs	bed surface layer
d	dynamic region
D	sediment dispersion
ds	dead-space
gs	ground surface
l	index for the sediment classes
ls	loose sediments on the ground surface
s	river water
sb	river bed
sd	newly deposited sediments
ss	free surface water combined with loose sediments on the ground surface
sw	free water lying on the ground surface
t	time derivative

SECTION VI

MODEL UTILITY

17

Catchment-Scale Solute Modelling in a Management Context

MALCOLM NEWSON

Department of Geography, University of Newcastle upon Tyne, UK

17.1 INTRODUCTION

Chapter 1 of this volume develops the theme that the solute models of current relevance have been developed *en masse* since 1980. Table 1.1 illustrates the domination of the field by models related to what might be called "headwater problems", notably acidification and nutrient enrichment. Table 1.2 shows how the facility to route solutes from an aerosol or agricultural application through plant and soil covers to groundwater or stream channel has ridden on the back of increasing knowledge of hydrological processes; again these are "headwater processes". The contributions from theoretical chemistry to these models cannot however be underestimated — initially through an understanding of reaction kinetics but latterly (and still in need of improvement) through a knowledge of biochemical processes. Biochemical cycling is a daunting field for modelling and measurement because of the relatively large size and controlling role of storages, rather than fluxes—as if a catchment water balance required a simultaneous knowledge of the volume of the world's oceans!

Nevertheless, progress in a decade and a half has been considerable; when not practical it has been heuristic as with all modelling, and there is no doubt that *headwater problems* have acquired a practical and political significance in land and water management during the last fifteen years of the model development. This importance has, indeed, helped finance model development. Nevertheless, those modelling solutes and those modelling pollutants (especially conservative, point-source chemicals) remain in separate camps, the former mainly probing or forming *catchment* policies, the latter a traditional "bread and butter" scientific role feeding directly into the management of more local, pressing problems. Natural solutes are often only of management significance if they are hyperconcentrated (e.g. mineralised groundwater) or enhanced/released through contamination by human wastes (e.g. heavy metals). However, solute processes and pathways are of huge relevance to all catchment management. Thus, while this book may appear to be mainly one of strategic relevance (e.g. for an era of anticipated global warming), it has also much to reveal about the basic processes that must be managed in all developed river basins.

Nevertheless, to conclude the pervasive review presented by this volume, it is fitting to ask questions about the relative standing of this body of scientific endeavour and the new demands

Solute Modelling in Catchment Systems. Edited by Stephen T. Trudgill
© 1995 John Wiley & Sons Ltd

being put on science by society under terms such as "wealth creation", "decision support", "rational management". The author's particular concerns (see Newson, 1992) are with the following problems faced by scientific hydrology (embracing, for the moment, solute modelling):

1. The scale issue posed by practical *problems*. At the scale of management units adopted by public policy, science can find itself maladjusted to human needs (e.g. wrong parameters, time-scales for prediction).
2. The scale issue raised by the need to extrapolate *solutions* from where they are based upon well-calibrated predictions to the larger and ungauged scale.
3. The *uncertainty* and "incertitude" (Chechile, 1991) issues considered normal to science but little understood by society, despite political terms such as "precautionary principle".

17.2 WHAT ARE MANAGED CATCHMENTS?

Although scientific hydrology has a relatively short history (*ca.* 300 years) and hydrochemistry of solutes an even shorter one (*ca.* 30 years) they provide irrefutable evidence of both the needs and opportunities for rational, sustainable resource management of whole river basins. At present it is hydrological, hydrochemical or biological *surveillance* studies that reveal the need for solute modelling in terms of harmful water quality trends; these revelations are often decades in advance of the "white box" studies of systems that support the best models. It is much more difficult to structure the chemical component of a catchment-scale model from first principles (i.e. without field studies) than it is to set up a deterministic hydrological routine. In the interim, surveillance may be intensified and/or *monitoring* added if precautionary restrictions are imposed on land users/dischargers.

The era of water quality problems began during the Industrial Revolution in Europe and centred initially on pathogens; disease slowly became reduced in the "sanitary age" of the late nineteenth century but the curative technology (i.e. sewerage and sewage treatment) itself led to radical changes in river chemistry. Relatively few variables were used to control and monitor these changes—suspended sediments, BOD, ammonia—despite the growing and diversifying "cocktail" of constituents emerging from industrial sewer connections. There is no doubt that river *contamination* is now considerable outside wilderness environments. The stage at which *pollution* occurs is culturally defined (as anyone who has bathed in the Ganges will know!), even if and where toxicological guidance is sought on "safe" limits. Nevertheless, the contemporary growth and spread of contamination is sufficient for warnings to be sounded and heeded about the finite environmental capacity of the fresh water resource, especially in relation to human needs for increasing volumes of unpolluted supplies. Figure 17.1 indicates the need for an international effort over the next few decades to avoid "fouling the nest" of our very stretched water resources — "dirty water" will limit the human carrying capacity of some basins and aquifers (L'vovitch and White, 1990). Extension of rights to other species means that there are sound reasons for protecting and restoring the ecosystems of our river basins, both on- and off-stream. Here studies of solutes are especially relevant because the levels of bases and nutrients in all waters will have a major say in the habitat provided. Ecotoxicological studies, fast replacing purely chemical tests, may yet give insights into human health links to solutes.

In one sense, therefore, all developed catchments are, or should be, managed catchments. The need for hydrochemical models has been extended by the rate of innovation in the chemical

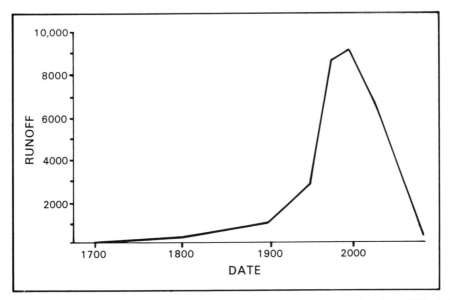

FIGURE 17.1 The component of waste water in global surface flows (modified from L'Vovitch and White, 1990)

industry but, more suited to the efforts described in this volume, by the increase in contamination of freshwater from *diffuse* sources. These may be diffuse (i.e. non-point) because they result from a well-spread activity, such as fertilising or controlling pests on crops, because they derive from a diffuse medium — such as acidification from aerosols or because a point source has been distributed to rivers by a regional aquifer.

Figure 17.2 indicates that, whilst point- and non-point (diffuse) sources of water pollution are convenient academic pidgeon holes, the practical needs of management must see the two together in terms of overall impact — though they may again need separation in terms of management. The point source pollutant, perhaps simple to model because of its "gulp" origins and conservative behaviour, can only be successfully controlled against a background in which upstream water quality i.e. the diluting capacity of the river (or groundwater) environment is optimised.

The problem for most, if not all, catchments is therefore that shown by Figure 17.2 in which a number of well-known point sources of pollution have been contained and managed but where "success" in doing so is compromised in three ways—from unknown sources or pathways, from diffuse pollution upstream or from pollution incidents. The range of such problems is shown in Figure 17.3, which also reveals the real-time challenge for future management models of point-source pollutants.

17.3 WHAT IS CATCHMENT MANAGEMENT?

As development proceeds in any nation state the needs of management become more complex and are represented as being more sophisticated (Table 17.1). Water quality concerns generally lag behind those of water resources and the two may not be as well linked initially as the

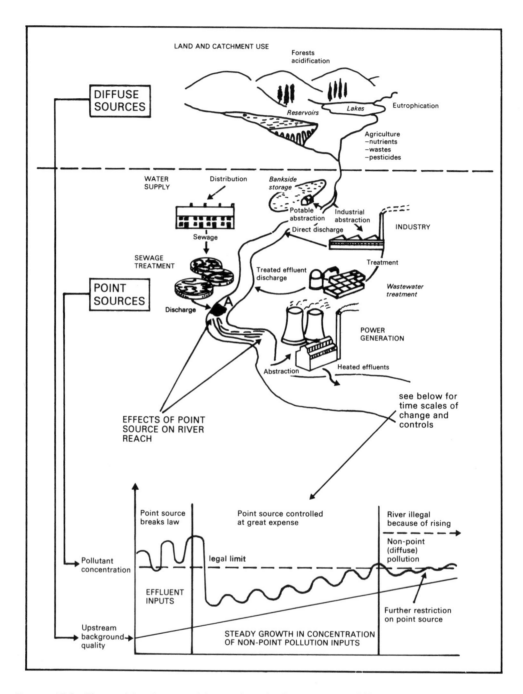

FIGURE 17.2   The spatial and temporal interaction of point-source and diffuse pollution; impacts on the setting and regulation of objectives

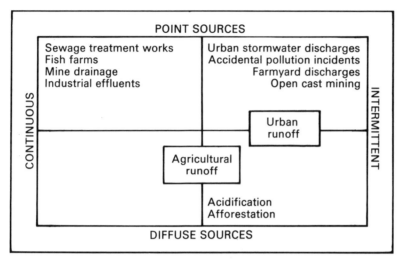

FIGURE 17.3   Water pollution sources: a simple classification

TABLE 17.1   River basin development: prioritising the issues of water-based schemes

	'Developed'		'Developing'
Life-permitting	Priorities often domestic and industrial supplies	Water resources	Priorities often irrigation and hydroelectric power
	Priorities normally urban centres	Flood protection	Priority is food security
	Major influence associated with property rights	Fisheries	Subsistence only: little enhancement
	Reacts to "chemophobia"	Pollution control	Eradication of disease
Life-enhancing	Increasingly	Recreation	Little known
	Increasingly	Conservation	Little known

scientific view would demand; quality concerns reach a peak coincident with an extension of popular demand to recreation (including fisheries), amenity and conservation in water bodies.

17.3.1   Tasks of catchment management

Management begins in a technocentric way, concentrating on "treatment" — the salvation of point-source pollution problems. However, it is important for solute modellers to consider the available options for management beyond treatment. If the river basin is to be managed as an ecosystem it is essential to manage *land and water together*; in small catchments (e.g. those of upland reservoirs) it is possible to control land-use/management activities by purchase or lease. This has been a tradition in the British uplands. However, elsewhere some form of indirect influence must be brought — normally through strategic or agricultural planning or "good practice" encouragements — on the use and management of land, e.g.British *Forests and Water Guidelines* (Forestry Commission, 1993).

A compromise is possible in the allocation of "protection zones" and this option is now deriving so much support that solute modellers should urgently consider it. In a protection zone prescriptions are made on land use, land management or the handling of potentially toxic materials (e.g. fuels, pesticides). Another approach is also becoming popular — the retention of natural or semi natural *buffer zones* of a surface-water catchment so that the products of an uncontrolled land system on the remainder of the catchment can be detained, diluted or removed from the flow. Table 17.2 shows that there are many aims for the widths of buffer zones.

Future trends in catchment management are likely to centre on *water quality objectives*, set by toxicological studies of popular choice (since aesthetic, recreational and other preference criteria enter the equation). In England and Wales, where there is a long tradition averse to emission standards (Newson, 1991), it is a logical corollary of the use of ambient standards (however fixed) that land-use planning should be used to attain those standards. Objectives are normally fixed in terms of mean or maximum allowable *concentrations*.

Similar in principle but vastly larger in scale is the approach to solute *loadings* through *critical loads* — where the solute source is atmospheric. Many of the models described in this volume are directly or indirectly relevant to this form of management by describing the pollutant transformations in plant/soil ecosystems or the impact of changing emissions to air. As such, their relevance peaks in systems for which ecological objectives are politically acceptable and where far-distant emission sources are amenable to control (e.g. within the European Union).

17.3.2 Problems of modelling for catchment management

The relationship between these management contexts is shown in Figure 17.4. Whatever the context, however, there remain major-scale problems (see Figure 17.5 and Newson, 1994); these will now be developed.

The *prediction gap* exists in all science, i.e. between controlled (or, in the case of field studies, "understood") conditions and those sites and conditions for which much less control (or information) exists. In hydrology this gap is generally one between small, headwater catchments (or aquifer recharge zones) and large, downstream areas where a population becomes dependent upon the resource system. Routing or the headwater output may be appropriate in isolation for some contaminants, but the effects of increasing scale on controlling processes remain poorly understood.

The *prescription gap* concerns the first part of the policy context of river basin or aquifer management. Whilst a well-established model may be one way to overcome the prediction gap (preferably backed by careful extension using the other methods shown in Table 17.3) it may be unable to deal with particular variants of a water quality problem and the available control tools. For example, science speaks of the acidification or the eutrophication problem; in fact, there are many variants of these problems and therefore of the potential solutions (e.g. between

TABLE 17.2 Summary of range of options for implementation of buffers in the United Kingdom (Lange and Petts, 1992)

Type	Function	Details	Size	Maintenance
Simple	Wildlife	Grazing retirement, strip of land adjacent to aquatic–terrestrial boundary. Land simply fenced off. Width variable and enhancement of habitat value for wildlife dependent on the range of growth forms left or introduced	1–2 m for small streams to 10–20 m for lowland rivers	Minimal, some weeding to maintain seral stage. Areas of remnant vegetation
Simple	Wildlife	Buffers in upland situations bounding coniferous plantations and upland streams	5–20 m	Minimal. Native scrub used, or broadleaves introduced
Simple	Wildlife, bank stability	Tree planting to provide bank-side habitat and bank stability, as well as shade and organic matter for aquatic system	1–2 m for small rivers to 10–20 m for larger areas	Little
Basic	Wildlife, bank stability	Grazing retirement coupled with tree planting to provide a wider variety of habitat. Native species used. Strategic mix of rooting types	10 m to 5 × stream width	Longer term for tree spp. Shorter term grazing and access restriction
Intermediate	Wildlife, water quality	Areas away from river bank included in buffer. Patch diversity begins to increase	10–50 m	Becomes less as natural development encouraged
Complex	Wildlife, water quality, recreation	Large areas of floodplain put over to riparian woodland and ponds. Reintroduction of wetlands on the floodplain. Autogenic input increases	20–200 m and larger patches	Maintenance of pioneer stages including marginal gravels bars, floodplain surfaces and a variety of aquatic habitats. Access permitted after establishment of buffer
Idea	Wildlife, water quality, recreation	Large areas of floodplain put over to riparian woodland. Reintroduction of wetlands and ponds on the floodplain. Removal of barrier to channel migration. Reintroduction of meanders. Variety of soil types/substrates included. Restoration of natural hydrological regime providing self-sustaining potential for recreation and flood storage. Self-cycling, mainly autogenic system, with regular input of allogenic material	50–300 m and patches of differing size	Maintenance of pioneer stages including marginal gravels bars, floodplain surfaces and a variety of aquatic habitats. Access permitted after establishment of buffer. Self-regulation and rejuvenation of succession encouraged

nutrient enrichment of conservation areas and nitrate pollution of drinking water). There are few "blanket solutions" in applied science and social scientists are now warning that solutions need local adjustments to be truly sustainable.

The *policy gap* cannot be fully addressed here, occurring as it does because administrative units often fail to coincide with aquifer or river basin boundaries; the ability and confidence of water management authorities to get involved with local and regional government is crucial to bridging this gap (e.g. the catchment management planning approach of authorities in England, Wales, Ontario and New Zealand, see Gardiner *et al.*, 1994).

17.3.3 Strategic management issues

So far we have stressed the catchment scale of management and have investigated the various "gaps" which open up between our normally small-scale scientific research and the management scale. However, we must necessarily go wider still if we are to provide truly scientific guidance and robust applied techniques. The world's rivers play a major part in *global cycles of nutrients*. In 1978 a SCOPE project, supported by UNEP, entitled "Transport of

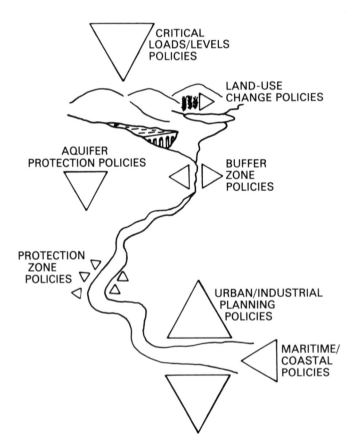

FIGURE 17.4 A typology of intervention policies seen in the catchment context

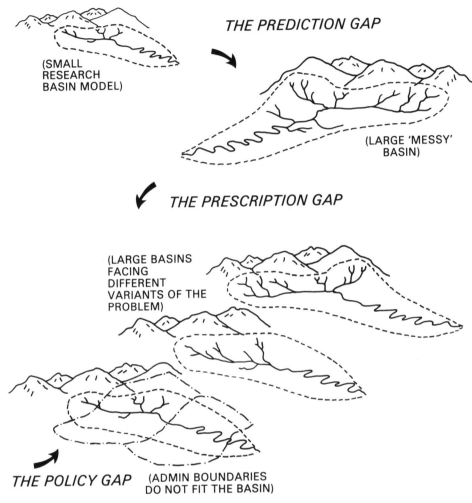

THE PREDICTION GAP

(SMALL
RESEARCH
BASIN MODEL)

(LARGE 'MESSY'
BASIN)

THE PRESCRIPTION GAP

(LARGE BASINS
FACING
DIFFERENT
VARIANTS OF THE
PROBLEM)

THE POLICY GAP (ADMIN BOUNDARIES
DO NOT FIT THE BASIN)

FIGURE 17.5 The problems of applying models at the catchment scale (modified from Newson, 1994)

carbon and minerals in major world rivers" began to assemble the albeit limited data on the solute fluxes on the world's major rivers (Degens *et al.*, 1991). A hundred scientists from 20 countries assembled what was known and carried out reconnaissance work on what was not known. For example, 18 stations were sampled nine times for 12 determinands on the River Amazon (Richey *et al.*, 1991 and Figure 17.6). At the other extreme, for Europe Kempe *et al.* (1991) established solute run off with the benefit of gauging station sites for 66% of the continent's runoff. Even so, with this data richness these authors concluded (p. 206) that "Data are too sparse to allow estimates of the percentages of dissolved and particulate nitrogen and phosphorous yet".

The point stressed by exercises such as the SCOPE project is that all environmental science now has an overtly global agenda; solute modelling is no exception. There is a need to answer profoundly strategic questions such as the role of dissolved organic carbon in the global cycle

TABLE 17.3 Linking small-area research to large-area application

Methods of extrapolation[a]	Methods of incorporation[a]
1 Replication of research in other environments	1 Education — broad approach
2 Pooling data from individual research efforts; synthesis	2 Technical education of practitioners — "good practice"
3 Use of geographical predictions	3 Fiscal manipulation of land-use financial support
4 Mathematical modelling of processes	4 Proscription of damaging operations (plus prescription of beneficial ones)
5 Natural "demonstration" of effect — hazardous event	5 Protection zones → whole-basin planning

[a] Whilst the columns do not cross-correlate, they both represent an ascending sequence of demonstration and action.

and the DOC's role as a carrier of nutrients, of metals and of pesticides. Esser and Kohlmaier (1991) take a modelling approach to this problem but they use only the crudest possible correlation model, being strapped for data. For the 18 major world catchments they correlate dissolved and particulate carbon loads against catchment characteristics (Figure 17.7); despite

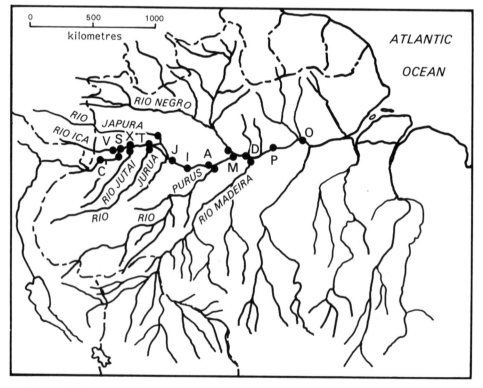

FIGURE 17.6 Sampling stations in the Amazon basin (modified from Richey *et al.*, 1992)

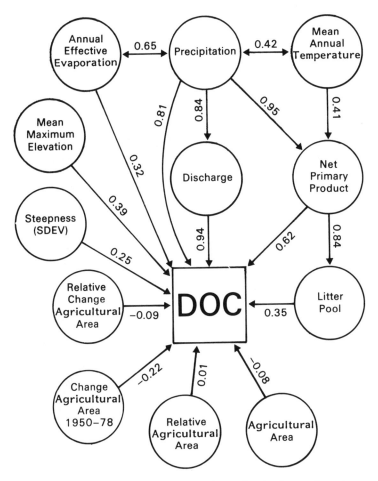

FIGURE 17.7 Stores, fluxes and control variables of the dissolved organic carbon system and their correlation in a study of 18 international watersheds (modified from Esser and Kohlmaier, 1991)

including major anthropogenic variables such as tropical deforestation the dominant effects *identifiable with these data* are of climate, topography and ecology. They muse that the impacts of deforestation must be "lost" in the processes operating upstream of the catchment outlet (the "prediction gap") and that the major impact could well be on mineral solutes during the 2−7 years it takes for tropical forest soils to lose their fertility after logging. However, there must be a high risk that the analysis has been frustrated by the agglomeration of land-use data of variable quality and this frustration will, despite the arrival of remotely sensed data, continue to dog our efforts to apply solute models in support of quantifying global chemical cycles.

The suggestion that we are witnessing the impact of scale problems is perhaps best supported by the fact that, in contrast, Kempe *et al.* point to considerable anthropogenic impact during the last century of industrial and agricultural development in the Rhine catchment; they make this observation after obtaining a rare historical analysis dating to 1854 (the work of Zorbrit and

Stumm). Figure 17.8 shows the degree of anthropogenic impact on solutes in the subsequent 140 years.

The SCOPE study pays considerable attention to the suspended sediment yield of the world's rivers but, predictably, makes little analysis of sediment-solute linkages. We all pay lip-service to physical and chemical bonds between sediments and solutes, but until recent years few have carried out phase analysis or even established the particulate load of rivers. There are no acceptable toxicity tests for sediments (Samoiloff, 1989) who admits that (p. 143) "In most aquatic ecosystems the sediments would be expected to contain both compounds." Perhaps the relative significance of this neglect is lessened when compared with the generally poor state of assessment for non-acute bioassays in general. The lethal concentration (LC_{50}) has been the stand-by technique for water quality for a very long period of time; ecotoxicology, focusing upon such impacts as those on physiology, biochemistry, morphology, genetics and behaviour, has been slow to develop.

17.3.4 Extension to groundwater solute models

It was from studies of limestone aquifers that dual porosity in the unsaturated zone first became acceptable for groundwater pathways. It was, for instance, only in the late 1960s that papers were published revealing the vulnerability of limestone aquifers to pollution incidents resulting in transmission through fissure flows. More detailed studies of the chalk aquifer have suggested a much wider application of dual porosity and therefore of preferential flow in the unsaturated zone (Foster *et al.*, 1991). Groundwater models abound and many incorporate the findings of empirical research; simple ones are being used to design surface protection zones for aquifers (Adams and Foster, 1992). Nevertheless, in the United Kingdom where one third of the

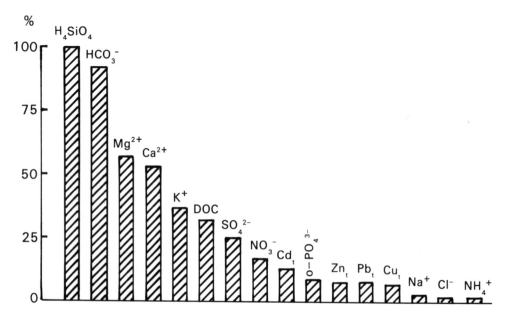

FIGURE 17.8 Changes of solute loads between the pristine River Rhine and its current loads

TABLE 17.4 Groundwater problems in the United Kingdom

Pollutant	Aquifer/region	Reference
Chlorinated solvents	Chalk (Luton)	Longstaff *et al.* (1992)
Pesticides	East Anglia	Croll (1991)
Metals, organics	Birmingham, Coventry	Leaner and Tellam (1992)
Chlorinated solvents	14 Uk sites	Rivett *et al.* (1990)

population depends for its water supply on major aquifers (mainly Chalk and Bunter Sandstone), it is surprising that we have not been more proactive in protection by making models more operational for pollution control. This is perhaps a good demonstration of the "policy gap" — only in the last three years has the United Kingdom adopted policies that delimit regional aquifers and their sensitive recharge zones with a view to implementing land-use controls.

Adams and Foster suggest that the extent of aquifer pollution around a surface source of pollution depends on the hydraulic loading of the pollutant, its chemical nature and the chemical nature of the unsaturated zone. However, we are less able to predict the mitigation potential of the unsaturated zone through processes such as:

- Interception, sorption of bacteria and viruses
- Precipitation, sorption or cation exchange for metals
- Sorption or biodegradation for organic compounds

Table 17.4 shows how severe the problem of groundwater pollution is becoming in the United Kingdom.

If the conventional applied model is one maximising on flow, the management model of the future is likely to have to deal much more closely with the effects of *storage* of water on chemical properties of the resulting flow. Storage locations that directly affect outflow chemistry include:

- "Old" water in soil
- Water stored in vegetation
- Groundwater
- Backwater and dead zones in channels
- Lakes and reservoirs
- Estuaries and coastal seas

17.4 SCIENTIFIC CHALLENGES FOR SOLUTE MODELLING AT THE CATCHMENT SCALE

We now return almost wholly to the catchment-scale challenge — i.e. to bridge the gaps identified in Figure 17.5. Given the division of the catchment water quality management problem outlined above (and in Figure 17.2) it is not surprising that there is a potential for two schools of thought to arise. The *traditional applied model* is well outlined by James (1980) who suggests the following logical steps:

- Formulation of problem

- Review of theory
- Formulation of equations
- Information flow
- Method of solution
- Preparation of programme
- Testing of model

These are used in the development of a model which may incorporate the hydraulic, chemical, physical and biological processes affecting a contaminant during its transport downstream. The biggest problem for this type of model is to decide between plug flow or mixing/reacting conditions and the means of describing the kinetics of the appropriate reactions. Applications require a simple prediction of the concentration of a contaminant at a key point of interest (e.g. abstraction), often as an average but, in these days of maximum allowable concentrations, as a range—and possibly in real time. This is the bread-and-butter model for calculating the conditions of a discharge licence or bringing a pollution prosecution. Nevertheless, we should not be too universal in applying such models. The European tradition of emission controls involves a much closer relationship between regulatory activity and toxicological studies—such that conservative measures to prevent harm can be taken "end of pipe", rather than in the medium concerned. The two philosophies are constantly battled out in the European Union (see Newson, 1991).

The requirements of *ecosystem protection models* are much more stringent and, as a direct result, more attention is applied to ecotoxicology *in vivo* as a means of identifying polluted locations and episodes. There is a considerable stimulus to develop such models for two main reasons:

1. To direct attention to the use and management of land in the case of diffuse pollutants (and to the use and management of remoter sites in the case of acidification).
2. To assess the time-scales of restoration of ecosystems through changes in land use or management or air pollution control.

The major problems of this type, in a broader framework of multi-functional basin management, are still treated as localised (e.g. acidified headwaters, eutrophic lakes), rather than in the conceptual framework of Figure 17.2, but the challenge remains that an integration of the local with the whole is essential in the future. This is particularly the case in semi-arid climates where the extensive use of irrigated agriculture, essential to maintain the booming populations of such regions, means that irrigation return flows must be the subject of such models.

Clearly there is great scope for scientific divisions to open up; as acronyms burgeon and modellers stress the particular benefits of their algorithms the point may be missed that the prescription gap is a serious obstacle to progress. Modelling may well be a case of "horses for courses" but it behoves researchers to appreciate both the context of the applied problems with which they identify and the policy gaps that may open up before remedies are contemplated. Perhaps the best demonstration of potential dilemmas will be for the reader to advise what happens on a stream emerging from granitic highlands where water quality standards are set by critical load concepts to a reach where local politicians have set an "industrial-use" water quality objective!

A clearer definition of our problems in research is set by the most recent SCOPE project in this area (Moldan and Cerny, 1994). To summarize the views of the editors of SCOPE 51 in pointing to "considerations for the future" (p. 23):

1. The "geo-bio" connection requires greater interdisciplinary effort.
2. The impact of climate change should be studied.
3. Local and global scales must be linked, requiring standard techniques and careful reporting of catchment land use for comparative purposes.
4. More manipulations of land use are required to give a truly experimental framework.
5. The biochemistry of heavy metals and organics requires more effort.
6. New focuses are required at time-scales of the event and long-term change.
7. Model catchment techniques should be refined to provide inexpensive monitoring tools.
8. More work is required in the tropical zone and in the developing countries generally.

Those who have read the SCOPE 51 report will detect a deliberate omission in our summary above. It is worth separating, for reasons of emphasis, the report's conclusions on the tasks for the field scientists. They include:

1. Long-term continuation of measurements in *existing research sites*.
2. Enhancing *data quality* of all collected data for small catchments.
3. Development and refinement of conceptual and mathematical *models*.
4. More extensive use of intersite comparison, *nested and paired catchments*.
5. *Regionalization*.

17.5 CONCLUSIONS

It is a reasonable question for a town planner to ask those concerned with solutes in surface water to suggest a future pattern of industrial development that optimises the purifying power of the river/estuarine/coastal environment. Such a town planner will be sensible if he or she also asked about precautions to be taken with land use and land management upstream of the town or city, again with a view to making maximum benefit of the natural "fluvial treatment works". Hopefully, the same questions will be asked for the aquifers too, since groundwater resources are more likely to sustain a population through a period of climate change than are surface resources that are less well modulated.

With this, the modern approach to resource management, the strategic knowledge of solute modellers becomes highly relevant, though it may need to expand rapidly into currently underestimated modulators of chemical throughput such as aquifers, lakes and wetlands. In fact it is likely that solute modellers will join a cross-media group sharing expertise and possibly model "shells" with those working on all aspects of environmental protection. The umbrella title of this grouping is provided by the title of a recent college text: *Chemical Fate and Transport in the Environment* (Hermond and Fechner, 1994).

REFERENCES

Adams, B. and Foster, S. S. D. (1992). "Land-surface zoning for groundwater protection", *J. Inst. of Water and Environ. Managemt*, **6**, 312–320.

Chechile, R. A. (1991). "Introduction to environmental decision-making", in *Environmental Decision Making, A Multidisciplinary Perspective*, (Eds. R. A. Chechile and S. Carlisle) pp. 1–13, Van Nostrand Reinhold, New York.

Croll, B. T. (1991). "Pesticides in surface waters and groundwaters", *J. Instn. of Water and Environ. Managemt.*, **5**(4), 389−395.

Degens, E. T.,, Kempe, S. and Richey, J. E. (Eds.) (1991). *Biogeochemistry of Major World Rivers*, John Wiley, Chichester.

Esser, G. and Kohlmaier, G. H. (1991). "Modelling terrestrial sources of nitrogen, phosphorus, sulphur and organic carbon to rivers", in *Biogeochemistry of Major World Rivers* (Eds. E.T., Degens, S. Kempe and J. E. Richey), pp. 297−322, John Wiley, Chichester.

Forestry Commission (1993). *Forests and Water Guidelines*, HMSO, London.

Foster, S. S. D., Chilton, P. J. and Stuart, M. E. (1991). "Mechanisms of groundwater pollution by pesticides", *J. Instn. of Water and Environ. Managemt*, **5**, 186−193.

Gardiner, J., Thompson, K. and Newson, M. D. (1994). "Integrated water/river catchment management. A comparison of selected Canadian and United Kingdom experiences", *J. Environ. Planning and Managemt.*, **7**(1), 53−67.

Hemond, H. F. and Fechner, E. J. (1994). *Chemical Fate and Transport in the Environment*, Academic Press, New York.

James, A. (1980). "Water quality modelling", in *Water Quality in Catchment Ecosystems* (Ed. A. M. Gower), pp. 265−284, John Wiley, Chichester.

Kempe, S., Pettine, M. and Cauwet, G. (1991). "Biogeochemistry of European rivers", in *Biogeochemistry of Major World Rivers* (Eds. E.T. Degens, S. Kempe and J. E. Richey), pp. 169−211, John Wiley, Chichester.

Lange, A. R. G. and Petts, G. E. (1992). *Buffer zones for conservation of rivers and bankside habitats.* National Rivers Authority, Bristol, UK., Research and Development Record, 340/5/Y.

Lerner, D. N. and Tellam, J. H. (1992). "The protection of urban groundwater from pollution", *J. Instn. of Water and Environ. Managemt.*, **6**(1), 28−37.

Longstaff, S. L., Aldous, P. J., Clark, L., Flavin, R. J. and Partington, J. (1992). "Contamination of the Chalk aquifer by chlorinated solvents: a case study of the Luton and Dunstable area", *J. Instn. of Water and Environ. Managemt.*, **6**(5), 541−559.

L'Vovitch, M. I. and White, G. F. (1990). "Use and transformation of terrestrial water systems", in *The Earth as Transformed by Human Action*, (Eds. B. L. Turner, W. C. Clark, R. W. Kates, J. F. Richards, J. T. Matthews and W. B. Meyer), Cambridge University Press.

Moldan, B. and Cerny, J. (Eds.) (1994). *Biochemistry of Small Catchments. A Tool for Environmental Research*, SCOPE 51, John Wiley, Chichester.

Newson, M. D. (1991). "Space, time and pollution control: geographical principles in UK public policy", *Area*, **23**(1), 5−10.

Newson, M. D. (1992). *Land, Water and Development*, Routledge, London.

Newson, M. D. (1994). "Scales and their appropriateness for integrating land use management: science, subsidiarity and sustainability", in *Land Use Science*, (Ed. J.A. Milne), pp. 1−14, Macaulay Land Use Research Institute.

Richey, J. E., Victoria, R. L., Salati, E. and Forsberg, B.R. (1991). "The biogeochemistry of a major river system: the Amazon case study", in *Biogeochemistry of Major World Rivers*, (Eds. E. T. Degens, S. Kempe, and J. E. Richey), pp. 57−74. John Wiley, Chichester.

Rivett, M. O., Lerner, D. N. and Lloyd, J. W., (1990). "Chlorinated solvents in UK aquifers", *J. Instn. of Water and Environ. Managemt.*, **4**(3), 242−250.

Royal Commission on Environmental Pollution (1992). *Sixteenth Report. Freshwater Quality*, HMSO, London.

Samoiloff, M. R. (1989). "Toxicity testing of sediments. Problems, trends, and solutions", in *Aquatic Toxicology and Water Quality Management* (Eds. J. O. Nriagu and J. S. S. Lakshminarayana), pp. 143−152, John Wiley, Chichester.

Welch, E. B. (1992). *Ecological Effects of Wastewater. Applied Limnology and Pollutant Effects*, Chapman and Hall, London.

Appendix

Conversion Factors for Solute Concentration

Species	mg l^{-1} to meq l^{-1} multiply by	mg l^{-1} to mmol l^{-1} multiply by
Al^{+3}	0.00009	0.03715
Ca^{+2}	0.04990	0.02495
Cl^-	0.02821	0.02821
CO_3^{-2}	0.03333	0.01666
Fe^{+2}	0.03581	0.01791
Fe^{+3}	0.00788	0.00788
H^+	0.99209	0.99209
$H_2PO_4^{-3}$	0.02084	0.01042
HCO_3^-	0.01639	0.01639
Mg^{+2}	0.08226	0.04113
Na^+	0.04350	0.04350
NH_4^+	0.05544	0.05544
NO_3^-	0.01613	0.01613
NO_2^-	0.02174	0.02174
OH^-	0.05880	0.05880
PO_4^{-3}	0.03159	0.01053
SiO_2		0.01664
SO_4^{-2}	0.02082	0.01041
S^{-2}	0.06238	0.03119

Source: Adapted from Hem, J. D. (1970). "Study and Interpretation of the chemical characteristics of natural water", *US Geological Survey Water-Supply Paper, 1473*, US Government Printing Office, Washington DC, 363 pp. (p. 83).

Author Index

Subject Index